Polymer Composites: Structure, Properties and Processing

Polymer Composites: Structure, Properties and Processing

Ana Pilipović
Phuong Nguyen-Tri
Mustafa Özcanli

Basel • Beijing • Wuhan • Barcelona • Belgrade • Novi Sad • Cluj • Manchester

Editors

Ana Pilipović
Department of Technology
University of Zagreb
Zagreb
Croatia

Phuong Nguyen-Tri
Department of Chemistry,
Biochemistry and Physics
Université du Québec à
Trois-Rivières (UQTR)
Quebec
Canada

Mustafa Özcanli
Department of Automotive
Engineering
Çukurova University
Adana
Turkey

Editorial Office
MDPI AG
Grosspeteranlage 5
4052 Basel, Switzerland

This is a reprint of articles from the Special Issue published online in the open access journal *Polymers* (ISSN 2073-4360) (available at: www.mdpi.com/journal/polymers/special_issues/LA75S57A7J).

For citation purposes, cite each article independently as indicated on the article page online and as indicated below:

Lastname, A.A.; Lastname, B.B. Article Title. *Journal Name* **Year**, *Volume Number*, Page Range.

ISBN 978-3-7258-2486-1 (Hbk)
ISBN 978-3-7258-2485-4 (PDF)
doi.org/10.3390/books978-3-7258-2485-4

© 2024 by the authors. Articles in this book are Open Access and distributed under the Creative Commons Attribution (CC BY) license. The book as a whole is distributed by MDPI under the terms and conditions of the Creative Commons Attribution-NonCommercial-NoDerivs (CC BY-NC-ND) license.

Contents

About the Editors vii

Preface ix

Naiara Lima Costa, Carlos Toshiyuki Hiranobe, Henrique Pina Cardim, Guilherme Dognani, Juan Camilo Sanchez and Jaime Alberto Jaramillo Carvalho et al.
A Review of EPDM (Ethylene Propylene Diene Monomer) Rubber-Based Nanocomposites: Properties and Progress
Reprinted from: *Polymers* **2024**, *16*, 1720, doi:10.3390/polym16121720 1

Mostafa Katouzian, Sorin Vlase, Marin Marin and Maria Luminita Scutaru
Modeling Study of the Creep Behavior of Carbon-Fiber-Reinforced Composites: A Review
Reprinted from: *Polymers* **2022**, *15*, 194, doi:10.3390/polym15010194 36

Adrian Korycki, Fabrice Carassus, Olivier Tramis, Christian Garnier, Toufik Djilali and France Chabert
Polyaryletherketone Based Blends: A Review
Reprinted from: *Polymers* **2023**, *15*, 3943, doi:10.3390/polym15193943 67

Chen Chen, Lifang Shen, Guang Liu, Yang Cui and Shubin Yan
Improved Energy Storage Performance of Composite Films Based on Linear/Ferroelectric Polarization Characteristics
Reprinted from: *Polymers* **2024**, *16*, 1058, doi:10.3390/polym16081058 99

Xiaoshuai Han, Hongyu Feng, Wei Tian, Kai Zhang, Lei Zhang and Jiangbo Wang et al.
A Sandwich Structural Filter Paper–AgNWs/MXene Composite for Superior Electromagnetic Interference Shielding
Reprinted from: *Polymers* **2024**, *16*, 760, doi:10.3390/polym16060760 111

Xinyuan Wan, Xiaojian Xia, Yunxiang Chen, Deyuan Lin, Yi Zhou and Rui Xiong
Bioinspired Thermal Conductive Cellulose Nanofibers/Boron Nitride Coating Enabled by Co-Exfoliation and Interfacial Engineering
Reprinted from: *Polymers* **2024**, *16*, 805, doi:10.3390/polym16060805 120

Lovro Travaš, Maja Rujnić Havstad and Ana Pilipović
Optimization of Thermal Conductivity and Tensile Properties of High-Density Polyethylene by Addition of Expanded Graphite and Boron Nitride
Reprinted from: *Polymers* **2023**, *15*, 3645, doi:10.3390/polym15173645 132

Von Clyde Jamora, Virginia Rauch, Sergii G. Kravchenko and Oleksandr G. Kravchenko
Effect of Resin Bleed Out on Compaction Behavior of the Fiber Tow Gap Region during Automated Fiber Placement Manufacturing
Reprinted from: *Polymers* **2023**, *16*, 31, doi:10.3390/polym16010031 153

Min Sun, Shuo Jing, Haibo Wu, Jun Zhong, Yongfu Yang and Ye Zhu et al.
Preparation Scheme Optimization of Thermosetting Polyurethane Modified Asphalt
Reprinted from: *Polymers* **2023**, *15*, 2327, doi:10.3390/polym15102327 175

Fan Pan, Lingling Sun and Shiben Li
Dynamic Processes and Mechanical Properties of Lipid–Nanoparticle Mixtures
Reprinted from: *Polymers* **2023**, *15*, 1828, doi:10.3390/polym15081828 193

Rifat Ara Nargis and David Abram Jack
Fiber Orientation Quantification for Large Area Additively Manufactured Parts Using SEM Imaging
Reprinted from: *Polymers* **2023**, *15*, 2871, doi:10.3390/polym15132871 . . . . **212**

Enrique Martínez-Franco, Victor Alfonzo Gomez Culebro and E. A. Franco-Urquiza
Technologies for Mechanical Recycling of Carbon Fiber-Reinforced Polymers (CFRP) Composites: End Mill, High-Energy Ball Milling, and Ultrasonication
Reprinted from: *Polymers* **2024**, *16*, 2350, doi:10.3390/polym16162350 . . . . **232**

Luísa Rosenstock Völtz, Linn Berglund and Kristiina Oksman
Use of Recycled Additive Materials to Promote Efficient Use of Resources While Acting as an Effective Toughness Modifier of Wood–Polymer Composites
Reprinted from: *Polymers* **2024**, *16*, 2549, doi:10.3390/polym16182549 . . . . **251**

About the Editors

Ana Pilipović

Ana Pilipović began to work at the University of Zagreb, Faculty of Mechanical Engineering and Naval Architecture as the Chair of Polymer Processing, Department of Technology in 2008. She completed her doctoral thesis in October 2012, focused on the topic of additive manufacturing: influence of processing parameters on the properties of polymer prototypes.

Since 2008, she has participated in several science international and domestic projects, like FP7, H2020, IPA, IRI, IRI2, Eco-Innovation, and the Croatian Science Foundation, as well as projects from the Croatian Ministry of Science and Education.

Her interests include additive manufacturing, polymer processing, composite processing, recycling of polymers, mechanical testing of polymers, biopolymers, etc. Up to now, she has published as an author or co-author 242 scientific and professional papers in journals and conference proceedings in Croatia and abroad.

She was a Visiting Professor at Tecnologico de Monterrey Santiago de Queretaro, San Luis Potosi and Guadalajara, Mexico, Çukurova University Adana, Turkey, Faculty of Mechanical Engineering in Maribor, Slovenia, and Faculty of Mechanical Engineering Sarajevo, Bosnia and Herzegovina. She has published chapters in three books, mentored forty masters and baccularius theses, two rector's award, and is currently mentoring eight doctorates in the fields of medicine, dentistry, and mechanical engineering. In addition to teaching at her home university, she lectures at the Polytechnic of Zagreb (Zagreb University of Applied Sciences) and the University of Lyon, France. She won the award "Vera Johanides" for young scientists in 2014 in the field of additive manufacturing awarded by the Croatian Academy of Engineering (HATZ). She is now an Associate Professor and Head of Chair of Polymer Processing in the University of Zagreb Faculty of Mechanical Engineering and Naval Architecture.

Phuong Nguyen-Tri

Professor Phuong Nguyen-Tri is a regular professor at the Department of Chemistry, Biochemistry and Physics, Université du Québec à Trois-Rivières (UQTR), Quebec, Canada. He is the founder of the Laboratory of Advanced Materials for Energy and Environment (Nguyen-Tri Lab) at UQTR and holds the UQTR Research Chair of Advanced materials for Health and Security at work. He obtained his M.Sc. degree from École Nationale Supérieure de Chimie de Mulhouse, France, and Ph.D. in Material Sciences from the Conservatoire National des Arts et Métiers in Paris, France, in 2009. His main research interests include nanomaterials, hybrid nanoparticles, smart coatings, polymer crystallization, polymer aging, and polymer blends and composites. He is also a Visiting Professor at UdeM and an Associate Professor at ETS of Montréal, as well as a regular member of I2E3, CQMF, and CREPEC of Quebec. Professor Phuong Nguyen-Tri has an impressive publication record, with over 200 scientific publications, including journal articles, conference papers, and book chapters. His work has garnered significant recognition, as reflected in his H-index of 38. His research has been cited over 5,5150 times, with 4,700 citations since 2019 alone. In addition to his publications, Professor Nguyen-Tri has edited 14 books and monographs and contributed as an Editor to 10 Special Issues. In the past two years, he has published over 30 articles, underscoring his continued impact and productivity in materials science, nanotechnology, and polymer chemistry.

Mustafa Özcanli

Professor Mustafa OZCANLI is a Professor at the Department of Automotive Engineering, Çukurova University, Turkiye. He is an expert in Internal Combustion Engines and currently serves as the Deputy Director at the Institute of Natural and Applied Sciences. He obtained his PhD degree from Cukurova University, Turkiye, in 2009. With a strong background in automotive engineering, Professor OZCANLI has also contributed significantly to the fields of materials and sustainability, focusing on the integration of composite materials and eco-friendly and energy-efficient technologies in public transportation.

Professor OZCANLI has a publication record of over 240 scientific publications, including journal articles and conference papers. His work has earned him an H-index of 24, reflecting the significant impact of his research on engineering.

Preface

Today, polymer products are used in various industries such as the transport industry, engineering, construction, medicine, dentistry, etc. Therefore, polymer materials are required to have properties that can compare with metal materials. This refers to the mechanical properties required for various applications in load-bearing structures, up to the improvement in compatibility with human tissue in medical applications. That is why polymer composites, which have an excellent ratio of strength and low mass, are increasingly being used. Whether fibers or particles are used as a reinforcement in thermoplastics, thermosets, and elastomers, it is necessary to establish their structure in the overall composite and how this affects their properties. When talking about classic fiber-reinforced polymer composites, hybridization (whether mixing with different types of fibers in the weave or in different layers) certainly plays a big role because better properties are achieved.

In addition to classic composite materials with the addition of the roving of glass, carbon and aramid fiber and natural reinforcements (particles/fibers) are increasingly used as reinforcement/fillers in polymer matrices due to their ecofriendly nature, lightweight, lifecycle superiority, biodegradability, low cost, and sustainability. With these natural polymer composites we are facing numerous challenges regarding their development, applications, and inferior properties (quality of the fiber, thermal stability, water absorption capacity, and incompatibility with the polymer matrices) in comparison with classical composites reinforced with carbon, aramid, or glass fibers.

However, for such materials to be used in industry, an important step is the production itself and how the processing parameters affect the composite products. In addition to the classic methods of processing composites regarding fibers (hand lay-up, vacuum bagging, filament winding, resin infusion, etc.), in recent years, this has also been achieved by additive manufacturing (AM), so-called 3D printing, and, for particles, by extrusion.

In this Special Issue, 13 papers were published, which include polymer composites with a matrix of thermoplastics, thermosets and elastomers reinforced with carbon fibers, boron nitride, and even natural fibers such as cellulose. Furthermore, it includes their production by classical procedures and additive manufacturing, as well as improvements and innovations in the field of recycling polymer composites.

Accordingly, as Guest Editors, we would like to thank all the authors and reviewers who contributed to this Special Issue of *Polymers*. We also wish to express thanks to the Editor in Chief, the Section Managing Editor, and all the assistants who helped to finish this Special Issue.

Ana Pilipović, Phuong Nguyen-Tri, and Mustafa Özcanli
Editors

Review

A Review of EPDM (Ethylene Propylene Diene Monomer) Rubber-Based Nanocomposites: Properties and Progress

Naiara Lima Costa [1,2], Carlos Toshiyuki Hiranobe [1], Henrique Pina Cardim [1], Guilherme Dognani [2], Juan Camilo Sanchez [3], Jaime Alberto Jaramillo Carvalho [3], Giovanni Barrera Torres [4], Leonardo Lataro Paim [1], Leandro Ferreira Pinto [1], Guilherme Pina Cardim [1], Flávio Camargo Cabrera [1,*], Renivaldo José dos Santos [1] and Michael Jones Silva [1,*]

1. School of Engineering and Science (FEC–UNESP), São Paulo State University, Rosana 19274-000, SP, Brazil; naiara.costa@unesp.br (N.L.C.); carlos.hiranobe@unesp.br (C.T.H.); henrique.cardim@unesp.br (H.P.C.); leonardo.paim@unesp.br (L.L.P.); leandro.f.pinto@unesp.br (L.F.P.); guilherme.cardim@unesp.br (G.P.C.); renivaldo.santos@unesp.br (R.J.d.S.)
2. School of Technology and Sciences (FCT–UNESP), São Paulo State University, Presidente Prudente 19060-900, SP, Brazil; dognanig@gmail.com
3. Mechanical Engineering Department, Pascual Bravo University Institution (IUPB), Medellín 050036, Colombia; juan.sancjezg@pascualbravo.edu.co (J.C.S.); jaime.jaramillo@pascualbravo.edu.co (J.A.J.C.)
4. Industrial Design Engineering Department, Arts and Humanities Faculty, Metropolitan Institute of Technology (ITM), Medellín 050036, Colombia; giovannibarrera@itm.edu.co

* Correspondence: fc.cabrera@unesp.br (F.C.C.); michael.silva@unesp.br (M.J.S.)

Citation: Costa, N.L.; Hiranobe, C.T.; Cardim, H.P.; Dognani, G.; Sanchez, J.C.; Carvalho, J.A.J.; Torres, G.B.; Paim, L.L.; Pinto, L.F.; Cardim, G.P.; et al. A Review of EPDM (Ethylene Propylene Diene Monomer) Rubber-Based Nanocomposites: Properties and Progress. *Polymers* **2024**, *16*, 1720. https://doi.org/10.3390/polym16121720

Academic Editors: Phuong Nguyen-Tri, Ana Pilipović and Mustafa Özcanli

Received: 23 April 2024
Revised: 3 June 2024
Accepted: 5 June 2024
Published: 17 June 2024

Copyright: © 2024 by the authors. Licensee MDPI, Basel, Switzerland. This article is an open access article distributed under the terms and conditions of the Creative Commons Attribution (CC BY) license (https://creativecommons.org/licenses/by/4.0/).

Abstract: Ethylene propylene diene monomer (EPDM) is a synthetic rubber widely used in industry and commerce due to its high thermal and chemical resistance. Nanotechnology has enabled the incorporation of nanomaterials into polymeric matrixes that maintain their flexibility and conformation, allowing them to achieve properties previously unattainable, such as improved tensile and chemical resistance. In this work, we summarize the influence of different nanostructures on the mechanical, thermal, and electrical properties of EPDM-based materials to keep up with current research and support future research into synthetic rubber nanocomposites.

Keywords: EPDM; synthetic rubber; nanocomposites; nanofillers; properties

1. Introduction

The search for versatile, lightweight materials with suitable hardness, a large area, appropriate coercivity, low production cost, and the ability to modify their chemical and physical properties when subjected to external stimuli has grown in recent decades [1,2]. With the recent developments in nanotechnology and nanomaterials, producing materials with excellent properties that meet their designed requirements has become possible. Among these classes of materials, polymer-based nanocomposites, especially those based on rubber matrices, such as ethylene propylene diene monomers (EPDM) [3,4], styrene–butadiene–styrene (SBS) [5,6], and natural rubber (NR) [7–9], are significant.

EPDM is one of the most studied and used synthetic rubbers by the scientific community and industry [10]. EPDM rubber is a member of the unsaturated polyolefin family; it is prepared via the polymerization of propylene and ethylene with a small quantity of non-conjugated diene (approximately 3–9%) [11–14]. Figure 1 illustrates the chemical structure of EPDM, which contains a saturated hydrocarbon backbone.

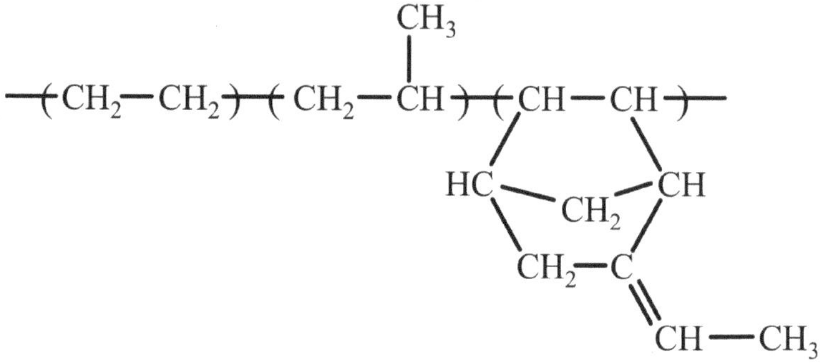

Figure 1. Representation of the chemical structure of EPDM.

In addition to its excellent heat resistance, elasticity and flexibility at low temperatures, weathering properties, oxidation resistance, ozone resistance, and aging resistance, EPDM has also been applied to a wide variety of applications, including weather-stripping for automobiles, roofing membranes, sealants, tubing, belts, radiators, thermal insulation, and electrical insulations [11,15–20].

Typically, the properties of EPDM rubber are improved through vulcanization by creating a crosslinked structure using sulfur as the vulcanizing agent and chemical accelerators [21]. However, nanofiller-based reinforcement has also been used as an alternative method for improving the physical and chemical properties of the EPDM matrix as well as reducing its production cost [22,23]. One advantage of EPDM compared to synthetic rubber and NR is its capacity to accept large amounts of nanofillers, which can significantly improve its properties [24]. Various nanofillers with inherent properties have been used to reinforce EPDM rubber and obtain nanocomposites with improved final properties. Among these nanofillers, organic-based nanoparticles, such as carbon nanotubes (CNTs) [25,26], multi-walled carbon nanotubes (MWCNTs) [27,28], graphene [29,30], and carbon black (CB) [31–33], and inorganic-based nanoparticles, such as nanoclay [34–36], nanosilica [37,38], montmorillonite [11,39], and polyhedral oligomeric silsesquioxane (POSS) [40,41], should be highlighted.

The incorporation of different nanofillers into the EPDM matrix can expand the application of EPDM in various industrial and technological areas, for example, EMI (electromagnetic interference), shielding effectiveness in electric devices [29,42,43], sensors [44,45], nuclear applications [46], thermal and electrical insulating materials [11], and solid rocket motor insulation [38,47], among others.

These diverse applications are due to the improved final mechanical and morphological properties caused by the dispersion of different nanofiller types in the polymeric matrices. For example, the dispersion of nanofillers in EPDM-based nanocomposites can improve the thermal stability and conductivity of the nanocomposites, allowing for application in the field of thermal energy storage and thermal management of electronic devices [48]. George et al. [38] evaluated the insulation behavior of nanosilica-filled EPDM/Kevlar fiber and established that a 220% improvement in char residue was observed with enhanced thermal stability and mechanical properties. The authors evaluated the insulation behavior of nanosilica-filled EPDM/Kevlar fiber and established that a 220% improvement in char residue was observed with enhanced thermal stability and mechanical properties [38]. These properties are attributed to the load of Kevlar in the composites, which maintains thermal stability and increases the char residue. However, nanosilica does not significantly improve the composite [38]. Guo et al. [49] studied the effects of the incorporation of MWCNTs on the char residue and carbothermal reduction reaction in EPDM-based nanocomposites. They observed that the MWCNT network structure in the char has a directional effect on in situ SiC (silicon carbide) formation, which improves the ablation resistance of the nanocomposite. Zhang et al. [50] investigated the impact of adding flame-retardant and dendrimer-modified organic montmorillonite (FR-DOMt) on the thermal and mechanical properties of EPDM-based nanocomposites. It was observed that an increase in the quantity of FR-DOMt in the nanocomposite enhanced the thermal stability and flame retardance of the EPDM matrix. Lu et al. [44] successfully prepared a highly stretchable and sensitive sensor based on a graphene nanoplatelet (GnPs)/EPDM nanocomposite with excellent heat dissipation performance. Owing to the good dispersion of the GnPs in the EPDM-based nanocomposite, the nanocomposite-based sensor had a low percolation threshold value when using approximately 2.9 wt.% GnPs and a thermal conductivity of 0.72 W/m K when using 7 wt.% GnPs. The results indicate that the GnPs/EPDM sensor shows excellent potential for monitoring deformation and motion in the human body [44].

Several types of nanofillers have been incorporated into EPDM for reinforcement purposes; one of the most important parameters used to characterize reinforcement materials is their specific surface area, which should be in the order of hundreds of square meters per gram, as it is directly related to the size of the material particles [51]. Another critical parameter used to describe these reinforcement materials is the relationship between the average length and the diameter, known as the aspect ratio [52]. In addition to the parameters above, the mechanical properties of the nanocomposites are improved through the good dispersion of the nanofillers, which can be achieved through different forms of processing [52].

The dispersion of nanofillers is intricately linked to three conditions. The first is related to nanofiller surface chemistry (presence or absence of active functional groups), which can compromise the compatibility between the filler and the polymeric matrix. The second is the interfacial interaction of the fillers with the polymeric matrix, which can be chemical or physical and depends on the formation of aggregates due to filler–filler interactions. The third condition is the correct choice of a polymeric matrix; there is a need for compatibility between the type of nanofiller and the polymeric matrix.

The parameters should be balanced to achieve the desired reinforcement properties in the nanocomposites. For example, when the nanofiller concentrations are relatively high, the quality of reinforcement and interactions between the matrix and the nanofillers can deteriorate owing to the smaller interfacial contact area [53]. Consequently, aggregates are formed that induce various localized stress concentrations, which leads to the initialization of defects, such as crack propagation and failure [54]. The relatively increased hardness and tension can improve the mechanical properties due to more nanofillers, but the abrasion resistance is reduced owing to concentrated stress points [55]. In such cases, it is necessary to improve the surface of the load with various treatments, for example, chemical treatments, so that the interfacial load–matrix interaction is optimized [56]. This causes multiple properties to improve, possibly reducing the volume of the incorporated load [56].

EPDM is an elastomer that exhibits acceptable and valuable mechanical and dielectric properties and shows specific resistance to several conditions, such as oxidation, chemical attack, and resistance to weather effects [57]. A modification using tungsten oxide nanoparticles was made by Sang et al. [58] and incorporated into EPDM foam to improve thermal storage capacity and seawater resistance; such properties could be potential applications in marine sportswear products, among other fields where these properties are desirable. In another study, Bianchi et al. [59] obtained EPDM foams that were filled with different amounts of paraffin, a typical phase change material (PCM) with a melting temperature of approximately 70 °C, to develop rubber foams with thermal energy storage (TES) capabilities [59]. This system utilizes a heat absorption/release principle, using materials called PCMs, which are excellent thermal conductors due to their high energy storage capacity [57,58,60–62]. The main characteristic of these materials is that they can be used for thermal insulation and heating and cooling purposes, especially in the field of Heating, Ventilation, and Air Conditioning (HVAC) [57,62,63]. Additionally, EPDM/paraffin compounds have been reported to be helpful in thermal energy storage applications in buildings [64,65]. As a positive consequence of materials with TES properties being used in construction, TES technology helps to balance the energy demands of buildings by reducing energy peaks for air conditioning processes [66,67].

EPDM rubbers can be classified as amorphous polymers, and the addition of reinforcing fillers is considered a significant factor in improving the mechanical properties and strength of rubber regarding unfilled EPDM rubber [68]. An essential application in industrial electric fields is related to the use of fillers and nanofillers in EPDM rubbers, which are used to suppress space charge in high-voltage direct current (HVDC) cables, which is a factor that promotes insulation breakdown. Yang et al. discussed using various types of nanofillers [69]. The first type of fillers includes inorganic nanofillers, such as ZnO and zeolite, inorganic nano-carbon, oxides such as SiO_2, and organic chemical modification, such as crosslinked polyethylene/organic nano-montmorillonite (XLPE/O-MMT), another polymer that exhibits a positive response in the presence of low electrical fields relative to suppressing space charge accumulation [70]. On the other hand, carbon series of nanofillers, such as graphene and graphene oxide (GO), including a carbon nanotube, exhibit a positive effect on space charge suppression, suggesting their use as potential filler materials for (HVDC) cables. Another critical problem associated with high-voltage power cables is related to humid environments, which can promote a water tree phenomenon responsible for the degradation of dielectric properties, consequently reducing the lifespan of XLPE when improving the properties of these cables [71]. Qingyue et al. [72] used three kinds of dopants, nano-montmorillonite (MMT), spherical nano-silicon dioxide (SiO_2), and polar ethylene-vinyl acetate copolymer (EVA), over XPLE samples using a melt blending process to analyze several properties. The results showed that final composites exhibit a better elastic modulus, better toughening, and the lowest growth of the water tree effect compared with XLPE without dopants [72].

Blends of EPDM/styrene butadiene rubber (SBR) have been analyzed by some authors [73–75] regarding the effect of different nanocomposite blends, such as reinforcing agents, and the consequent impact on mechanical properties. Broad engineering applications can be found using these kinds of (EPDM/SBR) rubber compounds, such as tires, seals, conveyor belts, and electrical cables, among others. The reinforced composite is essential for this (EPDM/SBR) composite. The addition of nanoclay (NC) and nanosilica (NS) can promote better results regarding tensile strength, tear strength, hardens, and abrasion resistance [73]. On the other hand, Badahar and Zawawi [75] used single-walled carbon nanotubes (SWCNTs) to reinforce EPDM/SBR nanocomposites, and the results showed values close to a 20% improvement in harder, better rheology, and viscoelastic response as well as thermomechanical performance, which are essential characteristics for improving shock absorption engineering applications. Deng et al. [76] designed and analyzed a double foaming system composed of EPDM/SBR and thermoplastic rubber/TPR composite foam to improve the mechanical properties of this polymeric foam, particularly in tear strength, which is considered an essential characteristic in terms of reducing costs and achieving greater toughness, with potential applications in automotive, aerospace, and construction. The authors suggest that the bimodal structure of the foam cells and the reduction in the crystallization of molecular chains are critical factors in achieving higher tear strength [76].

Given the importance of EPDM as a polymeric matrix and its wide application in industry and technological developments, this manuscript aims to present a brief review of the thermal, mechanical, electrical, morphological, and rheological properties of EPDM-based nanocomposites modified with nanofillers. Overall, this review emphasizes the role of nanofillers in improving EPDM-based nanocomposites across multiple dimensions, including mechanical, thermal, electrical, morphological, and rheological properties. As a result of these studies, a broader understanding of EPDM-based nanocomposite properties and their applications is provided.

2. Mechanical Properties

This section discusses improvements in the mechanical properties of EPDM-based nanocomposites incorporating nanofillers, including tensile strength, hardness, and abrasion resistance.

Kermani et al. [77] produced hybrid nanocomposites based on EPDM–butadiene rubber (XSBR) mixed with different concentrations of MWCNTs, using maleic anhydride grafted on EPDM rubber (EPDM-g-MA) as a compatibilizer. The effect of the MWCNT concentrations on the mechanical properties of the hybrid nanocomposites was studied. The authors noted that EPDM-g-MAH improved the distribution of the MWCNTs within the polymer matrix, resulting in a uniform distribution of MWCNTs with a small amount of aggregation. In addition, the mechanical properties, such as modulus, tensile strength, hardness, and elongation at break, of the compatible EPDM/XSBR nanocomposite were better than those of incompatible composites. Vayyaprontavida Kaliyathan et al. [78] showed that rubber blends are pivotal in tailoring the specific properties sought in rubber-based products. The performance of these blends is significantly influenced by their underlying morphology [78]. Detailed examination of this morphology is made possible by applying various microscopy techniques, allowing for a deeper understanding of their structure and behavior. Notably, achieving miscibility in rubber blends is a rarity; however, additives can be introduced to enhance compatibility between different rubber types. This approach serves as a method to bridge the gap in terms of properties and create more harmonious blends despite the inherent challenges of achieving complete miscibility in these mixtures [78]. As a result, Vayyaprontavida Kaliyathan et al. [78] analyzed the mechanical properties of rubber–rubber blends. A notable enhancement was observed in the mechanical characteristics of these blends throughout the blending process. Employing specific mixing equipment such as internal mixers and extruders proves instrumental in refining the formation of these blends, ensuring a more uniform and optimized composition [78]. Addressing incompatibility issues within blends and incorporating compatibilizers emerge as a solution to render these initially incompatible blends more harmonious. This addition bridges the gap between disparate rubber types, facilitating a more cohesive and compelling blend.

According to Sowińska et al. [79], Ionic Liquids (ILs) stand out in elastomer technology owing to their distinct and advantageous properties. Supported Ionic Liquid-Phase (SILP) materials serve to anchor ILs onto solid supports, a technique explored in this study to understand its impact on the properties of EPDM elastomers [79]. The investigation delves explicitly into how SILPs exert control over the vulcanization process without causing crosslink density or thermal stability alterations. SILP materials play a pivotal role in regulating the vulcanization process of EPDM without causing any degradation in crosslink density [79]. Moreover, they contribute positively by enhancing the tensile strength and hardness of the resulting vulcanizates. These modifications signify an improvement in the mechanical properties of the EPDM material facilitated by SILP incorporation. EPDM vulcanizates exhibit robust resistance to thermo-oxidative aging when subjected to a temperature of 100 °C. This underscores the material's durability and stability under harsh environmental conditions [79]. Furthermore, the analysis conducted through Dynamic Mechanical Analysis (DMA) sheds light on the influence exerted by SILPs on the mechanical loss factor, also known as tan delta (tan δ), demonstrating their impact on the viscoelastic behavior of EPDM [79]. The findings from this study [79] highlight that while employing SILPs, marginal enhancements are noted in the tensile strength and hardness of the resultant vulcanizates, suggesting a potential for SILPs to exert a subtle yet discernible influence on the mechanical characteristics of EPDM elastomers [79].

Vishvanathperumal and Anand [80] prepared EPDM/SBR hybrid composites reinforced with nanoclay (NC) and nanosilica (NS) and investigated the synergistic effect of NC and NS on the mechanical properties of the EPDM/SBR hybrid nanocomposites. Three different crosslinking systems were used: dicumyl peroxide, sulfur, and a combination of peroxide and sulfur. The best tensile strength was achieved using 4 phr of NC; using higher mass concentrations reduced the tensile strength [80]. This behavior was attributed to the adsorption of the MBTS and TMTD accelerator on the silica surface by the silanol groups. The authors found that sufficient Si-69 in filler reduces the adsorption process. In addition, the high NS content exhibits strong NS–NS interactions, compromising abrasion resistance, among other properties [80]. The concentrations of NS (4 phr) and NC (7.5 phr) in the sulfur-cured EPDM/SBR hybrid nanocomposites play an essential role in the microstructural and mechanical properties of the nanocomposites. The study revealed that nanocomposites containing 7.5 phr of NC and NS exhibited the best mechanical and abrasion resistance characteristics [80].

Within Rostami-Tapeh-Esmaeil et al. [81], comprehensive insights are offered into a spectrum of aspects concerning rubber foams. This includes detailed information encompassing formulation strategies, curing methodologies, diverse production techniques, an intricate study of morphology, an extensive examination of properties, and an exploration of the wide-ranging applications these rubber foams find across different industries [81]. Rubber foams exhibit superior attributes such as increased flexibility, abrasion resistance, and enhanced capabilities for energy absorption [81]. Formulation choices and various processing parameters significantly influence the physical and mechanical properties of these foams. Tensile testing is infrequently employed with foams primarily due to challenges in securely gripping the material for analysis. Instead, compressive loading finds more prevalence, particularly in applications involving cushioning and packaging [81]. The mechanical behavior of foams is significantly influenced by their morphology and the thickness of their cell walls [81]. In particular, an increase in the dispersion of cell wall thickness is observed to correspond with reductions in both Young's modulus and shear modulus. Introducing fillers such as rice husk powder and kenaf particles effectively enhances the tensile strength and stiffness of foams [81]. Moreover, the compression stress experienced by foams tends to be higher in instances where more giant cells and a higher content of foaming agents are present. Additionally, a higher content of foaming agents correlates with increased compression set observed in foams, indicating a greater degree of deformation or relaxation under sustained compressive loads [81].

To enhance the properties of an NR/EPDM blend, Lee et al. [82] introduced a phlogopite mineral filler into the mixture. This process involves the incorporation of compatibilizers, specifically Amino Ethyl Amino Propyltrimethoxy Silane (AEAPS) and Stearic Acid (SA). The resulting NR/EPDM/phlogopite/AEAPS composite displayed comparable properties to those observed in the NR/EPDM/CB composite, suggesting similar performance between the two materials [82]. This composite of NR/EPDM/phlogopite/AEAPS holds the potential to serve as a viable alternative to costly fillers like CB, offering similar properties and performance while potentially reducing production costs [82].

A novel approach has been introduced to enhance the toughness of polypropylene (PP) by incorporating nanostructured rubber into the material by Chang et al. [83]. This involves a scalable method for producing PP nanocomposites, leveraging crosslinked EPDM nanofibrils to bolster toughness. As EPDM loading increases, the yield strength and elastic modulus of the PP nanocomposites consistently decline. This trend suggests the incorporation of relatively softer inclusions into the matrix. In conventional rubber-toughened systems utilizing rubber microparticles, a notable decline is observed in both yield strength and modulus. This decrease surpasses what the Rule of Mixtures theoretically predicts. The measurement of the viscoelastic moduli (G', G'') and complex viscosity (η) in both PP nanocomposites and fibrillar composites illustrates the distinct contributions of individual polymeric constituents (PP and EPDM) to the overall material response [83].

Azizli et al. [84] produced several elastomeric nanocomposites by mixing NR with EPDM compatible with maleic anhydride (EPDM-g-MA) reinforced with different amounts of GO (1.0, 3.0, 5.0, 7.0, and 10.0 phr); the mechanical properties were evaluated theoretically and experimentally. The results showed that incorporating GO improved the nanocomposites' mechanical properties, such as hardness, tensile strength, elongation at break, and modulus [84]. These improvements are due to variations in the NR content that increase the crosslinking points, the size of the surface area, and the dispersion of GO in the nanocomposites [84].

A study by Burgaz and Goksuzoglu [85] synthesized isotropic Magnetorheological Elastomers (MREs) with potential use in automotive applications. Among the formulations tested, EPDM/CB/Carbonyl Iron Powder (CIP) MREs demonstrated superior properties compared to EPDM/CB/Bare Iron Powder (BIP) MREs. The particles known as CIP displayed distinct characteristics in contrast to BIP particles within these MRE formulations. CIP particles exhibited lower damping factor, Payne effect, elastic modulus, and hardness but demonstrated higher values in tensile strength and elongation at break. Moreover, the sample comprising 5 phr CIP and 60 phr CB showcased a substantial MR effect of 77%, signifying the material's responsiveness to a magnetic field [85].

Several reported studies demonstrate nanofiller dispersion's effect on the mechanical properties of EPDM-based nanocomposites. Nazir et al. [86] tested the effect of boron nitride (BN) on the breakdown, surface tracking, and mechanical performance of EPDM for high-voltage insulation. It was established that by obtaining a hybrid composite formed using BN-based micro and nanoparticles, better tensile strength (σ_t) and elongation at break (ε_b) values were achieved; for example, the EPDM matrix filled with BN micro-25 wt.% + nano-5 wt.% particles presented better mechanical properties (σ_t = 5.2 MPa and ε_b = 353%) than EPDM filled with BN micro-30 wt.% (σ_t = 4.06 MPa and ε_b = 302%) [86]. In rubber nanocomposites, the dispersion of nanofillers and the interfacial interactions between the rubber matrix and nanofillers have a significant impact on their properties. The limits are associated with the interaction of the filler with the polymeric phase, as specified by the author [86], and a combination of fillers with sizes on the order of micro and nano, for example, to EPDM-based nanocomposite with 25 wt.% microfillers and 5 wt.% nanofillers exhibited tensile strength of more than 5 MPa and elongation of more than 300%, but not for other combinations or individual load applications. It is also possible to use percolation theory to describe the mechanical behavior of composite materials [9]. This is because when nanofillers are dispersed in a polymeric matrix, they can form a three-dimensional network of geometric contact between them in the composite when their concentration is equal to the mechanical percolation threshold of the nanocomposite [9]. Consequently, when the nanofiller concentration is equal to or greater than the percolation threshold and mechanical tension is applied to the composite, part of the mechanical stress is efficiently transferred from the matrix to the nanofillers [9]. Consequently, when the concentration of nanofiller in the composite approaches the percolation threshold, the mechanical resistance and elastic modulus of the composite tend to increase. This behavior may be related to a significant reduction in the inter-particle distances and an increase in the total surface area of the particles in the co-filled composites [86]. Molavi et al. [87] evaluated the effect of MWCNTs on the mechanical and rheological properties of silane-modified EPDM rubber (VTMSg–EPDM). Even at deficient MWCNT concentrations, enhanced physical properties were observed for the VTMSg–EPDM matrix owing to the grafting reaction that improved the dispersion of the MWCNTs [87]. Nanocomposites with 1.5 wt.% MWCNTs exhibited remarkable elastic modulus and tensile strength improvements. Increasing the MWCNT content also positively affected the storage modulus (G') and complex viscosity (η^*), both of which were enhanced; however, the difference in the G' and η^* values decreased as the frequency increased [87]. As a result of the reduced interparticle distance, the total surface area of the particles increases, improving the processability of the material. In addition to the excellent dispersion and low nanofiller–nanofiller interaction, the ease of processing and incorporation of nanofillers into polymeric matrix increases the nanofiller–matrix interaction. This good dispersion resulted in improved mechanical and thermal properties for the nanocomposite.

3. Rheological Properties

The curing characteristics of EPDM-based nanocomposites filled with nanoparticles are established during the vulcanization process. They can be measured using the materials' rheometric parameters [13,84,88], including the minimum (ML) and maximum torque (MH), scorch time (ts_2), curing time (t_{90}), and cure rate index (CRI), which will be discussed in this section.

Khalaf et al. [22] produced a hybrid EPDM/chitin/nanoclay bionanocomposite using a binary loading system formed by nanoclay and chitin from shrimp shells. The purpose of this study was to reuse the residue from previously treated shrimp shells and to evaluate the rheological properties of the bionanocomposite with the addition of nanoclay. In addition, the physical–mechanical properties, swelling parameters, morphological analysis, and water absorption of the vulcanized polymers were studied. Sulfur was used as the crosslinking agent in the EPDM rubber containing 55% ethylene. The curing characteristics were measured using a Monsanto oscillating disc rheometer (ODR -100 s) at 152 °C according to ASTM D-2084-07 [22,89]. Different concentrations of nanoclay (3, 5, 7, and 10 phr) were incorporated into the EPDM/chitin mixtures (5 and 10 phr). The results showed that the ML increased with the nanoclay quantity. An increase influenced the viscosity of the EPDM-based nanocomposite; the decrease in fluidity was associated with processing [22]. By contrast, the MH value showed a contradictory trend as a decrease in the MH values with increased nanoclay load concentration was observed. The authors established that as the amount of nanoclay in the prepared vulcanized polymers increased, there was a reduction in t_{90} and an increase in the ts_2 and CRI values. This behavior was attributed to the polarity of the silicate layers of the nanoclay, which contributed to the formation of hydrogen bonds and accelerated the curing process at low nanofiller concentrations [22]. Therefore, the authors concluded that the ideal binary charge concentration for EPDM/chitin/nanoclay was 5 and 10 phr of chitin with 3 phr of nanoclay for EPDM rubber applications in industrial heat exchange processes. The different geometries, particle sizes, and surface activities of nanofiller influence the ML, which increases with an increase in nanoclay content on hybrid EPDM/chitin/nanoclay composites [22]. The increase in pre-curing time is a result of an increase in the specificity of the rubber compound and the decrease in fluid capacity associated with the processing and dispersion of the load. Contrary to the ML, the MH decreased with increasing nanoclay quantities and possible interactions with EPDM matrixes [22]. As the nanofiller content in the prepared vulcanizates increases, the curing speed index and optimal curing time decrease [22]. There is a possibility that this is due to the polarity of the clay silicate layers, which facilitate the formation of hydrogen bonds and thus accelerate the curing process at low nanoclay concentrations [22]. In addition, it is possible to reduce waste and promote sustainability by reusing materials such as chitin, from shrimp shells as nanofillers in EPDM. Incorporating recycled materials into EPDM-based nanocomposite can enhance their biodegradability, contributing to the preservation of the environment. Using recycled chitin as a nanofiller in EPDM-based nanocomposite allows for a more comprehensive lifecycle assessment of the rubber products, considering their environmental impact from the point of production to the end of disposal [22]. EPDM-based nanocomposites containing recycled chitin fillers may have improved durability and longevity, potentially extending their useful life and reducing their environmental impact [22]. The manufacturing process can be made more cost-effective using recycled materials by repurposing waste materials, such as chitin from shrimp shells. Using sustainable and cost-efficient production practices, EPDM-based nanocomposite containing recycled chitin fillers may provide a more economical alternative to traditional rubber formulations [22].

Sowińska et al. utilized a rotorless curemeter to analyze the rheological properties of rubber compounds [79]. The samples underwent vulcanization at 150 °C employing t_{95} and t_{80} (at the point when the torque increases by 95% and 80% of the maximum value, respectively) as the vulcanization times. This study aimed to determine the impact of SILPs on the curing characteristics of EPDM compounds. Figure 2 displays the rheometric curves representing the EPDM compounds' behavior during the process.

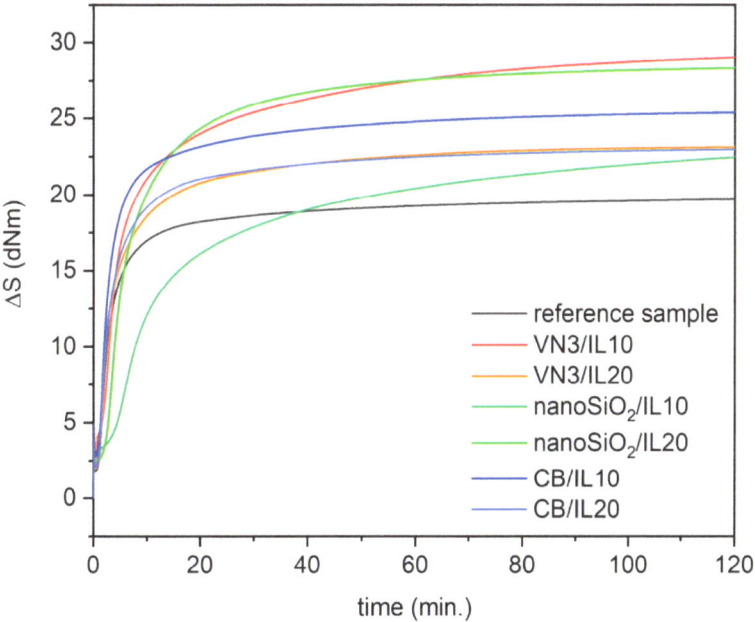

Figure 2. Rheometric curves of EPDM/SILPs compounds. (This figure is from ref. [79]).

As per the rheological analysis of the developed EPDM biocomposites by Chen et al. [90], both EPDM/WF and EPDM/Si-WF composite samples exhibited typical shear thinning behavior and reduced viscosity as the shear rate increased. The higher WF/Si-WF loadings increased the viscosity due to the filler-induced inhibition of molecular movement. Comparing EPDM/WF with EPDM/Si-WF samples, the former showed loosely stacked viscosities across different WF loading levels, notably with the 50 WF sample displaying significantly higher viscosity [90]. Conversely, EPDM composites with Si-WF displayed closely aligned viscosities across varying filler levels, attributed to improved interfacial compatibility through silane modification. This enhancement resulted in minimal viscosity variation among different filler levels. In the case of samples with DCP, their viscosity curves surpassed that of pure EPDM. The increased viscosity in Si-WF + DCP-based composites is anticipated to stem from crosslinking or molecular entanglement initiated by DCP [90].

Abdelsalam et al. [91] focused on the rheological parameters of *ML*, *MH*, tc_{90}, and scorch time (ts_2). *ML* gauges the stiffness and viscosity of an unvulcanized elastomer, signifying its processability, observed at the lowest point on the torque–time plot. *MH* represents the peak torque during curing, which is directly linked to the compound's modulus, reflecting its stiffness. The difference between *MH* and *ML*, Δ*M*, typically estimates the crosslink density. The tc_{90} marks the duration to reach 90% of *MH*, while ts_2 signals the vulcanization initiation as torque increases by two units [91]. The rise in *ML*, *MH*, and Δ*M* with escalating CB content reveals enhanced viscosity in the NR/SBR/NBR matrix, suggesting improved interactions between CB and the polymer matrix. S_1 (without filler) displays the lowest *ML*, indicating easier processing than CB-loaded samples [91]. Higher *MH* in filled rubber blends indicates robust interactions between the nanofiller and the polymer matrix, which is attributed to the smaller particle size promoting increased rubber-filler interactions and more excellent resistance to molecular motion, leading to elevated torque values. The rise in Δ*M* relates to increased crosslink density [91]. Moreover, as filler loading increases, both ts_5 and tc_{90} decrease [91]. Lower ts_5 signifies better processing ability associated with the heat history that the rubber can withstand before reaching a crosslinked state.

This potentially increases the CRI, showcasing the application of CB as a reinforcement and an effective accelerator of NR/SBR/NBR vulcanization, reducing process duration in tire manufacturing. The decline in ts_2 is attributed to increased crosslinking, while the reduction in tc_{90} is linked to heightened energy input and heat build up due to increased viscosity during mixing. The curing rate index elevation with filler addition stems from CB's basicity, expediting the vulcanization reaction [91].

4. Swelling Properties

The blending of the two materials has been a focus of interest for specific applications in several studies for more than 40 years, including the performance of polymer systems and their variations in elasticity and viscosity at different strain rates and temperatures [92–95]. An example of a polymer blend is EPDM, which is used in industrial applications owing to its high resistance to aging, heat, polar solvents, and other chemicals and its outstanding electrical characteristics and mechanical properties [96]. There has been considerable interest in the swelling properties of this type of material in several research studies [97–101]. In most cases, this property is associated with the absorption of liquid, which improves the material so that it can exhibit changes in dimension up to 100% of its initial volume [102].

Vishvanathperumal and Anand [103] showed that several parameters affect the swelling properties of reinforced EPDM with nanosilica, such as the matrix shape, type and geometry of filler, the reaction of the solvent on the matrix, temperature, and penetrant, among others, in addition, they identified that crosslinks restrict the swelling-induced extensibility of rubber polymeric chains by reducing the dispersion of the solvent on the matrix. Stelescu et al. [96] also identified parameters associated with swelling: reinforcement density, crosslink density, and solvent absorption. Additionally, it has been shown that the amount of oil used to improve the mixing and processing while maintaining the stiffness influences the swellability of the compounds due to the oil particles that impact the crosslinking process, affecting the stiffness and leading to softness [104].

On the other hand, Abdelsalam et al. showed that swelling could be improved by using a silane-like coupling agent when Al_2O_3 is added to the EPDM [105]. Likewise, it has been demonstrated that the swelling resistance is possibly improved via the addition of nanocomposites as nGO with amounts around 6 phr [106]. Colom et al. confirmed the inverse relationship between swelling and crosslink density by showing that increasing this last property reduces the possibility of toluene entering the tighter polymeric network, reducing swelling [107]. A similar phenomenon was demonstrated by Simet et al. when swelling samples within xylene over 24 h, achieving up to 30% variation in the swelling ratio, which is associated with the degradation of the structure due to interactions with the solvent [108]. To minimize dimensional changes, it is possible to use pre-swelling, as carried out by Nijibah et al., which significantly reduces the swelling or shrinking of the membranes when compared to the previous process [109].

5. Thermal Properties

Thermal properties refer to the characteristics of a material that describe its behavior in response to changes in temperature. The influence of different types of filler on the thermal behavior of EPDM-based nanocomposites has attracted the interest of researchers since the dispersion of nanoparticles can attribute new properties to the final composite. In this sense, properties such as thermal stability, thermal conductivity, melting point, residue, and glass transition temperature (T_g) can be studied. These analyses include differential scanning calorimetry (DSC), thermogravimetric analysis (TGA/DTGA), dynamic mechanical thermal analysis (DMTA), and thermal conductivity. This section focuses on studying the thermal properties of EPDM-based nanocomposites using some of the main thermal analysis techniques.

5.1. TGA Analysis

Thermogravimetric (TGA/DTGA) analysis evaluates the mass of a sample as a function of temperature using a controlled atmosphere, such as inert or oxidizing gas [110]. Using this technique, changes in the mass (loss or gain) and decomposition events of materials, including those based on EPDM, can be analyzed, and the effect of nanofiller dispersions on the thermal stability of the polymeric matrix can be evaluated [110,111].

Several studies have used TGA/DTGA analysis to evaluate the effect of nanofiller dispersions on EPDM-based nanocomposites, such as carbon-based materials (CB, non-functionalized and functionalized carbon nanotubes (fCNT), graphene, etc.) and inorganic-based particles (nanosilica, nanoclay, tungsten bronze nanorods, tungsten oxide, etc.), among others.

Introducing inorganic-based nanoparticles into EPDM-based nanocomposites influences the thermal profile and stability. Sang et al. [58] fabricated a bio-EPDM/tungsten oxide nanocomposite foam with improved thermal storage and seawater resistance and evaluated its thermal properties using TGA/DTGA analysis. Figure 3a shows the thermal profile of a bio-EPDM-based nanocomposite containing various mass ratios of TBNR (tungsten bronze nanorods). Two mass loss events are observed in the 25–700 °C temperature range. The first corresponds to a mass loss of 2% due to the decomposition of the hydrocarbon chain, and a second mass loss event occurs between 430 and 590 °C due to the decomposition of the EPDM polymer. Figure 3b indicates that the maximum decomposition rate temperature (T_m) of the bio-EPDM foams shifts to a higher temperature with an increase in the amount of TBNRs. According to the authors, this behavior indicates that the interfacial interaction between the chain on the surface of the TBNRs and the bio-EPDM matrix enhanced the chemical stability of the bio-EPDM matrix and improved the thermal stability of the nanocomposite [58].

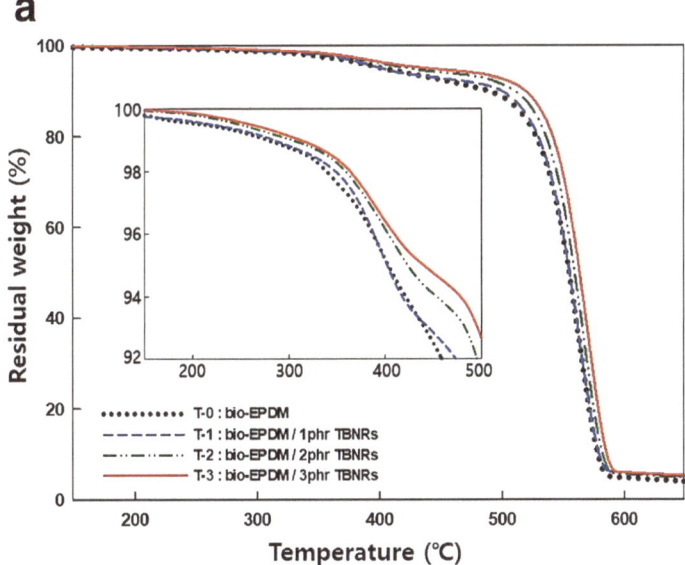

Figure 3. Cont.

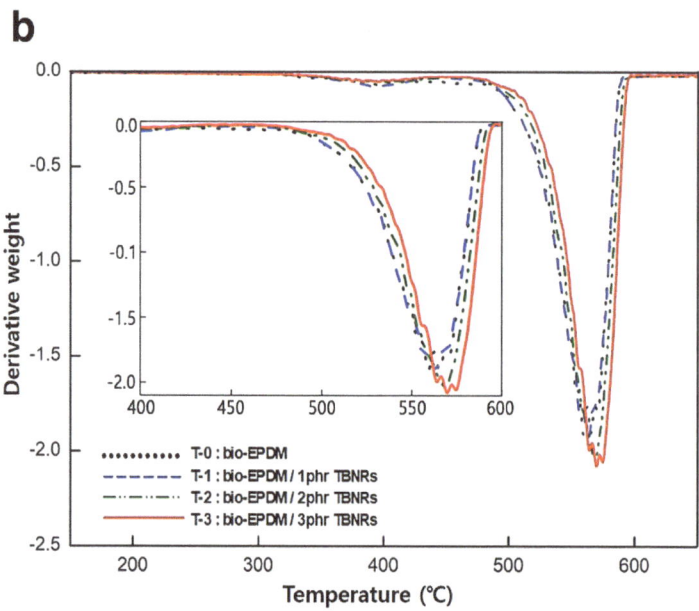

Figure 3. (**a**) TGA and (**b**) derivative TGA curves of bio-EPDM-based nanocomposite with various mass ratios of TBNRs. (This figure is from ref. [58] licensed under a Creative Commons Attribution 4.0 International License).

Rana et al. [112] observed similar behaviors for EPDM-based nanocomposites filled with varying amounts of nanoclay using TGA analysis. According to the authors, the mass loss in the main event (between 300 and 500 °C) is lower for neat nanoclay than for the nanocomposites containing 6, 4, and 2 wt.% nanoclay. In contrast, EPDM presented a higher mass loss between 30 and 690 °C. A shift in the T_m peak (DTG curve) to a higher temperature is observed with an increase in the nanoclay quantity relating to neat EPDM. This shift was attributed to the increased thermal stability of the EPDM-based nanocomposite with the increased amount of nanoclay [112]. Zhang et al. [113] also used TG/DTG analysis to evaluate the behavior of an EPDM-based nanocomposite with a crosslinked interfacial design containing nanoclay. The results obtained demonstrate that the incorporation of nanoclay did not influence the thermal profile of the EPDM-based nanocomposite; however, the residue percentage increased significantly with increasing nanoclay content, indicating that the nanoclay content influences the thermal stability of the EPDM matrix.

The influence of organic-based nanoparticles on the thermal properties of EPDM-based nanocomposites has also been investigated in several studies. To assess the dispersion effect of MWCNTs on the physicochemical properties of EPDM insulation for solid rocket motors, Guo et al. [114] used TG/DTG analysis to evaluate the thermal stability and ablation characteristics of two formulations of EPDM (with and without MWCNTs). The study indicated that adding MWCNTs increased the residue rate after thermal decomposition and suppressed the consumption reaction of the char layer under high-temperature conditions [114]. Guo et al. [49] studied the effects of the incorporation of MWCNTs on the char residue and carbothermal reduction reaction in EPDM-based nanocomposites. They observed that the MWCNT could significantly enhance the rate of carbothermal reduction due to their large specific surface areas and excellent activity, resulting in a rapid increase in the partial pressure of SiO and CO inside the char layer. In addition to that, the MWCNT network structure in the char has a directional effect on in situ SiC (silicon carbide) formation, which improves the ablation resistance of the nanocomposite. In this sense, higher amounts of SiC can contribute to some challenging applications in the automotive and aerospace industries. To tailor the thermomechanical properties of EPDM-based nanocomposites, Iqbal et al. [115] incorporated silane-treated MWCNTs (S-MWCNTs) into an EPDM matrix using five different mass concentrations: 0, 0.1, 0.3, 0.5, and 1.0%. Using TG/DTG analysis, the authors observed that the thermal stability increased as the amount of S-MWCNTs in the EPDM matrix increased. This is due to the high thermal endurance and nanoscale interaction of the nanotubes, which restricts the thermal motion of the molecular polymer chains in the heating environment [115]. In addition, two main mass loss events were observed for the nanocomposites in the temperature range of 200 to 520 °C. The first event in the 200–450 °C temperature range was associated with the evaporation of aromatic oil, wax, and other processing ingredients. The second event between 450 and 520 °C was attributed to EPDM pyrolysis [115].

5.2. DSC Analysis

Dynamical Scanning Calorimetry (DSC) is a type of thermal analysis in which the heat flow from the sample and the reference material is measured (energy absorbed or released) as a function of temperature during heating, cooling, or at constant temperatures [110]. The DSC curve shows the exothermic and endothermic effects, specific heat capacity, reaction and transition enthalpies, and other effects during temperature variations [110].

DSC measurements are widely used in the thermal analyses of materials to evaluate the transition temperature of exothermic and endothermic events and the specific heat capacity of materials. Jeon et al. [116] studied the effect of tungsten bronze nanorod and nanoparticle (TBNR or TBNP) loading on the T_g of EPDM using DSC analysis, as shown in Figure 4. Second-order transitions in the DSC thermogram with T_g values equal to −48.3 °C and −47.3 °C for neat EPDM and EPDM/TBNR (3 wt.%), respectively, were observed; for the 3 wt.% EPDM/TBNP, a T_g of −46.5 °C was observed [116]. Only the second-order transition peak was observed in the DSC thermogram, indicating that the EPDM-based nanocomposite filled with TBNRs or TBNPs has an amorphous structure [116].

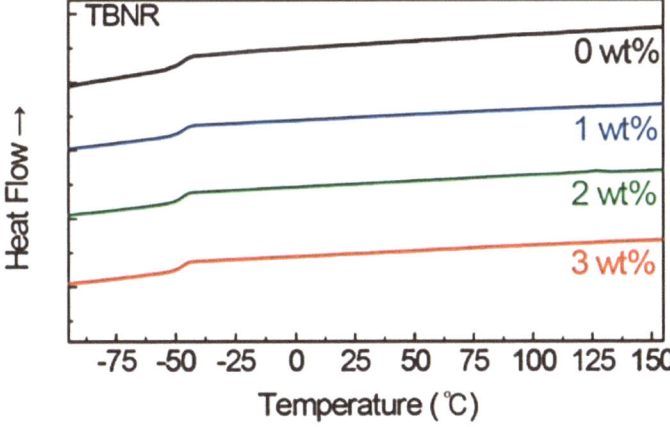

Figure 4. DSC curves of EPDM-based nanocomposite with various mass ratios of $Na_{0.33}WO_3$ TBNR. (This figure is from ref. [116] licensed under a Creative Commons Attribution 4.0 International License).

Iqbal et al. [115] investigated the influence of S-MWCNTs on the glass transition (T_g), crystallization (T_c), onset melting (T_{m1}), and peak melting (T_{m2}) temperatures of an EPDM-based nanocomposite. They observed that the T_g value was reduced with the incorporation of S-MWCNTs in the EPDM matrix; the T_g value of neat EPDM was −10.35 ± 0.10 °C and that of the EPDM-based nanocomposite with 1.0 wt.% S-MWCNT was −20.01 ± 0.10 °C [115]. By contrast, the opposite was observed for T_c; the T_c values were 4.55 ± 0.10 °C and 11.82 ± 0.10 °C for neat EPDM and the EPDM-based nanocomposite containing 1.0 wt.% S-MWCNT, respectively. These results were attributed to the formation of a nanofiller network in the EPDM matrix that restricts phase changes by interacting with the polymeric chains [115]. T_{m1} and T_{m2} were improved up to 58.28 °C and 11.21 °C, respectively, due to the incorporation of S-MWCNTs, which resisted polymer chain mobility by entrapping phonons within the nanofiller network. Additionally, incorporating S-MWCNTs resulted in higher thermal stability [115].

DSC analyses were used to investigate the T_g behavior of EPDM with incorporated montmorillonite (MMT) nanoclay [112]. The results obtained by Rana et al. [112] showed that an increase in the amount of nanoclay increased the T_g value of the EPDM-based nanocomposite, i.e., the T_g values were −62.7 °C, −58 °C, −47.3 °C, and −33 °C for neat EPDM, and 2 wt.%, 4 wt.%, and 6 wt.% for nanoclay, respectively. Zhang et al. [113] also obtained similar results for an EPDM/nanoclay nanocomposite with a crosslinking interfacial design. The study was conducted from −80 °C to 80 °C, and the authors observed an increase in the T_g value of the EPDM-based nanocomposite as the amount of nanoparticles amount increased [113]. The T_g value of neat EPDM was equal to −53.5 °C, which increased to approximately −49 °C with the incorporation of 10 phr (per hundred rubber) in the EPDM matrix [113].

5.3. DMTA Analysis

The DMTA technique is a type of thermal analysis that involves the application of periodic tension and evaluating the sampling effort with temperature variation, i.e., the study of stress can be performed as a function of temperature or frequency of minor variable mechanical stress [110]. From DMTA analysis, it is possible to calculate the complex modulus (E' and E''), tan δ, and the viscoelastic parameters of materials. During cyclic strain over a temperature range, the complex storage modulus E^* can be written as shown in Equation (1):

$$E^* = E' + E'' \tag{1}$$

E' represents the storage modulus, which can also be defined as the quantity of energy stored and retained during the load application cycle; E'' is classified as the loss modulus or energy dissipated during the cycle. tan δ can be calculated using the E''/E' ratio, as shown in Equation (2):

$$\tan \delta = \frac{E''}{E'} \tag{2}$$

Most of the materials developed by the industrial and technological sectors are subjected to dynamic deformations/strain and temperature variations during their life cycles. In this sense, it is essential to investigate the dynamic mechanical properties during the temperature variation of polymer nanocomposite and analyze the modes of cyclic deformation and relaxation processes similar to those of temperature.

DMTA measurements were carried out on neat EPDM and EPDM-based nanocomposite reinforced with silica (SiO_2) nanoparticles (Figure 5) to evaluate E', E'', and tan δ at temperatures ranging from −100 °C to 100 °C [117]. The behavior of E' and E'' as a function of temperature shows two well-defined regions, i.e., above (rubbery region) and below (glassy region) the T_g. For E' below the T_g (glassy region), an increase in of up to 10 wt.% of SiO_2 was observed in the EPDM matrix, which indicates an increase in stiffness and restricted polymeric chain movement; on the other hand, a decrease in the quantity of SiO_2 was observed [117]. In the T_g region, the E' value tends to decrease for all EPDM-based nanocomposite and neat EPDM, indicating decreased energy stored at temperatures superior to T_g [117]. Above T_g in the rubbery region, the E' value was higher for samples with 30 and 20 wt.% SiO_2; according to Mokhothu et al. [117], this behavior may be related to the degree of crosslinking in terms of the EPDM matrix and the content of the rigid dispersed phase. T_g was determined using tan δ as a function of temperature at the glass transition maximum peak; the values were −41.5 °C and −42.3 °C for neat EPDM and the EPDM/SiO_2 nanocomposite with 30 wt.% SiO_2, respectively [117]. A slight increase in the T_g value was observed for samples containing SiO_2 due to the reduced mobility of the

EPDM chain in the nanocomposites because the EPDM chains strongly interact with the SiO$_2$ nanoparticles, decreasing polymer chain mobility [117]. In another work, the authors observed similar behavior when studying in situ generated silica nanoparticles on EPDM, in which they evaluated the morphology, thermal, thermomechanical, and mechanical properties of EPDM-based nanocomposites [118].

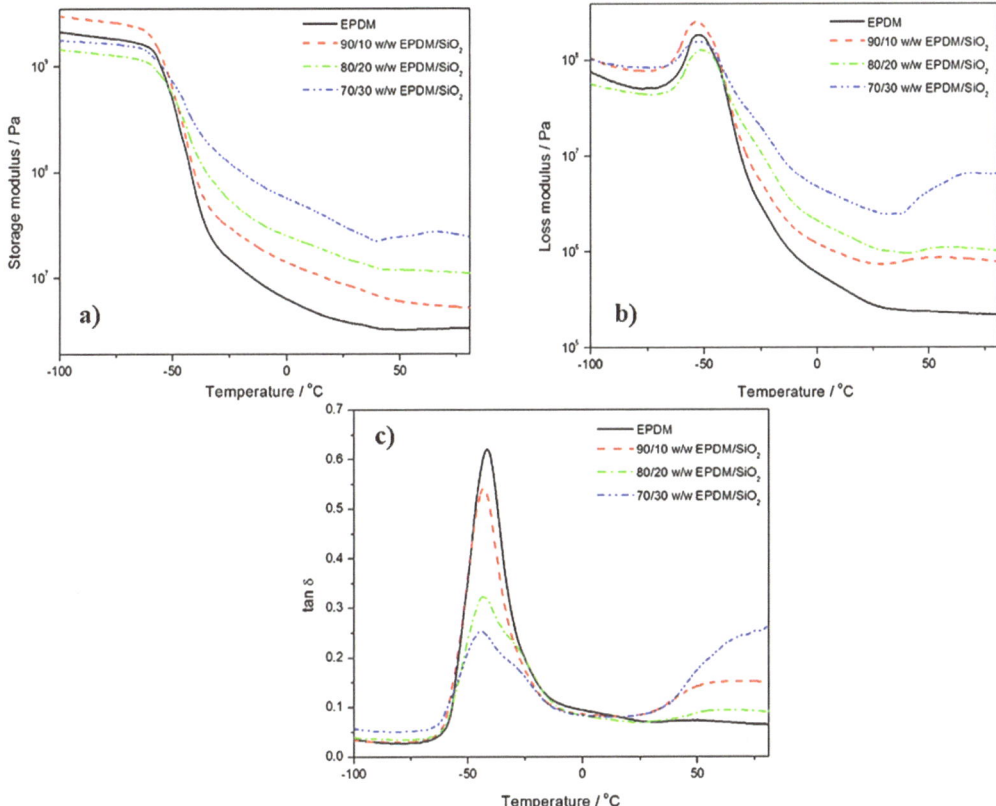

Figure 5. DMA curves of EPDM and EPDM/SiO$_2$ composites: (**a**) storage modulus (E'), (**b**) loss modulus (E''), and (**c**) damping factor (tan δ). (This figure is from ref. [116] licensed under a Creative Commons Attribution 4.0 International License).

Basu et al. [119] reported the effect of fire-safe and environmentally friendly nanocomposites based on layered double hydroxides (LDHs) and an EPDM matrix, which can be applied to many products, e.g., appliance hoses, radiator hoses in cars, washers, and insulation for water pipes and outdoor appliances. Through DMTA analysis, the authors observed good EPDM–nanofiller interactions due to a positive shift in the T_g (peak height of the tan δ vs. temperature) of the EPDM-based nanocomposite when the loading of LDH was increased from 4 to 100 phr, i.e., the T_g value was equal to -42.4 °C and -36.8 °C for EPDM—4 LDH (4 phr) and EPDM—100 LDH (100 phr) nanocomposites, respectively [119]. Additionally, this work showed a decrease in the peak height of tan δ with increasing LDH quantity. This behavior indicates the reinforcing characteristics of LDH in the EPDM-based nanocomposite [119]. The E' value as a function of temperature also presents an improvement with an increasing amount of LDH in the vitreous and rubbery regions; however, above T_g, the E' value decreased as the temperature increased for all samples [119]. This behavior of E' can be attributed to the excess amount of LDH, which provides a higher aspect ratio and a greater exposed surface area.

6. Morphological Properties

The morphological characterization of polymers is essential as studies on polymer structure and morphology provide insight into the macroscopic, morphological, and conformation properties that depend on the composites or nanocomposites used as fillers in their composition [120]. There are several microscopy techniques, such as scanning electron microscopy (SEM), transmission electron microscopy (TEM), and atomic force microscopy (AFM), that can be used to characterize polymers, such as EPDM. Morphological analysis allows for the determination of changes in the structures and properties of polymers caused by the addition of nanofillers. These changes include the dispersion of charges [2], micro-deformations [121,122], topographical changes [33,87], and modifications in the self-organization of a polymeric structure [114,123]. For example, Table 1 shows some observations obtained via the morphological analysis of EPDM matrices modified with various nanofillers or nanoparticles.

SEM is an analytical tool that enables the qualitative characterization of polymer surfaces and is used to identify characteristics such as the topography [35,87], morphology [35,77,124,125], and composition of materials. SEM studies are essential because the mechanical and thermal properties of a polymer depend on the type of filler dispersion used in the polymeric matrix [2,84].

Mohamed Bak et al. [126] studied the effect of nanoclay loading on blends of EPDM/silica–styrene–butadiene–rubber (S-SBR) nanocomposites. The fractured surface of the nanoclay-filled EPDM/S-SBR rubber blends was investigated using SEM images. The SEM observations showed that the fractured surfaces containing several tear lines, matrix crack lines, crack initiation, and crack propagation, as well as high nanoclay loading (10.0 phr), are the reasons for the formation of clusters and its contribution to the poor dispersion of the nanofiller in the EPDM matrix. Azizi et al. [123] studied EPDM and EPDM/silicone rubber composites with additives of modified fumed silica (MFS), titanium dioxide (TiO_2), and graphene. The SEM micrographs showed that the addition of graphene as an additive in the formulation of the composites facilitated the formation of MFS and TiO_2 particles that were homogeneously dispersed in the EPDM matrix.

Table 1. Some observations were obtained by morphological analyses with different nanocomposite charges added to the EPDM matrices.

Matrices	Nanofiller	Observation	Ref.
EPDM/Silica-SBR	Nanoclay	High loading of nanoclay (10 phr)—formation of clusters with poor dispersion	[2]
EPDM	Organoclay	3 wt.% of fillers to the matrix cause a substantial wettability of elastomer	[35]
EPDM	MWCNT	The densities of samples increased as a function of MWCNT content	[28]
PP/EPDM	MWCNT	Morphological changes induced by the dynamic vulcanization process	[127]
EPDM/MFS	Graphene	The addition of graphene improved the homogeneous dispersion with a decrease in agglomerates.	[123]
EPDM/PP	Nanoclay and nanosilica	AFM and SEM show two phases and the uniform distribution of the nanofillers	[125]
SAN/EPDM	Organically modified montmorillonite	Micrographs indicate two phases (SAN + EPDM) with different domain sizes and shapes.	[124]
PP/EPDM	MWCNT	SEM and TEM show MWCNT aggregates in the matrix	[128]

Bizhani et al. [28] developed nanocomposites of EPDM rubber and MWCNTs from hot compression molding using a chemical blowing agent as a foaming agent. Figure 6 shows the SEM micrographs of the EPDM/MWCNTs containing different amounts (phr) of MWCNTs. The presence of MWCNTs in EPDM caused an increase in density owing to a decrease in the cell diameters from 227 ± 58 µm to 59 ± 30 µm [28]. According to the authors, the MWCNTs acted as nucleating sites during the foaming process, increasing the number of cells and decreasing the size of the cells.

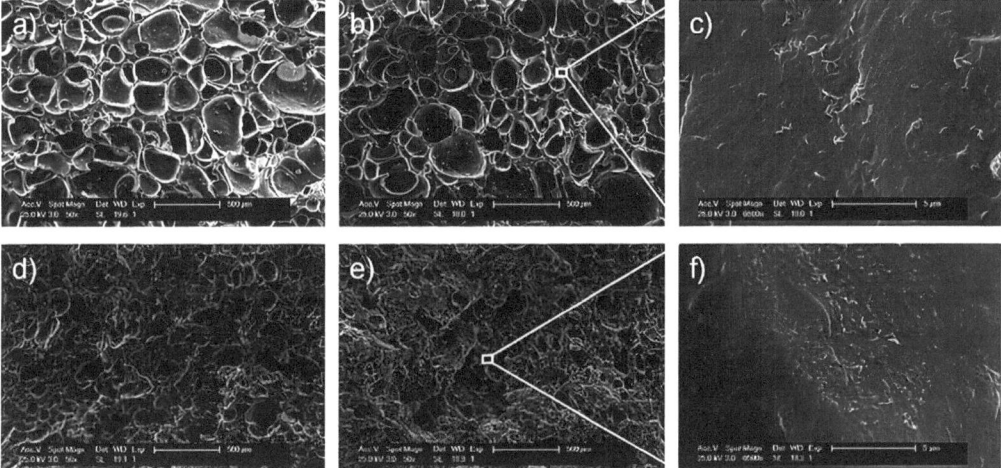

Figure 6. SEM micrographs of cryofractured EPDM/MWCNT nanocomposite foams containing (**a**) 0, (**b**) 2, (**d**) 6, and (**e**) 10 phr MWCNTs. (**c**,**f**) Micrographs of the MWCNTs within the morphology. (This figure is from ref. [28] licensed under a Creative Commons Attribution 4.0 International License).

TEM is an effective and adequate analytical method for conducting the morphological characterization of several types of polymer samples on a nanoscale, such as EPDM matrix polymers with nanometric fillers [45,128,129].

Jha and coauthors [45] studied oil-extended EPDM filled with nanographite. The morphological characterization conducted via SEM and TEM showed good dispersion when nanographite concentrations of up to 4% were utilized in the EPDM matrix; TEM was performed on the EPDM/nanographite samples with filler concentrations of 2%, 4%, and 6%. The nanofiller with up to 4% concentration was uniformly distributed in the EPDM matrix, and the EPDM elastomer was situated between the nanofiller.

Bhattacharya et al. [125] developed thermoplastic vulcanizates (TPVs) and TPV nanocomposites (TPVNs) based on EPDM and PP. The TPVNs were prepared with a fixed EPDM:PP ratio; the nanofiller content varied with the use of different concentrations. The AFM and SEM images showed that the nanofillers (nanosilica and nanoclay) were well distributed in the TPV matrix. Specifically, the AFM analysis made it possible to view the surface topography and surface heterogeneity of the different polymeric systems. The authors attributed the brighter regions to softer materials (lower nanofiller concentration) and the darker areas to harder materials (higher nanofiller concentration). The AFM images showed that the material exhibited a biphasic morphology, which corresponded with the SEM results of the same sample. A uniform distribution of nanofillers was observed in the samples filled with 1.0 phr nanofiller [125].

Vishvanathperumal and Anand [80] examined the microstructural characteristics of EPDM/SBR hybrid composites. Within these composites, reinforcement was achieved by incorporating nanoclay and nanosilica particles. This addition facilitated enhanced properties attributed to the improved interaction between the fillers and the rubber matrix. The concentration of nanosilica and nanoclay emerged as a crucial factor determining the composite's properties. Additionally, composites featuring agglomerated nanosilica exhibited superior wear resistance compared to other formulations within the study [80].

According to Rostami-Tapeh-Esmaeil et al. [81], rubber foams that undergo a more extensive procuring process exhibit thicker cell walls. As the procuring level increases, there is a noticeable trend toward a more confined range in cell sizes within the foam structure. Specifically, at a pre-curing level of 30, there is a greater quantity of cells than other pre-curing levels examined.

In Moustafa et al. [35], SEM images depicted proofed fabrics filled with different loads of EOC and OC. The effective dispersion of the organoclay nanofiller played a crucial role in preventing radiation-induced degradation of the polymer matrix. Additionally, incorporating organoclay notably enhanced the resistance of filled PEPA blends against photo-oxidation.

In an isotropic Magnetorheological Elastomer (MRE) presented by Kang et al. [130], carbonyl iron particles were evenly and haphazardly distributed within the elastomeric matrix. Conversely, in an anisotropic MRE, these carbonyl iron particles were aligned within the elastomeric medium in a specific direction or pattern.

The SEM images indicate a consistent and even distribution of MWCNTs within the EPDM composites in the work of Gou et al. [49]. Additionally, a TEM image validated this uniform dispersion of MWCNTs within the composite structure. Moreover, the SEM images illustrated the formation of a char layer, specifically in composite 3, which was within the high-speed area.

Chen et al. [90] proposed that EPDM biocomposites filled with silane-modified wood fibers (Si-WFs) should be manufactured. This study investigated the impacts of fiber loading, silane modification, and crosslinking. Silane modification notably enhanced water absorption resistance and thermal stability. Applying silane modification led to an enhancement in Si-WF dispersion within the elastomer matrix. This improved dispersion of Si-WFs within the matrix was confirmed through fractography analysis [90].

Observing the morphology in the work of Chang et al. [83] supports creating an EPDM network that percolates rheologically. Solid-state fibrils inhibit molecular mobility and constrain relaxation when subjected to shear flow. Additionally, the microstructure of blends was examined following partial etching. Burgaz and Goksuzoglu [85] noted a uniform dispersion of individual CIP particles alongside their clusters. Conversely, BIP particles were observed creating larger clusters. The EPDM/CB/CIP MREs displayed noteworthy enhancements in terms of their properties compared to EPDM/CB/BIP MREs.

Abdelsalam et al. [91] showed that fillers such as CB particles play a pivotal role in augmenting the physical attributes of rubber-based products. The dispersion of CB and its interaction within the rubber matrix play significant roles in determining the resulting mechanical properties. Increased CB loading correlates with elevated torque levels and enhancements in tensile strength, tear strength, and crosslink density. SEM images illustrated an adequate dispersion of CB particles within the rubber matrix. Furthermore, the surface achieved homogeneity upon reaching a CB loading of 45 phr [91].

7. Thermal Conductivity

The thermal conductivities of polymers, such as elastomers, are relatively low, with values of 0.1–0.2 W/(m K). Therefore, loading nanofillers, such as metal particles, graphene, and CNT, into polymeric matrices can improve thermal conductivities [131]. In a study reported by Bizhani et al. [43], a highly elastic foam with enhanced electromagnetic wave absorption made of EPDM filled with a barium titanate (BT)/MWCNT hybrid was produced. The EPDM-based nanocomposite foam demonstrated enhanced thermal conductivity with an increase in the quantity of MWCNTs, i.e., the hybrid EPDM foam exhibited a higher value [0.217 W/(m K)] compared to the sample composed of only 20 phr BT [0.077 W/(m K)], as illustrated in Figure 7 [43]. This behavior is attributed to an improvement in the thermal conduction through the solid and gas phases of the foam [43]. According to the authors, the EPDM-based nanocomposite can be a valuable asset in electromagnetic interference (EMI), shielding applications owing to its improved EM wave absorption due to heat dissipation and minimizing the impact of temperature on the electronic components [28,43].

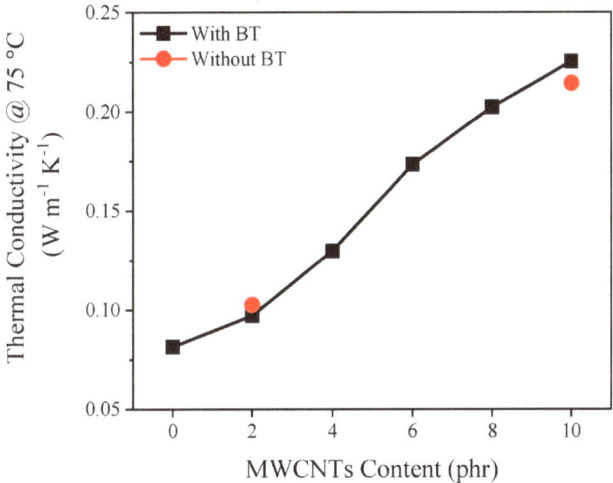

Figure 7. Thermal conductivity of cellular BT/MWCNTs and MWCNTs nanocomposites as a function of MWCNT loading at 25 °C. (This figure is from ref. [43] licensed under a Creative Commons Attribution 4.0 International License).

Lu et al. [29] developed a flexible GnPs/EPDM composite with excellent thermal conductivity and EMI shielding properties. The amount of GnPs incorporated into the EPDM matrix greatly affected its thermal conductivity, i.e., as the amount of GnPs in the EPDM-based nanocomposite increased, the thermal conductivity was enhanced. This behavior indicates that the GnPs assisted in improving the thermal conductivity inside the EPDM matrix; the thermal conductivity values were 0.23 W/(m K) and 0.79 W/(m K) for neat EPDM and the EPDM-based nanocomposite with 8 wt.% GnPs, respectively. The results are significant for potential applications in sensors, as heat dissipation affects the stability and sensitivity of sensors [44]. The percolation threshold was reached at 2.9 wt.% graphene platelets (GnPs), significantly increasing electrical conductivity. By increasing the concentration of GnPs to 7 wt.%, the thermal conductivity improved significantly, reaching 0.72 W/m K, nearly three times higher than at 2.9 wt.%. As a result of the improved dispersion of GnPs within the EPDM matrix, thermal conductivity is improved. In the sensor with 3 wt.% GnPs loading, the gauge factor was 129.33, indicating excellent sensitivity. With a 7 wt.% GnPs loading, the sensor exhibited enhanced sensitivity, with two linear change sensing stages, each with a different gauge factor, implying improved operating stability and sensitivity. Higher thermal conductivity resulting from the increased GnP content allows sensors to dissipate heat more efficiently, resulting in excellent stability and sensitivity.

8. Electrical Properties

As previously mentioned, dispersing nanofillers or nanoparticles in a polymeric matrix to produce nanocomposites offers several advantages regarding the morphology, thermal properties, and mechanical properties of materials. Additionally, the addition of nanofillers with excellent conductive properties can provide an additional benefit, improving the electrical properties of nanocomposites and, thereby, expanding their potential applications [132–134]. The combination of conductive nanofillers with polymeric matrixes provides unique and highly desirable electrical properties, making conductive nanocomposites a fascinating class of materials [135]. These conductive nanocomposites have several applications, ranging from flexible electronics to advanced sensors and energy storage devices, such as supercapacitors [135–137].

To develop new materials for desired applications, it is essential to understand the conduction processes and phenomena that govern electrical conductivity in conductive polymer nanocomposites [9,138,139]. The percolation theory is one of the widely used theories for understanding conduction processes and electrical conductivity in polymer nanocomposites [140,141]. According to this theory, the electrical conductivity of a composite material increases abruptly as the volume fraction of the conducting nanofillers reaches a certain critical fraction or percolation threshold [9,139,140]. Percolation thresholds may be defined as the critical concentration (mass or volume) of conductive nanofillers dispersed in an insulating matrix [9,139,140]. When the concentration of conductive nanofiller reaches the percolation threshold in the insulating matrix, the first conductive path is formed by which the charge carriers move under the action of the external electric field [9,139,140]. In other words, when concentrations of conductive nanofillers reach the percolation threshold, an insulator–conductor transition occurs in the nanocomposite, resulting in an abrupt increase in its electrical conductivity [9,139,140]. When the concentration of conducting nanofillers in nanocomposite is below the percolation threshold, they do not form conductive paths far away from each other, resulting in low electrical conductivity [9,139,140]. Above the percolation threshold, however, a continuous conductive path is formed that allows electrons to flow through the composite material [9,139,140].

In this sense, it is possible to understand how the dispersion of nanofillers influences the electrical conductivity of conductive nanocomposites by using the percolation theory [142,143]. For nanofillers to form effective conductive paths, homogeneous dispersion and good interconnectivity are essential [136,142,143]. Furthermore, the size, shape, and concentration of nanofillers are also crucial factors in determining the percolation threshold and the electrical conductivity of the nanocomposite [139,144,145]. Hence, the percolation threshold is defined as the critical volumetric fraction of conductive nanofillers in the nanocomposite, above which a continuous conductive path is formed [140,141,146]. Because of the considerable distance between the conducting nanofillers below the percolation threshold, no conductive paths are formed, resulting in an electrical conductivity like that of the insulating polymer matrix [9,138,139]. In contrast, above the percolation threshold, electrical conductivity increases dramatically due to the interconnection of the conductive paths, allowing electrons to flow efficiently, bringing the electrical conductivity of the nanocomposites closer to the conductive nanofillers [9,138,139].

Notably, dispersing conductive nanofillers in EPDM matrices has benefits. EPDM elastomers are widely used in applications requiring flexibility, chemical resistance, and durability [147]. Unfortunately, the low electrical conductivity of these materials often limits their use in electronic devices and in applications where good electrical conductivity is essential. It is possible to significantly improve the electrical properties of EPDM by incorporating conductive nanofillers, such as CB, CNT, graphene, metallic nanoparticles, etc. [145,148–150]. As a result of the uniform dispersion of these nanofillers, a three-dimensional conductive network is created within the material, allowing the charge carrier to transfer efficiently [148,150,151]. Increasing the volume fraction of the conductive nanofillers increases the electrical conductivity of the EPDM/nanofiller nanocomposite because a continuous conductive path is formed, as predicted by percolation theory [148,150,151]. In this way, EPDM can be modified and improved in terms of its electrical properties, making it suitable for a wide range of electronic and electrical applications, such as antistatic coatings and films, sensors, conducting adhesives, electromagnetic interference shielding materials, etc. [135–137,147].

The thermal and electrical conductivity of EPDM-based nanocomposites with GnPs and CB has been extensively studied by Koca et al. [148]. The electrical conductivity of EPDM/GnPs composites is slightly affected by the specific surface area or lateral size of GnPs up to 20 phr [148]. The electrical conductivity of EPDM-based nanocomposites was higher when hybrid fillers, 50 phr CBs, and 7 phr GnPs with higher lateral sizes were used. EPDM-based nanocomposites with a single CB have electrical conductivity increases of 4.4×10^{15} S/cm to 4.8×10^{-6} S/cm from 20 phr to 50 phr, according to the authors [148]. Electrical conductivity is enhanced by the increasing number of contacts between CB particles, which leads to more conductive pathways and improves conductivity. Moreover, the filler–filler interaction pathway facilitates electron transfer through the matrix, and hopping and tunneling are two mechanisms for electron conduction in polymer composites [148].

Zhang et al. [150] used CNT and hexagonal boron nitride nanosheets to improve the electrical properties of EPDM-based nanocomposites for cable accessory applications. Based on the results, it can be concluded that the non-linear conductivity of CNT/h-BN/EPDM composites becomes more prominent with increasing CNT content, followed by a reduction in threshold field strength and an increase in the non-linear coefficient [150]. In order to study the electrical properties of donor ZnO nanoparticles/EPDM composites, Chi et al. [149] developed EPDM-based nanocomposites that exhibit outstanding non-linear electrical conductivity. The results indicate that the non-linear conductivity becomes much more distinct with the increase in ZnO nanoparticles, along with an increase in the non-linear coefficient of conductance and a decrease in breakdown field strength [149]. Consequently, ZnO/EPDM nanocomposites demonstrate promise as an effective means of protecting the safe operation of power transmission systems [149].

9. EPDM-Based Nanocomposites: Advantages, Disadvantages, and Perspectives

EPDM rubber is a type of synthetic rubber that is used in a variety of applications. EPDM rubber is classified as a class M rubber by ASTM standard D-1418 [152], which includes elastomers of the saturated chain polyethylene type. In 2022, the market for EPDM is estimated to be worth USD 4.45 billion, and it is expected to grow at a compound annual growth rate (CAGR) of 5.7% between 2023 and 2030. The primary driver of the growth of the automotive industry is increasing investments in emerging economies such as China and India. There has been considerable expansion of application sectors, including automotive, building, and construction industries. In recent years, construction and automotive industries in the United States have experienced rapid growth, which has increased regional demand for products [153].

Because of their mechanical and thermal properties, EPDM rubber nanocomposites are highly promising for a variety of industrial applications. When nanoparticles are incorporated into nanocomposites, they significantly increase tensile strength, elastic modulus, and wear resistance, outperforming other nanocomposites under certain conditions [154,155]. Additionally, these nanoparticles enhance the thermal stability and oxidative degradation resistance of the materials, making them suitable for use in adverse environments [154,155].

Compared to SBR-based nanocomposites, EPDM-based nanocomposites exhibit superior heat and oxidation resistance, making them more suitable for extreme conditions [156,157]. EPDM is highly resistant to heat and oxidation, which makes it ideal for applications that require durability under severe conditions [158,159]. When compared to silicone rubber, EPDM offers a unique combination of flexibility and mechanical strength. Despite the fact that it cannot reach the extreme temperatures of silicone, EPDM is more cost-effective.

The versatility and weather resistance of EPDM nanocomposites make them stand out among synthetic elastomers. However, achieving a homogeneous dispersion of nanoparticles can be challenging, negatively impacting the final properties and causing production costs to increase, which limits the application of nanoparticles at a large scale. Rubber-based nanocomposites, such as EPDM-based nanocomposites, offer a balance between price and performance, but the synthesis and incorporation of nanoparticles can increase the production cost. Nanocomposites can have adverse environmental effects due to the presence of nanoparticles, which are difficult to recycle or degrade [160,161].

Depending on the specific application requirements, EPDM rubber nanocomposites are preferred over other materials. In general, these materials are selected for applications that require high flexibility, wear resistance, and thermal stability, such as automotive components and building materials. In contrast, epoxy- and polyamide-based nanocomposites may be more suitable for applications requiring high mechanical strength and rigidity [147,162]. Table 2 summarizes the advantages and disadvantages of specific elastomeric materials.

Table 2. Advantages and disadvantages of elastomeric materials.

Matrix	Advantage	Disadvantage	Ref.
EPDM	Increased tensile strength, elastic modulus, wear resistance; superior resistance to heat and oxidation; high mechanical strength and flexibility; cost-effectiveness	Nanoparticle dispersion challenges, increased production costs, environmental concerns	[126,163–167]
SBR	Cost-effective; good abrasion resistance; excellent adhesion to fabrics	Heat and ozone resistance are low; oil and solvent resistance is low	[168,169]
Natural Rubber	Excellent tensile strength; high elasticity; high resilience	The product has a low resistance to heat, ozone, oils, and solvents	[161,170,171]
Silicone Rubber	Heat and cold resistance; good electrical properties; chemically inert	High cost; low tensile strength	[172–174]
Synthetic Rubbers (General)	Adjustable properties: more excellent resistance to oils, solvents, and weather than natural rubber	It may be less elastic and resilient than natural rubber; some types may be more harmful to the environment.	[154,155]
Epoxy- or Polyamide-based Nanocomposites	Mechanical strength and rigidity are high	Comparatively less flexible than elastomeric materials	[147,162]

Figure 8a illustrates the effects that different types of nanofillers can have on EPDM-based nanocomposites and how they are intended to affect their properties. Figure 8b illustrates a simple representation of the microstructure of different types of nanofillers, including 0D, 1D, and 2D fillers.

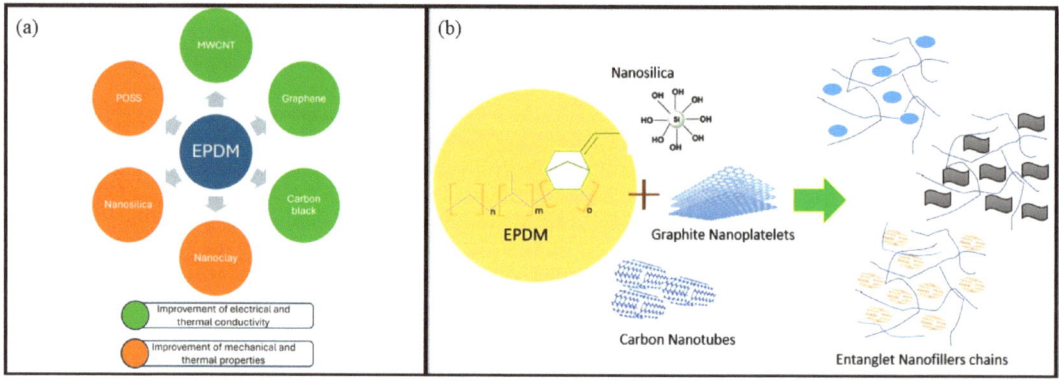

Figure 8. (**a**) Flowchart of EPDM-based nanocomposite with different nanofillers and their effect on its properties. (**b**) Microstructural representation of EPDM-based nanocomposite with 0D, 1D, and 2D nanofillers.

Furthermore, it is essential to be aware of the main trends in terms of improving the properties, processes, and load systems of this elastomer. In recent years, there has been a growing interest in improving the energy absorption properties of natural-fiber-reinforced polymer composites, especially in structural applications. This study focuses on developing PP/EPDM composites based on coconut fibers (10–40 phr), which have been plasma pretreated and coated with highly hydrophobic fluoroalkyl functional siloxanes. A natural fiber elastomer is an example of the trend toward proposing elastomers with high sustainability for technical applications [175]. In addition, a foam composite was developed with bamboo fiber, polypropylene, and EPDM, with an elastomer content of 10%. This material demonstrated the best foaming effects, and its impact toughness increased by 34.42% compared with pure polypropylene. This study expands the possibilities of using bamboo fiber-reinforced polypropylene microcellular foam materials [176].

This exercise illustrates how a mixture can be produced with the required properties and cost-effectively, as opposed to producing samples, which does not energy-consuming equipment without which development would not be possible and that has an ever-increasing value, without which the sample could not be produced and the results studied. According to the results of this study, the equivalent properties of the composite studied were 23.0% in terms of rebound resilience, 57.8% in terms of elongation at break, 7.8 MPa in terms of tensile strength, and 66.8% in terms of Shore A hardness. Particulate load optimization techniques and statistical design of experiments have been demonstrated to be effective methods for generating compositional variables when combined. As a result, there is potential for the development of new analytical processes for elastomers in the future, such as EPDM, as well as other applications that utilize machine learning tools to validate and predict the development of new elastomer blends, including [177–179]. Furthermore, in the development of thermoplastic elastomers (TPVs), the mixtures EPDM/PP, the higher crystallinity of the PP in the TPV blend, the higher Mooney viscosity, as well as the better compatibility between EPDM and PP, result in higher tensile strength and elongation at break, which are enhanced by the crystals present in EPDM [180]. Therefore, EPDM is an elastomer with a long way to go in terms of new applications relating to viscosity and rheology, as in this case for the application of TPV. Table 3 shows the correlation between the EPDM matrix or EPDM-polymer blend and how different nanofillers affect the microstructures, as well as their effect on the properties of EPDM-based nanocomposites.

Table 3. Effects of nanofiller loading on EPDM-based nanocomposite microstructure and properties.

Matrix	Nanofiller	Morphology	Properties	Ref.
EPDM-SBR	Nanoclay	A high loading of nanoclay (10 phr) results in the formation of clusters with poor dispersion	As the concentration of nanoclay (4–10 phr) is gradually increased in the EPDM/S-SBR rubber nanocomposites, the 100% modulus of elasticity increases. In the same manner, when nanoclay loading is up to 8 phr, the compression set increases, but as the load increases, the compression set decreases due to the afferent crosslinking density and mobility of the long rubber chains. From abrasion resistance to hardness, nanocomposites showed a gradual increase in both properties.	[2]
EPDM	Organoclay	The presence of 3 wt.% of fillers in the matrix makes the elastomer more wettable.	Thermal degradation of pure EPDM was observed using TGA. The maximum degradation temperature (T_{max}) at 10% weight loss of the material was 478 °C, indicating high thermal stability when compared to other studies. The onset temperature of degradation ($T_{10\%}$) at 10% weight loss of the material was 401 °C. Antimicrobial tests were conducted on tissue samples coated with 5% by weight of EOC or OC and 20% by weight of a long oil alkyd resin based on soybean oil. It was conducted against three different types of microbes: Staphylococcus aureus (G-ve bacteria), Pseudomonas aeruginosa (G-ve bacteria), and Candida albicans (fungi). LAR resin (fatty acid) with expanded organoclay is likely to serve as an essential component of microbial fatty acids that are a promising target for the development of antimicrobial-resistant tissues in the presence of 20 wt.% LAR resin (fatty acid).	[35]

Table 3. *Cont.*

Matrix	Nanofiller	Morphology	Properties	Ref.
EPDM	MWCNT	Sample density increased with MWCNT content	The development of EPDM/MWCNT foams has achieved superior EMI-specific SE and deformability. The use of EPDM should also ensure properties such as chemical, moisture, and ozone resistance. The developed foam samples exhibit high thermal and electrical conductivities of up to 2.7×10^{-4} S/cm and EMI shielding efficiencies of up to 45 dB, which do not degrade significantly after repeated bending. The EPDM matrix exhibits these properties because of the formation of a three-dimensional interconnected network.	[28]
PP/EPDM	MWCNT	Dynamic vulcanization causes morphological changes	Despite a high loading level, the samples have formed a dense, stable conductive network, which is difficult to destroy, resulting in excellent resistance stability. The cyclic loading test revealed no fractures and electrical shots, which indicates strong potential for practical applications. Approximately 0.1% of incident electromagnetic waves could penetrate the material without being reflected or absorbed, primarily as a result of the high density of the MWCNT network and the greater electrical conductivity of the samples compared to TPE without filler, indicating that the TPV elastomer composites may be helpful to as materials for stretchable conductors with electromagnetic radiation shielding capabilities.	[127]
EPDM/MFS	Graphene	With the addition of graphene, homogeneous dispersion was improved, and agglomerates were reduced.	EPDM/S/TiO_2 samples showed a 30% increase in dielectric strength over vulcanized EPDM rubber samples. The addition of MFS, TiO_2, and graphene to EPDM and EPDM/S rubbers resulted in a higher thermal conductivity for all composites. The EPDM composites containing TiO_2 had the highest thermal conductivity values. The thermal conductivity of the composites did not appear to be affected by the addition of graphene.	[123]
ultra-high molecular weight EPDM (UHMW-EPDM) and PP	Nanoclay and nanosilica	SEM shows the presence of two phases and a uniform distribution of nanofillers	Compared with the other blends, the tensile strength and modulus were 250% higher with the 7 phr nanofiller loading. Nanosilica-added TPVs are superior to nanoclay-added TPVs in terms of mechanical properties, crosslink density, and morphology.	[125]
SAN/EPDM	Organically modified montmorillonite	Micrographs indicate two phases (SAN-EPDM) with different domain sizes and shapes.	Incorporating nanoclay and EPDM improved the thermal stability of SAN. Due to the establishment provided by EPDM rubber (known for its high thermal stability) and the improved interaction between SAN and nanoclay, nanocomposite blends demonstrated higher thermal stability than pure SAN. Because the EPDM soft phase has a low modulus, the Young's modulus of the binary blends was decreased. As long as there is good phase interaction, adding a low-modulus EPDM dispersed phase to a high-modulus matrix (pure SAN) will tend to reduce its modulus.	[124]
PP/EPDM	MWCNT	TEM and SEM images show aggregates of MWCNTs	There was a significant difference between the unvulcanized mixtures and the vulcanized TPVs, which had superior tensile strength and deformation resistance. Dynamic vulcanization played a substantial role in determining the fracture toughness of vulcanized samples, although different curing mechanisms contributed to fracture toughness in various ways.	[128]

The development of porous systems has been one of the prospects for the application of EPDM-based nanocomposites, which have been applied to a variety of industrial contexts, ranging from household products, acoustic systems, thermal protection, and aerospace applications. Ohadi, Asghar, and Hosseinpour [181] have shown the application of MWCNT on EPDM foams to absorb low-frequency waves. Samples with low MWCNT loading showed higher damping factors, and smaller pore sizes exhibited high viscous motions with superior sound absorption characteristics. Thus, it was possible to know that the nanocomposite foam sample with only 0.1 phr (parts per hundred of rubber) of MWCNT offered the best sound absorption coefficient with a sharp peak (0.98) in the frequency range of 500–1000 Hz. This occurs due to the high porosity, the airflow resistance of the thin, and the partially crosslinked cell structure. With additional MWCNT loading, the cell size was drastically reduced, resulting in a decrease in porosity and an excessive increase in airflow resistance. It can be concluded that the MWCNTs improve the sound absorption and acoustic insulation performance of EPDM foam so that with the application of nanoparticles, a lightweight acoustic absorber with high resistance to adverse environmental conditions and excellent low-frequency sound absorption can be obtained.

To improve foams, biobased materials can reduce the cost of commercial fillers, change the degradation lifetime, and add environmentally safe credits that offset the price of the new material. EPDM rubber-based biocomposites filled with silane-modified wood fiber (SiWF) were foamed using the chemical foaming method by Chen, Gupta, and Mekonnen [182]. Using SiWF and dicumyl peroxide (DCP), the developed EPDM biocomposite foams showed much lower water absorption and improved thermomechanical properties. The tensile strength and modulus of the EPDM biocomposite foams exhibited up to 90% and 600% enhancements with SiWFs and in situ crosslinking [182]. Figure 9 illustrates the EPDM biocomposite in foam form, as shown by the authors [182]. There was a high degree of foaming in the unfilled EPDM. In contrast, biocomposites are less foamed due to the higher concentration of fillers, silane modification, or DCP crosslinking [182].

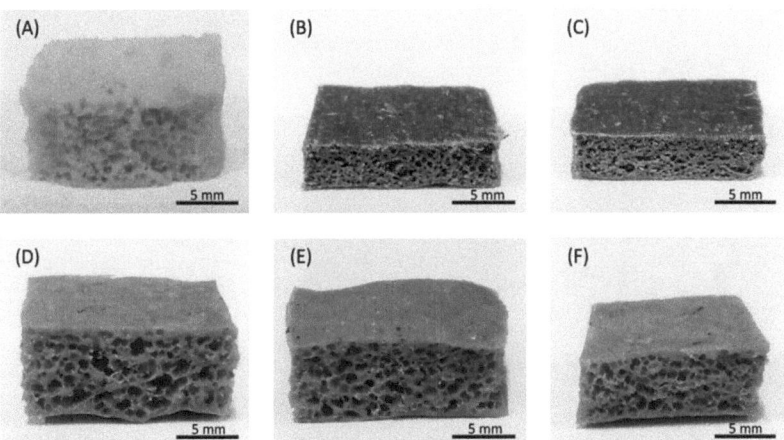

Figure 9. Images of the developed EPDM biocomposite foam. (**A**) Unfilled foam; (**B**) 40 WF; (**C**) 40 SiWF; (**D**) 10 WF; (**E**) 10 SiWF; (**F**) 10 SiWF-DCP. (This figure is from ref. [182] licensed under a Creative Commons Attribution 4.0 International License).

The developed biocomposite foams also displayed lower apparent thermal conductivity than the unfilled EPDM foam over the heating temperature range, showing possible applications as insulation materials. Developing lightweight insulating materials has also proven to be an effective means of improving the performance of solid rocket motors (SRMs) [183]. Through rubber foaming technology, Wang et al. [183] introduced a porous structure into EPDM insulating materials to prepare lightweight insulating materials [183].

When studying morphology formation in the various samples, it was found that with an increase in pre-curing temperature, the cell size decreased significantly, and the original cell was more uniform, which improved the mechanical properties [183]. With an increase in blowing agent content, the cell density gradually increased without a significant change in cell diameter. During this period, the mechanical properties of the insulation materials gradually decreased [183]. EPDM rubber foams have the highest rate of carbonization resistance when the pre-curing temperature is 120 °C and the blowing agent content is 6 phr [183]. As compared to the base formulation, their carbonization ablation rate is only 16% higher. As a result, CNTs and carbon fibers (CFs) were added to reinforce the EPDM rubber foams [183]. Compared to the foam formulation, the reinforced formulation exhibited a 36.4% increase in tensile strength, which is 19.2% lower than the base formulation [183]. In comparison to the foam formulation, the carbonization rate of the reinforced formulation decreases by 35.5%, which is only 19.2% lower than the base formulation [183].

Zhao et al. [184] showed that polylactic acid (PLA) is a promising biomass and biodegradable polymer, but it is brittle and has a poor foaming ability. Herein, a composite with UV-crosslinking-assisted in situ fibrillation of PLA reinforced with ethylene-propylene-diene terpolymer (EPDM) nanofibrils was developed. Due to the heterogeneous nucleation effect of UV-crosslinked EPDM nanofibrils, the crystallization rate of PLA was significantly increased. The crystalline morphology was refined, resulting in a refined crystalline morphology with fibrous structures, and the tensile and impact strength were significantly improved without sacrificing strength and stiffness. In addition, the expansion ratio and cell population density of the foams increased by 470% in the presence of UV-cured EPDM nanofibrils. Due to the very high expansion ratio and the unique micro/nanoscale structures of the cell walls, the 28-fold-expanded PLA/EPDM foams exhibited excellent thermal insulation performance, with a thermal conductivity of only 26.5 μm.

On the other hand, Sang et al. [185] studied the properties of a bio-EPDM where Keltan® Eco 6950C was produced from biological raw materials. In contrast, the ethylene used in the process is derived from ethanol produced from sugarcane. Foams prepared by mixing chemical and encapsulated foaming agents (OBSH (4,4′oxybis-(benzene sulfonyl hydrazide) at different mixing ratios were compared the results of sulfur crosslinking systems and dicumyl peroxide systems, with an evaluation of the mechanical properties, thermal stability, and saltwater resistance of the bio-EPDM foam. The mechanical and elastic properties of the bio-EPDM foam decreased with increasing amounts of encapsulated blowing agents. At the same time, the thermal stability and saltwater resistance improved significantly with increasing amounts of encapsulated blowing agents. The mechanical properties obtained with the peroxide system were superior to those obtained with sulfur. Further research focusing on sustainability, and no less important, involved the reuse of rubber (RR) in the development of closed-cell elastomeric foams based on ethylene-propylene-diene rubber (EPDM). Rheometric results showed that the introduction of RR up to 20 phr increased the cure rate from 11.7 to 13.48%/min, reduced the cure time from 12.21 to 9.3 min, and increased the final torque from 6.51 to 7.24 N·m. RR increased the cell density from 12 to 78 cells/mm^3 and reduced the average cell number size from 940 to 110 μm. Furthermore, the introduction of RR proved to be a viable alternative to produce high-performance EPDM foams with improved toughness and resilience and predicts opportunities for EPDM/RR foam composites to be used in the sealing and gasketing industry as an environmentally friendly substitute for virgin rubber [186].

Nowadays, with the approach of computational tools for the optimization of processes and materials, the study of predictions and projections towards the design of composite systems has been considered. Buchen et al. [187] performed a study involving the time-dependent modeling and experimental characterization of foamed EPDM rubber. In their research, they implemented a derived constitutive model in finite element software (ABAQUS) and calibrated it experimentally with multi-step relaxation tensile tests of foamed EPDM rubber [187]. It is possible to predict the time-dependent stress–strain relationship for EPDM foam rubber using the viscoelastic model as a first approximation. A homogenized hollow sphere constitutive model described the hyperelastic behavior, while rheological elements described time dependence. Analyses were conducted using the finite element program Abaqus. By using a material without pores, it was possible to obtain information regarding the material's behavior and mechanical characteristics, as well as multi-step relaxation experiments. In each case, the parameters were identified using least squares fitting. An equilibrium simulation resulted in the overestimation of stress even after optimizing parameter identification based on different deformation states. In addition to the analytical hollow sphere model, a numerical hollow sphere model was employed to evaluate the impact of matrix compressibility and to compare the predictions of the two models. Consequently, the numerical solution converges to the analytical model for increasing Poisson's ratio, which opens the door for designing porous systems applicable to industrial applications.

EPDM-based nanocomposites can also be applied to the treatment of wastewater from the petroleum industry by using cellulose nanofibers to obtain nanocomposite membranes. According to Bhuyan et al. [188], a superhydrophilic and organic-solvent-resistant nanocomposite membrane was designed using waste bottles made from poly(ethylene terephthalate) (PET) and cellulosic paper. According to the study, 98 $Lm^{-2}h^{-1}$ permeability was measured and applied under 1.5 bar pressure [188]. Additionally, the membrane was able to remove more than 97% of the organic substances from a crude oil–water emulsion [188]. Moreover, EPDM-based nanocomposite foams or fibrous membranes can be used as membrane-based desalination technologies for the treatment of water, opening new avenues for water recovery [189].

10. Conclusions

EPDM is one of the most studied and used synthetic rubbers in the industry, and incorporating different types of nanofillers can expand its application in several areas. The dispersion of these nanofillers into EPDM plays an important role in improving thermal stability and conductivity. It is important to consider three aspects of filler dispersion. First, the surface chemistry of the nanofiller (presence or absence of active functional groups) may compromise the compatibility between the filler and the polymeric matrix. The second involves the interaction between the fillers and the polymer matrix, which can be chemical or physical and is dependent upon the formation of aggregates because of the interactions between the filler and the polymer matrix. Furthermore, the morphology of the particle should be considered as it should have a specific surface area of hundreds of square meters per gram, which can affect its dispersion and interaction with the polymeric matrix. Thirdly, it is important to select the correct polymeric matrix with which nanofillers and polymeric matrices must be compatible. Nanocomposites based on EPDM have several advantages over other rubber matrices, including their ability to accept large amounts of nanofillers that can greatly enhance their properties and extend their application range.

Author Contributions: Conceptualization, N.L.C., C.T.H., L.L.P., F.C.C., R.J.d.S. and M.J.S.; methodology, C.T.H., L.L.P., F.C.C., R.J.d.S. and M.J.S.; formal analysis, C.T.H., H.P.C., G.D., J.C.S., J.A.J.C., G.B.T., L.L.P., G.P.C., F.C.C., R.J.d.S. and M.J.S.; investigation, N.L.C., C.T.H., H.P.C., G.D., J.C.S., J.A.J.C., G.B.T., L.L.P., L.F.P., G.P.C., F.C.C., R.J.d.S. and M.J.S.; resources, C.T.H., H.P.C., G.D., J.C.S., J.A.J.C., G.B.T., L.L.P., G.P.C., F.C.C., R.J.d.S. and M.J.S.; writing—original draft preparation, N.L.C., C.T.H., H.P.C., G.D., J.C.S., J.A.J.C., L.F.P., G.B.T., L.L.P., G.P.C., F.C.C., R.J.d.S. and M.J.S.; writing—review and editing, F.C.C., R.J.d.S. and M.J.S.; supervision, L.L.P., F.C.C., R.J.d.S. and M.J.S.; project administration, L.L.P., F.C.C., R.J.d.S. and M.J.S.; funding acquisition, M.J.S. All authors have read and agreed to the published version of the manuscript.

Funding: The APC was covered by PROPe (Pró-Reitoria de Pesquisa) of the São Paulo State University.

Institutional Review Board Statement: Not applicable.

Data Availability Statement: No data were used to support this study.

Conflicts of Interest: The authors declare no conflicts of interest.

References

1. Nazir, M.T.; Phung, B.T.; Sahoo, A.; Yu, S.; Zhang, Y.; Li, S. Surface Discharge Behaviours, Dielectric and Mechanical Properties of EPDM based Nanocomposites containing Nano-BN. *Appl. Nanosci.* **2019**, *9*, 1981–1989. [CrossRef]
2. Malas, A.; Das, C.K. Carbon black–clay hybrid nanocomposites based upon EPDM elastomer. *J. Mater. Sci.* **2012**, *47*, 2016–2024. [CrossRef]
3. Nazir, M.T.; Phung, B.T.; Kabir, I.; Yuen, A.C.Y.; Yeoh, G.H.; Zhang, Y.; Yu, S.; Li, S. Investigation on Dry Band Arcing Induced Tracking Failure on Nanocomposites of EPDM Matrix. In Proceedings of the 2019 2nd International Conference on Electrical Materials and Power Equipment (ICEMPE), Guangzhou, China, 7–10 April 2019; IEEE: Piscataway, NJ, USA, 2019; pp. 309–312.
4. Nazir, M.T.; Butt, F.T.; Phung, B.T.; Yeoh, G.H.; Yasin, G.; Akram, S.; Bhutta, M.S.; Hussain, S.; Nguyen, T.A. Simulation and Experimental Investigation on Carbonized Tracking Failure of EPDM/BN-Based Electrical Insulation. *Polymers* **2020**, *12*, 582. [CrossRef] [PubMed]
5. Attia, N.F.; Saleh, B.K. Novel Synthesis of Renewable and Green Flame-Retardant, Antibacterial and Reinforcement Material for Styrene–Butadiene Rubber Nanocomposites. *J. Therm. Anal. Calorim.* **2020**, *139*, 1817–1827. [CrossRef]
6. Ahamad, A.; Kumar, P. Effect of Reinforcing Ability of Halloysite Nanotubes in Styrene-Butadiene Rubber Nanocomposites. *Compos. Commun.* **2020**, *22*, 100440. [CrossRef]
7. Silva, M.J.; Cena, C.R.; Sanches, A.O.; Mattoso, L.H.C.; Malmonge, J.A. DBSA to Improve the Compatibility, Solubility, and Infusibility of Cellulose Nanowhiskers Modified by Polyaniline in Reinforcing a Natural Rubber-Based Nanocomposite. *Polym. Bull.* **2019**, *76*, 3517–3533. [CrossRef]
8. Silva, M.J.; Dias, Y.J.; Zaszczyńska, A.; Kołbuk, D.; Kowalczyk, T.; Sajkiewicz, P.Ł.; Yarin, A.L. Three-phase bio-nanocomposite natural-rubber-based microfibers reinforced with cellulose nanowhiskers and 45S5 bioglass obtained by solution blow spinning. *J. Appl. Polym. Sci.* **2023**, *140*, e54661. [CrossRef]
9. Silva, M.J.; Sanches, A.O.; Cena, C.R.; Nagashima, H.N.; Medeiros, E.S.; Malmonge, J.A. Study of the Electrical Conduction Process in Natural Rubber-Based Conductive Nanocomposites Filled with Cellulose Nanowhiskers Coated by Polyaniline. *Polym. Compos.* **2020**, *42*, 1519–1529. [CrossRef]
10. Stelescu, M.D.; Manaila, E.; Georgescu, M.; Nituica, M. New Materials Based on Ethylene Propylene Diene Terpolymer and Hemp Fibers Obtained by Green Reactive Processing. *Materials* **2020**, *13*, 2067. [CrossRef]
11. Abdel-Aziz, M.M.; Amer, H.A.; Atia, M.K.; Rabie, A.M. Effect of Gamma Radiation on the Physicomechanical Characters of EPDM Rubber/Modified Additives Nanocomposites. *J. Vinyl Addit. Technol.* **2017**, *23*, E188–E200. [CrossRef]
12. Jahed, M.; Naderi, G.; Hamid Reza Ghoreishy, M. Microstructure, Mechanical, and Rheological Properties of Natural Rubber/Ethylene Propylene Diene Monomer Nanocomposites Reinforced by Multi-wall Carbon Nanotubes. *Polym. Compos.* **2018**, *39*, E745–E753. [CrossRef]
13. Ismail, H.; Che Mat, N.S.; Othman, N. Curing Characteristics, Tear, Fatigue, and Aging Properties of Bentonite-filled Ethylene-propylene-diene (EPDM) Rubber Composites. *J. Vinyl Addit. Technol.* **2018**, *24*, E77–E84. [CrossRef]
14. Razak, J.; Mohamad, N.; Mahamood, M.A.; Jaafar, R.; Othman, I.S.; Ismail, M.M.; Tee, L.K.; Junid, R.; Mustafa, Z. On the Preparation of EPDM-g-MAH Compatibilizer via Melt-Blending Method. *J. Mech. Eng. Sci.* **2019**, *13*, 5424–5440. [CrossRef]
15. Valentini, L.; Bittolo Bon, S.; Lopez-Manchado, M.A.; Verdejo, R.; Pappalardo, L.; Bolognini, A.; Alvino, A.; Borsini, S.; Berardo, A.; Pugno, N.M. Synergistic Effect of Graphene Nanoplatelets and Carbon Black in Multifunctional EPDM Nanocomposites. *Compos. Sci. Technol.* **2016**, *128*, 123–130. [CrossRef]
16. Amin, A.; Kandil, H.; Rabia, A.M.; Nashar, D.E.E.; Ismail, M.N. Enhancing the Mechanical and Dielectric Properties of EPDM Filled with Nanosized Rice Husk Powder. *Polym. Plast. Technol. Eng.* **2018**, *57*, 1733–1742. [CrossRef]
17. Li, C.; Wang, Y.; Yuan, Z.; Ye, L. Construction of Sacrificial Bonds and Hybrid Networks in EPDM Rubber towards Mechanical Performance Enhancement. *Appl. Surf. Sci.* **2019**, *484*, 616–627. [CrossRef]
18. Lei, Y.; He, J.; Zhao, Q.; Liu, T. A Nitrile Functionalized Graphene Filled Ethylene Propylene Diene Terpolymer Rubber Composites with Improved Heat Resistance. *Compos. B Eng.* **2018**, *134*, 81–90. [CrossRef]
19. Güngör, A.; Akbay, I.K.; Özdemir, T. EPDM Rubber with Hexagonal Boron Nitride: A Thermal Neutron Shielding Composite. *Radiat. Phys. Chem.* **2019**, *165*, 108391. [CrossRef]
20. Ghoreishi, A.; Koosha, M.; Nasirizadeh, N. Modification of Bitumen by EPDM Blended with Hybrid Nanoparticles: Physical, Thermal, and Rheological Properties. *J. Thermoplast. Compos. Mater.* **2020**, *33*, 343–356. [CrossRef]
21. Restrepo-Zapata, N.C.; Osswald, T.A.; Hernández-Ortiz, J.P. Vulcanization of EPDM Rubber Compounds with and without Blowing Agents: Identification of Reaction Events and TTT-Diagram Using DSC Data. *Polym. Eng. Sci.* **2015**, *55*, 2073–2088. [CrossRef]
22. Khalaf, A.; Helaly, F.; El Sawy, S. Improvement Properties of EPDM Rubber Using Hybrid Chitin/Clay Filler for Industrial Products. *Egypt. J. Chem.* **2019**, *63*, 129–143. [CrossRef]
23. Manaila, E.; Airinei, A.; Stelescu, M.D.; Sonmez, M.; Alexandrescu, L.; Craciun, G.; Pamfil, D.; Fifere, N.; Varganici, C.-D.; Doroftei, F.; et al. Radiation Processing and Characterization of Some Ethylene-Propylene-Diene Terpolymer/Butyl (Halobutyl) Rubber/Nanosilica Composites. *Polymers* **2020**, *12*, 2431. [CrossRef] [PubMed]
24. Craciun, G.; Manaila, E.; Ighigeanu, D.; Stelescu, M.D. A Method to Improve the Characteristics of EPDM Rubber Based Eco-Composites with Electron Beam. *Polymers* **2020**, *12*, 215. [CrossRef] [PubMed]

25. Su, J.; Li, C.H. Effect of Carbon Nanotube Content on Properties of Ethylene Propylene Diene Rubber/CaCO$_3$ Carbon Nanotube Composites. *Adv. Mat. Res.* **2017**, *1142*, 206–210. [CrossRef]
26. Guo, M.; Li, J.; Li, K.; Zhu, G.; Hu, B.; Liu, Y.; Ji, J. Carbon Nanotube Reinforced Ablative Material for Thermal Protection System with Superior Resistance to High-Temperature Dense Particle Erosion. *Aerosp. Sci. Technol.* **2020**, *106*, 106234. [CrossRef]
27. Jha*, N.; Mahapatra, S.P. Morphology, Barrier, Mechanical and Electrical Conductivity Properties of Oil-Extended EPDM/MWCNT Nanocomposites. *Int. J. Eng. Adv. Technol.* **2019**, *8*, 4116–4121. [CrossRef]
28. Bizhani, H.; Katbab, A.A.; Lopez-Hernandez, E.; Miranda, J.M.; Verdejo, R. Highly Deformable Porous Electromagnetic Wave Absorber Based on Ethylene–Propylene–Diene Monomer/Multiwall Carbon Nanotube Nanocomposites. *Polymers* **2020**, *12*, 858. [CrossRef]
29. Lu, S.; Bai, Y.; Wang, J.; Chen, D.; Ma, K.; Meng, Q.; Liu, X. Flexible GnPs/EPDM with Excellent Thermal Conductivity and Electromagnetic Interference Shielding Properties. *Nano* **2019**, *14*, 1950075. [CrossRef]
30. Allahbakhsh, A.; Mazinani, S. Influences of Sodium Dodecyl Sulfate on Vulcanization Kinetics and Mechanical Performance of EPDM/Graphene Oxide Nanocomposites. *RSC Adv.* **2015**, *5*, 46694–46704. [CrossRef]
31. Han, S.-W.; Choi, N.-S.; Ryu, S.-R.; Lee, D.-J. Mechanical Property Behavior and Aging Mechanism of Carbon-Black-Filled EPDM Rubber Reinforced by Carbon Nano-Tubes Subjected to Electro-Chemical and Thermal Degradation. *J. Mech. Sci. Technol.* **2017**, *31*, 4073–4078. [CrossRef]
32. Sagar, M.; Nibedita, K.; Manohar, N.; Kumar, K.R.; Suchismita, S.; Pradnyesh, A.; Reddy, A.B.; Sadiku, E.R.; Gupta, U.N.; Lachit, P.; et al. A Potential Utilization of End-of-Life Tyres as Recycled Carbon Black in EPDM Rubber. *Waste Manag.* **2018**, *74*, 110–122. [CrossRef] [PubMed]
33. Zhang, G.; Zhou, X.; Liang, K.; Guo, B.; Li, X.; Wang, Z.; Zhang, L. Mechanically Robust and Recyclable EPDM Rubber Composites by a Green Cross-Linking Strategy. *ACS Sustain. Chem. Eng.* **2019**, *7*, 11712–11720. [CrossRef]
34. Ashok, N.; Balachandran, M. Effect of Nanoclay and Nanosilica on Carbon Black Reinforced EPDM/CIIR Blends for Nuclear Applications. *Mater. Res. Express* **2020**, *6*, 125364. [CrossRef]
35. Moustafa, H.; Lawandy, S.N.; Rabee, M.; Zahran, M.A.H. Effect of Green Modification of Nanoclay on the Adhesion Behavior of EPDM Rubber to Polyester Fabric. *Int. J. Adhes. Adhes.* **2020**, *100*, 102617. [CrossRef]
36. Wasfy, S.A.; Hegazi, E.M.; Abd El-megeed, A.A.; Mahmoud, T.S.; El-Kady, E.Y. Effect of Nanoclay on EPDM Composites under Gamma Irradiation. *Arab. J. Nucl. Sci. Appl.* **2018**, *51*, 168–176.
37. Mokhothu, T.H.; Luyt, A.S.; Messori, M. Preparation and Characterization of EPDM/Silica Composites Prepared through Non-Hydrolytic Sol-Gel Method in the Absence and Presence of a Coupling Agent. *Express Polym. Lett.* **2014**, *8*, 809–822. [CrossRef]
38. George, K.; Mohanty, S.; Biswal, M.; Nayak, S.K. Thermal Insulation Behaviour of Ethylene Propylene Diene Monomer Rubber/Kevlar Fiber Based Hybrid Composites Containing Nanosilica for Solid Rocket Motor Insulation. *J. Appl. Polym. Sci.* **2021**, *138*, 49934. [CrossRef]
39. Zygo, M.; Lipinska, M.; Lu, Z.; Ilcíková, M.; Bockstaller, M.R.; Mosnacek, J.; Pietrasik, J. New Type of Montmorillonite Compatibilizers and Their Influence on Viscoelastic Properties of Ethylene Propylene Diene and Methyl Vinyl Silicone Rubbers Blends. *Appl. Clay Sci.* **2019**, *183*, 105359. [CrossRef]
40. Lee, K.; Chang, Y. Peroxide Vulcanized EPDM Rubber/Polyhedral Oligomeric Silsesquioxane Nanocomposites: Vulcanization Behavior, Mechanical Properties, and Thermal Stability. *Polym. Eng. Sci.* **2015**, *55*, 2814–2820. [CrossRef]
41. Zaharescu, T.; Blanco, I.; Bottino, F.A. Antioxidant Activity Assisted by Modified Particle Surface in POSS/EPDM Hybrids. *Appl. Surf. Sci.* **2020**, *509*, 144702. [CrossRef]
42. Lu, S.; Li, B.; Ma, K.; Wang, S.; Liu, X.; Ma, Z.; Lin, L.; Zhou, G.; Zhang, D. Flexible MXene/EPDM Rubber with Excellent Thermal Conductivity and Electromagnetic Interference Performance. *Appl. Phys. A* **2020**, *126*, 513. [CrossRef]
43. Bizhani, H.; Katbab, A.A.; Lopez-Hernandez, E.; Miranda, J.M.; Lopez-Manchado, M.A.; Verdejo, R. Preparation and Characterization of Highly Elastic Foams with Enhanced Electromagnetic Wave Absorption Based on Ethylene-Propylene-Diene-Monomer Rubber Filled with Barium Titanate/Multiwall Carbon Nanotube Hybrid. *Polymers* **2020**, *12*, 2278. [CrossRef]
44. Lu, S.; Ma, J.; Chen, D.; Du, K.; Ma, K.; Bai, Y.; Lu, Z.; Wang, X. Highly Stretchable and Sensitive Sensor Based on GnPs/EPDM Composites with Excellent Heat Dissipation Performance. *Appl. Phys. A* **2019**, *125*, 425. [CrossRef]
45. Jha, N.; Sarkhel, G.; Mahapatra, S.P. Morphology, Barrier and Electrical Properties of Oil-Extended EPDM/Nanographite Nanocomposites. *Mater. Today Proc.* **2021**, *45*, 3850–3856. [CrossRef]
46. Neelesh, A.; Vidhyashree, S.; Meera, B. The Influence of MWCNT and Hybrid (MWCNT/Nanoclay) Fillers on Performance of EPDM-CIIR Blends in Nuclear Applications: Mechanical, Hydrocarbon Transport, and Gamma-Radiation Aging Characteristicscharacteristics. *J. Appl. Polym. Sci.* **2020**, *137*, 49271. [CrossRef]
47. Ahmed, A.F.; Hoa, S. Development and Analysis of a Hybrid Reinforced Composite Material for Solid Rocket Motor Insulation. In Proceedings of the American Society for Composites 23rd Annual Technical Conference, Memphis, TN, USA, 9–11 September 2008; Volume 3, pp. 131–143.
48. Ding, Z.; He, F.; Li, Y.; Jiang, Z.; Yan, H.; He, R.; Fan, J.; Zhang, K.; Yang, W. Novel Shape-Stabilized Phase Change Materials Based on Paraffin/EPDM@Graphene with High Thermal Conductivity and Low Leakage Rate. *Energy Fuels* **2020**, *34*, 5024–5031. [CrossRef]
49. Guo, M.; Li, J.; Xi, K.; Liu, Y.; Ji, J.; Ye, C. Effects of Multi-Walled Carbon Nanotubes on Char Residue and Carbothermal Reduction Reaction in Ethylene Propylene Diene Monomer Composites at High Temperature. *Compos. Sci. Technol.* **2020**, *186*, 107916. [CrossRef]

50. Zhang, C.; Wang, J.; Zhao, Y. Effect of Dendrimer Modified Montmorillonite on Structure and Properties of EPDM Nanocomposites. *Polym. Test.* **2017**, *62*, 41–50. [CrossRef]
51. Chazeau, L.; Gauthier, C.; Chenal, J.M. Mechanical Properties of Rubber Nanocomposites: How, Why ... and Then? In *Rubber Nanocomposites*; Wiley: Hoboken, NJ, USA, 2010; pp. 291–330.
52. Fu, S.-Y.; Feng, X.-Q.; Lauke, B.; Mai, Y.-W. Effects of Particle Size, Particle/Matrix Interface Adhesion and Particle Loading on Mechanical Properties of Particulate–Polymer Composites. *Compos. B Eng.* **2008**, *39*, 933–961. [CrossRef]
53. Mirjalili, F.; Chuah, L.; Salahi, E. Mechanical and Morphological Properties of Polypropylene/Nano α-Al_2O_3 Composites. *Sci. World J.* **2014**, *2014*, 718765. [CrossRef]
54. Kallungal, J.; Chazeau, L.; Chenal, J.-M.; Adrien, J.; Maire, E.; Barrès, C.; Cantaloube, B.; Heuillet, P.; Wilde, F.; Moosmann, J.; et al. Crack Propagation in Filled Elastomers: 3D Study of Mechanisms Involving the Filler Agglomerates. *Eng. Fract. Mech.* **2022**, *274*, 108771. [CrossRef]
55. Ragupathy, K.; Prabaharan, G.; Pragadish, N.; Vishvanathperumal, S. Effect of Silica Nanoparticles and Modified Silica Nanoparticles on the Mechanical and Swelling Properties of EPDM/SBR Blend Nanocomposites. *Silicon* **2023**, *15*, 6033–6046. [CrossRef]
56. Lin, C.; Kanstad, T.; Jacobsen, S.; Ji, G. Bonding Property between Fiber and Cementitious Matrix: A Critical Review. *Constr. Build. Mater.* **2023**, *378*, 131169. [CrossRef]
57. Da Silva, A.L.N.; Tavares, M.I.B.; Politano, D.P.; Coutinho, F.M.B.; Rocha, M.C.G. Polymer Blends Based on Polyolefin Elastomer and Polypropylene. *J. Appl. Polym. Sci.* **1997**, *66*, 2005–2014. [CrossRef]
58. Sang, J.S.; Kim, T.; Park, E.-Y.; Park, J.; Eum, Y.; Oh, K.W. Bio-EPDM/Tungsten Oxide Nanocomposite Foam with Improved Thermal Storage and Sea Water Resistance. *Fash. Text.* **2020**, *7*, 29. [CrossRef]
59. Bianchi, M.; Valentini, F.; Fredi, G.; Dorigato, A.; Pegoretti, A. Thermo-Mechanical Behavior of Novel EPDM Foams Containing a Phase Change Material for Thermal Energy Storage Applications. *Polymers* **2022**, *14*, 4058. [CrossRef] [PubMed]
60. Hu, H. Recent Advances of Polymeric Phase Change Composites for Flexible Electronics and Thermal Energy Storage System. *Compos. B Eng.* **2020**, *195*, 108094. [CrossRef]
61. Peng, S.; Fuchs, A.; Wirtz, R.A. Polymeric Phase Change Composites for Thermal Energy Storage. *J. Appl. Polym. Sci.* **2004**, *93*, 1240–1251. [CrossRef]
62. Liu, C.; Xiao, T.; Zhao, J.; Liu, Q.; Sun, W.; Guo, C.; Ali, H.M.; Chen, X.; Rao, Z.; Gu, Y. Polymer Engineering in Phase Change Thermal Storage Materials. *Renew. Sustain. Energy Rev.* **2023**, *188*, 113814. [CrossRef]
63. Dincer, I.; Rosen, M.A. Heat Storage Systems. In *Exergy Analysis of Heating, Refrigerating and Air Conditioning*; Elsevier: Amsterdam, The Netherlands, 2015; pp. 221–278.
64. Valentini, F.; Dorigato, A.; Fambri, L.; Bersani, M.; Grigiante, M.; Pegoretti, A. Production and Characterization of Novel EPDM/NBR Panels with Paraffin for Potential Thermal Energy Storage Applications. *Therm. Sci. Eng. Prog.* **2022**, *32*, 101309. [CrossRef]
65. Valentini, F.; Fredi, G.; Dorigato, A. Thermal Energy Storage (TES) for Sustainable Buildings: Addressing the Current Energetic Situation in the EU with TES-Enhanced Buildings. In *Natural Energy, Lighting, and Ventilation in Sustainable Buildings*; Nazari-Heris, M., Ed.; Springer Nature: Cham, Switzerland, 2024; pp. 191–224. ISBN 978-3-031-41148-9.
66. Mofijur, M.; Mahlia, T.M.I.; Silitonga, A.S.; Ong, H.C.; Silakhori, M.; Hasan, M.H.; Putra, N.; Rahman, S.M.A. Phase Change Materials (PCM) for Solar Energy Usages and Storage: An Overview. *Energies* **2019**, *12*, 3167. [CrossRef]
67. Li, G.; Zheng, X. Thermal Energy Storage System Integration Forms for a Sustainable Future. *Renew. Sustain. Energy Rev.* **2016**, *62*, 736–757. [CrossRef]
68. Chandrasekaran, C. Rubbers Mostly Used in Process Equipment Lining. In *Anticorrosive Rubber Lining*; Elsevier: Amsterdam, The Netherlands, 2017; pp. 87–101.
69. Yang, P.; He, M.; Ren, X.; Zhou, K. Effect of Carbon Nanotube on Space Charge Suppression in PP/EPDM/CNT Nanocomposites. *J. Polym. Res.* **2020**, *27*, 132. [CrossRef]
70. Gao, J.; Zhang, H.a.o.; Li, L.; Zhang, J.; Guo, N.; Zhang, X.H. Study on space charge characteristics of xlpe/o-mmt nanocomposites. *Acta Polym. Sin.* **2013**, *13*, 126–133. [CrossRef]
71. Resner, L.; Lesiak, P.; Taraghi, I.; Kochmanska, A.; Figiel, P.; Piesowicz, E.; Zenker, M.; Paszkiewicz, S. Polymer Hybrid Nanocomposites Based on Homo and Copolymer Xlpe Containing Mineral Nanofillers with Improved Functional Properties Intended for Insulation of Submarine Cables. *Polymers* **2022**, *14*, 3444. [CrossRef]
72. Qingyue, Y.; Xiufeng, L.; Peng, Z.; Peijie, Y.; Youfu, C. Properties of Water Tree Growing in XLPE and Composites. In Proceedings of the 2019 2nd International Conference on Electrical Materials and Power Equipment (ICEMPE), Guangzhou, China, 7–10 April 2019; IEEE: Piscataway, NJ, USA, 2019; pp. 409–412.
73. Vishvanathperumal, S.; Anand, G. Effect of Nanoclay/Nanosilica on the Mechanical Properties, Abrasion and Swelling Resistance of EPDM/SBR Composites. *Silicon* **2020**, *12*, 1925–1941. [CrossRef]
74. Abdelsalam, A.A.; Mohamed, W.S.; Abd El-Naeem, G.; El-Sabbagh, S.H. Effect of the Silane Coupling Agent on the Physicomechanical Properties of EPDM/SBR/AL_2O_3 Rubber Blend Nanocomposites. *J. Thermoplast. Compos. Mater.* **2023**, *36*, 1811–1832. [CrossRef]
75. Bahadar, A.; Zwawi, M. Development of SWCNTs-Reinforced EPDM/SBR Matrices for Shock Absorbing Applications Development of SWCNTs-Reinforced EPDM/SBR Matrices for Shock Absorbing Applications. *Mater. Res. Express* **2020**, *7*, 025310. [CrossRef]

76. Deng, F.; Jin, H.; Zhang, L.; He, Y. Higher Tear Strength of EPDM/SBR/TPR Composites Foam Based on Double Foaming System. *J. Elastomers Plast.* **2021**, *53*, 373–385. [CrossRef]
77. Hasanzadeh Kermani, H.; Mottaghitalab, V.; Mokhtary, M.; Alizadeh Dakhel, A. Morphological, Rheological, and Mechanical Properties of Ethylene Propylene Diene Monomer/Carboxylated Styrene-Butadiene Rubber/Multiwall Carbon Nanotube Nanocomposites. *Int. J. Polym. Anal. Charact.* **2020**, *25*, 479–498. [CrossRef]
78. Vayyaprontavida Kaliyathan, A.; Varghese, K.M.; Nair, A.S.; Thomas, S. Rubber–Rubber Blends: A Critical Review. *Prog. Rubber Plast. Recycl. Technol.* **2020**, *36*, 196–242. [CrossRef]
79. Sowińska, A.; Maciejewska, M.; Guo, L.; Delebecq, E. Effect of SILPs on the Vulcanization and Properties of Ethylene-Propylene-Diene Elastomer. *Polymers* **2020**, *12*, 1220. [CrossRef]
80. Vishvanathperumal, S.; Gopalakannan, S. Effects of the Nanoclay and Crosslinking Systems on the Mechanical Properties of Ethylene-Propylene-Diene Monomer/Styrene Butadiene Rubber Blends Nanocomposite. *Silicon* **2019**, *11*, 117–135. [CrossRef]
81. Rostami-Tapeh-esmaeil, E.; Vahidifar, A.; Esmizadeh, E.; Rodrigue, D. Chemistry, Processing, Properties, and Applications of Rubber Foams. *Polymers* **2021**, *13*, 1565. [CrossRef]
82. Lee, S.H.; Park, S.Y.; Chung, K.; Jang, K.S. Phlogopite-Reinforced Natural Rubber (Nr)/Ethylene-Propylene-Diene Monomer Rubber (EPDM) Composites with Aminosilane Compatibilizer. *Polymers* **2021**, *13*, 2318. [CrossRef] [PubMed]
83. Chang, E.; Zhao, J.; Zhao, C.; Li, G.; Lee, P.C.; Park, C.B. Scalable Production of Crosslinked Rubber Nanofibre Networks as Highly Efficient Toughening Agent for Isotactic Polypropylene: Toughening Mechanism of Non-Traditional Anisotropic Rubber Inclusion. *Chem. Eng. J.* **2022**, *438*, 134060. [CrossRef]
84. Azizli, M.J.; Barghamadi, M.; Rezaeeparto, K.; Mokhtary, M.; Parham, S. Compatibility, Mechanical and Rheological Properties of Hybrid Rubber NR/EPDM-g-MA/EPDM/Graphene Oxide Nanocomposites: Theoretical and Experimental Analyses. *Compos. Commun.* **2020**, *22*, 100442. [CrossRef]
85. Burgaz, E.; Goksuzoglu, M. Effects of Magnetic Particles and Carbon Black on Structure and Properties of Magnetorheological Elastomers. *Polym. Test.* **2020**, *81*, 106233. [CrossRef]
86. Nazir, M.T.; Phung, B.T.; Li, S.; Akram, S.; Mehmood, M.A.; Yeoh, G.H.; Hussain, S. Effect of Micro-Nano Additives on Breakdown, Surface Tracking and Mechanical Performance of Ethylene Propylene Diene Monomer for High Voltage Insulation. *J. Mater. Sci. Mater. Electron.* **2019**, *30*, 14061–14071. [CrossRef]
87. Khademeh Molavi, F.; Soltani, S.; Naderi, G.; Bagheri, R. Effect of Multi-Walled Carbon Nanotube on Mechanical and Rheological Properties of Silane Modified EPDM Rubber. *Polyolefins J.* **2016**, *3*, 69–77. [CrossRef]
88. Zvereva, U.G.; Solomatin, D.V.; Kuznetsova, O.P.; Prut, E.V. Rheological Properties of Ethylene-Propylene-Diene Elastomers. *Polym. Sci. Ser. D* **2016**, *9*, 234–237. [CrossRef]
89. ASTM D-2084-07; Test Standard Methods for Rubber Property-Vulacization Using Oscillating Disk Cure Meter. ASTM International: West Conshohocken, PA, USA, 2007.
90. Chen, G.; Gupta, A.; Mekonnen, T.H. Silane-Modified Wood Fiber Filled EPDM Bio-Composites with Improved Thermomechanical Properties. *Compos. Part A Appl. Sci. Manuf.* **2022**, *159*, 107029. [CrossRef]
91. Abdelsalam, A.A.; Araby, S.; El-Sabbagh, S.H.; Abdelmoneim, A.; Hassan, M.A. Effect of Carbon Black Loading on Mechanical and Rheological Properties of Natural Rubber/Styrene-Butadiene Rubber/Nitrile Butadiene Rubber Blends. *J. Thermoplast. Compos. Mater.* **2021**, *34*, 490–507. [CrossRef]
92. Utracki, L.A.; Walsh, D.J.; Weiss, R.A. *Multiphase Polymers: Blends and Ionomers*; American Chemical Society: Washington, DC, USA, 1989; pp. 1–35. [CrossRef]
93. Yang, Z.; Peng, H.; Wang, W.; Liu, T. Crosslink Density and Diffusion Mechanisms in Blend Vulcanizates Loaded with Carbon Black and Paraffin Wax. *J. Appl. Polym. Sci.* **2009**, *116*, 2658–2667. [CrossRef]
94. Monakhova, T.V.; Bogaevskaya, T.A.; Shlyapnikov, Y.A. Effect of Polydimethylsiloxane on Oxidation of Isotactic Polypropylene. *Polym. Degrad. Stab.* **1999**, *66*, 149–151. [CrossRef]
95. Robeson, L.M. Applications of Polymer Blends: Emphasis on Recent Advances. *Polym. Eng. Sci.* **1984**, *24*, 587–597. [CrossRef]
96. Stelescu, M.D.; Airinei, A.; Bargan, A.; Fifere, N.; Georgescu, M.; Sonmez, M.; Nituica, M.; Alexandrescu, L.; Stefan, A. Mechanical Properties and Equilibrium Swelling Characteristics of Some Polymer Composites Based on Ethylene Propylene Diene Terpolymer (EPDM) Reinforced with Hemp Fibers. *Materials* **2022**, *15*, 6838. [CrossRef]
97. Ghobashy, M.M.; Mousa, S.A.S.; Siddiq, A.; Nasr, H.M.D.; Nady, N.; Atalla, A.A. Optimal the Mechanical Properties of Bioplastic Blend Based Algae-(Lactic Acid-Starch) Using Gamma Irradiation and Their Possibility to Use as Compostable and Soil Conditioner. *Mater. Today Commun.* **2023**, *34*, 105472. [CrossRef]
98. Zarrinjooy Alvar, M.; Abdeali, G.; Bahramian, A.R. Influence of Graphite Nano Powder on Ethylene Propylene Diene Monomer/Paraffin Wax Phase Change Material Composite: Shape Stability and Thermal Applications. *J. Energy Storage* **2022**, *52*, 105065. [CrossRef]
99. Jiang, X.; Yuan, X.; Guo, X.; Zeng, F.; Wang, H.; Liu, G. Study on the Application of Flory–Huggins Interaction Parameters in Swelling Behavior and Crosslink Density of HNBR/EPDM Blend. *Fluid. Phase Equilib.* **2023**, *563*, 113589. [CrossRef]
100. Salimi, A.; Abbassi-Sourki, F.; Karrabi, M.; Reza Ghoreishy, M.H. Investigation on Viscoelastic Behavior of Virgin EPDM/Reclaimed Rubber Blends Using Generalized Maxwell Model (GMM). *Polym. Test.* **2021**, *93*, 106989. [CrossRef]
101. Evgin, T.; Mičušík, M.; Machata, P.; Peidayesh, H.; Preťo, J.; Omastová, M. Morphological, Mechanical and Gas Penetration Properties of Elastomer Composites with Hybrid Fillers. *Polymers* **2022**, *14*, 4043. [CrossRef] [PubMed]

102. Azaar, K.; Vergnaud, J.M.; Rosca, I.D. Anisotropic Swelling of EPDM Rubber Discs by Absorption of Toluene. *Rubber Chem. Technol.* **2002**, *76*, 1031–1044. [CrossRef]
103. Vishvanathperumal, S.; Anand, G. Effect of Nanosilica on the Mechanical Properties, Compression Set, Morphology, Abrasion and Swelling Resistance of Sulphur Cured EPDM/SBR Composites. *Silicon* **2022**, *14*, 3523–3534. [CrossRef]
104. Darko, C. The Link between Swelling Ratios and Physical Properties of EPDM Rubber Compound Having Different Oil Amounts. *J. Polym. Res.* **2022**, *29*, 325. [CrossRef]
105. Abdelsalam, A.A.; Ward, A.A.; Abdel-Naeem, G.; Mohamed, W.S.; El-Sabbagh, S.H. Effect of Alumina Modified by Silane on the Mechanical, Swelling and Dielectric Properties of Al_2O_3/EPDM/SBR Blend Composites. *Silicon* **2023**, *15*, 3609–3621. [CrossRef]
106. Prakash, P.C.; Srinivasan, D.; Navaneethakrishnan, V.; Vishvanathperumal, S. Effect of Modified Nanographene Oxide Loading on the Swelling and Compression Set Behavior of EPDM/SBR Nano-Composites. *J. Inorg. Organomet. Polym. Mater.* **2023**, *34*, 593–610. [CrossRef]
107. Colom, X.; Carrillo-Navarrete, F.; Saeb, M.R.; Marin, M.; Formela, K.; Cañavate, J. Evaluation and Rationale of the Performance of Several Elastomeric Composites Incorporating Devulcanized EPDM. *Polym. Test.* **2023**, *121*, 107976. [CrossRef]
108. Simet, C.; Mougin, K.; Moreau, M.; Perche, M.; Vaulot, C.; Ponche, A.; Bally-Le Gall, F. Investigation of Ethylene-Propylene-Diene Monomer (EPDM) Ageing Behaviour in PMDIs Environment by Surface NMR. *Polym. Test.* **2023**, *127*, 108171. [CrossRef]
109. Najibah, M.; Kong, J.; Khalid, H.; Hnát, J.; Park, H.S.; Bouzek, K.; Henkensmeier, D. Pre-Swelling of FAA3 Membranes with Water-Based Ethylene Glycol Solution to Minimize Dimensional Changes after Assembly into a Water Electrolyser: Effect on Properties and Performance. *J. Memb. Sci.* **2023**, *670*, 121344. [CrossRef]
110. Mothé, C.G.; Azevedo, A.D. *De Analise Termica de Materiais*, 1st ed.; Artliber Editora: São Paulo, Brazil, 2009; ISBN 8588098490.
111. Mohamad, N.; Yaakub, J.; Ab Maulod, H.E.; Jeefferie, A.R.; Yuhazri, M.Y.; Lau, K.T.; Ahsan, Q.; Shueb, M.I.; Othman, R. Vibrational Damping Behaviors of Graphene Nanoplatelets Reinforced NR/EPDM Nanocomposites. *J. Mech. Eng. Sci.* **2017**, *11*, 3274–3287. [CrossRef]
112. Rana, A.S.; Vamshi, M.K.; Naresh, K.; Velmurugan, R.; Sarathi, R. Mechanical, Thermal, Electrical and Crystallographic Behaviour of EPDM Rubber/Clay Nanocomposites for out-Door Insulation Applications. *Adv. Mater. Process. Technol.* **2020**, *6*, 54–74. [CrossRef]
113. Zhang, F.; Zhao, Q.; Liu, T.; Lei, Y.; Chen, C. Preparation and Relaxation Dynamics of Ethylene–Propylene–Diene Rubber/Clay Nanocomposites with Crosslinking Interfacial Design. *J. Appl. Polym. Sci.* **2018**, *135*, 45553. [CrossRef]
114. Guo, M.; Li, J.; Xi, K.; Liu, Y.; Ji, J. Effect of Multi-Walled Carbon Nanotubes on Thermal Stability and Ablation Properties of EPDM Insulation Materials for Solid Rocket Motors. *Acta Astronaut.* **2019**, *159*, 508–516. [CrossRef]
115. Iqbal, S.S.; Iqbal, N.; Jamil, T.; Bashir, A.; Khan, Z.M. Tailoring in Thermomechanical Properties of Ethylene Propylene Diene Monomer Elastomer with Silane Functionalized Multiwalled Carbon Nanotubes. *J. Appl. Polym. Sci.* **2016**, *133*, 43221. [CrossRef]
116. Jeon, B.; Kim, T.; Lee, D.; Shin, T.J.; Oh, K.W.; Park, J. Photothermal Polymer Nanocomposites of Tungsten Bronze Nanorods with Enhanced Tensile Elongation at Low Filler Contents. *Polymers* **2019**, *11*, 1740. [CrossRef]
117. Mokhothu, T.H.; Luyt, A.S.; Messori, M. Reinforcement of EPDM Rubber with in Situ Generated Silica Particles in the Presence of a Coupling Agent via a Sol-Gel Route. *Polym. Test.* **2014**, *33*, 97–106. [CrossRef]
118. Mokhothu, T.H.; Luyt, A.S.; Morselli, D.; Bondioli, F.; Messori, M. Influence of in Situ-Generated Silica Nanoparticles on EPDM Morphology, Thermal, Thermomechanical, and Mechanical Properties. *Polym. Compos.* **2015**, *36*, 825–833. [CrossRef]
119. Basu, D.; Das, A.; Wang, D.-Y.; George, J.J.; Stöckelhuber, K.W.; Boldt, R.; Leuteritz, A.; Heinrich, G. Fire-Safe and Environmentally Friendly Nanocomposites Based on Layered Double Hydroxides and Ethylene Propylene Diene Elastomer. *RSC Adv.* **2016**, *6*, 26425–26436. [CrossRef]
120. Drzeżdżon, J.; Jacewicz, D.; Sielicka, A.; Chmurzyński, L. A Review of New Approaches to Analytical Methods to Determine the Structure and Morphology of Polymers. *TrAC Trends Anal. Chem.* **2019**, *118*, 470–476. [CrossRef]
121. Haghnegahdar, M.; Naderi, G.; Ghoreishy, M.H.R. Fracture Toughness and Deformation Mechanism of Un-Vulcanized and Dynamically Vulcanized Polypropylene/Ethylene Propylene Diene Monomer/Graphene Nanocomposites. *Compos. Sci. Technol.* **2017**, *141*, 83–98. [CrossRef]
122. Hajibabazadeh, S.; Razavi Aghjeh, M.; Palahang, M. Study on the Fracture Toughness and Deformation Micro-Mechanisms of PP/EPDM/SiO_2 Ternary Blend-Nanocomposites. *J. Compos. Mater.* **2020**, *54*, 591–605. [CrossRef]
123. Azizi, S.; Momen, G.; Ouellet-Plamondon, C.; David, E. Performance Improvement of EPDM and EPDM/Silicone Rubber Composites Using Modified Fumed Silica, Titanium Dioxide and Graphene Additives. *Polym. Test.* **2020**, *84*, 106281. [CrossRef]
124. Benneghmouche, Z.; Benachour, D. Effect of Organophilic Clay Addition on Properties of SAN/EPDM Blends. *Compos. Interfaces* **2019**, *26*, 711–727. [CrossRef]
125. Bhattacharya, A.B.; Raju, A.T.; Chatterjee, T.; Naskar, K. Development and Characterizations of Ultra-High Molecular Weight EPDM/PP Based TPV Nanocomposites for Automotive Applications. *Polym. Compos.* **2020**, *41*, 4950–4962. [CrossRef]
126. Mohamed Bak, K.; Raj Kumar, G.; Ramasamy, N.; Vijayanandh, R. Experimental and Numerical Studies on the Mechanical Characterization of Epdm/S-Sbr Nano Clay Composites. *IOP Conf. Ser. Mater. Sci. Eng.* **2020**, *912*, 052016. [CrossRef]
127. Ma, L.; Yang, W.; Jiang, C. Stretchable Conductors of Multi-Walled Carbon Nanotubes (MWCNTs) Filled Thermoplastic Vulcanizate (TPV) Composites with Enhanced Electromagnetic Interference Shielding Performance. *Compos. Sci. Technol.* **2020**, *195*, 108195. [CrossRef]

128. Khodabandelou, M.; Razavi Aghjeh, M.K.; Khonakdar, H.A.; Mehrabi Mazidi, M. Effect of Localization of Carbon Nanotubes on Fracture Behavior of Un-Vulcanized and Dynamically Vulcanized PP/EPDM/MWCNT Blend-Nanocomposites. *Compos. Sci. Technol.* **2017**, *149*, 134–148. [CrossRef]
129. Haghnegahdar, M.; Naderi, G.; Ghoreishy, M.H.R. Electrical and Thermal Properties of a Thermoplastic Elastomer Nanocomposite Based on Polypropylene/Ethylene Propylene Diene Monomer/Graphene. *Soft Mater.* **2017**, *15*, 82–94. [CrossRef]
130. Kang, S.; Choi, K.; Nam, J.-D.; Choi, H. Magnetorheological Elastomers: Fabrication, Characteristics, and Applications. *Materials* **2020**, *13*, 4597. [CrossRef]
131. Cho, J.K.; Sun, H.; Seo, H.W.; Chung, J.-Y.; Seol, M.; Kim, S.-H.; Kim, R.-S.; Park, I.-K.; Suhr, J.; Park, J.C.; et al. Heat Dissipative Mechanical Damping Properties of EPDM Rubber Composites Including Hybrid Fillers of Aluminium Nitride and Boron Nitride. *Soft Matter* **2020**, *16*, 6812–6818. [CrossRef] [PubMed]
132. Melo, D.S.; Reis, I.C.; Queiroz, J.C.; Cena, C.R.; Nahime, B.O.; Malmonge, J.A.; Silva, M.J. Evaluation of Piezoresistive and Electrical Properties of Conductive Nanocomposite Based on Castor-Oil Polyurethane Filled with MWCNT and Carbon Black. *Materials* **2023**, *16*, 3223. [CrossRef] [PubMed]
133. Mittal, G.; Dhand, V.; Rhee, K.Y.; Park, S.-J.; Lee, W.R. A Review on Carbon Nanotubes and Graphene as Fillers in Reinforced Polymer Nanocomposites. *J. Ind. Eng. Chem.* **2015**, *21*, 11–25. [CrossRef]
134. Al Sheheri, S.Z.; Al-Amshany, Z.M.; Al Sulami, Q.A.; Tashkandi, N.Y.; Hussein, M.A.; El-Shishtawy, R.M. The Preparation of Carbon Nanofillers and Their Role on the Performance of Variable Polymer Nanocomposites. *Des. Monomers Polym.* **2019**, *22*, 8–53. [CrossRef] [PubMed]
135. Sharma, S.; Sudhakara, P.; Omran, A.A.B.; Singh, J.; Ilyas, R.A. Recent Trends and Developments in Conducting Polymer Nanocomposites for Multifunctional Applications. *Polymers* **2021**, *13*, 2898. [CrossRef] [PubMed]
136. Han, S.; Wei, H.; Han, L.; Li, Q. Durability and Electrical Conductivity of Carbon Fiber Cloth/Ethylene Propylene Diene Monomer Rubber Composite for Active Deicing and Snow Melting. *Polymers* **2019**, *11*, 2051. [CrossRef] [PubMed]
137. Peidayesh, H.; Špitalský, Z.; Chodák, I. Electrical Conductivity of Rubber Composites with Varying Crosslink Density under Cyclic Deformation. *Polymers* **2022**, *14*, 3640. [CrossRef] [PubMed]
138. Silva, M.J.D.; Kanda, D.H.F.; Nagashima, H.N. Mechanism of Charge Transport in Castor Oil-Based Polyurethane/Carbon Black Composite (PU/CB). *J. Non Cryst. Solids* **2012**, *358*, 270–275. [CrossRef]
139. Rebeque, P.V.; Silva, M.J.; Cena, C.R.; Nagashima, H.N.; Malmonge, J.A.; Kanda, D.H.F. Analysis of the Electrical Conduction in Percolative Nanocomposites Based on Castor-Oil Polyurethane with Carbon Black and Activated Carbon Nanopowder. *Polym. Compos.* **2019**, *40*, 7–15. [CrossRef]
140. Bunde, A.; Dieterich, W. Percolation in Composites. *J. Electroceram* **2000**, *5*, 81–92. [CrossRef]
141. Kirkpatrick, S. Percolation and Conduction. *Rev. Mod. Phys.* **1973**, *45*, 574. [CrossRef]
142. Xue, Q.; Sun, J. Electrical Conductivity and Percolation Behavior of Polymer Nanocomposites. In *Polymer Nanocomposites*; Springer: Cham, Switzerland, 2016; pp. 51–82.
143. Kwon, S.; Cho, H.W.; Gwon, G.; Kim, H.; Sung, B.J. Effects of Shape and Flexibility of Conductive Fillers in Nanocomposites on Percolating Network Formation and Electrical Conductivity. *Phys. Rev. E* **2016**, *93*, 032501. [CrossRef] [PubMed]
144. Sanches, A.O.; Kanda, D.H.F.; Malmonge, L.F.; da Silva, M.J.; Sakamoto, W.K.; Malmonge, J.A. Synergistic Effects on Polyurethane/Lead Zirconate Titanate/Carbon Black Three-Phase Composites. *Polym. Test.* **2017**, *60*, 253–259. [CrossRef]
145. Ma, J.; Wang, X.; Chen, Y.; Yu, J.; Zheng, W.; Zhao, Y. Largely Enhanced Thermal Conductivity of Ethylene-Propylene-Diene Monomer Composites by Addition of Graphene Ball. *Compos. Commun.* **2019**, *13*, 119–124. [CrossRef]
146. Balberg, I.; Azulay, D.; Goldstein, Y.; Jedrzejewski, J. Possible Origin of the Smaller-than-Universal Percolation-Conductivity Exponent in the Continuum. *Phys. Rev. E* **2016**, *93*, 062132. [CrossRef] [PubMed]
147. Athawale, A.A.; Joshi, A.M. *Electronic Applications of Ethylene Propylene Diene Monomer Rubber and Its Composites*; Springer: Cham, Switzerland, 2016; pp. 305–333.
148. Koca, H.D.; Turgut, A.; Evgin, T.; Ateş, İ.; Chirtoc, M.; Šlouf, M.; Omastová, M. A Comprehensive Study on the Thermal and Electrical Conductivity of EPDM Composites with Hybrid Carbon Fillers. *Diam. Relat. Mater.* **2023**, *139*, 110289. [CrossRef]
149. Chi, Q.; Jiang, L.; Zhang, T.; Zhang, C.; Zhang, Y.; Zhang, Y. Study on Electrical Properties of Donor ZnO Nanoparticles/EPDM Composites. *J. Mater. Sci. Mater. Electron.* **2021**, *32*, 26894–26904. [CrossRef]
150. Zhang, T.; Xu, H.; Dai, C.; Zhang, C.; Zhang, Y.; Chi, Q. Carbon Nanotubes and Hexagonal Boron Nitride Nanosheets Co-filled Ethylene Propylene Diene Monomer Composites: Improved Electrical Property for Cable Accessory Applications. *High Volt.* **2023**, 1–10. [CrossRef]
151. Ketikis, P.; Ketikis, I.; Klonos, P.; Giannakopoulou, T.; Kyritsis, A.; Trapalis, C.; Tarantili, P. The Effect of MWCNTs on the Properties of Peroxide Vulcanized Ethylene Propylene Diene Monomer (EPDM) Composites. *J. Compos. Mater.* **2023**, *57*, 2043–2058. [CrossRef]
152. *ASTM D-1418*; Standard Practice for Rubber and Rubber Latices-Nomenclature. ASTM International: West Conshohocken, PA, USA, 2022.
153. Grand View Research. *Ethylene Propylene Diene Monomer (EPDM) Market Size, Share & Trends Analysis Report by Product (Hoses, Seals & O-Rings), by Application (Building & Construction, Tires & Tubes, Automotive), and Segment Forecasts, 2023–2030*; Grand View Research: San Francisco, CA, USA, 2021.

154. Hussain, F.; Hojjati, M.; Okamoto, M.; Gorga, R.E. Review Article: Polymer-Matrix Nanocomposites, Processing, Manufacturing, and Application: An Overview. *J. Compos. Mater.* **2006**, *40*, 1511–1575. [CrossRef]
155. George, J.; Sreekala, M.S.; Thomas, S. A Review on Interface Modification and Characterization of Natural Fiber Reinforced Plastic Composites. *Polym. Eng. Sci.* **2001**, *41*, 1471–1485. [CrossRef]
156. Satterthwaite, K. Plastics Based on Styrene. In *Brydson's Plastics Materials*; Elsevier: Amsterdam, The Netherlands, 2017; pp. 311–328.
157. Akiba, M. Vulcanization and Crosslinking in Elastomers. *Prog. Polym. Sci.* **1997**, *22*, 475–521. [CrossRef]
158. Maiti, M.; Bhattacharya, M.; Bhowmick, A.K. Elastomer Nanocomposites. *Rubber Chem. Technol.* **2008**, *81*, 384–469. [CrossRef]
159. Li, W.; Huang, Y.D.; Ahmadi, S.J. Preparation and Properties of Ethylene–Propylene–Diene Rubber/Organomontmorillonite Nanocomposites. *J. Appl. Polym. Sci.* **2004**, *94*, 440–445. [CrossRef]
160. Koerner, R.M. *Designing with Geosynthetics*, 6th ed.; Xlibris Corporation: Bloomington, IL, USA, 2012; Volume 1.
161. Bokobza, L. Natural Rubber Nanocomposites: A Review. *Nanomaterials* **2018**, *9*, 12. [CrossRef] [PubMed]
162. Camargo, P.H.C.; Satyanarayana, K.G.; Wypych, F. Nanocomposites: Synthesis, Structure, Properties and New Application Opportunities. *Mater. Res.* **2009**, *12*, 1–39. [CrossRef]
163. Acharya, H.; Srivastava, S.K. Mechanical, Thermo-Mechanical, Thermal, and Swelling Properties of EPDM-Organically Modified Mesoporous Silica Nanocomposites. *Polym. Compos.* **2017**, *38*, E371–E380. [CrossRef]
164. Bartosik, D.; Szadkowski, B.; Kuśmierek, M.; Rybiński, P.; Mirkhodzhaev, U.; Marzec, A. Advanced Ethylene-Propylene-Diene (EPDM) Rubber Composites Filled with Raw Silicon Carbide or Hybrid Systems with Different Conventional Fillers. *Polymers* **2022**, *14*, 1383. [CrossRef]
165. Ateia, E.E.; El-Nashar, D.E.; Ramadan, R.; Shokry, M.F. Synthesis and Characterization of EPDM/Ferrite Nanocomposites. *J. Inorg. Organomet. Polym. Mater.* **2020**, *30*, 1041–1048. [CrossRef]
166. George, K.; Biswal, M.; Mohanty, S.; Nayak, S.K.; Panda, B.P. Nanosilica Filled EPDM/Kevlar Fiber Hybrid Nanocomposites: Mechanical and Thermal Properties. *Mater. Today Proc.* **2020**, *41*, 983–986. [CrossRef]
167. Azizli, M.J.; Mokhtary, M.; Khonakdar, H.A.; Goodarzi, V. Hybrid Rubber Nanocomposites Based on XNBR/EPDM: Select the Best Dispersion Type from Different Nanofillers in the Presence of a Compatibilizer. *J. Inorg. Organomet. Polym. Mater.* **2020**, *30*, 2533–2550. [CrossRef]
168. Peddini, S.K.; Bosnyak, C.P.; Henderson, N.M.; Ellison, C.J.; Paul, D.R. Nanocomposites from Styrene–Butadiene Rubber (SBR) and Multiwall Carbon Nanotubes (MWCNT) Part 2: Mechanical Properties. *Polymer* **2015**, *56*, 443–451. [CrossRef]
169. Rajasekar, R.; Pal, K.; Heinrich, G.; Das, A.; Das, C.K. Development of Nitrile Butadiene Rubber–Nanoclay Composites with Epoxidized Natural Rubber as Compatibilizer. *Mater. Des.* **2009**, *30*, 3839–3845. [CrossRef]
170. Santos, R.J.; Agostini, D.L.S.; Cabrera, F.C.; Budemberg, E.R.; Job, A.E. Recycling Leather Waste: Preparing and Studying on the Microstructure, Mechanical, and Rheological Properties of Leather Waste/Rubber Composite. *Polym. Compos.* **2015**, *36*, 2275–2281. [CrossRef]
171. Santos, R.J.; Hiranobe, C.T.; Dognani, G.; Silva, M.J.; Paim, L.L.; Cabrera, F.C.; Torres, G.B.; Job, A.E. Using the Lorenz–Park, Mooney–Rivlin, and Dynamic Mechanical Analysis Relationship on Natural Rubber/Leather Shavings Composites. *J. Appl. Polym. Sci.* **2021**, *139*, 51880. [CrossRef]
172. Mechanics, M.; Li, R.; Sun, L.Z. Dynamic Mechanical Analysis of Silicone Rubber Reinforced with Multi-Walled Carbon Nanotubes. *Interact. Multiscale Mech.* **2011**, *4*, 239–245. [CrossRef]
173. Avila Torrado, M.; Constantinescu, A.; Johlitz, M.; Lion, A. Viscoelastic Behavior of Filled Silicone Elastomers and Influence of Aging in Inert and Hermetic Environment. *Contin. Mech. Thermodyn.* **2022**, *36*, 333–350. [CrossRef]
174. Kumar, V.; Alam, M.N.; Manikkavel, A.; Song, M.; Lee, D.-J.; Park, S.-S. Silicone Rubber Composites Reinforced by Carbon Nanofillers and Their Hybrids for Various Applications: A Review. *Polymers* **2021**, *13*, 2322. [CrossRef]
175. Praveen, K.M.; Pillin, I.; Kervoelen, A.; Grohens, Y.; Thomas, S. Investigations on the Effect of Sol-Gel Coated Coir Fiber Reinforcement in PP/EPDM Composites. *Eur. Polym. J.* **2024**, *212*, 113081. [CrossRef]
176. Guo, W.; Zhao, J.; Zhao, F.; Feng, T.; Liu, L. Assessment of Properties of Bamboo Fiber and EPDM Reinforced Polypropylene Microcellular Foam Composites. *Polymer* **2024**, *304*, 127142. [CrossRef]
177. Scaglia, F.; Mazzini, A.; Perretta, C.; Regattieri, G.; Marchetti, G. Industrial EP(D)M Rubber Production Process: A First-Principle Data-Driven Modelling Approach. *Chem. Eng. Trans.* **2023**, *100*, 163–168. [CrossRef]
178. Kulkarni, S.D.; Manjunatha, B.; Chandrasekhar, U.; Siddesh, G.K.; Lenin, H.; Arul, S.J. Effect of Curing Temperature and Time on Mechanical Properties of Vinyl Polymer Material for Sealing Applications in Industry Using Machine Learning Techniques. *Adv. Polym. Technol.* **2023**, *2023*, 9964610. [CrossRef]
179. Correia, S.L.; Palaoro, D.; Segadães, A.M. Property Optimisation of EPDM Rubber Composites Using Mathematical and Statistical Strategies. *Adv. Mater. Sci. Eng.* **2017**, *2017*, 2730830. [CrossRef]
180. Song, L.-F.; Bai, N.; Shi, Y.; Wang, Y.-X.; Song, L.-X.; Liu, L.-Z. Effects of Ethylene-Propylene-Diene Monomers (EPDMs) with Different Mooney Viscosity on Crystallization Behavior, Structure, and Mechanical Properties of Thermoplastic Vulcanizates (TPVs). *Polymers* **2023**, *15*, 642. [CrossRef]
181. Hosseinpour, A.; Katbab, A.A.; Ohadi, A. Improving the Sound Absorption of a Highly Deformable Nanocomposite Foam Based on Ethylene-Propylene-Diene-Monomer (EPDM) Infused with Multi-Walled Carbon Nanotubes (MWCNTs) to Absorb Low-Frequency Waves. *Eur. Polym. J.* **2022**, *178*, 111522. [CrossRef]

182. Chen, G.; Gupta, A.; Mekonnen, T.H. Effects of Wood Fiber Loading, Silane Modification and Crosslinking on the Thermomechanical Properties and Thermal Conductivity of EPDM Biocomposite Foams. *Ind. Crops Prod.* **2023**, *200*, 116911. [CrossRef]
183. Wang, Y.; Li, J.; Wan, L.; Wang, L.; Li, K. A Lightweight Rubber Foaming Insulation Reinforced by Carbon Nanotubes and Carbon Fibers for Solid Rocket Motors. *Acta Astronaut.* **2023**, *208*, 270–280. [CrossRef]
184. Zhao, J.; Wang, G.; Chai, J.; Chang, E.; Wang, S.; Zhang, A.; Park, C.B. Polylactic Acid/UV-Crosslinked in-Situ Ethylene-Propylene-Diene Terpolymer Nanofibril Composites with Outstanding Mechanical and Foaming Performance. *Chem. Eng. J.* **2022**, *447*, 137509. [CrossRef]
185. Sang, J.S.; Park, E.-Y.; Lim, S.-W.; Park, S.; Oh, K.W. Performance of Bio-Ethylene Propylene Diene Monomer (Bio-EPDM) Foam with Mixed Chemical and Encapsulated Blowing Agents. *Fash. Text.* **2019**, *6*, 21. [CrossRef]
186. Kouhi, F.; Vahidifar, A.; Naderi, G.; Esmizadeh, E. Tire-Derived Reclaimed Rubber as a Secondary Raw Material for Rubber Foams: In the Framework of Circular Economy Strategy. *J. Polym. Res.* **2023**, *30*, 97. [CrossRef]
187. Buchen, S.; Kröger, N.H.; Reppel, T.; Weinberg, K. Time-Dependent Modeling and Experimental Characterization of Foamed EPDM Rubber. *Contin. Mech. Thermodyn.* **2021**, *33*, 1747–1764. [CrossRef]
188. Bhuyan, C.; Konwar, A.; Bora, P.; Rajguru, P.; Hazarika, S. Cellulose Nanofiber-Poly(Ethylene Terephthalate) Nanocomposite Membrane from Waste Materials for Treatment of Petroleum Industry Wastewater. *J. Hazard. Mater.* **2023**, *442*, 129955. [CrossRef] [PubMed]
189. Shahrim, N.A.; Abounahia, N.M.; El-Sayed, A.M.A.; Saleem, H.; Zaidi, S.J. An Overview on the Progress in Produced Water Desalination by Membrane-Based Technology. *J. Water Process Eng.* **2023**, *51*, 103479. [CrossRef]

Disclaimer/Publisher's Note: The statements, opinions and data contained in all publications are solely those of the individual author(s) and contributor(s) and not of MDPI and/or the editor(s). MDPI and/or the editor(s) disclaim responsibility for any injury to people or property resulting from any ideas, methods, instructions or products referred to in the content.

Review

Modeling Study of the Creep Behavior of Carbon-Fiber-Reinforced Composites: A Review

Mostafa Katouzian [1], Sorin Vlase [1,2,*], Marin Marin [3] and Maria Luminita Scutaru [1]

1. Department of Mechanical Engineering, Transilvania University of Brasov, 500036 Brasov, Romania
2. Romanian Academy of Technical Sciences, 700506 Bucharest, Romania
3. Department of Mathematics and Computer Science, Transilvania University of Brasov, 500036 Brasov, Romania
* Correspondence: svlase@unitbv.ro; Tel.: +40-7232-643-020

Abstract: The aim of this paper is to present some important practical cases in the analysis of the creep response of unidirectional fiber-reinforced composites. Some of the currently used models are described: the micromechanical model, homogenization technics, the Mori–Tanaka method, and the finite element method (FEM). Each method was analyzed to determine its advantages and disadvantages. Regarding the accuracy of the obtained results, comparisons are made with experimental tests. The methods presented here are applied to carbon-fiber-reinforced composites, but these considerations can also be applied to other types of composite materials.

Keywords: FRP; composite; viscoelasticity; generalized Hooke law; creep response; constitutive low; carbon

Citation: Katouzian, M.; Vlase, S.; Marin, M.; Scutaru, M.L. Modeling Study of the Creep Behavior of Carbon-Fiber-Reinforced Composites: A Review. *Polymers* **2023**, *15*, 194. https://doi.org/10.3390/polym15010194

Academic Editors: Ana Pilipović, Phuong Nguyen-Tri and Mustafa Özcanli

Received: 17 November 2022
Revised: 14 December 2022
Accepted: 26 December 2022
Published: 30 December 2022

Copyright: © 2022 by the authors. Licensee MDPI, Basel, Switzerland. This article is an open access article distributed under the terms and conditions of the Creative Commons Attribution (CC BY) license (https:// creativecommons.org/licenses/by/ 4.0/).

1. Introduction

The creep phenomenon that can occur in viscoelastic materials is defined as manifesting in three hypostases: primary, secondary, and tertiary. The creep phenomenon is defined as a deformation in time of the studied material, if it is loaded with a known force [1] (Figure 1). Creep phenomena usually manifest at high temperatures. However, there are situations in which the creep can appear at lower temperatures, for example, at room temperature, for some types of materials.

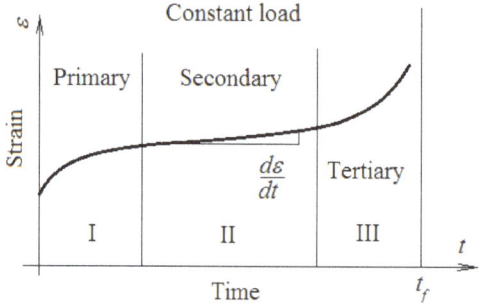

Figure 1. Usual creep behavior of a material (adapted from [2]).

Of course, this phenomenon, which manifests in the elongation of the material over time, can become dangerous in the operation of a machine. Figure 1 shows the three intervals of creep behavior. Current applications refer mostly to the first two stages of creep, when the deformation rate is relatively high. In the primary creep stage, a high rate is observed at the beginning, which slows down over time. The aspect of the creep curve depends on the material, load, and time. In the secondary creep stage, there is a relatively constant rate. A high rate of deformation characterizes the third creep stage. The time interval in which this high increase is observed is short and is associated with the destruction of the material. In engineering practice, it is not necessary to reach this stage; as a result, the study of behavior in this area has not attracted much attention. Designers must know the rate of deformation. This can be determined using measurements or by using a verified calculus model. The paper presents such models, which are useful for design activities [2,3]. Creep behavior is interesting for engineers and studies on this phenomenon are numerous [4–6].

The technology of advanced composites has developed to the point where these materials are being increasingly utilized in the commercial, military, and aerospace industries, among others. Composite materials are ideal for structural applications where high strength-to-weight and stiffness-to-weight ratios, improved fatigue resistance, and improved dimensional stability are required. Reinforced fiber polymers date back to the early years of the last century. There are two major steps in the manufacturing of polymer-based laminated composites, namely layup and curing. In the layup stage, continuous filaments are arranged in unidirectional laminae or are interwoven. The fibers are often impregnated with resinous material, such as polyester resin, which later serves as the matrix material. The next step, thermal curing, involves the drying or polymerization of the resinous matrix material and is accomplished in suitable autoclaves. The aim is to form a permanent bond between the fibers and the matrix, as well as between the laminae, in order to obtain lightweight, stiff panels [2].

The materials used in engineering have different purposes and are manufactured according to different technologies; as a result, they have a variety of properties. The creep diagrams of these materials can be very different, even under the same loading and temperature conditions. The simplest way to construct a creep diagram is to perform experimental measurements. However, such an approach is expensive and time consuming. Loads with different constant loads must be considered, and tests must be performed at different temperatures.

In [7], a scheme for accelerated characterization is proposed to analyze the viscoelastic response of general laminated composites. The use of this scheme allows a small number of experimental measurements to be performed. The measurements allow for short-term tests at high temperature, to predict the long-term response [8–10].

It would be much more advantageous for designers to have useful creep models which could be used to obtain creep diagrams by calculation.

To study the nonlinear viscoelastic behavior of a unidirectional composite, the well-known FEM method is applied. The symmetry properties of the composite allow for the simplification of such an analysis. A good correlation with the FEM micromechanics models developed in [11] is obtained. The method can also be used to study a composite with a complex topology [12–14]. Such a description also offers the possibility of studying the material in a wide range of boundary conditions. Thus, the thermal effects and the expansion due to humidity were included through the initial conditions. In [15], the above equations were used for unidirectional composites reinforced with graphite and glass.

The works [16–18] improve the classic models used in the case of nonlinear behavior. An empirical model was developed to achieve this. A method that can be easily implemented using a numerical procedure was thus obtained.

Based on the previously presented studies [16–18], a nonlinear viscoelastic model was developed in [19,20]. The developed model and the experimental measurements taken for test specimens allowed for an orthotropic material. The presented procedure can also be applied to study the long-term nonlinear viscoelastic response of laminates.

Other research [21] has shown that a law moisture concentration (at about 1%) can be a critical limit for carbon epoxy laminates. When this limit is exceeded, the viscoelastic rate of deformation occurs faster.

The study of a material made of an epoxy resin reinforced with unidirectional aramid fibers by tests and measurements at high temperatures is presented in [22]. An appropriate mathematical model for this study proved to be the "power law" which can describe behavior in both the linear and nonlinear domains, so that it can model viscoelastic behavior. To study the behavior in the nonlinear field, some nonlinear viscoelastic coefficients are introduced (these coefficients depend both on the stresses to which the materials are subjected and on the temperature). This method of analysis was proven to concur with the nonlinear model presented in [12].

In [23–26], a variational principle is used in which the time variable also appears, using a relatively simple mathematical description. In [27], the heat-induced stress field in the components of a polymer composite at low temperatures is studied (one application is considered for spacecraft). The geometry of the composite microstructure proves to be important in terms of the field of stresses and the deformation of this type of material under the conditions described above.

In [28], all the engineering constants that define one orthotropic and one transverse isotropic composite are determined. For a transverse isotropic material, the results [29–32] provide us with the upper and lower limits of engineering constants. In [33], the Mori–Tanaka method presented in [34] is extended.

Paper [35] shows the non-linear viscoelastic/viscoplastic behaviors of graphite/bismaleimide. Paper [12] presents a nonlinear formulation used to study materials at temperatures above 93 °C. A micromechanical analysis for the study of the behavior of a fiber-reinforced composite is described in [36,37]. These studies show good concordance with the findings presented in [11]. Biphasic materials and their mechanical properties have been extensively studied in numerous papers published in recent years [38–49]. New results are presented in [50–54].

In this review, the authors present more factors related to the analysis of the creep behavior of a composite material reinforced with fibers. The model's proposed offer results and a creep curve in the case of different loads. The results presented in this review are mainly based on the results obtained in [55–59].

The creep calculation of composite materials represents an important step in the process of designing a new material. A series of methods are therefore developed to achieve this objective. The problem remains an important one in the context of unprecedented advances in the development of new materials, with increasing numbers of properties that are useful in various applications. To the knowledge of the authors, the systematization and unitary presentation of these methods has not yet been achieved. This study thus makes a significant contribution to the field. The methods based on the homogenization theory are presented in Sections 2–4, and those based on the FEM theory are presented in Section 5.

2. The Micromechanical Model in Homogenization Theory

2.1. Model and Constitutive Law

The method of the micromechanical model aims to obtain the overall mechanical parameters of a composite based on models that use the parameters of the individual constituents of the composite and the interaction that exists between them. Consider a unidirectional composite with randomly distributed fibers in the matrix material. In the models used, it is normal to take into account a periodicity in the distribution of fibers. In this way, the existing periodicities allow us to simplify the analysis. Figure 2 presents such a model. The following assumptions can be formulated:

- The fibers are continuous and circular, and oriented in the X_1 direction. They are positioned regularly in a rectangular array in the transversal X_2–X_3 plane;
- The fibers are linearly elastic and anisotropic. The matrix is isotropic and nonlinearly viscoelastic;
- No cracks or holes appear or develop, and the contact fiber matrix is mechanical.

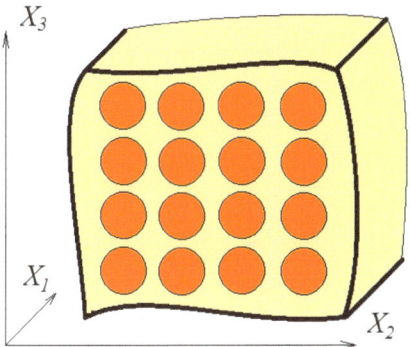

Figure 2. A micromechanical model of a one-dimensional fiber material.

Using the proposed model, it is possible to determine the response of the material if only a single repeating unit cell (RUC) is studied—see Figure 3. As such, the complexity of the problem can be significantly reduced. For this analysis, it is sufficient to study only a quarter of a fiber, as in Figure 3b. The main hypothesis of the theory is that the RUCs are very small, reported to be equal to the dimensions of the studied material.

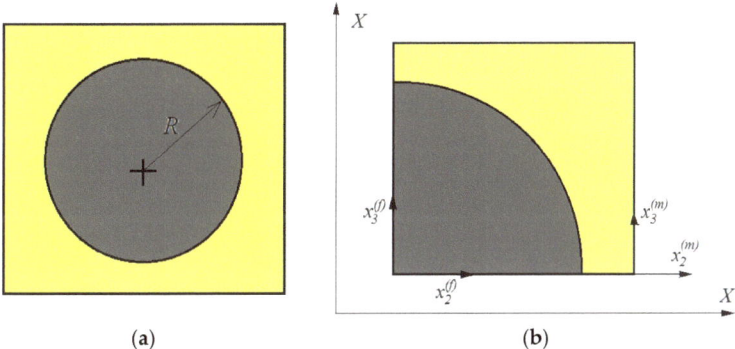

Figure 3. (a) A cell of the microstructure. (b) The representative unit cell (RUC).

The RUC refers to a local coordinate system $(X_1, x_2^{(\lambda)}, x_3^{(\lambda)})$ (Figure 4). The displacement in each subcell is defined through the following formulas [27,28]:

$$u_i^{(\lambda)} = u_i^{o(\lambda)} + x_2^{(\lambda)} \xi_i^{(\lambda)} + x_3^{(\lambda)} \zeta_i^{(\lambda)} \; ; \; i = 1, 2, 3. \qquad (1)$$

Here, u_i^o is the displacement component of the origin and "λ" represents both the fiber (when $\lambda = f$) and the matrix (when $\lambda = m$).

Considering the material to be linear, the strain–displacement relations are as follows:

$$\varepsilon_{ij} = \frac{1}{2}\left(\frac{\partial u_i}{\partial x_j} + \frac{\partial u_j}{\partial x_i}\right) \; ; \; i,j = 1,2,3, \qquad (2)$$

Equation (2) can be written for fiber and matrix in the unified form:

$$\varepsilon_{ij}^{(\lambda)} = \frac{1}{2}\left(\frac{\partial u_i^{(\lambda)}}{\partial x_j} + \frac{\partial u_j^{(\lambda)}}{\partial x_i}\right) \; ; \; i,j = 1,2,3 \qquad (3)$$

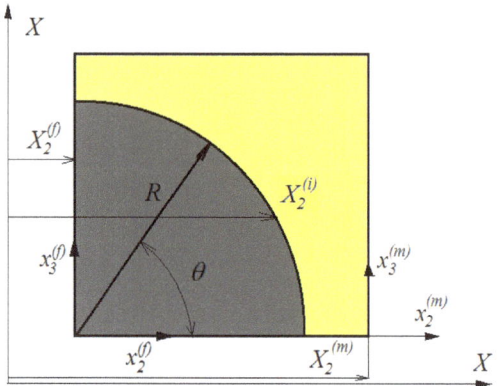

Figure 4. Continuity conditions in the RUC.

If $i \neq j$, it can be written as

$$\varepsilon_{ij}^{(\lambda)} = \frac{1}{2}\left(\frac{\partial u_i^{(\lambda)}}{\partial x_j} + \frac{\partial u_j^{(\lambda)}}{\partial x_i}\right) = \frac{\gamma_{ij}^{(\lambda)}}{2} \; ; \; i,j = 1,2,3 \; ; \; i \neq j \qquad (4)$$

The engineering shear strain is denoted as $\gamma_{ij}^{(\lambda)} = 2\varepsilon_{ij}^{(\lambda)} \; ; \; i,j = 1,2,3 \; ; \; i \neq j$.

Using Equation (1) into Equation (2) and considering Equations (3) and (4), the following equations can be obtained:

$$\varepsilon_{11}^{(\lambda)} = \frac{\partial u_i^{o(\lambda)}}{\partial X_1} \qquad (5)$$

$$\varepsilon_{11}^{(\lambda)} = \frac{\partial u_i^{o(\lambda)}}{\partial X_1} \qquad (6)$$

$$\varepsilon_{33}^{(\lambda)} = \zeta_3^{(\lambda)} \qquad (7)$$

$$\gamma_{23}^{(\lambda)} = \left[\xi_3^{(\lambda)} + \zeta_2^{(\lambda)}\right] \qquad (8)$$

$$\gamma_{31}^{(\lambda)} = \left[\frac{\partial u_3^{o(\lambda)}}{\partial X_1} + \zeta_1^{(\lambda)}\right] \qquad (9)$$

$$\gamma_{12}^{(\lambda)} = \left[\frac{\partial u_2^{o(\lambda)}}{\partial X_1} + \xi_1^{(\lambda)}\right] \qquad (10)$$

Considering a linear and transversely isotropic composite, the constitutive equation can be written as

$$\begin{Bmatrix} \varepsilon_{11} \\ \varepsilon_{22} \\ \varepsilon_{33} \\ \gamma_{23} \\ \gamma_{31} \\ \gamma_{12} \end{Bmatrix}^{(\lambda)} = \begin{bmatrix} S_{11} & S_{12} & S_{12} & 0 & 0 & 0 \\ S_{12} & S_{22} & S_{23} & 0 & 0 & 0 \\ S_{12} & S_{23} & S_{22} & 0 & 0 & 0 \\ 0 & 0 & 0 & S_{44} & 0 & 0 \\ 0 & 0 & 0 & 0 & S_{66} & 0 \\ 0 & 0 & 0 & 0 & 0 & S_{66} \end{bmatrix}^{(\lambda)} \begin{Bmatrix} \sigma_{11} \\ \sigma_{11} \\ \sigma_{11} \\ \tau_{23} \\ \tau_{31} \\ \tau_{12} \end{Bmatrix}^{(\lambda)} \qquad (11)$$

or, considering the expression of the engineering constant for this type of material:

$$\begin{Bmatrix} \varepsilon_{11} \\ \varepsilon_{22} \\ \varepsilon_{33} \\ \gamma_{23} \\ \gamma_{31} \\ \gamma_{12} \end{Bmatrix}^{(\lambda)} = \begin{bmatrix} \frac{1}{E_{11}} & -\frac{\nu_{12}}{E_{22}} & -\frac{\nu_{12}}{E_{22}} & 0 & 0 & 0 \\ -\frac{\nu_{12}}{E_{22}} & \frac{1}{E_{22}} & -\frac{\nu_{23}}{E_{22}} & 0 & 0 & 0 \\ -\frac{\nu_{12}}{E_{22}} & -\frac{\nu_{23}}{E_{22}} & \frac{1}{E_{22}} & 0 & 0 & 0 \\ 0 & 0 & 0 & \frac{1}{G_{23}} & 0 & 0 \\ 0 & 0 & 0 & 0 & \frac{1}{G_{12}} & 0 \\ 0 & 0 & 0 & 0 & 0 & \frac{1}{G_{12}} \end{bmatrix}^{(\lambda)} \begin{Bmatrix} \sigma_{11} \\ \sigma_{22} \\ \sigma_{33} \\ \tau_{23} \\ \tau_{31} \\ \tau_{12} \end{Bmatrix}^{(\lambda)} \quad (12)$$

Here, E_{11} and $E_{22} = E_{33}$ are Young's moduli, G_{23} and $G_{12} = G_{13}$ are the shear moduli, and ν_{23} and $\nu_{12} = \nu_{13}$ are the Poisson ratios. The direction of anisotropy is, in our model, X_1, and the plane of isotropy is X_2–X_3. From Equation (12), the following formula is obtained:

$$\begin{Bmatrix} \sigma_{11} \\ \sigma_{22} \\ \sigma_{33} \\ \tau_{23} \\ \tau_{31} \\ \tau_{12} \end{Bmatrix}^{(f)} = \begin{bmatrix} C_{11} & C_{12} & C_{12} & 0 & 0 & 0 \\ C_{12} & C_{22} & C_{23} & 0 & 0 & 0 \\ C_{12} & C_{23} & C_{22} & 0 & 0 & 0 \\ 0 & 0 & 0 & C_{66} & 0 & 0 \\ 0 & 0 & 0 & 0 & C_{44} & 0 \\ 0 & 0 & 0 & 0 & 0 & C_{44} \end{bmatrix}^{(f)} \begin{Bmatrix} \varepsilon_{11} \\ \varepsilon_{22} \\ \varepsilon_{33} \\ \gamma_{23} \\ \gamma_{31} \\ \gamma_{12} \end{Bmatrix}^{(f)} \quad (13)$$

with

$$C_{66} = \frac{C_{22} - C_{23}}{2} \quad (14)$$

Equation (13) can be written as

$$\sigma^{(f)} = C^{(f)} \varepsilon^{(f)} \quad (15)$$

The behavior of a viscoelastic material can be described using Boltzmann's superposition principle and the results presented in [12]:

$$\varepsilon(t) = D_n \sigma_n \quad (16)$$

with

$$D_n = g_0 D_o + g \sum_{j=1}^{m} D_j \left(1 - e^{-t/r_j}\right) \quad (17)$$

The paper [12] presents us with the possibility of writing the constitutive equations as

$$\varepsilon_{ij}^{(m)} = D_n [1 + \nu(t)] \sigma_{ij}^{(m)} - D_n \nu(t) \sigma_{kk}^{(m)} \delta_{ij} \quad (18)$$

In Equation (18), D_n is obtained using Equation (17), $\nu(t)$ is the Poisson ratio (in our study, this is considered to be independent of time), and δ_{ij} is Kronecker's delta.

The first step must be to determine the average stresses, and then the average strains. Thus, the general behavior of the material is obtained based on the average stresses and average strains in a RUC.

2.2. Average Stress

In the proposed model, the RUC is considered to be a rectangular parallelepiped with parallel edges. The reference frame axes are (X_1, X_2, X_3) of the volume V. This will be determined as the average stress $\overline{\sigma}_{ij}$ in V. This can be obtained via the following relation:

$$\overline{\sigma}_{ij} = \frac{1}{V} \int_V S_{ij} dV \quad (19)$$

Considering one-quarter of a cell, this relation can be written as

$$\overline{\sigma}_{ij} = \frac{1}{A}\left(\overline{S}_{ij}^{(f)} A_f + \overline{S}_{ij}^{(m)} A_m\right) \qquad (20)$$

where $\overline{S}_{ij}^{(\lambda)}$ are the average stresses. Now, consider a unit depth of the RUC, i.e., $V = A \times 1$. Using the notation presented in Figure 3b, the following is obtained:

$$A = (R+h/2)^2; \quad A_f = \pi R^2/4; \quad A_m = (R+h/2)^2 - \pi R^2/4 \qquad (21)$$

$$\overline{\sigma}_{ij} = \frac{1}{\left(R+\frac{h}{2}\right)^2}\left\{\frac{\pi R^2}{4}\overline{S}_{ij}^{(f)} + \left[\left(R+\frac{h}{2}\right)^2 - \frac{\pi R^2}{4}\right]\overline{S}_{ij}^{(m)}\right\} \qquad (22)$$

The partial average stress $\overline{S}_{ij}^{(\lambda)}$ is obtained with

$$\overline{S}_{ij}^{(\lambda)} = \frac{1}{A_v}\int_A \sigma_{ij}^{(\lambda)} dA = \frac{1}{A_v}\iint \sigma_{ij}^{(\lambda)} dx_2 dx_3 \qquad (23)$$

Using polar coordinates, the Jacobian can be obtained:

$$J = \frac{\partial(x_2, x_3)}{\partial(r, \theta)} = \begin{vmatrix} \cos\theta & \sin\theta \\ -r\sin\theta & r\cos\theta \end{vmatrix} = r \qquad (24)$$

Moreover, Equation (23) for fiber (subcell "f") becomes

$$e\overline{S}_{ij}^{(f)} = \frac{4}{\pi R^2}\int_0^{\pi/2}\int_0^R \sigma_{ij}^{(f)} r\, dr\, d\theta \qquad (25)$$

where $\sigma_{ij}^{(f)}$ is given by Equation (13). Introducing Equations (5)–(10) and (11) into Equation (25) leads to

$$\overline{S}_{ij}^{(f)} = \begin{cases} C_{11}\varepsilon_{11}^{(f)} + C_{12}\left(\varepsilon_{22}^{(f)} + \varepsilon_{33}^{(f)}\right) \\ C_{12}\varepsilon_{11}^{(f)} + C_{22}\varepsilon_{22}^{(f)} + C_{23}\varepsilon_{33}^{(f)} \\ C_{12}\varepsilon_{11}^{(f)} + C_{23}\varepsilon_{22}^{(f)} + C_{22}\varepsilon_{33}^{(f)} \\ C_{66}\gamma_{23}^{(f)} \\ C_{44}\gamma_{31}^{(f)} \\ C_{44}\gamma_{12}^{(f)} \end{cases} \qquad (26)$$

or

$$\overline{S}_{ij}^{(f)} = \begin{cases} C_{11}\frac{\partial u_1^{o(f)}}{\partial X_1} + C_{12}\left(\zeta_2^{(f)} + \zeta_3^{(f)}\right) \\ C_{12}\frac{\partial u_1^{o(f)}}{\partial X_1} + C_{22}\zeta_2^{(f)} + C_{23}\zeta_3^{(f)} \\ C_{12}\frac{\partial u_1^{o(f)}}{\partial X_1} + C_{23}\zeta_2^{(f)} + C_{22}\zeta_3^{(f)} \\ C_{66}\left[\zeta_3^{(f)} + \zeta_2^{(f)}\right] \\ C_{44}\left[\frac{\partial u_3^{o(f)}}{\partial X_1} + \zeta_1^{(f)}\right] \\ C_{44}\left[\frac{\partial u_2^{o(f)}}{\partial X_1} + \zeta_1^{(f)}\right] \end{cases} \qquad (27)$$

The average stresses in the matrix (subcell "m") are determined using the following relations:

$$\overline{S}_{ij}^{(m)} = \frac{1}{\left(R+\frac{h}{2}\right)^2 - \frac{\pi R^2}{4}}\left(\int_0^{R+h/2}\int_0^{R+h/2} \sigma_{ij}^{(m)} dx_2 dx_3 - \int_0^{\pi/2}\int_0^R \sigma_{ij}^{(m)} r\, dr\, d\theta\right) \qquad (28)$$

Equation (15) together with Equations (5)–(10) yields

$$\frac{\partial u_1^o}{\partial X_1} = D_n[1+\nu(t)]\overline{S}_{11}^{(m)} - D_n[\nu(t)]\overline{S}_{kk}^{(m)} \tag{29}$$

$$\xi_2^{(m)} = D_n[1+\nu(t)]\overline{S}_{22}^{(m)} - D_n[\nu(t)]\overline{S}_{kk}^{(m)} \tag{30}$$

$$\zeta_3^{(m)} = D_n[1+\nu(t)]\overline{S}_{33}^{(m)} - D_n[\nu(t)]\overline{S}_{kk}^{(m)} \tag{31}$$

$$\xi_3^{(m)} + \zeta_2^{(m)} = 2D_n[1+\nu(t)]\overline{S}_{23}^{(m)} \tag{32}$$

$$\frac{\partial u_3^{o(m)}}{\partial X_1} + \zeta_1^{(m)} = 2D_n[1+\nu(t)]\overline{S}_{31}^{(m)} \tag{33}$$

$$\frac{\partial u_2^{o(m)}}{\partial X_1} + \xi_1^{(m)} = 2D_n[1+\nu(t)]\overline{S}_{12}^{(m)} \tag{34}$$

2.3. Continuity Conditions

In a RUC, the conditions of continuity of the movements at the interface between subcells must be ensured. These conditions must be assured in both the X_2 and X_3 directions. From Figure 4, the following relations hold:

$$x_2^{(f)} = X_2^i \mp X_2^{(f)} = \mp R\cos\theta \tag{35}$$

$$x_2^{(m)} = X_2^i \mp X_2^{(m)} = \pm\left(\frac{h}{2} + R - R\cos\theta\right) \tag{36}$$

where there are θ located points on the interface.

Introducing Equations (35) and (36) into Equation (1) for the cases when $\lambda = f$ and $\lambda = m$ results in

$$u_i^{(f)} = u_i^{o(f)} \mp R\cos\theta\,\xi_i^{(f)} + x_3^{(f)}\zeta_i^{(f)} \tag{37}$$

$$u_i^{(m)} = u_i^{o(m)} \pm \left(\frac{h}{2} + R - R\cos\theta\right)\xi_i^{(m)} + x_3^{(m)}\zeta_i^{(m)} \tag{38}$$

where

$$u_i^{o(f)} = u_i^{o(f)}(X_2^{(f)}) \tag{39}$$

$$u_i^{o(m)} = u_i^{o(m)}(X_2^{(m)}) \tag{40}$$

The continuity of the displacements at the interface is considered in the average sense. This is expressed by the following relations:

$$\int_{-\pi/2}^{\pi/2}\left[u_i^{o(f)} \mp R\cos\theta\,\xi_i^{(f)} + x_3^{(f)}\zeta_i^{(f)}\right]R\cos\theta\,d\theta =$$

$$\int_{-\pi/2}^{\pi/2}\left[u_i^{o(m)} \pm \left(\frac{h}{2}+R-R\cos\theta\right)\xi_i^{(m)} + x_3^{(m)}\zeta_i^{(m)}\right]R\cos\theta\,d\theta \tag{41}$$

Equation (41) produces

$$u_i^{o(f)}\left(X_2^{(f)}\right) \pm \frac{\pi R}{2}\xi_i^{(f)} = u_i^{o(m)}\left(X_2^{(m)}\right) \mp \left(h + 2R - \frac{\pi R}{2}\right)\xi_i^{(m)} \tag{42}$$

The addition of the two Equations (42) offers us

$$u_i^{o(f)} = u_i^{o(m)} = u_i^o \tag{43}$$

which represent the continuity conditions for displacement.

2.4. Average Strain

Considering the composite specimen presented earlier, and the continuity conditions, the average of the strains over volume is

$$\bar{\varepsilon}_{ij} = \frac{1}{V} \int_V \varepsilon_{ij} dV \qquad (44)$$

For the representative cell studied, the following formula is obtained:

$$\bar{\varepsilon}_{ij} = \frac{1}{A} \sum_{\lambda=f,m} \bar{\varepsilon}_{ij}^{(\lambda)} A_\lambda \qquad (45)$$

Here $A = A_m + A_f$ and $\bar{\varepsilon}_{ij}^{(\lambda)}$ are the strains obtained using Equations (5)–(10) ($\lambda = f, m$). Considering Equations (1) and (3) (considering $i = j = 1$), we obtain

$$\bar{\varepsilon}_{ij}^{(\lambda)} = \frac{\partial u_1^o}{\partial X_1} \qquad (46)$$

Equation (46) together with Equation (45) yields

$$\bar{\varepsilon}_{11} = \frac{1}{\left(R + \frac{h}{2}\right)^2} \left\{ \frac{\pi R^2}{4} \frac{\partial u_1^o}{\partial X_1} + \left[\left(R + \frac{h}{2}\right)^2 - \frac{\pi R^2}{4} \right] \frac{\partial u_1^o}{\partial X_1} \right\} \qquad (47)$$

or

$$\bar{\varepsilon}_{11} = \frac{\partial u_1^o}{\partial X_1} \qquad (48)$$

The octahedral shear stress in the matrix is

$$S_{oct}^{(m)} = \left\{ \frac{1}{2} \left[\left(\bar{S}_{11}^{(m)} - \bar{S}_{22}^{(m)}\right)^2 + \left(\bar{S}_{22}^{(m)} - \bar{S}_{33}^{(m)}\right)^2 + \left(\bar{S}_{33}^{(m)} - \bar{S}_{11}^{(m)}\right)^2 \right] + 3 \left[\left(\bar{S}_{12}^{(m)}\right)^2 + \left(\bar{S}_{23}^{(m)}\right)^2 + \left(\bar{S}_{31}^{(m)}\right)^2 \right] \right\}^{\frac{1}{2}} \qquad (49)$$

The unknowns are calculated using an incremental procedure. The unknowns are

- the six values of stresses in the two subcells: $\bar{S}_{11}^{(\lambda)}, \bar{S}_{22}^{(\lambda)}$, and $\bar{S}_{33}^{(\lambda)}$;
- the four micro-variables in the two subcells: $\zeta_2^{(\lambda)}$ and $\zeta_3^{(\lambda)}$;
- the three strains in the composite: $\bar{\varepsilon}_{11}, \bar{\varepsilon}_{22}$, and $\bar{\varepsilon}_{33}$.

Now, the connection between the stress in the matrix and in the fiber of a RUC must be determined. Using the assumptions proposed in [36,37], the shear stress in the X_2 direction is

$$\bar{S}_{22}^{(f)} = \alpha_f \bar{\sigma}_{22} \qquad (50)$$

for the fiber and

$$\bar{S}_{22}^{(m)} = \alpha_m \bar{\sigma}_{22} \qquad (51)$$

for the matrix from where it results:

$$\bar{S}_{22}^{(f)} = \frac{\alpha_f}{\alpha_m} \bar{S}_{22}^{(m)} \qquad (52)$$

In the X_3 direction, a similar equation is obtained:

$$\bar{S}_{33}^{(f)} = \frac{\beta_f}{\beta_m} \bar{S}_{33}^{(m)} \qquad (53)$$

The concentration factors α_λ and β_λ are weighting coefficients and should satisfy the following relations:

$$\alpha_f v_f + \alpha_m v_m = 1 \qquad (54)$$

and
$$\beta_f v_f + \beta_m v_m = 1 \tag{55}$$

In a particular case, considering that the composite loaded is in only one of the directions X_2 or X_3, the relation (38) in direction X_3 (unloaded) becomes

$$\overline{S}_{33}^{(f)} = -\frac{v_m}{v_f}\overline{S}_{33}^{(m)} \tag{56}$$

and
$$\overline{S}_{22}^{(f)} = -\frac{v_m}{v_f}\overline{S}_{22}^{(m)} \tag{57}$$

Consider now the case of a uniaxial load. Therefore, we obtain a linear system with 13 equations and 13 unknowns:

$$\overline{S}_{11}^{(f)} = C_{11}\bar{\varepsilon}_{11} + C_{12}\left(\zeta_2^{(f)} + \zeta_3^{(f)}\right) \tag{58}$$

$$\overline{S}_{22}^{(f)} = C_{12}\bar{\varepsilon}_{11} + C_{22}\zeta_2^{(f)} + C_{23}\zeta_3^{(f)} \tag{59}$$

$$\overline{S}_{33}^{(f)} = C_{12}\bar{\varepsilon}_{11} + C_{23}\zeta_2^{(f)} + C_{22}\zeta_3^{(f)} \tag{60}$$

$$\bar{\varepsilon}_{11} = D_n[1+\nu(t)]\overline{S}_{11}^{(m)} - D_n[\nu(t)]\overline{S}_{kk}^{(m)} \tag{61}$$

$$\zeta_2^{(m)} = D_n[1+\nu(t)]\overline{S}_{22}^{(m)} - D_n[\nu(t)]\overline{S}_{kk}^{(m)} \tag{62}$$

$$\zeta_3^{(m)} = D_n[1+\nu(t)]\overline{S}_{33}^{(m)} - D_n[\nu(t)]\overline{S}_{kk}^{(m)} \tag{63}$$

$$\bar{\varepsilon}_{22} = \frac{1}{A}\left[A_f\zeta_2^{(f)} + A_m\zeta_2^{(m)}\right] \tag{64}$$

$$\bar{\varepsilon}_{33} = \frac{1}{A}\left[A_f\zeta_3^{(f)} + A_m\zeta_3^{(m)}\right] \tag{65}$$

$$\overline{S}_{22}^{(f)} = \frac{\alpha_f}{\alpha_m}\overline{S}_{22}^{(m)} \tag{66}$$

$$\overline{S}_{33}^{(f)} = \frac{\beta_f}{\beta_m}\overline{S}_{33}^{(m)} \tag{67}$$

$$\bar{\sigma}_{11} = \frac{1}{A}\left[A_f\overline{S}_{11}^{(f)} + A_m\overline{S}_{11}^{(m)}\right] \tag{68}$$

$$\bar{\sigma}_{22} = \frac{1}{A}\left[A_f\overline{S}_{22}^{(f)} + A_m\overline{S}_{22}^{(m)}\right] \tag{69}$$

$$\bar{\sigma}_{33} = \frac{1}{A}\left[A_f\overline{S}_{33}^{(f)} + A_m\overline{S}_{33}^{(m)}\right] \tag{70}$$

The analysis presented in this section shows that a micromechanical model for the study of a unidirectional composite can provide good results. Thus, analytical relations are obtained, which then allow for the calculation of the mechanical constants of such a composite and for the study of its behavior in a range of applications. Schapery's nonlinear constitutive equation for isothermal uniaxial loading conditions is used in the analysis, thus allowing us to consider the nonlinear viscoelastic response of the material. Papers that present many experimental results [55–59] demonstrate the potential of the method.

3. Description of Homogenization Theory
3.1. Overview

The theory of homogenization is a mathematical method used to average the physical properties of inhomogeneous materials. This method has been developed over the last eight decades and is used to analyze and solve differential equations with periodic coefficients.

As essentially inhomogeneous materials that have a periodicity or certain symmetries in their structure, composite materials lend themselves very well to the application of these methods that determine the mechanical characteristics of a material. The experimental results validate the methods used by the theory of homogenization [55–59]. For this reason, the homogenization method has been used in numerous cases and engineering applications [33,34,36,37,39–41,59] to determine the mechanical properties of multiphase composites. In this method, a transition is made, through homogenization, from a periodic structure to a homogeneous and isotropic or transversely isotropic material throughout its structure [44].

In the research, several analytical and numerical methods have been proposed to solve the problems generated by the application of this method. Experimental results have always shown the predominant acceptance of such methods [45,60]. The interaction between the phases of the composite is modeled by unifying the homogenization problems for heterogeneous elasto-plastic and elasto-viscoplastic materials [61,62]. Other works address the improvement of the method, using the experience gained by different engineering applications [61–69]. Other related methods are considered in [70,71]. The object of this research is the development of reliable procedures that can be easily applied by designers. The following section presents the homogenization theory used to determine the mechanical quantities that characterize the viscoelastic material in question. An application is suggested for a composite reinforced with carbon fibers.

3.2. Homogenized Model

One of the advantages offered by the homogenization theory is the possibility of studying differential equations in which the coefficients have rapid variations or periodic variations. Engineering constants, which are useful in engineering practice, are obtained following averaging processes. Thus, a material with a periodic structure can be treated as a homogeneous material. A differential equation with periodic coefficients with large variations is thus replaced in the modeling with an equation with constant coefficients. This is how the continuum concept is extended to micro-structured materials (composite materials also belong to this class). The bases of this mathematical theory are presented in [72–77]. In this application, the calculation method is used to analyze the creep response of a unidirectional composite reinforced with carbon fibers.

The stress field σ^δ for repeating unit cells of size δ must obey the following equations:

$$\begin{aligned}\frac{\partial \sigma_{11}^\delta}{\partial x_1} + \frac{\partial \tau_{12}^\delta}{\partial x_2} + \frac{\partial \tau_{13}^\delta}{\partial x_3} &= f_1(x) \\ \frac{\partial \tau_{21}^\delta}{\partial x_1} + \frac{\partial \sigma_{22}^\delta}{\partial x_2} + \frac{\partial \tau_{23}^\delta}{\partial x_3} &= f_2(x) \\ \frac{\partial \tau_{31}^\delta}{\partial x_1} + \frac{\partial \tau_{32}^\delta}{\partial x_2} + \frac{\partial \sigma_{33}^\delta}{\partial x_3} &= f_3(x) \end{aligned} \quad (71)$$

where $\sigma_{ij}^\delta = \sigma_{ji}^\delta$, for $i,j = 1,2,3$.

The contour conditions that must be respected by the displacements are

$$\left. u^\delta \right|_{\partial_1 \Omega} = \widetilde{u} \quad (72)$$

The boundary conditions are

$$\begin{aligned}\sigma_{11}^\delta n_1 + \tau_{12}^\delta n_2 + \tau_{13}^\delta n_3 &= T_1(x) \\ \tau_{21}^\delta n_1 + \sigma_{22}^\delta n_2 + \tau_{23}^\delta n_3 &= T_2(x) \\ \tau_{31}^\delta n_1 + \tau_{32}^\delta n_2 + \sigma_{33}^\delta n_3 &= T_3(x) \end{aligned} \quad (73)$$

on the contour $\partial_2 \Omega$, $(\partial_1 \Omega \cup \partial_2 \Omega = \partial \Omega)$. Hook's Law is

$$\begin{Bmatrix} \sigma_{11} \\ \sigma_{22} \\ \sigma_{33} \\ \tau_{23} \\ \tau_{31} \\ \tau_{12} \end{Bmatrix}^{\delta} = \begin{bmatrix} C_{1111} & C_{1122} & C_{1133} & & & \\ C_{2211} & C_{2222} & C_{2233} & & 0 & \\ C_{3311} & C_{3322} & C_{3333} & & & \\ & & & C_{2323} & & \\ & 0 & & & C_{3131} & \\ & & & & & C_{1212} \end{bmatrix} \begin{Bmatrix} \varepsilon_{11} \\ \varepsilon_{22} \\ \varepsilon_{33} \\ \gamma_{23} \\ \gamma_{31} \\ \gamma_{13} \end{Bmatrix}^{\delta} \quad (74)$$

Or, using a compact notation,

$$\sigma^{\delta} = C\varepsilon^{\delta} \quad (75)$$

The elasticity matrix C is semi-positive definite:

$$C_{ijkh} x_{ij} x_{kh} \geq \alpha\, x_{ij} x_{kh} \quad (76)$$

for $\alpha > 0$ and $\forall x_{ij}, x_{kh} \in \mathbf{R}$, $C_{ijkh}(x)$ is a periodical function of x with the period equal with the dimension δ of the unit cell. Considering a new function y, $y = x/\delta$:

$$C_{ijkh}(x) = C_{ijkh}(y\delta) = C_{ijkh}(y) \quad (77)$$

and the stress is

$$\sigma_{ij}^{\delta} = \sigma_{ij}^{0}(x,y) + \sigma_{ij}^{1}(x,y)\delta + \ldots \quad (78)$$

The dependence of stress on y is "quasi-periodical". Introducing Equation (78) in Equation (71) produces

$$\delta^{-1} \frac{\partial \sigma_{ij}^{0}}{\partial y_j} + \left(\frac{\partial \sigma_{ij}^{0}}{\partial x_j} + \frac{\partial \sigma_{ij}^{1}}{\partial y_j} \right) \delta^0 + \left(\frac{\partial \sigma_{ij}^{1}}{\partial x_j} + \frac{\partial \sigma_{ij}^{2}}{\partial y_j} \right) \delta^1 + \ldots = f_i(x) \quad (79)$$

The following relation is used:

$$\frac{d}{dx}(f) = \frac{\partial f}{\partial x} dx + \frac{\partial f}{\partial x} dy \quad (80)$$

but $y = x/\delta$, and, thus, $dy = dx/\delta$; so,

$$\frac{d}{dx}(f) = \frac{\partial f}{\partial x}(f) + \frac{1}{\delta} \frac{\partial}{\partial y}(f) \quad (81)$$

The coefficients of δ^{-1} in Equation (79) must be 0; therefore,

$$\frac{\partial \sigma_{ij}^{0}}{\partial y_j} = 0 \quad (82)$$

Equation (82) is called the "local equation". Identifying the terms of δ^0 produces

$$\frac{\partial \sigma_{ij}^{0}}{\partial x_j} + \frac{\partial \sigma_{ij}^{1}}{\partial y_j} = f_i(x)\ i = 1,2,3 \quad (83)$$

Applying the average operator to Equation (83) results in the following equation:

$$\frac{\partial \left\langle \sigma_{ij}^{0} \right\rangle}{\partial x_j} + \left\langle \frac{\partial \sigma_{ij}^{1}}{\partial y_j} \right\rangle = f_i(x)\ i = 1,2,3 \quad (84)$$

but

$$\left\langle \frac{\partial \sigma_{ij}^{1}}{\partial y_j} \right\rangle = \frac{1}{V} \int_V \partial \sigma_{ij}^{1} dV = \frac{1}{V} \int_{\partial V} \sigma_{ij}^{1} n_j dS = 0 \quad (85)$$

The stresses σ_{ij}^1 take equal values on the corresponding points of the boundary of the cell Γ (due to the property of periodicity), so

$$\frac{\partial \langle \sigma_{ij}^o \rangle}{\partial x_j} = f_i(x) \ i = 1,2,3 \tag{86}$$

We state that

$$\varepsilon_{ij,x}(w) = \frac{1}{2}\left(\frac{\partial w_i}{\partial x_j} + \frac{\partial w_j}{\partial x_i}\right); i,j = 1,2,3 \tag{87}$$

$$\varepsilon_{ij,y}(w) = \frac{1}{2}\left(\frac{\partial w_i}{\partial y_j} + \frac{\partial w_j}{\partial y_i}\right); i,j = 1,2,3 \tag{88}$$

The displacement field can be expressed by the series:

$$u(x,y) = u^o(x) + u^1(x,y)\,\delta + u^2(x,y)\,\delta^2 + \ldots \tag{89}$$

$u^o(x)$ is a function on x (only). The terms $u^1(x,y)$, $u^2(x,y)$ are considered to be quasi-periodical. Using (87)–(89), it can be written as

$$\begin{aligned}\varepsilon_{kh,x}(u) &= \tfrac{1}{2}\left(\tfrac{\partial u_k}{\partial x_h} + \tfrac{\partial u_h}{\partial x_k}\right) \\ &= \tfrac{1}{2}\left(\tfrac{\partial u_k^o}{\partial x_h} + \tfrac{\partial u_h^o}{\partial x_k}\right) + \tfrac{\delta}{2}\left(\tfrac{\partial u_k^1}{\partial x_h} + \tfrac{\partial u_h^1}{\partial x_k}\right) + \tfrac{\delta}{2}\left(\tfrac{\partial u_k^2}{\partial y_h} + \tfrac{\partial u_h^2}{\partial y_k}\right) + \ldots \\ &= \varepsilon_{kh,x}(u^o) + \varepsilon_{kh,y}(u^1) + \delta[\varepsilon_{kh,x}(u^1) + \varepsilon_{kh,y}(u^2)] + \delta^2[\ldots] + \ldots \ ; k,h = 1,2,3\end{aligned} \tag{90}$$

or

$$\varepsilon_{kh,x}(u) = \varepsilon_{kh}^o + \delta\,\varepsilon_{kh}^1 + \ldots \ ; k,h = 1,2,3 \tag{91}$$

where

$$\varepsilon_{kh}^o = \varepsilon_{kh,x}(u^o) + \varepsilon_{kh,y}(u^1) \ k,h = 1,2,3 \tag{92}$$

$$\varepsilon_{kh}^1 = [\varepsilon_{kh,x}(u^1) + \varepsilon_{kh,y}(u^2)] \ ; k,h = 1,2,3 \tag{93}$$

Applying the linear Hooke's law results in

$$\sigma_{ij}^o = C_{ijkh}\varepsilon_{kh}^o \ , \ i,j,k,h = 1,2,3 \tag{94}$$

From Equation (94), it follows that

$$\frac{\partial\left(C_{ijkh}\varepsilon_{kh}^o\right)}{\partial y_j} = 0 \ , \ i,j,k,h = 1,2,3 \tag{95}$$

or

$$\frac{\partial\left[C_{ijkh}\left(\varepsilon_{kh,x}(u^o) + \varepsilon_{kh,y}(u^1)\right)\right]}{\partial y_j} = 0 \ , \ i,j,k,h = 1,2,3 \tag{96}$$

The terms $\varepsilon_{kh,x}(u^o)$ depend only on x. Equation (96) can be written as

$$-\frac{\partial\left[C_{ijkh}\,\varepsilon_{kh,y}(u^1)\right]}{\partial y_j} = \varepsilon_{kh,x}(u^o)\frac{\partial C_{ijkh}}{\partial y_j} \ , \ i,j,k,h = 1,2,3 \tag{97}$$

introducing

$$u^1 = w^{kh}\varepsilon_{kh,x}(u^o) + k(x) \tag{98}$$

Using Equations (97) and (98) with $k(x)$ an arbitrary function on x, we can obtain

$$\varepsilon_{lm,y}(u^1) = \varepsilon_{kh,x}(u^o)\frac{1}{2}\left(\frac{\partial w_l^{kh}}{\partial y_m} + \frac{\partial w_m^{kh}}{\partial y_l}\right) = \varepsilon_{kh,x}(u^o)\varepsilon_{lm,y}(w^{kh}) \tag{99}$$

Equation (99) becomes

$$-\varepsilon_{kh,x}(u^o)\frac{\partial\left[C_{ijlm}\varepsilon_{lm,y}(w^{kh})\right]}{\partial y_j} = \varepsilon_{kh,x}(u^o)\frac{\partial C_{ijkh}}{\partial y_j}, \; i,j,k,h = 1,2,3 = 1 \tag{100}$$

Equation (100) is valid for any strain field $\varepsilon_{kh,x}(u^o)$, so Equation (100) becomes

$$-\frac{\partial[C_{ijlm}\varepsilon_{lm,y}(w^{kh})]}{\partial y_j} = \frac{\partial C_{ijkh}}{\partial y_j}, \; i,j,k,h = 1,2,3 \tag{101}$$

Using Green's theorem, we obtain

$$\int_\Gamma \frac{\partial[C_{ijlm}\varepsilon_{lm,y}(w^{kh})]}{\partial y_j}v_i dV + \int_\Gamma C_{ijlm}\varepsilon_{lm,y}(w^{kh})\frac{\partial v_i}{\partial y_j}dV \\ = \int_{\partial\Gamma} u_i C_{ijlm}\varepsilon_{lm,y}(w^{kh})v_i dS = 0, \; i,j,k,h = 1,2,3 \tag{102}$$

$$\int_\Gamma \frac{\partial[C_{ijlm}\varepsilon_{lm,y}(w^{kh})]}{\partial y_i}v_j dV + \int_\Gamma C_{ijlm}\varepsilon_{lm,y}(w^{kh})\frac{\partial v_j}{\partial y_i}dV \\ = \int_{\partial\Gamma} u_j C_{ijlm}\varepsilon_{lm,y}(w^{kh})v_j dS = 0, \; i,j,k,h = 1,2,3 \tag{103}$$

In Equation (103), the indices i and j have been interchanged and the property $C_{ijlm} = C_{jilm}$ has been considered. From (102) and (103), it follows that

$$\int_\Gamma \frac{\partial[C_{ijlm}\varepsilon_{lm,y}(w^{kh})]}{\partial y_j}v_i dV + \int_\Gamma \frac{\partial[C_{ijlm}\varepsilon_{lm,y}(w^{kh})]}{\partial y_i}v_j dV \\ = 2\int_\Gamma C_{ijlm}\varepsilon_{lm,y}(w^{kh})\frac{1}{2}\left(\frac{\partial v_i}{\partial y_j} + \frac{\partial v_j}{\partial y_i}\right)dV = 2\int_\Gamma C_{ijlm}\varepsilon_{ij,y}(v)\varepsilon_{lm,y}(w^{kh})dV, \; i,j,k,h = 1,2,3 \tag{104}$$

Because $C_{ijlm} = C_{jilm}$, multiplying Equation (104) by v results in

$$\frac{\partial[C_{ijlm}\varepsilon_{lm,y}(w^{kh})]}{\partial y_j}v_i = \frac{\partial C_{ijkh}}{\partial y_j}v_i \tag{105}$$

Interchanging the indices i and j results in

$$\frac{\partial\left[C_{ijlm}\varepsilon_{lm,y}(w^{kh})\right]}{\partial y_i}v_j = \frac{\partial C_{ijkh}}{\partial y_i}v_j \tag{106}$$

The integration and addition of Equations (105) and (106) using Equation (104) offer

$$\int_\Gamma C_{ijlm}\varepsilon_{ij,y}(v)\varepsilon_{lm,y}(w^{kh})dV = \int_\Gamma \frac{\partial C_{ijkh}}{\partial y_j}v_i dV \tag{107}$$

We must then find w^{kh} in V_y such that $\forall v \in V_y$, which verifies Equation (107). If w^{kh} is obtained, then

$$\sigma_{ij}^o = C_{ijkh}[\varepsilon_{kh,x}(u^o) + \varepsilon_{kh,y}(u^1)] \\ = C_{ijkh}[\varepsilon_{kh,x}(u^o) + \varepsilon_{kh,x}(u^o)\varepsilon_{lm,y}(w^{kh})] \tag{108}$$

By applying the average operator, it results in

$$\langle \sigma_{ij}^o \rangle = \langle C_{ijkh}\varepsilon_{kh,x}(u^o) \rangle + \langle C_{ijkh}\varepsilon_{kh,x}(u^o)\varepsilon_{lm,y}(u^{kh}) \rangle$$
$$= \langle C_{ijkh} \rangle \varepsilon_{kh,x}(u^o) + \langle C_{ijkh}\varepsilon_{lm,y}(u^{kh}) \rangle \varepsilon_{kh,x}(u^o) \tag{109}$$

This produces

$$\langle \sigma_{ij}^o \rangle = [\langle C_{ijkh} \rangle + \langle C_{ijkh}\varepsilon_{lm,y}(u^{kh}) \rangle]\varepsilon_{kh,x}(u^o) \tag{110}$$

A comparison of Equation (110) can be made with

$$\langle \sigma_{ij}^o \rangle = C_{ijkh}^o \tilde{\varepsilon}_{kh}(u^o) \tag{111}$$

and, if we denote $\varepsilon_{kh,x}(u^o) \equiv \tilde{\varepsilon}_{kh}(u^o)$, the homogenized coefficients can be obtained:

$$C_{ijkh}^o = \langle C_{ijkh} \rangle + \langle C_{ijkh}\varepsilon_{lm,y}(u^{kh}) \rangle \tag{112}$$

Therefore, there are two ways to obtain the homogenized coefficients:
- Using the local equations, the strain and stress field and the averages are determined, obtaining the homogenized coefficients;
- Using the variational formulation and determining the function w^{kh} can also help us to determine the homogenized coefficients.

For the fiber-reinforced composite, there is a class of solutions w^{kh}, with $k,h = 1,2,3$ satisfying

$$-\frac{\partial[C_{ijlm}\varepsilon_{lm,y}(w)]}{\partial y_j} = \frac{\partial C_{ijkh}}{\partial y_j}, i = 1,2,3 \tag{113}$$

with the boundary conditions

$$w^{kh}\big|_{\partial\Gamma} = 0 \tag{114}$$

and

$$\langle w^{kh} \rangle = 0 \tag{115}$$

If (x_1, x_2, x_3) are the principal material axes, we state

$$C_{1111} = C_{11}; C_{2222} = C_{22}; C_{1122} = C_{1133} = C_{12}; C_{2211} = C_{3311} = C_{21}$$
$$C_{3322} = C_{2233} = C_{23}; C_{3333} = C_{33}; C_{4444} = (C_{22} - C_{23})/2; C_{5555} = C_{44}; \tag{116}$$
$$C_{6666} = C_{44}$$

The other components of C_{ijkl} are zero (we work with a transversely isotropic material). The stress–strain relation becomes

$$\begin{Bmatrix} \sigma_{22} \\ \sigma_{33} \\ \tau_{23} \end{Bmatrix} = \begin{bmatrix} C_{22} & C_{23} & 0 \\ C_{23} & C_{33} & 0 \\ 0 & 0 & \frac{C_{22}-C_{23}}{2} \end{bmatrix} \begin{Bmatrix} \varepsilon_{22} \\ \varepsilon_{33} \\ \gamma_{23} \end{Bmatrix} \tag{117}$$

or

$$\{\sigma\} = [C]\{\varepsilon\}^o \tag{118}$$

The equilibrium conditions in Equation (71) are

$$\begin{bmatrix} \frac{\partial}{\partial y_2} & 0 & \frac{\partial}{\partial y_3} \\ 0 & \frac{\partial}{\partial y_3} & \frac{\partial}{\partial y_2} \end{bmatrix} \begin{Bmatrix} \sigma_{22} \\ \sigma_{33} \\ \tau_{23} \end{Bmatrix} = 0 \tag{119}$$

Equation (119) can be expressed in compact form:

$$[\partial]\{\sigma\} = [\partial][C]\{\varepsilon\}^o = 0 \qquad (120)$$

Equation (92) becomes

$$\{\varepsilon\}^o = \{\varepsilon\}^{u^o}_{,x} + \{\varepsilon\}^{u^1}_{,y} \qquad (121)$$

and (120) becomes

$$-\begin{bmatrix} \frac{\partial}{\partial y_2} & 0 & \frac{\partial}{\partial y_3} \\ 0 & \frac{\partial}{\partial y_3} & \frac{\partial}{\partial y_2} \end{bmatrix} \begin{bmatrix} C^{(\lambda)}_{22} & C^{(\lambda)}_{23} & 0 \\ C^{(\lambda)}_{23} & C^{(\lambda)}_{33} & 0 \\ 0 & 0 & \frac{C^{(\lambda)}_{22}-C^{(\lambda)}_{23}}{2} \end{bmatrix} \{\varepsilon\}^{u^1}_{,y}$$
$$= \begin{bmatrix} \frac{\partial}{\partial y_2} & 0 & \frac{\partial}{\partial y_3} \\ 0 & \frac{\partial}{\partial y_3} & \frac{\partial}{\partial y_2} \end{bmatrix} \begin{bmatrix} C^{(\lambda)}_{22} & C^{(\lambda)}_{23} & 0 \\ C^{(\lambda)}_{23} & C^{(\lambda)}_{33} & 0 \\ 0 & 0 & \frac{C^{(\lambda)}_{22}-C^{(\lambda)}_{23}}{2} \end{bmatrix} \{\varepsilon\}^{u^o}_{,x} \qquad (122)$$

Considering the plane strain loading conditions, we obtain

$$\begin{bmatrix} \frac{\partial}{\partial y_2} & 0 & \frac{\partial}{\partial y_3} \\ 0 & \frac{\partial}{\partial y_3} & \frac{\partial}{\partial y_2} \end{bmatrix} \begin{bmatrix} C^{(\lambda)}_{22} & C^{(\lambda)}_{23} & 0 \\ C^{(\lambda)}_{23} & C^{(\lambda)}_{33} & 0 \\ 0 & 0 & \frac{C^{(\lambda)}_{22}-C^{(\lambda)}_{23}}{2} \end{bmatrix} \{\varepsilon\}^{u^1}_{,y} = 0 \qquad (123)$$

Using the determined functions w^{kh}, it can be deduced that

$$\{\varepsilon\}^{u^1}_{,y} = \begin{bmatrix} \varepsilon_{22}(w^{22}) & \varepsilon_{22}(w^{33}) & \varepsilon_{22}(w^{23}) \\ \varepsilon_{33}(w^{22}) & \varepsilon_{33}(w^{33}) & \varepsilon_{33}(w^{23}) \\ \varepsilon_{23}(w^{22}) & \varepsilon_{23}(w^{33}) & \varepsilon_{23}(w^{23}) \end{bmatrix} \{\varepsilon\}^{u^o}_{,x} \qquad (124)$$

or, in an alternative form,

$$\{\varepsilon\}^{u^1}_{,y} = [\{\varepsilon(w^{22})\} \quad \{\varepsilon(w^{33})\} \quad \{\varepsilon(w^{23})\}]\{\varepsilon\}^{u^o}_{,x} \qquad (125)$$

Additionally,

$$\begin{bmatrix} \frac{\partial}{\partial y_2} & 0 & \frac{\partial}{\partial y_3} \\ 0 & \frac{\partial}{\partial y_3} & \frac{\partial}{\partial y_2} \end{bmatrix} \begin{bmatrix} C^{(\lambda)}_{22} & C^{(\lambda)}_{23} & 0 \\ C^{(\lambda)}_{23} & C^{(\lambda)}_{33} & 0 \\ 0 & 0 & \frac{C^{(\lambda)}_{22}-C^{(\lambda)}_{23}}{2} \end{bmatrix} [\{\varepsilon(w^{22})\} \quad \{\varepsilon(w^{33})\} \quad \{\varepsilon(w^{23})\}]\{\varepsilon\}^{u^o}_{,x} = 0 \qquad (126)$$

$$[\{\varepsilon(w^{22})\} \quad \{\varepsilon(w^{33})\} \quad \{\varepsilon(w^{23})\}]\{\varepsilon\}^{u^o}_{,x} = 0 \qquad (127)$$

Equation (127) should remain valid for all $\{\varepsilon\}^{u^o}_{,x}$. Equation (113) becomes

$$-\frac{C^{(\lambda)}_{ijlm}\partial\varepsilon^{(\lambda)}_{lm,y}(w^{kh})}{\partial y_j} = 0, \ i = 1,2,3 \qquad (128)$$

In we consider the case of plane strain $i = 2, 3$ and $j = 2, 3$,

$$C^{(\lambda)}_{22}\frac{\partial\varepsilon_{22}(w^{kh})}{\partial y_2} + C^{(\lambda)}_{23}\frac{\partial\varepsilon_{33}(w^{kh})}{\partial y_2} + \frac{1}{2}\left[C^{(\lambda)}_{22} - C^{(\lambda)}_{23}\right]\frac{\partial\varepsilon_{23}(w^{kh})}{\partial y_3} = 0 \qquad (129)$$

and

$$C^{(\lambda)}_{23}\frac{\partial\varepsilon_{22}(w^{kh})}{\partial y_3} + C^{(\lambda)}_{22}\frac{\partial\varepsilon_{33}(w^{kh})}{\partial y_3} + \frac{1}{2}\left[C^{(\lambda)}_{22} - C^{(\lambda)}_{23}\right]\frac{\partial\varepsilon_{23}(w^{kh})}{\partial y_2} = 0 \qquad (130)$$

The solution is

$$w^{kh} = \begin{cases} w^{kh,(f)} \text{ for } y \in V_f \\ w^{kh,(m)} \text{ for } y \in V_m \end{cases} \quad (131)$$

satisfying the boundary conditions

$$w^{kh,(f)}\big|_{\partial\Gamma} = w^{kh,(m)}\big|_{\partial\Gamma}; \; \sigma_{ij}^{(f)} n_j = \sigma_{ij}^{(m)} n_j \quad (132)$$

The boundary conditions for the RUC are: $u_i = \alpha_{ij} y_j$. It is possible to show that the average strain is: $\langle \varepsilon_{ij} \rangle = \bar{\varepsilon}_{ij} = \alpha_{ij}$. Let us denote the displacement field by w^* having the property $w^*|_{\partial\Gamma} = u|_{\partial\Gamma}$ and $\bar{\varepsilon}_{kh}(w^*) = \alpha_{ij}$. Due to the existing symmetry in the distribution of the unit cell it can be concluded that $\langle w^* \rangle = 0$. The field w is introduced as

$$w = w^* - u \quad (133)$$

with the boundary conditions

$$w|_{\partial\Gamma} = 0 \quad (134)$$

$$\langle w \rangle = \langle w^* - u \rangle = \langle w^* \rangle - \langle u \rangle = 0 \quad (135)$$

This function (w) verifies the condition of zero average and value zero on the contour, and it verifies Equation (130). For the "quasi-periodical fields" u_1, it follows that

$$u_1 = \alpha_{22} \begin{Bmatrix} \frac{w_{22}^*}{\alpha_{22}} - y_2 \\ 0 \end{Bmatrix} + \alpha_{33} \begin{Bmatrix} 0 \\ \frac{w_{22}^*}{\alpha_{22}} - y_2 \end{Bmatrix} \quad (136)$$

The strain field is

$$\varepsilon_{22}\left(w^{22}\right) = \frac{\varepsilon_{22}(w^*)}{\alpha_{22}} - 1 \; ; \; \varepsilon_{33}\left(w^{22}\right) = 0 \quad (137)$$

$$\varepsilon_{22}\left(w^{33}\right) = \frac{\varepsilon_{22}(w^*)}{\alpha_{33}} - 1 \; ; \; \varepsilon_{33}\left(w^{33}\right) = 0 \quad (138)$$

and

$$\varepsilon_{22}(u_1) = \varepsilon_{22}(w^*) - \alpha_{22} \; ; \; \varepsilon_{33}(u_1) = \varepsilon_{33}(w^*) - \alpha_{33} \quad (139)$$

or

$$\bar{\varepsilon}_{22}\left(w^{22}\right) = \frac{\bar{\varepsilon}_{22}(w^*)}{\alpha_{22}} - 1 = 0 \; ; \; \bar{\varepsilon}_{33}\left(w^{33}\right) = \frac{\bar{\varepsilon}_{33}(w^*)}{\alpha_{33}} - 1 = 0 \quad (140)$$

For the fiber-reinforced unidirectional composite, the homogenized coefficients can be obtained with the following relations:

$$\begin{aligned} C^o_{ijkh} &= \langle C_{ijkh} \rangle + \langle C_{ijkh}\varepsilon_{lm,y}\left(w^{kh}\right) \rangle \\ &= \frac{1}{V}\int_\Gamma C_{ijkh} dV + \frac{1}{V}\int_\Gamma C_{ijkh}\varepsilon_{lm}(w^*)dV \\ &= \frac{1}{V}\left(C^{(f)}_{ijkh}V_f + C^{(m)}_{ijkh}V_m\right) + \frac{1}{V}\left(C^{(f)}_{ijlm}\bar{\varepsilon}^{(f)}_{lm}(w)V_f + C^{(m)}_{ijlm}\bar{\varepsilon}^{(m)}_{lm}(w)V_m\right) \end{aligned} \quad (141)$$

where

$$\alpha_{lm}\bar{\varepsilon}^{(f)}_{lm}(w) = \bar{\varepsilon}^{(f)}_{lm}(w^*) - \alpha_{lm} \; ; \; \alpha_{lm}\bar{\varepsilon}^{(m)}_{lm}(w) = \bar{\varepsilon}^{(m)}_{lm}(w^*) - \alpha_{lm} \quad (142)$$

Thus, we have

$$C^o_{ijkh} = v_f C^{(f)}_{ijkh} + v_m C^{(m)}_{ijkh} + v_f C^{(f)}_{ijkh}\left[\frac{\bar{\varepsilon}^{(f)}_{lm}(w*)}{\alpha_{lm}} - 1\right] + v_m C^{(m)}_{ijkh}\left[\frac{\bar{\varepsilon}^{(m)}_{lm}(w*)}{\alpha_{lm}} - 1\right] \quad (143)$$

As a result (considering the plane strain loading conditions),

$$\begin{aligned} C^o_{22} &= v_f C^{(f)}_{22} + v_m C^{(m)}_{22} + v_f C^{(f)}_{22} \left[\frac{\bar{\varepsilon}^{(f)}_{22}(w*)}{\alpha_{22}} - 1 \right] + v_m C^{(m)}_{22} \left[\frac{\bar{\varepsilon}^{(m)}_{22}(w*)}{\alpha_{22}} - 1 \right] \\ &= v_f C^{(f)}_{22} \frac{\bar{\varepsilon}^{(f)}_{22}(w*)}{\alpha_{22}} + v_m C^{(m)}_{22} \frac{\bar{\varepsilon}^{(m)}_{22}(w*)}{\alpha_{22}} \end{aligned} \quad (144)$$

and

$$\begin{aligned} C^o_{23} &= v_f C^{(f)}_{23} + v_m C^{(m)}_{23} + v_f C^{(f)}_{23} \left[\frac{\bar{\varepsilon}^{(f)}_{33}(w*)}{\alpha_{33}} - 1 \right] + v_m C^{(m)}_{23} \left[\frac{\bar{\varepsilon}^{(m)}_{33}(w*)}{\alpha_{33}} - 1 \right] \\ &= v_f C^{(f)}_{23} \frac{\bar{\varepsilon}^{(f)}_{33}(w*)}{\alpha_{33}} + v_m C^{(m)}_{23} \frac{\bar{\varepsilon}^{(m)}_{33}(w*)}{\alpha_{33}} \end{aligned} \quad (145)$$

4. The Mori–Tanaka Model

4.1. Mathematical Model

In the following section, the mathematical model proposed by Mori and Tanaka is applied to obtain the engineering parameters that define Hooke's law for a one-dimensional fiber-reinforced composite [34]. We consider an epoxy matrix with a visco-elastic response, reinforced with monotonous and parallel aligned carbon fibers that are uniformly distributed inside the resin (Figure 5). The resulting material has an orthotropic behavior. However, there are applications where the fibers are elliptical cylinders. These cylinders are randomly distributed, and the behavior of the material is a transverse isotropic.

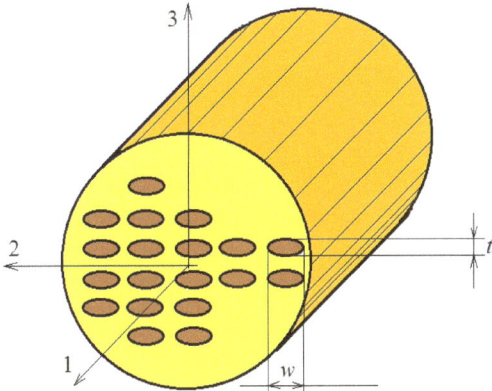

Figure 5. Reinforced composite.

The theory developed in [78] is applied in [28] for a reinforced material with continuous cylindrical fibers with an elliptic section. To solve this problem, Mori–Tanaka's [34] mean-field theory is used. In [79,80], the two phases of the composite are two isotropic materials.

We consider a comparison material (CM). In the CM, there is a linear relation between the mean strain field ε^o and the mean stress field $\bar{\sigma}$:

$$\bar{\sigma} = C_m \varepsilon^0 \quad (146)$$

The average strain field in the RUC is $\varepsilon^m = \varepsilon^0 + \tilde{\varepsilon}$ and the mean stress field is $\sigma^m = \bar{\sigma} + \tilde{\sigma}$. This results in

$$\sigma^m = \bar{\sigma} + \tilde{\sigma} = C_m(\varepsilon^0 + \tilde{\varepsilon}) \quad (147)$$

The mean strain fields in the fiber and in the matrix are differentiated through an additional term ε^{pt}, hence $\varepsilon^f = \varepsilon^m + \varepsilon^{pt} = \varepsilon^0 + \tilde{\varepsilon} + \varepsilon^{pt}$. In a similar way, the average stress field differs by the term σ^{pt} and, therefore, $\sigma^f = \overline{\sigma} + \widetilde{\sigma} + \sigma^{pt}$. The generalized Hooke law becomes

$$\sigma^f = \overline{\sigma} + \widetilde{\sigma} + \sigma^{pt} = C_f(\varepsilon^0 + \widetilde{\varepsilon} + \varepsilon^{pt}) \tag{148}$$

or

$$\sigma^f = \overline{\sigma} + \widetilde{\sigma} + \sigma^{pt} = C_f(\varepsilon^0 + \widetilde{\varepsilon} + \varepsilon^{pt}) = C_m(\varepsilon^0 + \widetilde{\varepsilon} + \varepsilon^{pt} - \varepsilon^*) \tag{149}$$

We introduce ε^{pt} in Equation (149).

$$\varepsilon^{pt} = P\varepsilon^* \tag{150}$$

The Eshelby transformation tensor P from Equation (150) is presented in Appendix A (where $P_{ikjl} = P_{jikl} = P_{ijtk}$). The average stress in the whole RUC is

$$\begin{aligned}\overline{\sigma} &= v_f\sigma_f + v_m\sigma_m = v_f(\overline{\sigma} + \widetilde{\sigma} + \sigma^{pt}) + v_m(\overline{\sigma} + \widetilde{\sigma}) \\ &= (v_f + v_m)\overline{\sigma} + (v_f + v_m)\widetilde{\sigma} + v_f\sigma^{pt} = \overline{\sigma} + \widetilde{\sigma} + v_f\sigma^{pt}\end{aligned} \tag{151}$$

which reduces to

$$\widetilde{\sigma} = -v_f\sigma^{pt} \tag{152}$$

In a similar way, we can obtain

$$\overline{\varepsilon} = -v_f(\varepsilon^{pt} - \varepsilon^*) = -v_f(P\varepsilon^* - \varepsilon^*) = -v_f(P - I)\varepsilon^* \tag{153}$$

I denotes the unit tensor. Equations (147) and (149) yield

$$C_f[\varepsilon^0 - v_f(P - I)\varepsilon^* + P\varepsilon^*] = C_m[\varepsilon^0 - v_f(P - I)\varepsilon^* + P\varepsilon^* - \varepsilon^*] \tag{154}$$

or

$$\left[C_f\Big(-v_f(P - I) + P\Big) + C_m\Big(v_f(P - I) - P + I\Big)\right]\varepsilon^* + \Big(C_f - C_m\Big)\varepsilon^0 = 0 \tag{155}$$

or

$$\left[C_f\Big(v_mP + v_fI\Big) - C_mv_m(P - I)\right]\varepsilon^* + \Big(C_f - C_m\Big)\varepsilon^0 = 0 \tag{156}$$

and

$$\left[v_m\Big(C_f - C_m\Big)P + v_f\Big(C_f - C_m\Big) + C_m\right]\varepsilon^* + \Big(C_f - C_m\Big)\varepsilon^0 = 0 \tag{157}$$

The final form is

$$\left[\Big(C_f - C_m\Big)\Big(v_mP + v_fI\Big) + C_m\right]\varepsilon^* + \Big(C_f - C_m\Big)\varepsilon^0 = 0 \tag{158}$$

This offers

$$\begin{aligned}\varepsilon^*_{11} &= \tfrac{1}{A}\left(A_{11}\varepsilon^0_{11} + A_{12}\varepsilon^0_{22} + A_{13}\varepsilon^0_{33}\right); \\ \varepsilon^*_{11} &= \tfrac{1}{A}\left(A_{21}\varepsilon^0_{11} + A_{22}\varepsilon^0_{22} + A_{23}\varepsilon^0_{33}\right); \\ \varepsilon^*_{11} &= \tfrac{1}{A}\left(A_{31}\varepsilon^0_{11} + A_{32}\varepsilon^0_{22} + A_{33}\varepsilon^0_{33}\right).\end{aligned} \tag{159}$$

The coefficients A_{ij} are presented in Appendix A [30]. The shear strain is [30]

$$\varepsilon^*_{12} = \frac{\Big(G_{12,f} - G_m\Big)}{\Big(G_{12,f} - G_m\Big)\Big(2v_mP_{1212} + v_f\Big) + G_m}\varepsilon^0_{12} \tag{160}$$

$$\varepsilon^*_{23} = \frac{\Big(G_{23,f} - G_m\Big)}{\Big(G_{23,f} - G_m\Big)\Big(2v_mP_{2323} + v_f\Big) + G_m}\varepsilon^0_{23} \tag{161}$$

$$\varepsilon_{31}^* = \frac{\left(G_{31,f} - G_m\right)}{\left(G_{31,f} - G_m\right)\left(2v_m P_{3131} + v_f\right) + G_m} \varepsilon_{31}^0 \quad (162)$$

Equations (158)–(162) can now be used to determine the elastic/viscoelastic parameters of a composite, which is considered as an orthotropic body. To compute the Young's modulus E_m, the composite specimen is subjected to a pure traction $\bar{\sigma}_{11}$. This results in the following equation: $\bar{\sigma}_{11} = E_{11}\bar{\varepsilon}_{11}$ and $\bar{\sigma}_{11} = E_m\bar{\varepsilon}_{11}^0$; $\bar{\varepsilon}_{22}^0 = \bar{\varepsilon}_{33}^0 = -v_m\bar{\varepsilon}_{11}^0$.

Equation (158) produces

$$\begin{aligned}\bar{\varepsilon}_{11} &= \bar{\varepsilon}_{11}^0 + v_f\bar{\varepsilon}_{11}^* = \bar{\varepsilon}_{11}^0 + v_f\left(\frac{A_{11}}{A}\bar{\varepsilon}_{11}^0 + \frac{A_{12}}{A}\bar{\varepsilon}_{22}^0 + \frac{A_{13}}{A}\bar{\varepsilon}_{33}^0\right) \\ &= \bar{\varepsilon}_{11}^0\left(1 + v_f a_{11}\right) - v_f a_{12} v_m \bar{\varepsilon}_{11}^0 - v_f a_{13} v_m \bar{\varepsilon}_{11}^0 \\ &= \bar{\varepsilon}_{11}^0\left[1 + v_f[a_{11} - v_m(a_{12} + a_{13})]\right]\end{aligned} \quad (163)$$

Here, we show that $a_{ij} = A_{ij}/A$, A_{ij}, and A is presented in Appendix A; see rel. (A6). This results in

$$E_{11} = \frac{\bar{\varepsilon}_{11}^0}{\bar{\varepsilon}_{11}} E_m = \frac{E_m}{1 + v_f[a_{11} - v_m(a_{12} + a_{13})]} \quad (164)$$

For the other directions, in a similar way, we obtain the following equations:

$$E_{22} = \frac{\bar{\varepsilon}_{22}^0}{\bar{\varepsilon}_{22}} E_m = \frac{E_m}{1 + v_f[a_{22} - v_m(a_{21} + a_{23})]} \quad (165)$$

and

$$E_{33} = \frac{\bar{\varepsilon}_{33}^0}{\bar{\varepsilon}_{33}} E_m = \frac{E_m}{1 + v_f[a_{33} - v_m(a_{31} + a_{32})]} \quad (166)$$

Considering the shear moduli, we have

$$\bar{\sigma}_{12} = 2G_{12}\bar{\varepsilon}_{12}\ ;\ \bar{\sigma}_{12} = 2G_m\bar{\varepsilon}_{12}^0 \quad (167)$$

However,

$$\bar{\varepsilon}_{12} = \bar{\varepsilon}_{12}^0 + v_f\bar{\varepsilon}_{12}^* = \varepsilon_{12}^* - v_f \frac{G_{12,f} - G_m}{\left(G_{12,f} - G_m\right)\left(2v_m P_{1212} + v_f\right) + G_m} \varepsilon_{12}^0 \quad (168)$$

Using Equations (167) and (168) produces G_{12}:

$$G_{12} = G_m\left(1 + \frac{v_f}{\frac{G_m}{G_{12,f} - G_m} + 2v_m P_{1212}}\right) \quad (169)$$

In the same way, we obtain

$$G_{23} = G_m\left(1 + \frac{v_f}{\frac{G_m}{G_{23,f} - G_m} + 2v_m P_{2323}}\right) \quad (170)$$

and

$$G_{31} = G_m\left(1 + \frac{v_f}{\frac{G_m}{G_{31,f} - G_m} + 2v_m P_{3131}}\right) \quad (171)$$

The Poisson ratio is computed using the formulas

$$\bar{\varepsilon}_{22} = -v_m\bar{\varepsilon}_{11}\ ;\ \bar{\varepsilon}_{22}^0 = \bar{\varepsilon}_{33}^0 = -v_m\bar{\varepsilon}_{11}^0 \quad (172)$$

Note that

$$\bar{\varepsilon}_{11} = \bar{\varepsilon}_{11}^0 + v_f \bar{\varepsilon}_{11}^* = \bar{\varepsilon}_{11}^0 + v_f a_{11} \bar{\varepsilon}_{11}^0 + v_f a_{12} \bar{\varepsilon}_{22}^0 + v_f a_{13} \bar{\varepsilon}_{33}^0$$
$$= \bar{\varepsilon}_{11}^0 \left(1 + v_f a_{11}\right) + v_f a_{12} \bar{\varepsilon}_{22}^0 + v_f a_{13} \bar{\varepsilon}_{33}^0 \tag{173}$$

and

$$\bar{\varepsilon}_{22} = \bar{\varepsilon}_{22}^0 + v_f \bar{\varepsilon}_{22}^* = v_f a_{21} \bar{\varepsilon}_{11}^0 + \bar{\varepsilon}_{22}^0 \left(1 + v_f a_{22}\right) + v_f a_{23} \bar{\varepsilon}_{33}^0 \tag{174}$$

or

$$\bar{\varepsilon}_{11} = \left[\left(1 + v_f a_{11}\right) - v_f a_{12} v_m - v_f a_{13} v_m\right] \bar{\varepsilon}_{11}^0 = \left[v_f a_{21} - v_m \left(1 + v_f a_{22}\right) - v_m v_f a_{23}\right] \bar{\varepsilon}_{11}^0 \tag{175}$$

Introducing Equation (174) into Equation (175) produces

$$v_{12} = -\frac{\bar{\varepsilon}_{22}}{\bar{\varepsilon}_{11}} = -\frac{v_f a_{21} - v_m \left(1 + v_f a_{22}\right) - v_m v_f a_{23}}{1 + v_f a_{11} - v_f a_{12} v_m - v_f a_{13} v_m} \tag{176}$$

which can be written as

$$v_{12} = \frac{v_m - v_f [a_{22} - v_m (a_{21} + a_{23})]}{1 + v_f [a_{11} - v_m (a_{12} + a_{13})]} \tag{177}$$

In the same way, it produces

$$v_{23} = \frac{v_m - v_f [a_{22} - v_m (a_{21} + a_{23})]}{1 + v_f [a_{33} - v_m (a_7 + a_8)]} \tag{178}$$

and

$$v_{31} = \frac{v_m - v_f [a_{33} - v_m (a_{31} + a_{32})]}{1 + v_f [a_{11} - v_m (a_{12} + a_{13})]} \tag{179}$$

5. The Finite Element Method Used to Obtain the Creep Response

Recently, FEM has become the main method used for the study of elastic systems, as it is able to address a multitude of situations and types of materials, including composite materials [81]. Specialized problems are also studied, such as the influence of temperature on the stresses that appear in the analyzed structures [82]. In [83], a model is presented for the study of a composite reinforced with silicon carbide fibers. A similar model is addressed in [84]. Bodies with transverse isotropy were also studied, as in [11,85]. If we are dealing with microstructured systems, where a unit cell can be identified, the geometric symmetry allows the analysis to be conducted only on a quarter or half of the unit cell, on a unit previously defined as the "representative unit cell" (RUC). The unit cell model with finite elements is presented in Figures 6 and 7; two models of a RUC that are used in various applications are also presented (Models 1 and 2).

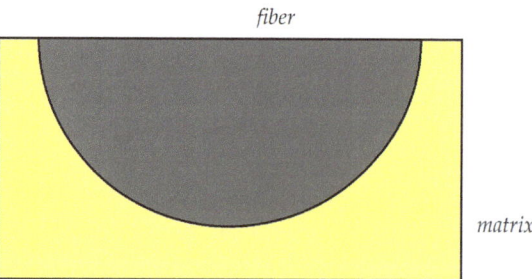

Figure 6. Finite element model 1 for a RUC.

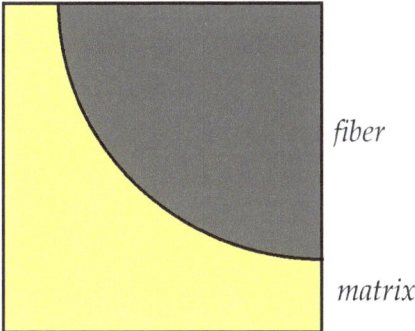

Figure 7. Finite element model 2 for a RUC.

The mechanical constants used in the application are

$$E_m = 4.14 \text{ GPa}; \ v_m = 0.22; \ E_f = 86.90 \text{ GPa}; \ v_f = 0.34 \tag{180}$$

The results of the analysis are shown in Tables 1–6 (in these tables, $\bar{\sigma}$ is the average stress and $\bar{\varepsilon}$ is the average strain).

Table 1. Average values of stress and strain (Case 1).

$\bar{\sigma}$	Fiber	Matrix	RUC	$\bar{\varepsilon}$	Fiber	Matrix	RUC
$\bar{\sigma}_{22}$	0.146×10^3	0.122×10^3	0.138×10^3	$\bar{\varepsilon}_{22}$	0.16×10^0	0.263×10^1	0.106×10^1
$\bar{\sigma}_{33}$	0.71×10^0	-0.906×10^0	0.118×10^0	$\bar{\varepsilon}_{33}$	-0.445×10^{-1}	-0.137×10^1	-0.529×10^0
$\bar{\sigma}_{11}$	0.324×10^2	0.413×10^3	0.357×10^2	$\bar{\varepsilon}_{11}$	0.0×10^0	0.0×10^0	0.0×10^0
$\bar{\sigma}_{23}$	0.194×10^{-4}	0.412×10^2	0.359×10^{-4}	$\bar{\varepsilon}_{23}$	0.539×10^{-7}	0.447×10^{-5}	0.167×10^{-5}

Table 2. Computed values of elastic moduli (Case 1).

Modulus [MPa]	Matrix	Fiber	Average
E_{11}	4140.0	86,900.0	56,278.0
$E_{23} = E_{13}$	4140.0	86,899.0	12,741.0
v_1	0.34	0.22	0.259
v_{23}	0.34	0.22	0.475
G_{23}	1544.0	35,614.7	4318.2
K_{23}	4827.4	63,597.7	12,886.2

Table 3. Average values of stress and strain (Case 2).

$\bar{\sigma}$	Fiber	Matrix	RUC	$\bar{\varepsilon}$	Fiber	Matrix	RUC
$\bar{\sigma}_{22}$	0.147×10^3	0.122×10^3	0.138×10^3	$\bar{\varepsilon}_{22}$	0.134×10^0	0.183×10^1	0.757×10^0
$\bar{\sigma}_{33}$	0.857×10^2	0.702×10^2	0.801×10^2	$\bar{\varepsilon}_{33}$	0.485×10^{-1}	0.157×10^0	0.881×10^{-1}
$\bar{\sigma}_{11}$	0.512×10^2	0.653×10^2	0.564×10^2	$\bar{\varepsilon}_{11}$	0.0×10^0	0.0×10^0	0.0×10^0
$\bar{\sigma}_{23}$	0.992×10^{-5}	0.322×10^{-4}	0.181×10^{-4}	$\bar{\varepsilon}_{23}$	0.306×10^{-7}	0.190×10^{-5}	0.717×10^{-6}

In this paper, we used a three-dimensional model to obtain the shear modulus and Poisson's ratios in a plane perpendicular to x_2x_3.

A few of the foregoing models are listed in Table 7, for which the results using finite element analysis are obtained.

Table 4. Computed values of the elastic moduli (Case 2).

Modulus	Matrix	Fiber	Average
E_{11}	4140.0	86,900.0	56,278.0
$E_{23} = E_{13}$	4140.0	86,900.0	12,741.0
ν_1	0.34	0.22	0.259
ν_{23}	0.34	0.22	0.475
G_{23}	1544.0	35,614.8	4318.2
K_{23}	4827.4	63,597.8	12,886.2

Table 5. Average values of stress and strain (Case 3).

$\bar{\sigma}$	Fiber	Matrix	RUC	$\bar{\varepsilon}$	Fiber	Matrix	RUC
$\bar{\sigma}_{22}$	0.147×10^3	0.123×10^3	0.138×10^3	$\bar{\varepsilon}_{22}$	0.160×10^0	0.263×10^1	0.106×10^{-1}
$\bar{\sigma}_{33}$	0.644×10^0	-0.983×10^0	0.049×10^{-1}	$\bar{\varepsilon}_{33}$	-0.446×10^{-1}	-0.137×10^1	0.530×10^0
$\bar{\sigma}_{11}$	0.324×10^2	0.414×10^2	0.357×10^2	$\bar{\varepsilon}_{11}$	0.0×10^0	0.0×10^0	0.0×10^0
$\bar{\sigma}_{23}$	0.899×10^{-5}	-0.979×10^{-4}	-0.301×10^{-4}	$\bar{\varepsilon}_{23}$	0.258×10^{-7}	-0.638×10^{-5}	0.232×10^{-5}

Table 6. Computed values of the elastic moduli (Case 3).

Modulus	Matrix	Fiber	Average
E_{11}	4140.0	86,900.0	56,279.0
$E_{23} = E_{13}$	4140.0	86,899.0	12,754.0
ν_1	0.34	0.22	0.259
ν_{23}	0.34	0.22	0.475
G_{23}	1544.0	35,614.7	4322.2
K_{23}	4827.4	63,597.7	12,900.8

Table 7. Finite element models and associated boundary conditions (BCs).

Case	Model	B.C. (x_2 Direction)	B.C. (x_3 Direction)
1	Model 1-a	p_x = 137.90 (MPa)	p_y = 0.00 (MPa)
2	Model 1-b	p_x = 137.90 (MPa)	p_y = 80.0 (MPa)
3	Model 2-a	p_x = 137.90 (MPa)	p_y = 0.00 (MPa)
4	Model 2-b	u_x = 0.01 (mm)	u_y = 0.01 (mm)
5	Model 2-c	u_x = 0.01 (mm)	u_y = 0.01 (mm)

There are some discrepancies between the present FE results and those presented in [49]. With respect to these discrepancies, the following verification should be considered. If the boundary condition for the FE model is taken as $u_i = \alpha_{ij} x_j$ (where $\alpha_{ij} = \alpha_{ji}$), the average strain should be equal to $\bar{\varepsilon}_{ij} = \alpha_{ij}$. This can be demonstrated as follows:

$$\bar{\varepsilon}_{ij} = \frac{1}{V} \int_\Gamma \varepsilon_{ij} dV = \frac{1}{2V} \int_\Gamma \left(\frac{\partial u_j}{\partial x_i} + \frac{\partial u_i}{\partial x_j} \right) dV \tag{181}$$

By applying Green's theorem, it follows that

$$\begin{aligned} \bar{\varepsilon}_{ij} &= \tfrac{1}{2V} \int_{\partial \Gamma} (n_i u_j + n_j u_i) ds \\ &= \tfrac{1}{2V} \left(\int_{\partial \Gamma} n_i \alpha_{jk} x_k ds + \int_{\partial \Gamma} n_j \alpha_{il} x_l ds \right) \\ &= \tfrac{1}{2V} \left(\alpha_{jk} \int_{\partial \Gamma} n_i x_k ds + \alpha_{il} \int_{\partial \Gamma} n_j x_l ds \right) \end{aligned} \tag{182}$$

or

$$\begin{aligned} \bar{\varepsilon}_{ij} &= \tfrac{1}{2V} \left(\alpha_{jk} \int_\Gamma \tfrac{\partial x_k}{\partial x_i} dV + \alpha_{il} \int_\Gamma \tfrac{\partial x_l}{\partial x_j} dV \right) \\ &= \tfrac{1}{2V} \left(\alpha_{jk} \int_\Gamma \delta_{ki} dV + \alpha_{il} \int_\Gamma \delta_{lj} dV \right) = \tfrac{1}{2V} (\alpha_{ji} + \alpha_{ij}) = \alpha_{ij} \end{aligned} \tag{183}$$

The discrepancy identified with the results of [49] can be attributed to the different type of finite elements used.

Therefore, we obtain average strains and stresses, viz., $\bar{\sigma}_{22}, \bar{\sigma}_{33}, \bar{\sigma}_{11}, \bar{\sigma}_{23} = \bar{\tau}_{23}, \bar{\varepsilon}_{22}, \bar{\varepsilon}_{33}, \bar{\varepsilon}_{11}, \bar{\varepsilon}_{23} = 1/2\,\gamma_{23}$. Using these values, it is now possible to obtain the mechanical constants of the studied composite [56]. To determine the longitudinal elastic modulus E_{11}, we use the well-known rule of mixture:

$$E_{11} = E_f v_f + E_m v_m \qquad (184)$$

where

$$v_f = \frac{A_f}{A}\ ;\ v_f = \frac{A_m}{A} \qquad (185)$$

The following relations exist:

$$\begin{aligned}\bar{\sigma}_{22} &= C_{22}\bar{\varepsilon}_{22} + C_{23}\bar{\varepsilon}_{33};\\ \bar{\sigma}_{33} &= C_{23}\bar{\varepsilon}_{22} + C_{22}\bar{\varepsilon}_{33};\\ \bar{\sigma}_{11} &= C_{12}(\bar{\varepsilon}_{22} + \bar{\varepsilon}_{33});\\ \bar{\tau}_{23} &= C_{66}\bar{\gamma}_{23}\end{aligned} \qquad (186)$$

from which results

$$\begin{bmatrix}\bar{\varepsilon}_{22} & \bar{\varepsilon}_{33}\\ \bar{\varepsilon}_{33} & \bar{\varepsilon}_{22}\end{bmatrix}\begin{Bmatrix}C_{22}\\ C_{23}\end{Bmatrix} = \begin{Bmatrix}\bar{\sigma}_{22}\\ \bar{\sigma}_{33}\end{Bmatrix} \qquad (187)$$

and

$$\begin{Bmatrix}C_{22}\\ C_{23}\end{Bmatrix} = \frac{1}{\bar{\varepsilon}_{22}^2 - \bar{\varepsilon}_{33}^2}\begin{bmatrix}\bar{\varepsilon}_{22} & -\bar{\varepsilon}_{33}\\ -\bar{\varepsilon}_{33} & \bar{\varepsilon}_{22}\end{bmatrix}\begin{Bmatrix}\bar{\sigma}_{22}\\ \bar{\sigma}_{33}\end{Bmatrix} \qquad (188)$$

This results in the following:

$$C_{22} = \frac{\bar{\sigma}_{22}\bar{\varepsilon}_{22} - \bar{\sigma}_{33}\bar{\varepsilon}_{33}}{\bar{\varepsilon}_{22}^2 - \bar{\varepsilon}_{33}^2}\ ;\ C_{23} = \frac{\bar{\sigma}_{33}\bar{\varepsilon}_{22} - \bar{\sigma}_{22}\bar{\varepsilon}_{33}}{\bar{\varepsilon}_{22}^2 - \bar{\varepsilon}_{33}^2} \qquad (189)$$

For C_{12} and C_{66},

$$C_{12} = \frac{\bar{\sigma}_{11}}{\bar{\varepsilon}_{22} + \bar{\varepsilon}_{33}}\ ;\ C_{66} = \frac{\bar{\tau}_{23}}{\bar{\gamma}_{23}} \qquad (190)$$

To determine the bulk modulus K_{23} is used in the relation:

$$K_{23} = \frac{C_{22} + C_{33}}{2} = \frac{\bar{\sigma}_{22} + \bar{\sigma}_{33}}{2(\bar{\varepsilon}_{22} + \bar{\varepsilon}_{33})}\ . \qquad (191)$$

The longitudinal Poisson's ratio is calculated via the following relation:

$$\nu_1 = \nu_{21} = \nu_{31} = \frac{1}{2}\left(\frac{C_{11} - E_{11}}{K_{23}}\right)^{1/2} = \frac{C_{12}}{C_{22} + C_{33}} = \frac{\bar{\sigma}_{11}}{(\bar{\sigma}_{22} + \bar{\sigma}_{33})}\ . \qquad (192)$$

and the shear modulus

$$G_{23} = \frac{C_{22} - C_{33}}{2} = \frac{\bar{\sigma}_{22} - \bar{\sigma}_{33}}{2(\bar{\varepsilon}_{22} - \bar{\varepsilon}_{33})}\ . \qquad (193)$$

or from

$$G_{23} = C_{66} = \frac{\bar{\sigma}_{23}}{2\bar{\varepsilon}_{23}}\ . \qquad (194)$$

By introducing the following parameter,

$$\psi = 1 + \frac{4\nu_1^2 K_{23}}{E_{11}}, \qquad (195)$$

the transverse moduli and Poisson's ratio are obtained using the following relations:

$$E_{22} = E_{33} = \frac{4G_{23}K_{23}}{K_{23} + \psi G_{23}} \qquad (196)$$

and

$$\nu_{23} = \frac{K_{23} - \psi G_{23}}{K_{23} + \psi G_{23}} \qquad (197)$$

As such, the expressions for E_{11}, $E_{22} = E_{33}$, $\nu_{12} = \nu_{13}$, ν_{23}, G_{23}, K_{23} were determined. From

$$C_{22} + C_{33} = 2K_{23}\ ;\ C_{22} - C_{33} = 2G_{23} \qquad (198)$$

one may obtain

$$C_{22} = K_{23} + G_{23}\ ;\ C_{23} = K_{23} - G_{23}\ . \qquad (199)$$

Recall that

$$C_{44} = G_1 = G_{12} = G_{13};\ C_{12} = \nu_1(C_{22} + C_{23}) = 2\nu_1 K_{23} \qquad (200)$$

and

$$C_{11} = E_{11} + \frac{2C_{12}^2}{C_{22} + C_{23}} = E_{11} + 4\nu_1^2 K_{23} = \psi E_{11}\ . \qquad (201)$$

In a similar way, FEM was used to determine the average stresses and strains in a 3D elastic solid. These are: $\overline{\sigma}_{11}, \overline{\sigma}_{22}, \overline{\sigma}_{33}, \overline{\sigma}_{12} = \overline{\tau}_{12}, \overline{\sigma}_{23} = \overline{\tau}_{23}, \overline{\sigma}_{31} = \overline{\tau}_{31}, \overline{\varepsilon}_{11}, \overline{\varepsilon}_{22}, \overline{\varepsilon}_{33}, \overline{\varepsilon}_{12} = 1/2\,\gamma_{12}, \overline{\varepsilon}_{23} = 1/2\,\gamma_{23}, \overline{\varepsilon}_{31} = 1/2\,\gamma_{31}$. The general Hooke's Law can be written as follows:

$$\begin{aligned}
\overline{\sigma}_{11} &= C_{11}\overline{\varepsilon}_{11} + C_{12}\overline{\varepsilon}_{22} + C_{12}\overline{\varepsilon}_{33} \\
\overline{\sigma}_{22} &= C_{12}\overline{\varepsilon}_{11} + C_{22}\overline{\varepsilon}_{22} + C_{23}\overline{\varepsilon}_{33} \\
\overline{\sigma}_{33} &= C_{12}\overline{\varepsilon}_{11} + C_{23}\overline{\varepsilon}_{22} + C_{22}\overline{\varepsilon}_{33} \\
\overline{\sigma}_{23} &= \overline{\tau}_{23} = (C_{11} - C_{23})\overline{\varepsilon}_{23} = \tfrac{1}{2}(C_{11} - C_{23})\overline{\gamma}_{23} \\
\overline{\sigma}_{31} &= \overline{\tau}_{31} = 2C_{44}\overline{\varepsilon}_{31} = C_{66}\overline{\gamma}_{31} \\
\overline{\sigma}_{12} &= \overline{\tau}_{12} = 2C_{44}\overline{\varepsilon}_{12} = C_{66}\overline{\gamma}_{12}
\end{aligned} \qquad (202)$$

From the last part of Equation (202), we can obtain

$$C_{44} = \frac{\overline{\sigma}_{12}}{2\overline{\varepsilon}_{12}} = \frac{\overline{\tau}_{12}}{\overline{\gamma}_{12}} = G_{12} = G_{13} = G_1 \qquad (203)$$

Equation (202) yields

$$\overline{\sigma}_{22} - \overline{\sigma}_{33} = (C_{22} - C_{33})(\overline{\varepsilon}_{22} - \overline{\varepsilon}_{33}) \qquad (204)$$

The law of mixture offers us

$$E_{11} = E_f v_f + E_m v_m \qquad (205)$$

Using Equation (205), one can replace the redundant fourth relation from Equation (201) with

$$E_{11} = C_{11} - \frac{2C_{12}^2}{C_{22} + C_{23}} \qquad (206)$$

The addition of the second and third equations in Equation (202) yields

$$\overline{\sigma}_{22} + \overline{\sigma}_{33} - \frac{2C_{12}\overline{\varepsilon}_{11}}{\overline{\varepsilon}_{22} + \overline{\varepsilon}_{33}} = C_{22} + C_{23} \qquad (207)$$

From the first part of Equation (202), one can show that

$$C_{11} = \frac{\overline{\sigma}_{11} - C_{12}(\overline{\varepsilon}_{22} + \overline{\varepsilon}_{33})}{\overline{\varepsilon}_{11}} \tag{208}$$

The substitution of Equation (208) into E_{11} yields

$$E_{11} = \frac{\overline{\sigma}_{11}}{\overline{\varepsilon}_{11}} + C_{12}\frac{\overline{\varepsilon}_{22} + \overline{\varepsilon}_{33}}{\overline{\varepsilon}_{11}} - \frac{2C_{12}^2(\overline{\varepsilon}_{22} + \overline{\varepsilon}_{33})}{(\overline{\sigma}_{22} + \overline{\sigma}_{33} - 2C_{12}\overline{\varepsilon}_{11})} \tag{209}$$

from which it is possible to compute C_{12}.

Figures 8 and 9 present two creep curves for a composite carbon/epoxy at two different temperatures [56,57].

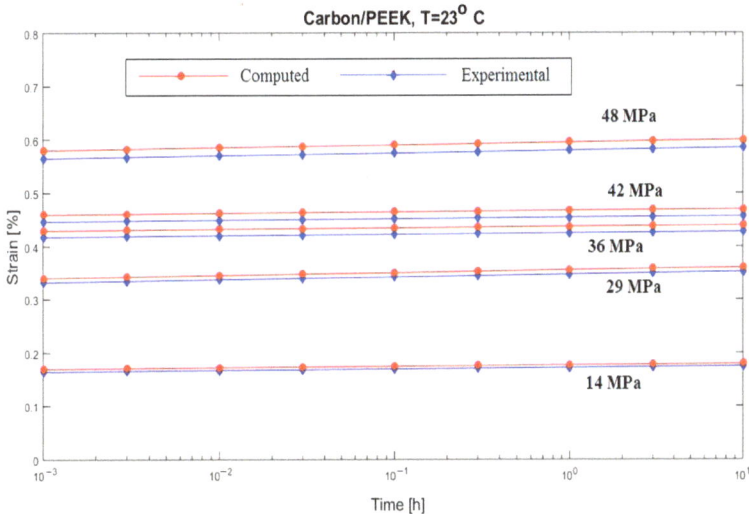

Figure 8. Creep response of a carbon/epoxy {90}$_{4s}$ at T = 23 °C.

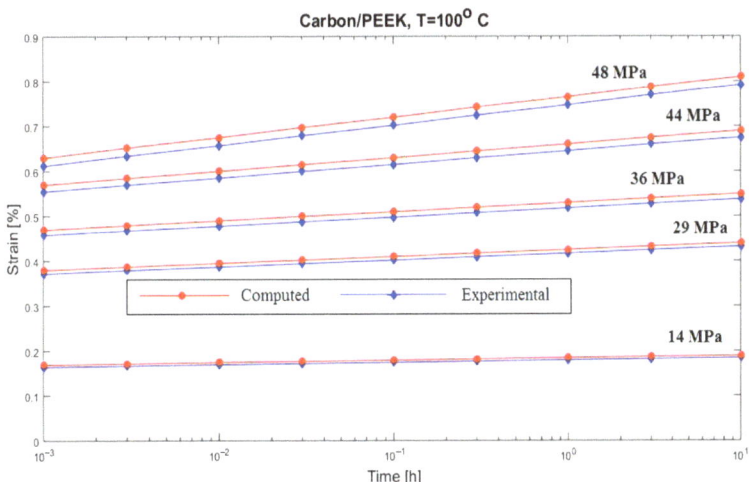

Figure 9. Creep response of a carbon/epoxy {90}$_{4s}$ at T = 100 °C.

6. Conclusions

The method presented in this paper proves to be a calculus method suitable for obtaining the general mechanical constants of a multiphase composite material. The material constants required by designers are obtained using the average of the values obtained by applying FEM. The results obtained experimentally verified the models proposed by different researchers. The tests and measurements conducted here show a good concordance between the results obtained using the proposed models and the experimental verifications. Thus, FEM proves to be a powerful tool for determining the engineering constants of composite materials. Compared to the methods described in the other sections, this method proves to be a useful and relatively simple means of identifying the constitutive laws. The results were also applied to a study of the creep behavior of a composite material. This case is more complicated because, in the case of creep phenomena, the influences of temperature prove to be nonlinear. All the presented models can replace expensive methods of determining the engineering constants of a viscoelastic material by experimental measurements with calculation-based methods.

This review focuses on the behavior of unidirectional fiber composites.

Author Contributions: Conceptualization, M.K., M.L.S. and M.M; methodology, M.K., M.L.S. and M.M.; software, M.K., M.L.S., M.M. and S.V.; validation, M.K., M.L.S., M.M. and S.V.; formal analysis, M.K., M.L.S., M.M. and S.V.; investigation, M.K., M.L.S., M.M. and S.V.; resources, M.K., M.L.S., M.M. and S.V.; data curation, M.K., M.L.S., M.M. and S.V.; writing—original draft preparation M.L.S.; writing—review and editing, M.K., M.L.S., M.M. and S.V.; visualization, M.K., M.L.S., M.M. and S.V.; supervision, M.L.S., M.M. and S.V.; project administration, M.K., M.L.S., M.M. and S.V.; funding acquisition, M.K., M.L.S., M.M. and S.V. All authors have read and agreed to the published version of the manuscript.

Funding: This research received no external funding. The APC was funded by the Transilvania University of Brasov.

Institutional Review Board Statement: Not applicable.

Informed Consent Statement: Not applicable.

Data Availability Statement: Not applicable.

Conflicts of Interest: The authors declare no conflict of interest.

Appendix A

Eshelby's tensor for an elliptic cylinder:

$$\begin{aligned} P_{2222} &= \frac{1}{2(1-v_m)} \left[\frac{1+2\alpha}{(1+\alpha)^2} + \frac{1-2v_m}{1+\alpha} \right] \\ P_{3333} &= \frac{\alpha}{2(1-v_m)} \left[\frac{\alpha+2}{(1+\alpha)^2} + \frac{1-2v_m}{1+\alpha} \right] \\ P_{2211} &= \frac{v_m}{1-v_m} \frac{1}{1+\alpha} \\ P_{2233} &= \frac{1}{2(1-v_m)} \left[\frac{1}{(1+\alpha)^2} + \frac{1-2v_m}{1+\alpha} \right] \\ P_{3311} &= \frac{v_m}{1-v_m} \frac{\alpha}{1+\alpha} \\ P_{3322} &= \frac{\alpha}{2(1-v_m)} \left[\frac{\alpha}{(1+\alpha)^2} + \frac{1-2v_m}{1+\alpha} \right] \\ P_{1212} &= \frac{1}{2(1+\alpha)} \\ P_{1313} &= \frac{\alpha}{2(1+\alpha)} \\ P_{2323} &= \frac{1}{4(1-v_m)} \left[\frac{1+\alpha^2}{(1+\alpha)^2} + (1-2v_m) \right] \end{aligned} \quad \text{(A1)}$$

and

$$P_{1111} = 0; \ P_{1122} = 0; \ P_{1133} = 0 \quad \text{(A2)}$$

where all other $P_{ijkl} = 0$. Relation (9) presents the link between ε_{ij}^0 and ε_{ij}^*. It must use the following relations (see Reference [25]):

$$\begin{bmatrix} M_1 & M_2 & M_3 \\ M_4 & M_5 & M_6 \\ M_7 & M_8 & M_9 \end{bmatrix} \begin{Bmatrix} \varepsilon_{11}^* \\ \varepsilon_{22}^* \\ \varepsilon_{33}^* \end{Bmatrix} + \begin{bmatrix} N_1 & 1 & 1 \\ 1 & N_2 & 1 \\ 1 & 1 & N_3 \end{bmatrix} \begin{Bmatrix} \varepsilon_{11}^0 \\ \varepsilon_{22}^0 \\ \varepsilon_{33}^0 \end{Bmatrix} = 0 \qquad (A3)$$

where

$$\begin{aligned}
M_1 &= v_f N_1 + N_2 + v_m(P_{2211} + P_{3311}); \\
M_2 &= v_f + N_3 + v_m(P_{2222} + P_{3322}); \\
M_3 &= v_f + N_3 + v_m(P_{2233} + P_{3333}); \\
M_4 &= v_f + N_3 + v_m(P_{2211} + P_{3311}); \\
M_5 &= v_f N_1 + N_2 + v_m(N_1 P_{2222} + P_{3322}); \\
M_6 &= v_f + N_3 + v_m(P_{2233} + P_{3333}); \\
M_7 &= v_f + N_3 + v_m(N_1 P_{3311} + P_{2211}); \\
M_8 &= v_f + N_3 + v_m(N_1 P_{3322} + P_{2222}); \\
M_9 &= v_f N_1 + N_2 + v_m(P_{3333} + P_{2233})
\end{aligned} \qquad (A4)$$

and

$$\begin{aligned}
N_1 &= 1 + 2\big(G_f - G_m\big)/\big(\lambda_f - \lambda_m\big); \\
N_2 &= (\lambda_m + 2G_m)/\big(\lambda_f - \lambda_m\big); \\
N_3 &= \lambda_m/\big(\lambda_f - \lambda_m\big)
\end{aligned} \qquad (A5)$$

where λ_f and λ_m are the Lamé constants (for the fiber and the matrix).

From (A3), we obtain the following:

$$\begin{Bmatrix} \varepsilon_{11}^* \\ \varepsilon_{22}^* \\ \varepsilon_{33}^* \end{Bmatrix} = \frac{1}{A} \begin{bmatrix} A_{11} & A_{12} & A_{13} \\ A_{21} & A_{22} & A_{23} \\ A_{31} & A_{32} & A_{33} \end{bmatrix} \begin{Bmatrix} \varepsilon_{11}^0 \\ \varepsilon_{22}^0 \\ \varepsilon_{33}^0 \end{Bmatrix} \qquad (A6)$$

$$\begin{aligned}
A_{11} &= A \cdot a_{11} = N_1(M_6 M_8 - M_5 M_9) + M_3(M_5 - M_8) + M_2(M_9 - M_6); \\
A_{12} &= A \cdot a_{12} = N_1(M_2 M_9 - M_3 M_8) + M_6(M_8 - M_2) + M_5(M_3 - M_9); \\
A_{13} &= A \cdot a_{13} = N_1(M_3 M_5 - M_2 M_6) + M_8(M_6 - M_3) + M_9(M_2 - M_5); \\
A_{21} &= A \cdot a_{21} = N_1(M_4 M_9 - M_6 M_7) + M_1(M_6 - M_9) + M_3(M_7 - M_4); \\
A_{22} &= A \cdot a_{22} = N_1(M_3 M_7 - M_1 M_9) + M_4(M_9 - M_3) + M_6(M_1 - M_7); \\
A_{23} &= A \cdot a_{23} = N_1(M_1 M_6 - M_3 M_4) + M_9(M_4 - M_1) + M_7(M_3 - M_6); \\
A_{31} &= A \cdot a_{31} = N_1(M_5 M_7 - M_4 M_8) + M_2(M_4 - M_7) + M_1(M_8 - M_5); \\
A_{32} &= A \cdot a_{32} = N_1(M_1 M_8 - M_2 M_7) + M_5(M_7 - M_1) + M_4(M_2 - M_8); \\
A_{33} &= A \cdot a_{33} = N_1(M_2 M_4 - M_1 M_5) + M_7(M_5 - M_2) + M_8(M_1 - M_4)
\end{aligned} \qquad (A7)$$

References

1. Cristescu, N.; Craciun, E.-M.; Gaunaurd, G. Mechanics of Elastic Composites (CRC Series in Modern Mechanics and Mathematics). *Appl. Mech. Rev.* **2004**, *57*, B27. [CrossRef]
2. Katouzian, M. On the Effect of Temperature on Creep Behavior of Neat and Carbon Fiber Reinforced PEEK and Epoxy—A Micromechanical Approach. Ph.D. Thesis, University of München, München, Germany, 1994.
3. Garajeu, M. Contribution à L'étude du Comportement non Lineaire de Milieu Poreaux Avec ou Sans Renfort. Ph.D. Thesis, Aix-Marseille University, Marseille, France, 1995.
4. Brauner, C.; Herrmann, A.S.; Niemeier, P.M.; Schubert, K. Analysis of the non-linear load and temperature-dependent creep behaviour of thermoplastic composite materials. *J. Thermoplast. Compos. Mater.* **2016**, *30*, 302–317. [CrossRef]
5. Fett, T. Review on Creep-Behavior of Simple Structures. *Res. Mech.* **1988**, *24*, 359–375.
6. Sá, M.F.; Gomes, A.; Correia, J.; Silvestre, N. Creep behavior of pultruded GFRP elements—Part 1: Literature review and experimental study. *Compos. Struct.* **2011**, *93*, 2450–2459. [CrossRef]
7. Brinson, H.F.; Morris, D.H.; Yeow, Y.I. A New Method for the Accelerated Characterization of Composite Materials. In Proceedings of the Sixth International Conference on Experimental Stress Analysis, Munich, Germany, 18–22 September 1978.
8. Xu, J.; Wang, H.; Yang, X.; Han, L.; Zhou, C. Application of TTSP to non-linear deformation in composite propellant. *Emerg. Mater. Res.* **2018**, *7*, 19–24. [CrossRef]

9. Nakano, T. Applicability condition of time–temperature superposition principle (TTSP) to a multi-phase system. *Mech. Time-Depend. Mater.* **2012**, *17*, 439–447. [CrossRef]
10. Achereiner, F.; Engelsing, K.; Bastian, M. Accelerated Measurement of the Long-Term Creep Behaviour of Plastics. *Superconductivity* **2017**, *247*, 389–402.
11. Schaffer, B.G.; Adams, D.F. *Nonlinear Viscoelastic Behavior of a Composite Material Using a Finite Element Micromechanical Analysis*; Dept. Report UWME-DR-001-101-1, Dep. Of Mech. Eng.; University of Wyoming: Laramie, WY, USA, 1980.
12. Schapery, R. Nonlinear viscoelastic solids. *Int. J. Solids Struct.* **2000**, *37*, 359–366. [CrossRef]
13. Violette, M.G.; Schapery, R. Time-Dependent Compressive Strength of Unidirectional Viscoelastic Composite Materials. *Mech. Time-Depend. Mater.* **2002**, *6*, 133–145. [CrossRef]
14. Hinterhoelzl, R.; Schapery, R. FEM Implementation of a Three-Dimensional Viscoelastic Constitutive Model for Particulate Composites with Damage Growth. *Mech. Time-Depend. Mater.* **2004**, *8*, 65–94. [CrossRef]
15. Mohan, R.; Adams, D.F. Nonlinear creep-recovery response of a polymer matrix and its composites. *Exp. Mech.* **1985**, *25*, 262–271. [CrossRef]
16. Findley, W.N.; Adams, C.H.; Worley, W.J. The Effect of Temperature on the Creep of Two Laminated Plastics as Interpreted by the Hyperbolic Sine Law and Activation Energy Theory. In Proceedings of the Proceedings-American Society for Testing and Materials, Conshohocken, PA, USA, 1 January 1948; Volume 48, pp. 1217–1239.
17. Findley, W.N.; Khosla, G. Application of the Superposition Principle and Theories of Mechanical Equation of State, Strain, and Time Hardening to Creep of Plastics under Changing Loads. *J. Appl. Phys.* **1955**, *26*, 821. [CrossRef]
18. Findley, W.N.; Peterson, D.B. Prediction of Long-Time Creep with Ten-Year Creep Data on Four Plastics Laminates. In Proceedings of the American Society for Testing and Materials, Sixty-First (61th) Annual Meeting, Boston, MA, USA, 26–27 June 1958; Volume 58.
19. Dillard, D.A.; Brinson, H.F. A Nonlinear Viscoelastic Characterization of Graphite Epoxy Composites. In Proceedings of the 1982 Joint Conference on Experimental Mechanics, Oahu, HI, USA, 23–28 May 1982.
20. Dillard, D.A.; Morris, D.H.; Brinson, H.F. Creep and Creep Rupture of Laminated Hraphite/Epoxy Composites. Ph.D. Thesis, Virginia Polytechnic Institute and State University, Blacksburg, VA, USA, 30 September 1980.
21. Charentenary, F.X.; Zaidi, M.A. Creep Behavior of Carbon-Epoxy (+/-45o)2s Laminates. In *Progess in Sciences and Composites*; ICCM-IV; Hayashi, K., Umekawa, S., Eds.; The Japan Society for Composite Materials: Tokyo, Japan, 1982.
22. Walrath, D.E. Viscoelastic response of a unidirectional composite containing two viscoelastic constituents. *Exp. Mech.* **1991**, *31*, 111–117. [CrossRef]
23. Hashin, Z. On Elastic Behavior of Fibre Reinforced Materials of Arbitrary Transverse Phase Geometry. *J. Mech. Phys. Solids* **1965**, *13*, 119–134. [CrossRef]
24. Hashin, Z.; Shtrikman, S. On some variational principles in anisotropic and nonhomogeneous elasticity. *J. Mech. Phys. Solids* **1962**, *10*, 335–342. [CrossRef]
25. Hashin, Z.; Shtrikman, S. A Variational Approach to the Theory of the Elastic Behavior of Multiphase Materials. *J. Mech. Phyds. Solids* **1963**, *11*, 127–140. [CrossRef]
26. Hashin, Z.; Rosen, B.W. The Elastic Moduli of Fiber-Reinforced Materials. *J. Appl. Mech.* **1964**, *31*, 223–232. [CrossRef]
27. Bowles, D.E.; Griffin, O.H., Jr. Micromecjanics Analysis of Space Simulated Thermal Stresses in Composites. Part I: Theory and Unidirectional Laminates. *J. Reinf. Plast. Compos.* **1991**, *10*, 504–521. [CrossRef]
28. Zhao, Y.H.; Weng, G.J. Effective Elastic Moduli of Ribbon-Reinforced Composites. *J. Appl. Mech.* **1990**, *57*, 158–167. [CrossRef]
29. Hill, R. Theory of Mechanical Properties of Fiber-strengthened Materials: I Elastic Behavior. *J. Mech. Phys. Solids* **1964**, *12*, 199–212. [CrossRef]
30. Hill, R. Theory of Mechanical Properties of Fiber-strengthened Materials: II Inelastic Behavior. *J. Mech. Phys. Solids* **1964**, *12*, 213–218. [CrossRef]
31. Hill, R. Theory of Mechanical Properties of Fiber-strengthened Materials: III Self-Consistent Model. *J. Mech. Phys. Solids* **1965**, *13*, 189–198. [CrossRef]
32. Hill, R. Continuum Micro-Mechanics of Elastoplastic Polycrystals. *J. Mech. Phys. Solids* **1965**, *13*, 89–101. [CrossRef]
33. Weng, Y.M.; Wang, G.J. The Influence of Inclusion Shape on the Overall Viscoelastic Behavior of Compoisites. *J. Appl. Mech.* **1992**, *59*, 510–518. [CrossRef]
34. Mori, T.; Tanaka, K. Average Stress in the Matrix and Average Elastic Energy of Materials with Misfitting Inclusions. *Acta Metal.* **1973**, *21*, 571–574. [CrossRef]
35. Pasricha, A.; Van Duster, P.; Tuttle, M.E.; Emery, A.F. The Nonlinear Viscoelastic/Viscoplastic Behavior of IM6/5260 Graphite/Bismaleimide. In Proceedings of the VII International Congress on Experimental Mechanics, Las Vegas, NV, USA, 8–11 June 1992.
36. Aboudi, J. Micromechanical characterization of the non-linear viscoelastic behavior of resin matrix composites. *Compos. Sci. Technol.* **1990**, *38*, 371–386. [CrossRef]
37. Aboudi, J. *Mechanics of Composite Materials—A Unified Micromechanical Approach*; Elsevier: Amsterdam, The Netherlands, 1991.
38. Abd-Elaziz, E.M.; Marin, M.; Othman, M.I. On the Effect of Thomson and Initial Stress in a Thermo-Porous Elastic Solid under G-N Electromagnetic Theory. *Symmetry* **2019**, *11*, 413. [CrossRef]

39. Abbas, I.A.; Marin, M. Analytical solution of thermoelastic interaction in a half-space by pulsed laser heating. *Phys. E Low-Dimens. Syst. Nanostruct.* **2017**, *87*, 254–260. [CrossRef]
40. Vlase, S.; Teodorescu-Draghicescu, H.; Motoc, D.L.; Scutaru, M.L.; Serbina, L.; Calin, M.R. Behavior of Multiphase Fiber-Reinforced Polymers Under Short Time Cyclic Loading. *Optoelectron. Adv. Mater. Rapid Commun.* **2011**, *5*, 419–423.
41. Teodorescu-Draghicescu, H.; Stanciu, A.; Vlase, S.; Scutaru, L.; Calin, M.R.; Serbina, L. Finite Element Method Analysis of Some Fibre-Reinforced Composite Laminates. *Optoelectron. Adv. Mater. Rapid Commun.* **2011**, *5*, 782–785.
42. Stanciu, A.; Teodorescu-Drăghicescu, H.; Vlase, S.; Scutaru, M.L.; Călin, M.R. Mechanical behavior of CSM450 and RT800 laminates subjected to four-point bend tests. *Optoelectron. Adv. Mater. Rapid Commun.* **2012**, *6*, 495–497.
43. Niculiță, C.; Vlase, S.; Bencze, A.; Mihălcică, M.; Calin, M.R.; Serbina, L. Optimum stacking in a multi-ply laminate used for the skin of adaptive wings. *Optoelectron. Adv. Mater. Rapid Commun.* **2011**, *5*, 1233–1236.
44. Katouzian, M.; Vlase, S.; Calin, M.R. Experimental procedures to determine the viscoelastic parameters of laminated composites. *J. Optoelectron. Adv. Mater.* **2011**, *13*, 1185–1188.
45. Teodorescu-Draghicescu, H.; Vlase, S.; Stanciu, M.D.; Curtu, I.; Mihalcica, M. Advanced Pultruded Glass Fibers-Reinforced Isophtalic Polyester Resin. *Mater. Plast.* **2015**, *52*, 62–64.
46. Fliegener, S.; Hohe, J. An anisotropic creep model for continuously and discontinuously fiber reinforced thermoplastics. *Compos. Sci. Technol.* **2020**, *194*, 108168. [CrossRef]
47. Xu, B.; Xu, W.; Guo, F. Creep behavior due to interface diffusion in unidirectional fiber-reinforced metal matrix composites under general loading conditions: A micromechanics analysis. *Acta Mech.* **2020**, *231*, 1321–1335. [CrossRef]
48. Lal, H.M.M.; Xian, G.-J.; Thomas, S.; Zhang, L.; Zhang, Z.; Wang, H. Experimental Study on the Flexural Creep Behaviors of Pultruded Unidirectional Carbon/Glass Fiber-Reinforced Hybrid Bars. *Materials* **2020**, *13*, 976. [CrossRef]
49. Wang, Z.; Smith, D.E. Numerical analysis on viscoelastic creep responses of aligned short fiber reinforced composites. *Compos. Struct.* **2019**, *229*, 111394. [CrossRef]
50. Fattahi, A.M.; Mondali, M. Theoretical study of stress transfer in platelet reinforced composites. *J. Theor. Appl. Mech.* **2014**, *52*, 3–14.
51. Fattahi, A.M.; Moaddab, E.; Bibishahrbanoei, N. Thermo-mechanical stress analysis in platelet reinforced composites with bonded and debonded platelet end. *J. Mech. Sci. Technol.* **2015**, *29*, 2067–2072. [CrossRef]
52. Tebeta, R.T.; Fattahi, A.M.; Ahmed, N.A. Experimental and numerical study on HDPE/SWCNT nanocomposite elastic properties considering the processing techniques effect. *Microsyst. Technol.* **2021**, *26*, 2423–2441. [CrossRef]
53. Selmi, A.; Friebel, C.; Doghri, I.; Hassis, H. Prediction of the elastic properties of single walled carbon nanotube reinforced polymers: A comparative study of several micromechanical models. *Compos. Sci. Technol.* **2007**, *67*, 2071–2084. [CrossRef]
54. Selmi, A. Void Effect on Carbon Fiber Epoxy Composites. In Proceedings of the 2nd International Conference on Emerging Trends in Engineering and Technology, London, UK, 30–31 May 2014.
55. Katouzian, M.; Vlase, S.; Scutaru, M.L. A Mixed Iteration Method to Determine the Linear Material Parameters in the Study of Creep Behavior of the Composites. *Polymers* **2021**, *13*, 2907. [CrossRef]
56. Katouzian, M.; Vlase, S.; Scutaru, M.L. Finite Element Method-Based Simulation Creep Behavior of Viscoelastic Carbon-Fiber Composite. *Polymers* **2021**, *13*, 1017. [CrossRef] [PubMed]
57. Katouzian, M.; Vlase, S. Creep Response of Carbon-Fiber-Reinforced Composite Using Homogenization Method. *Polymers* **2021**, *13*, 867. [CrossRef] [PubMed]
58. Katouzian, M.; Vlase, S. Mori–Tanaka Formalism-Based Method Used to Estimate the Viscoelastic Parameters of Laminated Composites. *Polymers* **2020**, *12*, 2481. [CrossRef]
59. Katouzian, M.; Vlase, S. Creep Response of Neat and Carbon-Fiber-Reinforced PEEK and Epoxy Determined Using a Micromechanical Model. *Symmetry* **2020**, *12*, 1680. [CrossRef]
60. Teodorescu-Draghicescu, H.; Vlase, S.; Scutaru, L.; Serbina, L.; Calin, M.R. Hysteresis effect in a three-phase polymer matrix composite subjected to static cyclic loadings. *Optoelectron. Adv. Mater. Rapid Commun.* **2011**, *5*, 273–277.
61. Jain, A. Micro and mesomechanics of fibre reinforced composites using mean field homogenization formulations: A review. *Mater. Today Commun.* **2019**, *21*, 100552. [CrossRef]
62. Lee, J.; Choi, C.W.; Jin, J.W. Homogenization-based multiscale analysis for equivalent mechanical properties of nonwoven carbon-fiber fabric composites. *J. Mech. Sci. Technol.* **2019**, *33*, 4761–4770. [CrossRef]
63. Koley, S.; Mohite, P.M.; Upadhyay, C.S. Boundary layer effect at the edge of fibrous composites using homogenization theory. *Compos. Part B Eng.* **2019**, *173*, 106815. [CrossRef]
64. Xin, H.H.; Mosallam, A.; Liu, Y.Q. Mechanical characterization of a unidirectional pultruded composite lamina using micromechanics and numerical homogenization. *Constr. Build. Mater.* **2019**, *216*, 101–118. [CrossRef]
65. Chao, Y.; Zheng, K.G.; Ning, F.D. Mean-field homogenization of elasto-viscoplastic composites based on a new mapping-tangent linearization approach. *Sci. China-Technol. Sci.* **2019**, *62*, 736–746. [CrossRef]
66. Sokołowski, D.; Kamiński, M. Computational Homogenization of Anisotropic Carbon/RubberComposites with Stochastic Interface Defects. In *Carbon-Based Nanofillers and Their Rubber Nanocomposites*; Elsevier: Amsterdam, The Netherlands, 2019; Chapter 11; pp. 323–353.
67. Dellepiani, M.G.; Vega, C.R.; Pina, J.C. Numerical investigation on the creep response of concrete structures by means of a multi-scale strategy. *Constr. Build. Mater.* **2020**, *263*, 119867. [CrossRef]

68. Choo, J.; Semnani, S.J.; White, J.A. An anisotropic viscoplasticity model for shale based on layered microstructure homogenization. *Int. J. Numer. Anal. Methods Geomech.* **2021**, *45*, 502–520. [CrossRef]
69. Cruz-Gonzalez, O.L.; Rodriguez-Ramos, R.; Otero, J.A. On the effective behavior of viscoelastic composites in three dimensions. *Int. J. Eng. Sci.* **2020**, *157*, 103377. [CrossRef]
70. Chen, Y.; Yang, P.P.; Zhou, Y.X. A micromechanics-based constitutive model for linear viscoelastic particle-reinforced composites. *Mech. Mater.* **2020**, *140*, 103228. [CrossRef]
71. Kotha, S.; Ozturk, D.; Ghosh, S. Parametrically homogenized constitutive models (PHCMs) from micromechanical crystal plasticity FE simulations, part I: Sensitivity analysis and parameter identification for Titanium alloys. *Int. J. Plast.* **2019**, *120*, 296–319. [CrossRef]
72. Sanchez-Palencia, E. Homogenization method for the study of composite media. In *Asymptotic Analysis II Lecture Notes in Mathematics*; Verhulst, F., Ed.; Springer: Berlin/Heidelberg, Germany, 1983; Volume 985. [CrossRef]
73. Sanchez-Palencia, E. Non-homogeneous media and vibration theory. In *Lecture Notes in Physics*; Springer: Berlin/Heidelberg, Germany, 1980. [CrossRef]
74. Xu, W.; Nobutada, O. A Homogenization Theory for Time-Dependent Deformation of Composites with Periodic Internal Structures. *JSME Int. J. Ser. A Solid Mech. Mater. Eng.* **1998**, *41*, 309–317.
75. Duvaut, G. Homogénéisation des plaques à structure périodique en théorie non linéaire de Von Karman. *Lect. Notes Math.* **1977**, *665*, 56–69.
76. Caillerie, D. Homogénisation d'un corps élastique renforcé par des fibres minces de grande rigidité et réparties périodiquement. *Compt. Rend. Acad. Sci. Paris Sér.* **1981**, *292*, 477–480.
77. Bensoussan, A.; Lions, J.L.; Papanicolaou, G. *Asymptotic Analysis for Periodic Structures*; American Mathematical Soc.: Amsterdam, The Netherlands, 1978.
78. Khodadadian, A.; Noii, N.; Parvizi, M.; Abbaszadeh, M.; Wick, T.; Heitzinger, C. A Bayesian estimation method for variational phase-field fracture problems. *Comput. Mech.* **2020**, *66*, 827–849. [CrossRef] [PubMed]
79. Eshelby, J.D. The determination of the elastic field of an ellipsoidal inclusion, and related problems. *Proc. R. Soc. Lond. Ser. A Math. Phys. Sci.* **1957**, *241*, 376–396.
80. Lou, Y.C.; Schapery, R.A. Viscoelastic Characterization of a Nonlinear Fiber-Reinforced Plastic. *J. Compos. Mater.* **1971**, *5*, 208–234. [CrossRef]
81. Bowles, D.; Griffin, O.H. Micromechanics Analysis of Space Simulated Thermal Stresses in Composites. Part II: Multidirectional Laminates and Failure Predictions. *J. Reinf. Plast. Compos.* **1991**, *10*, 522–539. [CrossRef]
82. Adams, D.F.; Miller, A.K. Hygrothermal Microstresses in a Unidirectional Composite Exhibiting Inelastic Material Behavior. *J. Compos. Mater.* **1977**, *11*, 285–299. [CrossRef]
83. Wisnom, M.R. Factors Affecting the Transverse Tensile Strength of Unidirectional Continuous Silicon Carbide Fiber Reinforced 6061 Aluminum. *J. Compos. Mater.* **1990**, *24*, 707–726. [CrossRef]
84. Brinson, L.C.; Knauss, W.G. Finite Element Analysis of Multiphase Viscoelastic Solids. *J. Appl. Mech.* **1992**, *59*, 730–737. [CrossRef]
85. Hahn, H.G. *Methode der Finiten Elemente in der Festigkeitslehre*; Akademische Verlagsgesellschaft: Franfurt am Main, Germany, 1975.

Disclaimer/Publisher's Note: The statements, opinions and data contained in all publications are solely those of the individual author(s) and contributor(s) and not of MDPI and/or the editor(s). MDPI and/or the editor(s) disclaim responsibility for any injury to people or property resulting from any ideas, methods, instructions or products referred to in the content.

Review

Polyaryletherketone Based Blends: A Review

Adrian Korycki [1,2], Fabrice Carassus [1,2], Olivier Tramis [1], Christian Garnier [1], Toufik Djilali [2] and France Chabert [1,*]

[1] LGP-ENIT-INPT, Université de Toulouse, 47 Avenue d'Azereix, 65016 Tarbes, France; adrian.korycki@enit.fr (A.K.); fabrice.carassus@groupe-lauak.com (F.C.); olivier.tramis@outlook.com (O.T.); christian.garnier@enit.fr (C.G.)

[2] LAUAK Service Innovation, 8 Rue Louis Caddau, 65000 Tarbes, France; toufik.djilali@groupe-lauak.com

* Correspondence: france.chabert@enit.fr

Citation: Korycki, A.; Carassus, F.; Tramis, O.; Garnier, C.; Djilali, T.; Chabert, F. Polyaryletherketone Based Blends: A Review. *Polymers* **2023**, *15*, 3943. https://doi.org/10.3390/polym15193943

Academic Editors: Ana Pilipović, Phuong Nguyen-Tri and Mustafa Özcanli

Received: 29 July 2023
Revised: 19 September 2023
Accepted: 27 September 2023
Published: 29 September 2023

Copyright: © 2023 by the authors. Licensee MDPI, Basel, Switzerland. This article is an open access article distributed under the terms and conditions of the Creative Commons Attribution (CC BY) license (https:// creativecommons.org/licenses/by/ 4.0/).

Abstract: This review aims to report the status of the research on polyaryletherketone-based thermoplastic blends (PAEK). PAEK are high-performance copolymers able to replace metals in many applications including those related to the environmental and energy transition. PAEK lead to the extension of high-performance multifunctional materials to target embedded electronics, robotics, aerospace, medical devices and prostheses. Blending PAEK with other thermostable thermoplastic polymers is a viable option to obtain materials with new affordable properties. First, this study investigates the miscibility of each couple. Due to different types of interactions, PAEK-based thermoplastic blends go from fully miscible (with some polyetherimides) to immiscible (with polytetrafluoroethylene). Depending on the ether-to-ketone ratio of PAEK as well as the nature of the second component, a large range of crystalline structures and blend morphologies are reported. The PAEK-based thermoplastic blends are elaborated by melt-mixing or solution blending. Then, the effect of the composition and blending preparation on the mechanical properties are investigated. PAEK-based thermoplastic blends give rise to the possibility of tuning their properties to design novel materials. However, we demonstrate hereby that significant research effort is needed to overcome the lack of knowledge on the structure/morphology/property relationships for those types of high-performance thermoplastic blends.

Keywords: polymer blend; miscibility; thermal transition; crystallization; mechanical properties

1. Introduction

High-performance thermoplastic (HPT) polymers lead the extension of advanced tailored applications such as embedded electronics, robotics, aerospace, medical devices and prostheses. In addition, HPT research and growth are driven by their use in structural composite materials, in which thermoplastics are an option for the matrix, offering significant weight savings and time-to-market reduction compared to thermoset composites or incumbent steel. Other widespread applications of HPT are as membranes from the barrier, filtration, osmosis and as templates for chemical reactions, such as proton exchange membrane fuel cells.

Thermostable polymers are defined as materials with heat resistance in continuous uses at 200 °C or above and resistance to aging in a thermo-oxidative environment. Only a few thermoplastics reach these outstanding properties, their common feature being the presence of aromatic groups in their chemical backbone. Among them, polyaryletherketones have been developed since the 1980s and have demonstrated some of the longest lifespans when submitted to thermo-oxidative aging [1]. PAEK are semicrystalline polymers displaying a large range of melting temperatures (T_m) from 320 °C for polyetheretheretherketone (PEEEK) to 390 °C for polyetherketoneketone (PEKK), with their crystalline morphology, the kinetics of crystallization and properties. The most popular is polyetheretherketone (PEEK). PEEK has a glass transition temperature (T_g) of 143 °C and a melting temperature of 335 °C. Its maximum operating temperature is from 250 °C to 260 °C and its processing temperature is from 370 °C to 400 °C [2]. The maximum achievable crystallinity is 48% [3]. Due to the fast kinetics of crystallization, it was reported that the amorphous state of pure PEEK can be obtained only when cooled at very high cooling rates, nearly 1000 K·min^{-1} [4].

All PAEKs offer a compromise between thermal stability, mechanical properties, chemical resistance and durability. The enhancement of these features can lead to the development of cutting-edge applications [5]. Ideally, one material should fit all the targeted properties, namely "multi-functional" material. Blending two polymers may be the easiest option to design such new materials. Indeed, two or more thermoplastics, often chosen such as their properties complement one another, yield a new material with synergetic properties.

When blending thermoplastics, the miscibility of the two polymers is often required, especially in load-carrying applications, since no miscibility gives a weak interfacial adhesion, resulting in a poor stress transfer from one phase to the other, leading to early fracture. In some specific cases, blends of immiscible polymers may be used, such as in chemical environments or tribological applications. The first polymer has poor chemical resistance or wear but it serves as the mechanical structure and the added polymer acts as a barrier or coating, giving a blend with improved targeted properties [6].

Blending PAEK to another HPT is a possible way to obtain materials with new affordable properties. However, the incorporation of an amorphous polymer in a semicrystalline matrix could induce an increase in the amorphous phase, thus reducing the amount of the crystalline phase. In most cases, the amorphous phase is considered as a diluent for the crystalline phase. Thus, the more amorphous content, the more difficult the crystallization of the crystalline phase. It is then possible to affect the crystallization kinetics by blending an amorphous polymer with a semicrystalline one.

This review focuses on blends of polyaryletherketone-based thermoplastics with various high-performance thermoplastic polymers. Despite their uses and rising interest, no review article has addressed the topic of high-performance thermoplastics blends until now. This work reviews research works started over 30 years ago. It is important to point out that some of these studies are former. Scientific techniques and knowledge could have evolved since then. We have carefully examined some hypotheses put forward in these papers. More specifically, the main topics of this review will be the miscibility of PAEK with HPT, their effect on PAEK crystallization, thermal transitions (glass transition and melting) and the mechanical properties of the blends.

2. Presentation of Blends Components

In this section, the polymer blends are classified according to their miscibility and then the chemical structures and properties of PAEKs and other high-performance thermoplastics are briefly presented.

2.1. Classes of Blends

Many attempts have been constructed to classify the polymer blends: miscible, immiscible, compatible, incompatible, partially miscible, etc. Our goal is not to review terminology or to discuss which definition is best suited to each situation. We rather define the framework of this review in comprehensive terms, for which we will use the following terms to classify the PAEK/HPT blends: miscible, partially miscible and immiscible. Miscible blends behave in a single phase and exhibit a single glass transition temperature. Partially miscible blends exhibit miscibility to an extent, as a function of blend composition or process parameters. It is worth mentioning that, whatever the polymer added to PAEK, the miscibility depends on the molecular mass. However, this information is barely provided in the reviewed articles. Finally, the blends that do not fall into the previous categories are classified as immiscible blends. Table 1 shows the common PAEK/HPT blends found in the literature. It is noteworthy that PEI is the only HPT that forms a fully miscible blend with various PAEK in the melted state. However, PEI comes in different conformations, and some of them are not miscible with PAEK. This point will be developed in Section 6.

Table 1. Classification of the most common HPT blended with PAEK as a function of their miscibility and method of blending.

Blending Method	Miscible Blends	Partially Miscible Blends	Immiscible Blends
Melt-mixing	PAEK/*meta*-PEI	PAEK/PAI PAEK/PES PEKK/TPI PEK/TPI PAEK/LCP	PEEK/*para*-PEI PEEK/*ortho*-PEI PEEK/TPI PEEK/PBI PAEK/PTFE
Solution blending	SPEEK/PAI SPEEK/PEI SPEEK/PBI SPEEK/PES	PEKK/PBI PEEK/PES	

2.2. Presentation of PAEK

Polyaryletherketones are semicrystalline high-performance thermoplastics with strong molecular rigidity of their repeating units. They demonstrate high-temperature stability, chemical resistance and high mechanical strength over a wide temperature range. PEEK (Figure 1), PEKK (Figure 2), PEK (Figure 3) and others have similar crystalline structures of two-chain orthorhombic packing [7] but not the same ether/ketone content [8].

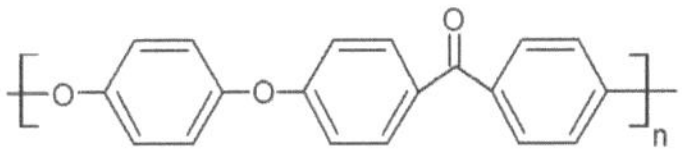

Figure 1. Structural formula of PEEK.

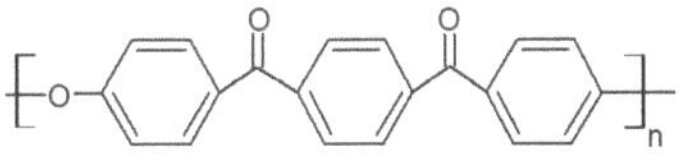

Figure 2. Chemical structure of PEKK.

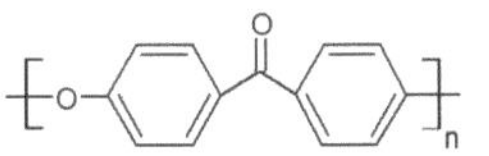

Figure 3. Scheme of PEK unit.

Polyetheretherketone combines the strength of 98 MPa and the stiffness of 125 MPa with a very good tensile fatigue of 97 MPa, thermal and chemical resistance (including the majority of organic solvents, oils and acids). Its mechanical properties remain stable up to temperatures of about 240 °C [9].

Polyetherketoneketone is a polymer with high heat resistance above 300 °C, chemical resistance and an ability to withstand high mechanical loads from 88 MPa to 112 MPa [10–12]. This polymer is synthesized in various formulations with individually unique properties. The PEKK formulations are expressed by the ratio of the percent of terephthaloyl (T) to isophthaloyl (I) moieties used during the synthesis that created the polymer. The T/I ratio affects the melting point ranging from 305 °C to 360 °C, the glass transition temperature from 160 °C to 165 °C and the crystallization kinetics [13].

Polyetherketone is characterized by very good material properties such as a tensile strength of 110 MPa, an elastic modulus of 4200 MPa and an elongation at break of 35%. It has a high resistance to abrasion and increased compression strength of 180 MPa at higher temperatures [14].

The properties of polyaryletherketones are due to the occurrence of phenylene rings linked via oxygen bridges (ether, R–O–R) and carbonyl groups (ketone, R-CO-R) in different configurations and proportions. The glass transition temperature, T_g, and the melting temperature, T_m, of the polymer depend on the ratio and sequence of ethers and ketones. They also affect its heat resistance and processing temperature. The lower the ratio of ether/ketone, the more rigid the polymer chain is and the higher the T_g and T_m as seen in Figure 4. As a consequence, the processing temperature ranges from 350 °C to 400 °C [15]. The main suppliers are Victrex, Arkema, Solvay, Evonik, Sabic and Gharda.

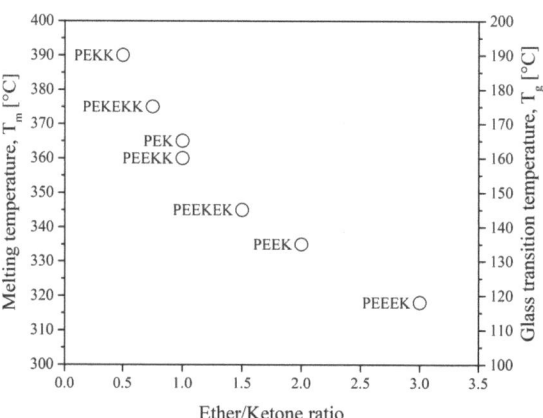

Figure 4. The melting temperature and glass transition temperature of PAEK as a function of ether/ketone ratio [16].

Another type of PAEK was launched recently on the market by Victrex, with a lower melting temperature while keeping the same glass transition, maned as LM-PAEK (LM: low melting). The typical trend for T_g and T_m as seen in Figure 4 is not valid anymore, due to the higher rigidity of its monomers. Furthermore, to increase the hydrophilicity of PEEK materials, charged groups are introduced using sulfuric acid into the polymer chains to make them ion-exchangeable. Sulfonation aids in the transport of cations and increases the hydrophilicity of the polymer. Sulfonated PEEK is presented in Figure 5.

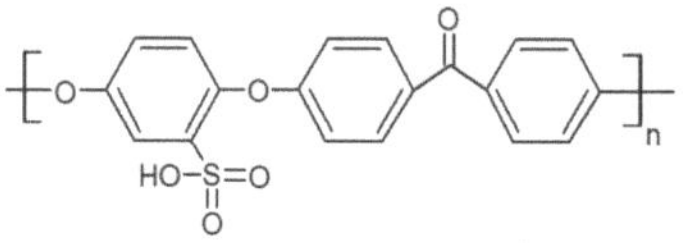

Figure 5. Sulfonated PEEK structure.

2.3. Presentation of HPT Usually Blended with PAEK

Polyetherimide is an amorphous polymer whose chemical structure of grade Ultem 1000 from Sabic is seen in Figure 6. It is renowned for being inherently flame retardant. Ultem 1000 resin in an unreinforced general-purpose grade offers a high strength of 105 MPa and a modulus of 3.2 GPa and a broad chemical resistance up to high temperatures of 170 °C while maintaining stable electrical properties over a wide range of frequencies [17,18]. The ketone groups in its backbone render it more flexible than polyimides, hence better processability. Similar to other amorphous thermoplastics, its mechanical strength decreases fast above its T_g at 215 °C. PAEK/PEI blends have been reported to be miscible, especially the binary blend PEEK/PEI. The latter is commonly used in PEEK composite parts [19–21], where the PEI is a joining agent, a method referred to as "Thermabond" [22]. Also, PEI is used as an energy director, meaning interfacial film, to assemble carbon fiber/PAEK composites through ultrasonic welding [23]. PEEK/PEI blends have also found application as tribological material [24], biomedical implants [25] and lightweight foam structures [26]. Since this blend has been reported to be miscible at all compositions in the amorphous state, it has attracted considerable attention to further the fundamental understanding of miscible blends [27–31].

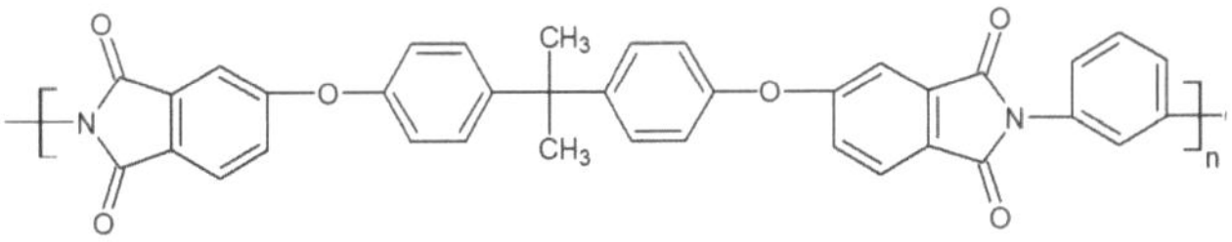

Figure 6. Schematic structure of PEI Ultem 1000.

Most studies with polyamideimides mention Torlon® 4000T from Solvay, whose chemical structure is presented in Figure 7, as well as other PAI Torlon®. Torlon® 4000T is the unfilled PAI powder mainly for adhesive applications [32]. Some authors have synthesized their PAI [33] with specific properties. Their mechanical, thermal and oxidative properties make them suitable for various applications thanks to a high glass transition temperature of around 275 °C [34].

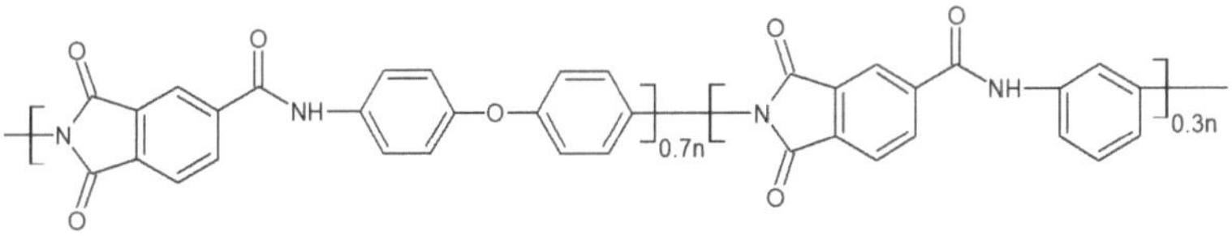

Figure 7. Chemical structure of the PAI Torlon® 4000T.

The chemical structure of polybenzimidazole is presented in Figure 8. Its rigid structure gives excellent thermal and mechanical properties, such as a melting point above 600 °C [35] and a glass transition of 435 °C [36]. It has the highest tensile strength among high-performance polymers, up to 145 MPa, and offers good chemical resistance [35,37–39]. However, it has a large water uptake of 15 wt.%, and the imidazole ring in the repeat unit may be subjected to hydrolysis, reducing its lifetime in applications such as fuel cells [40].

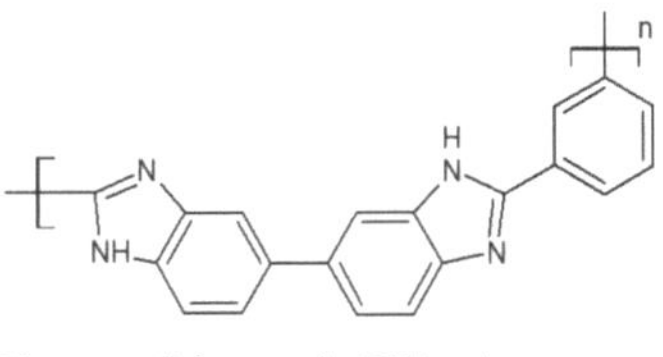

Figure 8. Scheme of a PBI unit.

Polyethersulfone, whose chemical structure is sketched in Figure 9, is an amorphous thermoplastic with a T_g between 190 °C and 230 °C, yielding high thermal stability. It also offers high mechanical rigidity and creep resistance, which make it a good candidate to blend with PAEK [41].

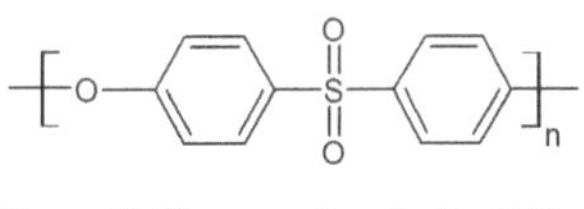

Figure 9. The repeat unit of a PES.

Thermoplastic polyimides, whose structure is depicted in Figure 10, display high mechanical properties of 128.7 MPa at tensile strength and 14.2% at elongation at break for TPI-4 [42]. They have one of the highest continuous operating temperatures for an unfiled thermoplastic of 240 °C [43]. Their T_g is at 250 °C. The main drawbacks of TPI are their high melt viscosity and low chemical resistance, making their processing challenging. Modification of the aromatic backbone by the inclusion of flexible functional bonds has enabled the synthesis of polyimide with a lower melt viscosity [44,45]. While TPIs are amorphous (Matrimid 5218, LaRC-TPI, Extem XH/UH), semicrystalline TPIs (with a crystallinity at 20%) were developed as an alternative, such as the so-called "New-TPI" by Mitsui Toatsu Chemical, Inc. (Tokyo, Japan) [46], later renamed "Aurum" or "Regulus" [47] depending on its end-use. Both amorphous and semicrystalline TPI suffer from low processability, even though the N-TPI has better overall mechanical properties and chemical resistance than TPI. Blending TPI or N-TPI with PAEK is an interesting way to improve the processability of the former, despite their immiscibility with PAEK.

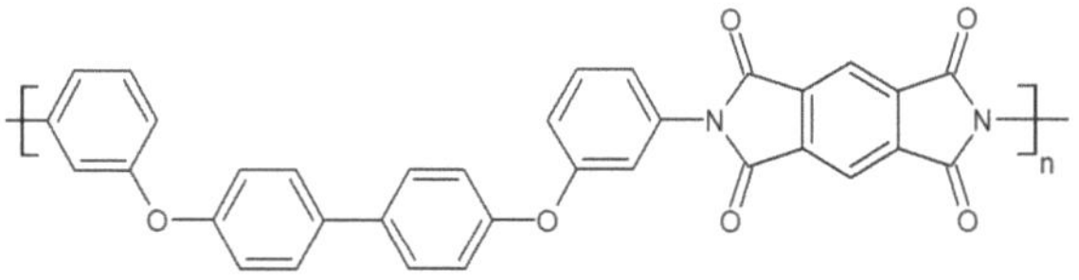

Figure 10. TPI repeat unit.

Polytetrafluoroethylene is a synthetic fluorocarbon with a high molecular weight compound consisting of carbon and fluorine. Their structure is shown in Figure 11. It is highly hydrophobic, biocompatible and widely used as a solid lubricant thanks to its low friction coefficient. Concerning engineering applications, PEEK is wear-resistant but may suffer from wear loss or high friction coefficient at elevated temperatures. Blending PEEK with PTFE is a way to improve the tribological performances of PAEK-based blends. But PTFE is extremely viscous in the melted state, which complicates its processability. Its glass transition temperature is around 114 °C and its melting point is 320 °C and it starts deteriorating above 260 °C, making it difficult to blend with PAEK.

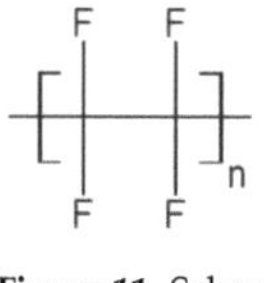

Figure 11. Schematic structure of the PTFE unit.

Liquid crystal polymers are characterized by the properties of liquid crystal, which can be those of conventional liquids or those of solid crystals. For example, a liquid crystal may flow like a liquid, but its molecules may be oriented in a crystal-like way. Those thermoplastics showed up on the market in the 1980s. Thermotropic LCP, shown in Figure 12, is melt processable, while lyotropic LCP with a higher chain rigidity and intermolecular bonding is spun from a solution such as sulfuric acid. During melt processing, the rigid molecules with units disrupt chain linearity and pack slightly to reduce the melting point. It has been reported that the blends with LCP often tend to be immiscible but have a range of useful properties. TLCP have been blended with many thermoplastics including PEEK [48].

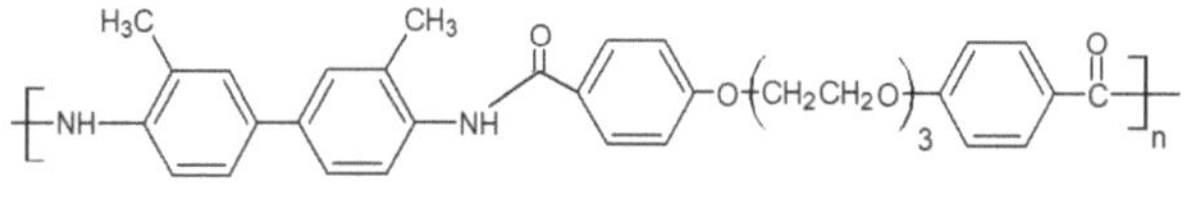

Figure 12. Repeat unit of TLCP.

3. Thermodynamics of Miscibility and Morphologies

Polymer–polymer theory of miscibility is based on the Flory–Huggins theory for a solution, for which, ideally, each part of the solute fills a space in the solution that it is soluble in, e.g., the so-called "lattice model" [49]. Polymers, and especially HPT, are at best semirigid due to their aromatic backbone. The Flory–Huggins theory still serves to describe the polymer–polymer mixture. The free energy of mixing depends on combinatorial entropy since the polymer of the mixture cannot fill all the space offered by the lattice due to its restricted degree of freedom. This effect is taken into account by considering the volume of the polymer, i.e., steric effects [50]. The free energy of mixing depends also on the interaction between like and unlike pairs of monomers, which is an enthalpic contribution. Usually, since macromolecules contain a large number of monomers, the repulsion between macromolecules is large [51]. The strength of this interaction, as the difference between like and unlike pairs, is defined as the interaction Flory–Huggins parameter, χ. The free energy of mixing is usually positive, i.e., unfavorable to mixing. However, strong specific interactions (dipole–dipole, ion–dipole, hydrogen bonding, acid-base or charge transfer) [52] may occur, and these will counterbalance the unfavorable entropic (steric) and enthalpic (van der Walls) interactions to give miscibility. Additionally, the free energy of mixing must have a positive curvature (in other words, increased differential growth at the center, for example, the surface of the sphere) at any concentration to give a miscible blend concerning the negative enthalpic term. Usually, this is accomplished by specific interactions, such as dipole–dipole and hydrogen bonding [53]. Such interactions are highly dependent on temperature; as a consequence, the miscibility depends on the temperature. This dependency is evidenced by miscibility diagrams depicting an upper critical solution temperature (UCST) or a lower critical solution temperature (LCST).

Partially miscible blends have a negative curvature (in other words, increased differential growth at the edges, for example, "saddle") for some concentrations on a miscibility diagram [54]. For example, PEEK/PES blends are partially miscible; their miscibility depends on the processing method. A probable cause, in view of the discussion above, would be that their respective conformation does not allow the macromolecules to interact attractively. However, sulfonated PEEK is fully miscible with PES, mainly due to the presence of sulfonate groups, which would specifically interact with the likes on the PES backbone. Another example would be PEEK/PBI blends, where steric repulsion would be the main reason for their immiscibility. Nevertheless, SPEEK is fully miscible with PBI, mainly due to the acid-base interaction between the basic, amide groups of the PBI and the acid, sulfonate groups of the SPEEK.

When functionalization is not possible, immiscible polymers may be compatibilized. The addition of a compatibilizer is the main route to obtain melt-mixed polymer blends [55,56]. The compatibilizers may be classified into two categories:

- Nanoparticles, which are partially or fully miscible with the polymers in the blend [57]. Usually, they are amphiphilic, "Janus" nanoparticles [58] aimed at the interface to reinforce it and promote the stress transfer between the two phases;
- Block copolymers: For instance, PET/PP blends are often compatibilized by PP-*b*-MA (MA: maleic anhydride) copolymers, where the *block* MA is miscible with PET. Gao et al. [59] compatibilized PEEK/PI blends with PEEK-*b*-PI block copolymers. In any case, the length of the blocks and the number of copolymers control the stability and the final morphology of the blends. It is worth noting that the choice of copolymers available to compatibilize PAEK with another HPT is very narrow.

Immiscible blends exhibit a two-phase morphology, where each phase is a rich phase of one component, whereas miscible blends should behave as a single-phase material. Compatibilization leads to fine dispersion of one of the phases into the other, depending on the composition. Partially miscible blends have a spectrum of morphologies depending on the composition and processing parameters, as in Figure 13. Of interest is obtaining a co-continuous structure [60]. Indeed, even if the blend has an immiscible state, a co-continuous morphology may give a blend with mechanical properties better than expected.

Thus, if blending two polymers leads to an immiscible blend, but they have complementary properties (for instance, thermal stability or solvent resistance), one may want to process them with a set that gives a co-continuous morphology to obtain enhanced properties. Such morphologies are expected when blending PAEK with other thermoplastics, likewise with conventional polymer blends.

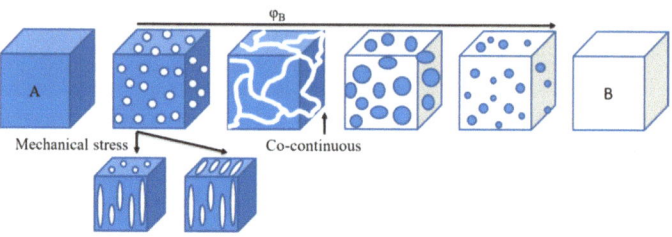

Figure 13. Basic types of phase structures in polymer blends.

Moreover, due to the crystallization of PAEK, the resulting morphology depends on the crystallinity generated in the blend. In the PAEK/HPT blends, the HPT are mostly amorphous, so the semicrystalline/amorphous blends may develop three main configurations [28,29,61,62] as shown in Figure 14:

- Interlamellar, where the semicrystalline polymer forms pure lamellas, alternating with an amorphous-rich phase (Figure 14A, left or Figure 14b, right);
- (Inter)spherulitic, where the spherulites develop in an amorphous-rich matrix (Figure 14B, left or Figure 14a, right);
- Interfibrillar or inter(lamellar-bundle), where the amorphous polymer may be trapped between bundles of the semicrystalline polymer (Figure 14C, left or Figure 14b, right).

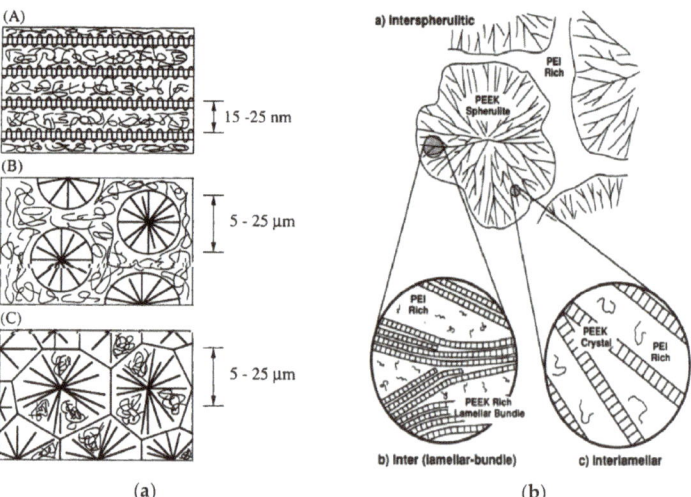

Figure 14. Schemes of the morphologies in the semicrystalline/amorphous blends. On the left (**a**), the spherolite sizes, with permission from Wiley [29] and right (**b**), macromolecular organization inside a spherulite, with permission from ACS Publications [28].

The competition between the diffusion of the amorphous component out of the crystal domains and the crystal growth is the controlling factor, while strong segmental interactions may hinder such competition. In the case of weak interactions, interlamellar morphologies (cases Figures 14A and 14C) may be favored if the amorphous component diffuses fast and the crystal growth is slow. Conversely, trapping of the amorphous component may occur, leading to spherulitic morphologies.

4. Evaluation of the Miscibility

Citing Olabis et al. [52], "The most commonly used method for establishing miscibility in polymer-polymer blends or partial phase mixing in such blends is through the determination of the glass transition (or transitions) in the blend versus those of the each separated constituent. A miscible polymer blend will exhibit a single glass transition between the T_g's of the components with a sharpness of the transition similar to that of the components. In cases of the borderline, miscibility broadening of the transitions will occur", supported by a citation from MacKnight et al. [63], "Perhaps the most unambiguous criterion of polymer compatibility is the detection of a single glass transition whose temperature is intermediate between those corresponding to the two-component polymers". Thus, the miscibility of polymer blends is most often checked by the measurement of the T_g of the blends. The blends of amorphous and semicrystalline polymers have a broader T_g than the pure components, attributed to a wider distribution of the amorphous domains. From thermodynamical considerations, such as those developed by Couchman [64], T_g was found to depend on the composition of the blend and usually increases with it. Several equations have been developed on those assumptions to describe the T_g dependency over the composition. The well-known Fox equation [65] (Equation (1), where w_i is the weight fraction) may be the most widely used for polymeric blends,

$$1/T_{g,mix} = \Sigma_i(w_i/T_{g,i}) \tag{1}$$

whereas more sophisticated equations, such as the Gordon–Taylor equation [66], have been developed based on the same theoretical basis, but with a different hypothesis where the equation (Equation (2)) is reduced by Wood [67] for a two-component system. Utracki and Jukes proposed an equation that follows T_g versus weight fraction dependencies for miscible blends and plasticized systems [68]. For partially miscible blends, they introduced an empirical parameter, K, describing the deviation from the assumed negligibility of entropy of mixing.

$$T_{g,mix} = (w_1 T_{g1} + K w_2 T_{g2})/(w_1 + K w_2) \tag{2}$$

More equations derived from those considerations may be found in the report of Aubin and Prud'homme [69].

The miscibility of the blends is often assessed by differential scanning calorimetry (DSC). This technique is quite accurate and it requires a small piece of samples. The blends are considered miscible if only one T_g appears on the DSC thermogram. In other cases, the immiscibility is proved by the presence of unchanged T_g of the pure components, in partially miscible blends, two T_g dependent on the composition appear [70]. Dynamic mechanical analysis (DMTA) may also be used to assess glass transitions (referred to as T_α) at a macroscopic scale. The shape at the T_α point in the loss factor (E'') signal may be an indicator of miscibility. In this case, a sharp T_α will be observed. A broader peak of the T_α would indicate phase separation to an extent, as the broad transition would encompass two distinct amorphous phases. If the two components display T_α that are close to each other, its measurement would yield a single T_α even if the polymers are not miscible. In this case, the sharpness of the E'' peak signal may help to distinguish miscible from immiscible blends. Another method of assessing miscibility would be the measure of the lower critical solution temperature (LCST). Its simplest measurement would be the detection of the cloud point, where phase separation would produce opacity in an otherwise transparent blend in the melt state. This method, however, may not be applicable for semicrystalline blends if crystallization occurs before crossing the LCST, i.e., for blends with a high percentage of a semicrystalline component.

Citing Wang and Cooper [71], "It is well known that compatible polymer blends are rare". Due to a lack of thermodynamic compatibility, only a few polymer blends are intimately miscible, i.e., miscible at a segmental level [72].

5. Elaboration of PAEK Blends

Two methods exist to obtain polymer blends: mixing the components of the blend in the melt state or by dissolution. In the latter case, both components are dissolved in a common solvent. PAEK, however, may suffer chemical modification if dissolved in strong acids, a mainly undesired functionalization. Therefore, controlled functionalization of PAEK is performed before dissolution. Both methods are explained in the following.

5.1. Melt-Mixing

Components of the blends are mechanically processed in the melted state, either by conventional mixing using an internal mixer or double-screw extrusion. Further, the so-obtained blends are processed through compression molding or injection molding to obtain specimens or final parts. Melt-mixing enables the independent control of various experimental parameters, such as melt temperature, pressure, residence time, shear rate and cooling rate. This method is appropriated for the industry as an environmentally friendly process because of the absence of the solvent. Additionally, melt-mixing may produce a finely dispersed phase [73] or co-continuous structure [74] depending on the parameters chosen. However, in the case of HPT blends, the residence time and melt temperature have to be carefully selected to prevent the degradation of one or both components upon processing.

During the polymer processing of blends, the viscosity mixing rules should be considered [75]. Some equations give the viscosity of the blend as a function of component viscosities, blend composition and shear rate. The flowing ability is associated with the viscosity of the molten polymers. The viscosity is highly influenced by the temperature and shear rate of molten polymers [76]. Molten polymers generally exhibit shear-thinning behavior with a Newtonian plateau at a low shear rate, standard polymer processing such as injection or extrusion involves shear rates above 1000 s^{-1}. Bakrani Balani et al. [77] reported a decrease in the viscosity from 11,000 Pa·s to 1200 Pa·s for PEEK at 0.03 rad·s^{-1} and 100 rad·s^{-1}, respectively, at 350 °C. Moreover, the viscosity of thermoplastics is highly influenced by the temperature: Bakrani Balani et al. [78] also reported a decrease in the viscosity from 11,000 Pa·s to 5300 Pa·s for PEEK from 350 °C to 400 °C.

The rheological behavior of PAEK and their blends have been barely published [79–81]. Molten PEEK demonstrates a classical behavior similar to conventional thermoplastics. The behavior of blends is more complicated; only a few articles analyze their rheological properties [19,82–85]. The blends do not reach a Newtonian plateau in the experimental frequency range, continuously increasing with decreasing frequency. This may be due to the morphological organization of immiscible domains [86]. Like any copolymer, it self-organizes into spheres, cylinders, lamellae or network-like morphology, depending on the relative polymer-block chain lengths, Figure 13. Such morphologies strongly impede the flow behavior and relaxation characteristics of the material.

Processing PAEK through melt-mixing requires knowledge and practice. Indeed, molten PAEK is subjected to thermo-oxidative degradation, resulting in viscosity increases for the highest residence times. As an example, the viscosity of PEEK slowly increases from 1600 Pa·s to 1750 Pa·s at 330 °C; at the higher temperature of 390 °C, the viscosity increases from 750 Pa·s to 900 Pa·s. It was observed that at 370 °C, the PEEK modulus increases after less than 4 min if under air atmosphere [87].

For these reasons, understanding the thermal decomposition mechanisms of aromatic polyketone, consisting of ketone and aromatic moieties, is essential. The kinetic parameters of the mechanisms involved during decompositions were first studied by Day et al. [88,89]. An isothermal weight-loss method in air and nitrogen atmosphere was investigated and the degradation was observed between 575 °C and 580 °C. Char yields of this polymer are above 40% [90]. The thermal degradation of PEEK occurs in a two-step decomposition. The first step is a random chain scission of the ether and ketone bonds [91]. Carbonyl bonds create more stable radical intermediates, which would be expected to predominate. The second step is due to the oxidation of the carbonaceous char formed. Oxidation of pure PEEK occurs at around 700 °C.

5.2. Solution Blending—Sulfonation of PAEK

One of the main advantages of PEEK, the chemical resistance, becomes one of its drawbacks when it comes to being dissolved. Indeed, its poor solubility is due to crystallinity and side reactions such as interpolymer crosslinking and degradation [92,93]. One of the methods to increase the solubility of PEEK is to modify its chemical structure to reduce its intrinsic crystallinity. PEEK is only soluble in strong solvents at room temperature and in other solvents (e.g., diphenyl sulfone) at temperatures approaching the melting temperature at 330 (°C). Chemical modifications (e.g., sulfonation) and protonation are the main causes of the solubility of PEEK in strong acids. Sulfonation is a method of choice since it reduces the degree of crystallinity of the PEEK, thus enhancing its solubility. Sulfonation substitutes some atoms of the phenyl rings with sulfonated groups and the position of the substituted carbon depends on the acid used [94]. In the case of sulfuric acid, Shibuya et al. [95] noted that sulfonation only takes place on the phenyl ring sandwiched by two ether groups. The packing and conformation of the backbone are thus modified by the introduction of sulfonate groups as the main cause of the loss of crystallinity. However, the T_g of the sulfonated PEEK is increased compared to the neat PEEK. Arigonda et al. [96] reported an increase in the T_g from 143 °C for PEEK to 216 °C after sulfonation. This is due to an ionomeric effect, where a polar ionic site can increase intermolecular association [97]. Since sulfonated PEEK is soluble in strong acids (dimethylacetamide, methanesulfonic, dimethylformamide) [97,98], it may be blended with various polymers, such as polyethersulfone, poly(4-vinylpyridine) or polybenzimidazole, as reported by Kerres et al. [99]. These miscible blends are prepared by the dissolution of both components in a common solvent.

However, the same blends prepared via melt-mixing are immiscible. This difference may be explained by other interactions since the backbone of PEEK is different from SPEEK, meaning different miscibility for a given polymer.

6. PAEK/PEI Blends Obtained by Melt-Mixing

This section is dedicated to the PAEK/PEI blends. Among all the blends reviewed, they are the only ones falling in the category "miscible in the melt state" and the wider number of articles available in the literature.

6.1. PAEK/PEI Blends in the Amorphous State

Polyetheretherketone/polyetherimide blends are the most studied among the PAEK/PEI blends since they are easily obtained by melt-mixing. They give fully miscible blends at all compositions in the amorphous state. Indeed, all reports reviewed measured a single T_g on quenched samples, either using DSC or DMTA [19,20,25,27–30,61,62,100–104], over the whole range of composition. Good agreement between the values of the measured T_g and Fox equation is found in those works, as illustrated in Figure 15.

When having a deeper look, PEI displays several conformations with a great influence on the miscibility with PAEK.

Nemoto et al. [31] showed that the conformation of the phenylenediamines in the PEI backbone greatly influences its miscibility with PEEK. They reported that PEI containing diamines in the meta conformation (i.e., the so-called "Ultem 1000" (*m*-PEI)) were fully miscible, while PEI containing para diamines were immiscible. According to Dingeman et al. [105], the inclusion of meta-linked aryl ether spacer results in a large increase in free volume compared to para linkages. Para-PEI (*p*-PEI) would have a linear macromolecular chain structure with all the junctions between repeat units aligned as if on a straight line. As a consequence, the only degree of freedom would be a rotation around this axis. Oppositely, m-PEI would offer a more hooked chain with an overall higher degree of conformation. Thus, the free energy of mixing would be lower for PEEK/m-PEI than PEEK/p-PEI to the point of giving miscible blends for the former and immiscible blends for the latter. Additionally, phenyl rings in the PEEK are in the para conformation. Thus, in the PEEK/p-PEI blend, like pair (p-phenyl/p-phenyl) interactions would increase and thus increase the enthalpic contribution, such as favoring PEI–PEI chain interaction over PEEK–PEI. On the other hand, the blend of PEEK/m-PEI would have an increase in unlike pair interactions, more favorable to mixing. Kong et al. [106] suggested that charge transfer between PEEK and PEI may be the specific interaction that would change the balance toward (im)miscibility. They reported a difference in miscibility between the two PEI conformations. It shall be noted that the PEI may also exist with ortho linkages [105], but such PEI was not investigated further. In the vast majority of scientific studies, m-PEI is the chosen conformation, commercially known as Ultem 1000. Overall, it may be supposed that the specific interactions responsible for its miscibility with PEEK would be the same for other PAEK. Indeed, PEK [27,29,107], PEEKK [108,109] and PEKK [27,29,30,107] have been reported to be fully miscible at all compositions in the amorphous state.

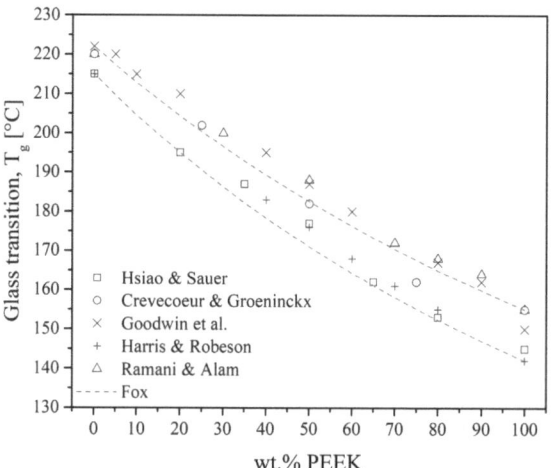

Figure 15. Glass transition temperature as a function of PEEK content in PEEK/PEI amorphous blends. Points represent experimental data from the literature [20,27,29,62,100] and solid lines represent the Fox trend using Equation (1).

6.2. Crystallization of PAEK/PEI Blends

Depending on the thermal conditions, the PAEK crystallizes in different morphologies and gives various spherulite sizes.

(1) When cooling PAEK/PEI blends from the melt, only PAEK partially crystallizes. The most significant effect of PEI on crystallization is that chains near the crystals in the amorphous phase are probably less mobile than those farther away, which increases the T_g of the amorphous phase in the blend [20,29,62]. Chen and Porter [61] recorded, 30 years ago, a slight negative variation from linearity when measuring the specific volume of the existence of PEI in the interlamellar zone of PEEK crystals, indicating favorable intermolecular interactions between PEEK and PEI. While amorphous blends agree with Fox law, crystalline blends deviate from it, with T_g superior to the predicted value. In the range of all blends, the broadening of T_g was observed up to 40 wt.% of PEI [61]. The broadening was attributed to a change in the amorphous phase distribution since, upon crystallization, the PEEK has to diffuse to the crystal domains, depleting the amorphous phase [29]. At the same time, the PEI has to diffuse

from the crystallizing PEEK to the amorphous phase, creating an enrichment of the amorphous phase in PEI. Thus, the broadening observed could be due to a broad distribution of amorphous PEI, which may comprise free chains (i.e., in the amorphous phase) and trapped chains (i.e., chains at the crystal-amorphous boundary). As already mentioned, blends of PEEK/PEI are reported to be amorphous upon preparation. However, the semicrystalline state is the most likely to be encountered for practical applications. Partial miscibility was revealed in the semicrystalline state by the presence of two distinct glass transitions between 40 wt.% and 90 wt.% of PEEK in the blend [110].

(2) Cold crystallization of PAEK blends was reported not to occur for PEI content from 75 wt.% [61,101], for which the PEI impacts and hinders the PEEK crystallization. Below this concentration, the PEI does not influence the PEEK crystallinity [20]. A decrease in the crystallization with increasing PEI content was measured [20,21,26,61,62,100,104,111,112], attributed to a rejection of PEI into the amorphous domains of PEEK [20]. The final degree of crystallinity of the PEEK reached in the blends did not change with the PEI content and in some cases slightly increased [20,21,26,62,101,111]. The PEEK crystallized similarly to that of pure PEEK, and the melting temperature was not affected by the PEI content.

(3) Isothermally crystallized PEEK/PEI blends exhibited a double melting behavior [62,111,113], seen by two melting peaks on DSC scans. The first peak corresponds to the melting of secondary PEEK crystals, while the second peak, occurring at a higher temperature, corresponded to primary PEEK crystals [29]. Both T_ms did not depend on the PEI content [26,27,29,62]. As the PEI content increases, the crystal growth of PEEK decreases [29,61], indicating a disruption of PEEK crystallization by the PEI matrix, which induced different nucleation mechanisms [29,111]. Similar effects were observed in PEKK/PEI blends. As PEKK crystallizes slower than PEEK, the PEI may, even more, reduce the nucleation site density, disrupting the PEKK crystallization [29]. The lamellar thickness in all blends was independent of the PEI content; however, it differs for PEKK/PEI blends with about 85 Å and it is higher for PEEK/PEI blends at 100 Å. Torre and Kenny [101] noted that for 50/50 PEEK/PEI blends, the PEI acted as a diluent for the PEEK crystallization, suggesting that PEEK crystallized as if in neat PEEK, since the PEI, in this case, may be completely expelled from the PEEK crystals. This means that blended PEEK and neat PEEK have a similar crystalline structure. However, they may have different growth rates. In another former study [62] on the isothermal crystallization temperature, it can be noticed that deviations observed with T_g of crystallized blends containing up to 50 wt.% of PEI were equal to pure PEI, which truly demonstrates the presence of almost 100% of PEI phase in the blend. Some thermograms are indeed incomplete, and the second peak could be assumed to be present on some broader DMTA signals.

Hsiao and Sauer [29] described the general trends for the morphology of PAEK/PEI blends. They supposed that an increase in PEI concentration and a decrease in the degree of undercooling both favor interspherulitic morphology. Good agreement with experimental data was found, as PEEK/PEI blends with a PEI content below 50 wt.% exhibited interlamellar bundle morphologies, revealed by transmission electron microscopy (TEM) by Hudson et al. [28] and by optical microscopy by Hsiao and Sauer [29], as presented in Figure 16. For PEI equal and above 50 wt.%, interspherulitic morphologies appeared, where the PEI was almost entirely excluded from the crystals [29,61,62]. Goodwin et al. [100] described those trends in terms of Avrami analysis. They found that the Avrami exponent shifted downward with an increase in the PEEK concentration, denoting a change from 3D to 2D crystal growth.

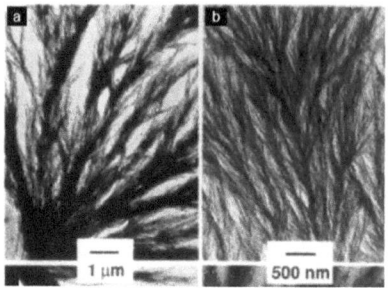
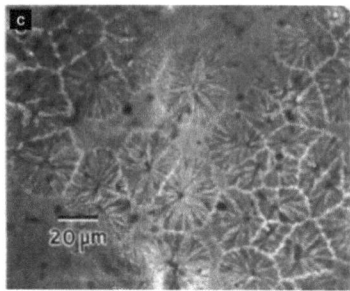

Figure 16. PEI/PEEK morphologies. Left: transmission electron micrographs of inter (lamellar-bundle) segregation, for 25/75 PEI/PEEK blends. Right: optical micrographs of 50/50 PEI/PEEK blends showing interspherulitic morphology, with permission from ACS Publications [28].

The PEEK/PEI blends have been successfully used for the preparation of PEEK hollow fiber membranes [114,115]. Ding and Bikson [114] removed the polyimide phase utilizing a primary amine reagent monoethanolamine (MEA). Porous PEEK membranes, from a 50/50 PEEK/PEI blend, were largely cylindrically shaped and interconnected, quite uniform with a pore diameter of around 10 nm. The degree of crystallinity was determined to be 34% based on the heat of fusion with T_m at 325 °C. Huang et al. [115] studied the influence of solvent-induced crystallization during extraction for PEEK/PEI blends with 60 wt.% of PEI. The average crystallite size for membranes before extraction was 5.7 nm with a degree of crystallinity of 39%, while a weak polar solvent dichloromethane and a composite extractant have caused an increase in the crystallite size to 6.2 nm and degree of crystallinity to 46.6%, which shows a strong ability to induce the crystallization of PEEK. A strong polar solvent N-methyl-2-pyrrolidone (NMP) did not change the crystallization parameters. In addition, a composite extractant (80 vol.% NMP, 10 vol.% ethanolamine, 10 vol.% water) had the most dominant distribution of small pores, indicating the strongest extraction ability for interlamellar or interfibrillar PEI. Solvent-induced crystallization can promote further phase separation of PEEK and PEI to form larger pores of membranes.

The degree of crystallinity of PEEKK/PEI blends decreased as the PEI content increased. The crystallinity of PEEKK was delayed by the presence of PEI and was attributed to an increased surface energy of the PEEKK crystals in the blends; the latter was supposed more defect-free than in pure PEEKK. However, the melting behavior of the PEEKK crystalline domains was unaffected by the blending, as the equilibrium T_m for the PEEKK in the blends was found to be close to the one for PEEKK alone, as measured by Zimmermann and Könnecke [108] and Wang et al. [116] on pure PEEKK.

7. Other PAEK Blends Obtained by Melt-Mixing

This section describes blends with other high-performance thermoplastics. Their miscibility depends on several factors, including the polymer's structure, the mixing parameters or the blend composition. The whole literature reviewed is gathered in Table 1. PAEK/PAI and PAEK/PES give partially miscible blends in the melted state. The partial miscibility of PAEK/TPI and PAEK/LCP blends is discussed. PAEK/PBI and PAEK/PTFE blends are immiscible.

7.1. PAEK/PAI Blends

The miscibility of PEEK with PAI seems to depend on its molecular structure. Karcha and Porter [117] measured two distinct T_g at all compositions for PEEK/PAI blends, proving their immiscibility. Smyser and Brooks [118] patented, over 30 years ago, a method to compatibilize PEEK and PAI by adding various inorganic hydrates, such as zinc sulfate ($ZnSO_4·7H_2O$) or iron II sulfate ($FeSO_4·7H_2O$) and aluminum hydroxide ($Al(OH)_3$) or cuprous hydroxide (CuOH). The hydrated blends may improve the injection molding process by preventing crosslinking between PEEK and PAI. They may also help to improve the mechanical properties to achieve better results than pure molded constituents. This method has been patented with other PAEK [119,120].

7.2. PAEK/PES Blends

Blends of PEEK and PES were reported to be either immiscible [121–125] or miscible [41,124,126], depending on how they were processed. Melt-mixed blends had two T_g for a processing temperature of 360 °C [121] or 350 °C [83,122,123] while Malik [126] measured single T_g for melt-mixed blends at 335 °C. This blend may undergo a low critical soluble temperature (LCST) at a temperature of around 340 °C [41,127]. Since the T_m of the PEEK is around 340 °C, the PEEK-rich phase begins to flow, hence having higher mobility, facilitating phase separation. Then, the high mobility of the PEEK overcomes the specific interaction between the sulfonate functions of the PES and the ether functions of the PEEK. Yu et al. [41] noted that postprocessing of the blends may alter their compatibility. Indeed, a single T_g was measured if the blends were processed using compression molding at 310 °C, while two T_g, each close to that of the pure component, were measured for a postprocessing temperature of 350 °C.

It was reported in 30-year-old studies that an increase in the PES content reduced the crystallinity of PEEK/PES blends. Moreover, no crystallization of PEEK was observed above 70 wt.% of PES. For a wide range of compositions, no crystallization was noticed on quenched blends [121,124,126]. In recent work, Korycki et al. [125] showed the crystallinity in PEEK/PES blends above 70 wt.% of PEEK was unchanged, revealing that PES as a minor phase has no impact on the PEEK crystalline phase.

Malik [126] showed, using scanning electron microscopy (SEM), that PEEK forms well-dispersed spherical submicron domains in a PES matrix (20/80 PEEK/PES blend), which is the result of phase separation. However, the PEEK domains remained bonded during the fracture of films broken under nitrogen, supporting a good adhesion between phases. Nandan et al. [83,127] gave in-depth information about the morphology of those blends. For PEEK-rich blends (Figure 17a) before the phase inversion point, which is assumed to be at a 75/25 PEEK/PES composition, the PES is dispersed within the PEEK matrix in the form of submicron droplets. As the PES content increases (Figure 17c), the morphology changes to a two-phase structure, with PEEK droplets larger than a micron forming the dispersion. At the phase inversion point (Figure 17b), an almost co-continuous phase was observed. For cryogenically fractured blends at any composition, the dispersed phase was debonded from the matrix and deformed. The latter evidenced a good adhesion between the two phases. It shall be noted that Malik used a PEEK 380G, while all other studies used a PEEK 450G. Arzak et al. [128] noted a difference in the structure between slowly cooled and quenched blends. However, the structure of the quenched blend seems to account for a better phase adhesion than the slowly cooled blends, as fewer PEEK droplets debonded from the PES matrix are seen in the structure of the latter (Figure 18).

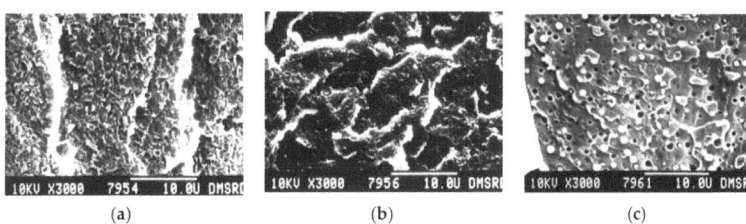

Figure 17. SEM fractography images of PEEK/PES blends obtained through melt-mixing then compression molding (**a**) 90/10 PEEK/PES (**b**) 75/25 PEEK/PES (**c**) 25/75 PEEK/PES, with permission from Wiley [127].

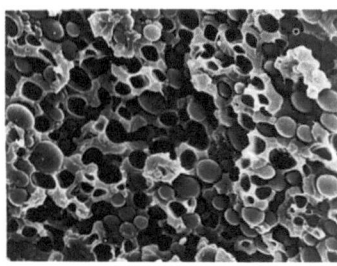 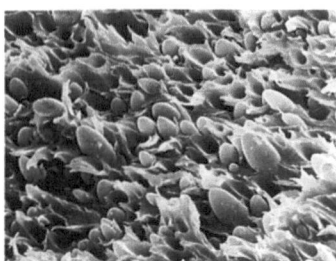

Figure 18. Morphologies of tensile fracture surface for 30/70 PEEK/PES compression-molded blends. Left: slowly cooled. Right: quenched, with permission from Wiley [128].

In order to compatibilize PEEK/PES blends, Korycki et al. [125] used phenolphthalein as a compatibilizer. The cardo side groups of phenolphthalein are supposed to be chemically bonded to PEEK and PES chains, increasing the interfacial adhesion between phases. The SEM analysis of fractured specimens displayed nano- to microsized PES spherical domains in a continuous PEEK phase, presented in Figure 19. The small pieces hooked on the surface of the PES droplets confirm the adhesion between both phases. It was noted that the crystallinity of PEEK was increased with PES content in the blends, demonstrating that despite phenolphthalein as a compatibilizer, the blends keep enough mobility to favor miscibility and PEEK crystallization. Among the two grades of PES tested with PEEK and phenolphthalein, the lowest molecular weight (PES 3010G) led to the highest crystallinity of PEEK, above 50%. Thus, the PEEK phase may be softened by the PES phase in the melted state, which gives macromolecules more mobility to self-organize into crystalline structures.

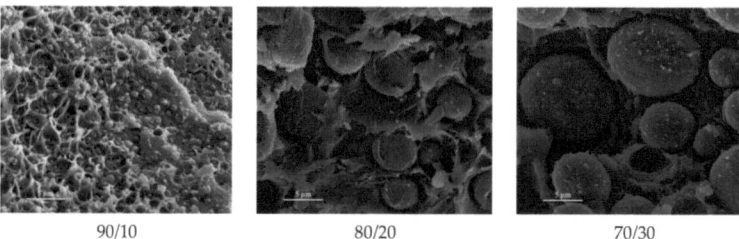

Figure 19. SEM images of the microstructure of PEEK/PES blends [wt.%] with phenolphthalein [125].

7.3. PAEK/TPI Blends

Blends of PEKK and TPI were defined as miscible at high and low TPI concentrations while being immiscible at intermediate (30 wt.% and 50 wt.%) compositions [30,129]. It was found that higher content of TPI favors miscibility, as the interaction parameter decreases with the TPI content increase. Moreover, the Fox equation was verified in miscible blends by the compositions and the T_α measured by DMTA [107,129]. Sauer et al. [18] provided a better understanding of the compatibility reasons for PAEK/TPI blends. By comparing PEEK, PEK and PEKK, it was shown that ketone linkages had an influence on miscibility with TPI. Indeed, it was hypothesized that the higher content of flexible ether linkages in PAEK, or their 120° angle, altered the ability to interact with the rigid polyimides. As a result, PEEK was immiscible with TPI but PEK and PEKK showed partial to full miscibility for high PAEK content. Looking further, a comparison of several PEKK with different terephthalic/isophthalic (T/I) content highlighted that a high fraction of 1.3 ketone linkage was thought to be more compatible with TPI, making PEKK (60T/40I) the most suitable for blending. Increasing the isophthalate moieties decreases the melting temperature and crystallization rate while increasing chain flexibility and retaining glass transition temperature at a high level. For PEKK content up to 20 wt.%, it acted as a high molecular diluent for strongly viscous polyimides [130]. More recently, Dominguez et al. [107] noted that semicrystalline blend miscibility undergoes the same phenomena as the mixtures of a semicrystalline polymer with an amorphous polymer. Thus, as PEEK crystallizes fast, the expelled TPI does not have enough time to reach the amorphous phase, creating phase separation. Also, PEKK crystallizes more slowly, giving more time for TPI to diffuse away from the crystal domains.

From a polarized optical microscope, the polyetheretherketone in PEEK/TPI blends seemed to crystallize as in neat PEEK, i.e., its crystalline structure is similar to that of neat PEEK. However, for high TPI content (above 60 wt.%), no PEEK crystallites were seen, indicating a hindering effect of the TPI on PEEK crystallization [131]. PEEK cold crystallization was also hindered by the TPI-rich phase [129]. Depending on the blend composition, both PEKK and TPI were able to crystallize during cooling from the melted state. Phase separation may also occur to some extent, as DMTA indicated the presence of two distinct amorphous phases in the melt crystallized blends, one rich in PEKK and the other rich in TPI.

To increase the compatibility between PEEK and TPI, Gao et al. [59] melt-mixed PEEK and TPI by adding synthesized optimized block length PEEK-*b*-TPI copolymers. The so-obtained compatibilized blends displayed a decrease in the interfacial tension and a narrowing of the T_g of both components. The impact of compatibilization on the crystallization of PEEK was not addressed, but since the segregation is reduced, it may hinder the PEEK crystallization.

7.4. PAEK/LCP Blends

Several studies have been conducted on melt-mixed PEEK/LCP blends. While Mehta and Isayev [132] obtained a blend behaving as pure PEEK, LCP seemed to promote PEEK hot crystallization, identically to a nucleating agent. Another work obtained nearly miscible-looking blends for more than 80 wt.% of PEEK but tending more toward a broadening of the T_g due to LCP crystal melting, which increased the amorphous population [133]. Partially or nonmiscible blends showed two distinct T_g affected by the other component, indicating the slight influence of LCP on PEEK [133,134].

Overall, the crystallinity of PAEK in immiscible blends was not affected by the presence of the other polymer for all PAEK/LCP blends reviewed. Thus, in immiscible blends, blended PAEK may crystallize as pure PAEK. However, Carvalho et al. [133] noted that in annealed PEEK/LCP, the melted LCP may act as an amorphous polymer and delay the PEEK crystallization.

7.5. PAEK/PBI Blends

Blends of PEEK and PBI have been shown to form immiscible blends [36,135]. PEEK can hardly be blended at temperatures above 420 °C because of its degradation, its T_m being around 340 °C. The glass transition temperature of PBI is 435 °C, so it remains solid during mixing. The obtained morphology is depicted in Figure 20a.

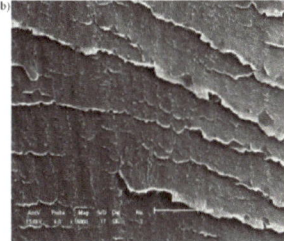

Figure 20. (**a**) Transmittance optical microscopy image of the melt-mixed PEEK/PBI blend with permission from ACS Publications [135]. (**b**) Scanning electron microscopy image of solution-blended SPEEK/PBI blend with permission from Elsevier [39].

The presence of PBI affects the crystallization of the PEEK, as no spherulites are formed upon cooling [135]. While it was supposed that PEEK crystallization may be delayed by the PBI, no experiment supports this hypothesis, which is a question of interest that needs to be taken on.

7.6. PAEK/PTFE Blends

Some studies have reported the immiscibility of PEEK with PTFE [136,137]. Melt processable PTFE (MP-PTFE) is easier to process by melting thanks to the incorporation of perfluoropropylvinylether. Two distinct T_g were measured from its blend with PEEK (110 °C for the PTFE, 166 °C for the PEEK). SEM observation revealed a dispersed morphology of the minor phase in the rich phase and a bad adhesion between them. In the case of the PEEK-rich phase, MP-PTFE droplets had a size of over 100 microns. MP-PTFE copolymers irradiation using an electron beam created -COOH and -COF functional groups and enabled the compatibility of the blend components [137]. Indeed, the T_g of the PEEK phase was shifted toward a higher temperature (from 164 °C to 167 °C) for 50/50 compatibilized blends: the evidence of a better degree of miscibility. SEM images showed both a finer dispersion of PTFE and better adhesion between phases.

Stuart and Briscoe [6,136] showed that PTFE increased PEEK crystallinity. Using Raman spectroscopy, they found that the frequency of the carbonyl band shifted and that this band became narrower, hinting at PEEK macromolecules being more ordered. They later confirmed it [138] as a higher degree of order in the crystalline phase was discovered through X-ray diffraction.

8. Solution-Blending through Chemical Modification of Components

8.1. Solution-Blending with Sulfonated PEEK

Among the blends reviewed, solution blended SPEEK/PEI [139,140], SPEEK/PAI [117,140], SPEEK/PES [141] and SPEEK/PBI [39,142,143] were reported to be miscible at all compositions. For all the blends, a single T_g with a positive deviation from the additivity law was measured, suggesting the presence of specific interactions.

Karcha and Porter [117,140] measured a single, sharp T_g at all compositions and sulfonation levels, presented in Figure 21. However, the T_g has deviated from the Fox equation. They suggested that the sulfonate functions of the sulfonated PEEK added strong intermolecular interactions. The same authors also studied SPEEK/PAI blends, which were found to be also miscible at all compositions. Using a Fourier transform infrared (FTIR) spectrometer, they identified that the specific interaction is the formation of intermolecular electron donor-acceptor complexes between the substituted phenylene rings of the SPEEK and the N-phenylene units of the PAI. The sulfonated group of the SPEEK acts as the electron acceptor, which helps in the formation of the complexes. Indeed, PEEK and PAI are not miscible in the melt state. Thus, the formation of such complexes by sulfonation of the PEEK decreases the free energy of mixing to the point of giving a miscible SPEEK/PAI blend. In the case of SPEEK/PBI blends, the underlying specific interaction is reported to be an acid-base interaction. Specifically, sulfonated groups of the SPEEK are acidic. They can form proton exchange complexes with the amino units of the PBI, the latter with the role of proton donor. Since SPEEK/PBI blends are mostly prepared for use in exchange membranes, most research focuses on their improved thermal and chemical stability [39,99,142,144–146]. These features may be explained by the morphology of the blends. As shown in Figure 20b, the PBI is well dispersed in the SPEEK matrix. Figure 20a shows the morphology of a melt-mixed PEEK/PBI blend, where immiscibility appears through different separated phases. Thus, good dispersion arises from the acid-base complexes formation.

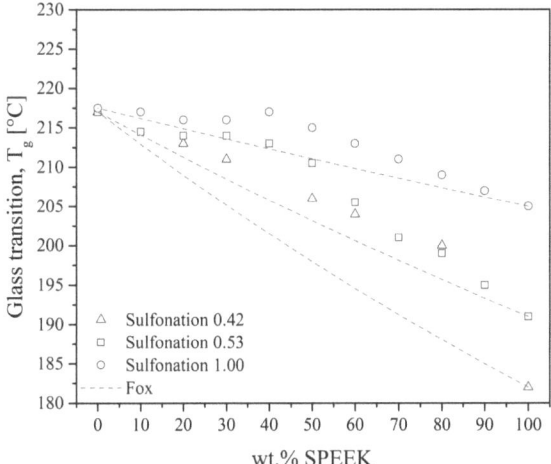

Figure 21. Glass transition temperature as a function of SPEEK content in SPEEK/PEI blends. Points represent experimental data from the literature [117,140], solid lines represent the Fox equation's trend.

8.2. Crystallization of SPEEK Blends

Reports about SPEEK/PEI blends have not assessed the crystallinity of the blends. Nevertheless, nanoscale transparency has been reported for SPEEK/PAI blends [33,141]. DSC traces of SPEEK/PBI [142] show no melting peak, hinting at a reduction in SPEEK crystallization. Wu et al. [33,141] supposed that SPEEK crystallinity would be greatly reduced when blended with PES while investigating the water and solvent uptake of SPEEK/PES blends.

Sulfonation has a twofold effect in SPEEK-based blends: (i) on the one hand, it reduces the intrinsic PEEK crystallinity [94], thus decreasing intermolecular (i.e., PEEK-PEEK) interactions; (ii) on the other hand, it increases the specific interactions through the formation of electron/proton donor-acceptor complexes. Both cases point toward the same direction, which is a reduction in crystallinity.

Overall, in-depth investigations would be required to make the distinction between sulfonation and the presence of an amorphous HPT on SPEEK crystallinity (such as blending para-PEI with SPEEK), to elucidate the structure–properties relationship behind the absence of crystallinity in miscible SPEEK/HPT blends, in order to enlighten and add to a deeper understanding of the nature of miscible blends.

9. Effect of Blends on PAEK's Mechanical Properties

9.1. Blends Obtained by Melt-Mixing

Mechanical properties of blends display various results depending on their thermomechanical history, whether they have been cooled rapidly to get amorphous blends or crystallized through annealing. Additivity rules are expected to fit well the experimental data for miscible blends, such as the properties that would follow a mixing law. Deviations from additivity may indicate that the phases in the blends, either:

- do not adhere to each other, leading to a negative deviation;
- or, strongly adhere, yielding a positive deviation, mainly attributed to specific interactions in the blend.

Semicrystalline PAEKs are more brittle than amorphous PAEKs due to the higher stiffness of the crystalline network compared to amorphous ones [27]. This section reports the results of the mechanical characterization of PAEK/HPT blends. However, among all the articles reviewed, a wider number of studies deal with PEEK/PEI blends. The blend preparation for some examples is presented in Table 2. It is known that the processing temperature may affect the miscibility properties of blends. The mechanical properties of the other types of blends are mentioned at the end of this section.

Table 2. Summary of the PAEK-based blend preparation with different additives.

Grade of PAEK	Tensile Strength [MPa]	Grade of HPT	Tensile Strength [MPa]	Blend and Sample Preparation	Processing Parameters	Max Tensile Strength of Blends [MPa]	Ref.
PEKK (synthesized)	93.7	PEI Ultem 1000	113	Extrusion (Ex) and injection molding (IM) or compression molding (CM) Annealed	360–380 °C / 370–380 °C / 370–390 °C / 200 °C/2 h	106.8 (60/40)	[27]
PEEK 450G	76.1	PEI Ultem 1000	92.4	Mixing (Mx) Annealed	365 °C/5′ / 300 °C	78.4 (20/80)	[111]
PEEK 450G	97	PEI Ultem 1000	120	Dry Mx IM Annealed	370 °C / 185 °C/24 h	116 (15/85)	[147]
PEEK 450G	135	PES 4100	72	Mx CM	355 °C/10′ / 355 °C	129 (90/10)	[126]
PEEK 380G	72.5	PES Ultrason E-2000	89	Mx IM	360 °C/12′ / 370 °C	/	[121]
PEKK (synthesized)	89	PES Radel A-300	90	Mx Annealed	350 °C/30″ / 185 °C/24 h	95 (50/50)	[123]
PEEK	84.1	LCP polyester	235.1	Ex IM	/ /	71.1 (70/30)	[48]
PEEK	80.8	LCP coPAEK (synthesized)	/	Ex IM	350 °C / 350 °C	110 (98/2)	[148]
PEEK	/	PBI	/	Ex CM	385–425 °C / 420 °C/30′	127 (50/50)	[135]
PEEK	97	PBI	/	Ex IM	455–510 °C / 385–455 °C	125 (50/50)	[149]
PEEK Gatone 5400	87	PTFE	/	Ex IM	330–350 °C / /	84 (92.5/7.5)	[150]

The general trend for PEEK/PEI blends follows that of PEEK, i.e., annealed blends are more brittle than amorphous ones. Harris and Robeson [27] noted that the toughness of PEEK/PEI blends passed through a maximum of 60 wt.% of PEEK when measured by tensile impact strength. As-molded pure PEEK had an estimated toughness of 279 kJ·m^{-2}, whereas blends with 60 wt.% of PEEK had more than a 60% increase with a 414 kJ·m^{-2} toughness. Annealed blends followed the same trend, with a toughness of 254 kJ·m^{-2} for blends with 80 wt.% of PEEK compared to 170 kJ·m^{-2} for neat PEEK. Moreover, tensile strength goes through a maximum of 106.8 MPa for 60 wt.% of PEEK for an annealed blend. As-molded (i.e., crystallinity is not controlled) shows negative deviation and annealed shows positive deviation (probably causes crystalline network reinforcement, annealing relieves stress according to the authors) [27,111,147]. Goodwin et al. [100] also noted some deviation from linearity when measuring the flexural modulus as obtained from DMTA for PEEK/PEI blends. The 40 wt.% of PEEK had a modulus of 2.8 GPa, while pure PEEK had 4 GPa of modulus. Going through a minimum of 80 wt.% of PEEK indicates specific interactions may take place at the interface, not explained by the authors. The mechanical properties of PEEK/PEI blends are complicated by the crystallinity of molded samples containing greater than 70 wt.% of PEEK [27]. Above 70 wt.% of PEEK, the authors highlight the brittleness of the blends. Harris et al. [151] noted such deviation for both PEEK/PEI (minimum at 40 wt.% of PEEK) and PEK/PEI (minimum at 60 wt.% of PEK) blends. Frigione et al. [111] reported that crystallized PEEK/PEI 20/80 blends had similar or slightly weaker mechanical properties to the tensile strength of 78.4 MPa and modulus of 1.47 GPa compared to pure PEEK, while the as-molded blend was much lower with 19.3 MPa and 1.18 GPa, respectively, Figure 22.

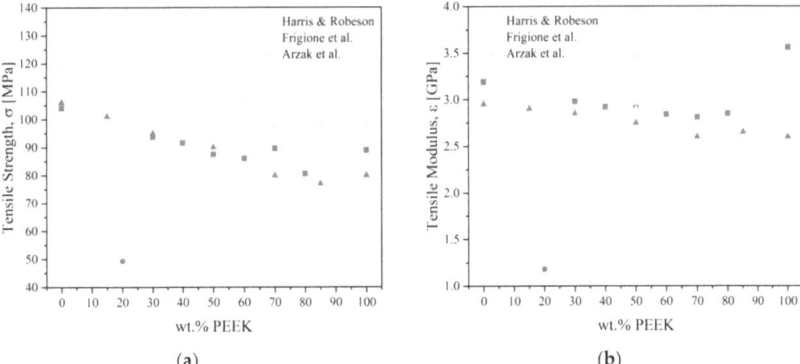

Figure 22. The tensile strength (**a**) and modulus (**b**) as a function of the volume fraction of PEEK for PEEK/PEI blends from the literature [27,111,147]. Empty symbols for annealed blends, full symbols for as-molded blends.

Additionally, after annealing, blends have better mechanical properties (Young's modulus and tensile strength) than as-molded blends, as noted by Arzak et al. [147]. Their tensile modulus and tensile strength deviated from the linearity predicted by the mixing rule because of crystallinity development. Though crystallinity gives an overall increase in rigidity, the incorporation of PEI diminished the overall quantity of PEEK crystals, hence a negative deviation from additivity. Also, the ductility of the blends has a more complex behavior. Pure annealed PEI and PEEK have a ductility of 10% and 25%, respectively. The maximum 55% is observed at 50/50 PEEK/PEI composition in annealed blends, while the ductility of as-molded blends appears fairly uniform and slightly enhanced concerning the additive rule of mixtures. PEI increases the ductility of the blends from 45% to 55%, balancing the loss of properties. This balance of opposite trends was of greater importance at temperatures close to the T_g of the blends. Annealed blends tested at 125 °C exhibited strong deviations, while as-molded blends showed no variation or followed additivity. However, the ductility was more pronounced in annealed blends, thus rendering them usable at this range of temperature. Thus, blending with PEI has the additional effect that some crystalline PEEK is converted to the more ductile and less stiff amorphous PEEK, which gives rise to higher ductilities and a smaller modulus of elasticity and tensile strength in the blends than the pure materials. Gensler et al. [152] noted for thin PEEK/PEI films a brittle-ductile transition with increasing temperature for blends containing up to 60 wt.% of PEEK. However, blends containing more PEEK showed no transition, as PEEK crystallization was strain-induced, preventing disentanglement and further crazing. The 50/50 PEEK/PEI blends were also reported to have reduced solvent (acetone) uptake compared to pure components [153]. The acetone uptake induces a phase swelling, resulting in the recrystallization of PEEK. The mechanical properties (tensile strength, yield stress, Young's modulus and strain at break) of the blends were also affected by the presence of the solvent. Overall, Young's modulus was more sensitive to plasticization and dropped from 1340 MPa to 500 MPa, while the tensile strength (decrease from 70 MPa to 16 MPa) and the strain at failure (decrease from 50% to 17%) were more sensitive to the degree of crystallinity.

To sum up, PAEK and PEI are miscible; all the authors presented better tensile strength and Young's modulus for annealed blends than for as-molded ones. The balance of properties changed in annealed blends because of the change in crystallinity with composition, giving clear trend changes at the intermediate composition, where crystallinity content also clearly changed.

As a reminder, PEEK is not miscible with PES. Young's modulus of PEEK/PES blends increases with increasing PEEK content [121,123,126,128]. All authors agreed on the fact that blends containing more than 60 wt.% of PEEK followed a mixing rule, despite using a different grade of PEEK (450G and 380G, respectively, by Azrak et al. [121,128] and Malik [126], while Nandan et al. [83,122,123,127] synthesized their own PEEK). Quenched [121] blends followed a mixing rule, while annealed [128] or slowly cooled blends [121] exhibited a negative deviation, Figure 23. The deviation was the result of the development of crystallinity. Malik [126] and Nandan et al. [123] measured a positive deviation from linearity, with a maximal modulus at 40–50 wt.% of PEEK content. Good adhesion and synergistic interaction, such as the increased density of chain entanglement [16,123] between the PEEK and the PES were thought to be responsible for this positive deviation. Azrak et al. [121] measured the yield stress for the blends that followed a linear behavior with the concentration in the case of quenched blends. Slowly cooled blends were brittle; for example, they broke before reaching the yield point. The crystallized PEEK increases the brittleness of the blends. Nandan et al. [123] measured elongation at break that deviated negatively for all compositions except for 90/10 PEEK/PES, which showed a positive deviation due to an important synergistic effect at this composition with ductile behavior.

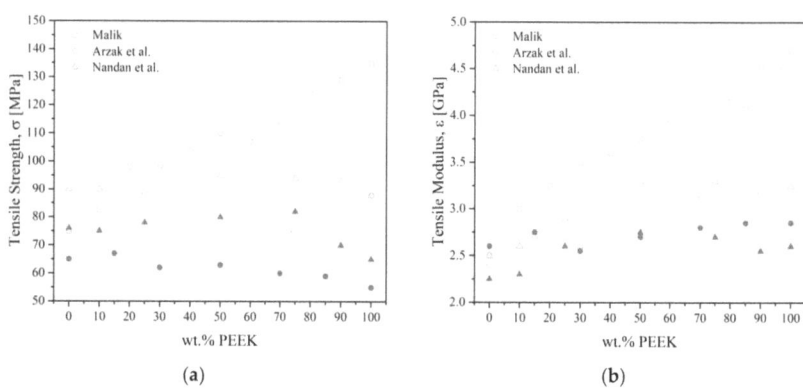

Figure 23. The tensile strength (**a**) and modulus (**b**) as a function of the volume fraction of PEEK for PEEK/PES blends from the literature [121,123,126]. Empty symbols for annealed or slowly cooled blends, full symbols for quenched blends.

PEEK and PES are not miscible from a thermodynamic point of view. However, some compatibility may be achieved for particular blending conditions and cooling phases. Compatibility arises from specific interaction between the PES sulfonated function and the PEEK backbone, hence the mechanical properties slightly deviating from additivity. Annealed or slowly cooled blends result in a higher tensile strength and Young's modulus than quenched blends.

Next, blends of TLCP with PEEK reported by Kiss [48] show that PEEK modulus increased from 3.5 GPa to 4.3 GPa with the addition of 30 wt.% of TLCP; however, the tensile strength and strain to failure decreased from 84.1 MPa to 71.7 Mpa and from 48% to 2.5%. Mehta and Isayev [132] noted that LCP may enhance PEEK's mechanical properties. The flexural modulus as obtained from DMTA, increased by an increasing LCP concentration, indicating the reinforcement of PEEK with LCP. Miao et al. [148] reported an increase in the tensile modulus and tensile strength from 1.85 GPa to 2.42 GPa and from 80.8 MPa to 110 MPa with only 2 wt.% of LCP.

Liu et al. [135] measured Young's modulus of 5.4 GPa for PEEK/PBI blends, which was close to the value reported by Alvarez and DiSano [149] of 5.1 GPa. In their work, they produced PEEK/PBI blends through extrusion and compression molding. PBI and PEEK are known to be incompatible with each other in the dry state. The possible influence of hygrothermal exposure on interfacial bonding between PEEK and PBI has been studied. The blends of PEEK/PEI 50/50 were treated under different conditions. Dry and water-saturated at 60 °C blends demonstrate an increase in the tensile strength from 122 MPa to 127 MPa, while the blend treated in hot water at 288 °C reached only 56 MPa. Alvarez and DiSano [149] reported that both compressive and tensile properties were improved by the incorporation of PBI into the PAEK. The tensile strength of pure PEK and PEEK were 102 MPa and 97 MPa while their blends with 50 wt.% of PBI reached 119 MPa and 125 MPa, respectively.

PTFE is usually incorporated into PEEK to target tribological applications. Indeed, PTFE implied a reduction in the coefficient of friction compared to pure PEEK [1,150,154] and it implied a loss of wear. The PTFE is generally thought to migrate to the surface of the blend [6,155]. Thus, if the blends contain too much PTFE, its surface consists only of PTFE. An optimal PTFE concentration has been found between 10 wt.% and 20 wt.%, depending on the molecular weight of the PEEK used [136,154,155] and the process used to elaborate the blends. Plasma treatment before blending may improve the tribological properties, as plasma-treated blends had a decreased specific wear rate and sliding friction coefficient [156]. Blends of irradiated MP-PTFE showed better mechanical properties than PEEK/PTFE blends. Indeed, a tripled strain at break and a doubled stress at break were reported for the former blend [137]. Bijwe et al. [150] noted that the incorporation of PTFE may deteriorate the mechanical properties of PEEK, rendering the PEEK/PTFE blend relevant for only a few tribological applications. So, the tensile strength for pure PEEK of 87 MPa decreased to 64.7 MPa for a blend with 30 wt.% of PTFE, while the tensile modulus changed from 3.9 GPa to 1.23 GPa, respectively.

Thus, TLCP and PBI, despite the lack of miscibility with PAEK, can be compatible in some blend compositions, achieving a higher tensile strength and modulus as blends than pure PAEK. The special case of PAEK/PTFE blends attracted our attention, in terms of that those polymers are highly immiscible but their blends give the foreseen properties, especially a reduced friction coefficient. Indeed, the incorporated PTFE migrates to the PEEK surface acting as a lubricant. PTFE was also found to increase PEEK crystallinity but an in-depth reason was not given to explain the observed trend.

9.2. Performances of Blends Obtained by Solution Blending

SPEEK is mainly blended with HPT to elaborate exchange membranes. A proton exchange membrane fuel cell transforms the energy liberated during the hydrogen and oxygen reaction from chemical to electrical energy. The sulfonation of PEEK allows its proton conductivity to be increased [35,40]. Electrochemical properties (i.e., water uptake, thermal stability, proton exchange, dielectric conductivity) are the most reported properties. Most SPEEK/HPT blends aim to create a cost-effective solution as an alternative membrane to Nafion®. The mechanical properties of such membranes have barely been studied.

Daud et al. [157] reported that SPEEK/PES had performances comparable to commercial ones. The presence of the ionic sulfonate groups increased the hydrophilic nature of the blends and increased the surface charges [158]. The mechanical properties reported in such studies cannot be compared due to varying water uptake. According to Arigonda et al. [96], SPEEK/PES demonstrates a lower Young's modulus for dry blends of 3.6 MPa and 2.2 MPa for 20 wt.% and 40 wt.% of PES, respectively, than pure SPEEK of 4.8 MPa. Hydrated SPEEK had a three-fold decrease in its modulus of 1.5 MPa, whereas the SPEEK/PES blends retained their mechanical properties of 2.6 MPa and 2.4 MPa for 20 wt.% and 40 wt.% of PES, respectively.

10. Ternary and Compatibilized Blends

Ternary blends may be relevant as the third component and may act as a link between two incompatible materials, provided it is miscible with both.

The properties of PAEK arise mainly from a difference in the molecular arrangement of the phenyl unit, which in turn influences their melting point and crystalline structure. The advantages of blending PAEK altogether reside in the fact that PAEK blends are isomorphic. They exhibit a single T_g and T_m. The high ethers-to-ketones ratio of PAEK has limited use in injection molding, as their T_m is too close to their degradation temperature. Blending those PAEK with lower T_m PAEK enables them to be processed at a lower temperature while keeping their outstanding properties.

Harris et al. [159] blended various PAEK (such as PEK/PEKK) and found a single T_g, a single T_m and mechanical compatibility, indicating isomorphism. They also prepared blends with random copolymers composed of different PAEK, such as a PEKK-g-PEEK copolymer, which was also isomorphic when blended with other PAEK. Some blends were not isomorphic when in a binary mixture but were isomorphic when a third PAEK was added, the latter forming binary isomorphic blends with the two other components. Thus, the PEKK-g-PEEK copolymer acted as a compatibilizer. Still keeping with the same idea, Gao et al. [59] observed an increased flexural modulus (from 3280 MPa to 3314 MPa) and elongation at break (from 8.6% to 27.4%) for the compatibilized blends compared to uncompatibilized ones. For an optimized block length copolymer, they recorded a 200% improvement in terms of elongation at break. Dominguez et al. [107] showed that an increase in the crystallinity in such ternary blends induced an increase in T_g but a decrease in rubbery moduli.

Dawkins et al. [160] prepared PBI/PAEK/PEI blends through both solution-blending and mix-melting. They measured only two T_g as the PAEK would be miscible with the PEI but not with the PBI. The second T_g was always close to the T_g of the PBI. Interestingly, both methods yielded a first T_g in agreement with that of binary PEEK/PEI blends, highlighting incompatibility between the PEEK and the PBI and a possible incompatibility of PEI with PBI.

Chen et al. [161] prepared miscible PEEK/PEI/PPS blends and they reported a homogeneous morphology [104]. The unique T_g of the blends was higher than that of pure PEEK. The crystallinity of PEEK was also higher than that of pure PEEK, with a maximum crystallinity of 37.8% for 60/30/10 PEEK/PEI/PES blends. Regarding the mechanical properties, they focused on tribology: higher coefficients of friction and wear rates were recorded for the blends compared to pure PEEK.

PAI/SPEEK/PEI blends were investigated by Karcha and Porter [162] since they studied the miscible binary blends SPEEK/PAI and SPEEK/PEI. The immiscibility of PEI and PAI makes SPEEK a suitable candidate as a compatibilizer. Only a few compositions lead to miscible blends both before and after annealing. The composition with up to 10 wt.% of PAI was miscible for the degree of sulfonation of 0.53, whereas, with the degree of sulfonation of 1.00, the blends containing up to 20 wt.% of PAI and PEI were miscible (i.e., 20/40/20 PAI/SPEEK/PEI).

Based on the consideration that PEI/LCP blends exhibit compatibility and synergistic effects and that PEEK and PEI are miscible, Bretas and Baird [163] explored the possibility of blending LCP with PEEK with PEI as compatibilizer. The ternary PEEK/PEI/LCP blend exhibited compatibility both before and after annealing. In the latter case, the blends were more stable and some compositions showed miscibility. Either blend with a high content of LCP (such as 10/10/80 PEEK/PEI/LCP blends) or with a high content of PEI (such as 10/80/10 PEEK/PEI/LCP blends) showed compatibility. Figure 24 shows the complexity of these blends, where it is easy to see that miscibility is not related to composition in a straightforward way, even though blends with low LCP content tend to be miscible [164] as they were mostly composed of PEEK and PEI. Specific interactions such as dipole–dipole interaction between some carbonyl groups were given causes for the observed miscibility. The link between the mechanical properties and composition was complex but overall high LCP content blends had a higher modulus and high PEI and PEEK contents were more ductile (i.e., high tensile strength and high strain at failure), with more PEI leading to higher values.

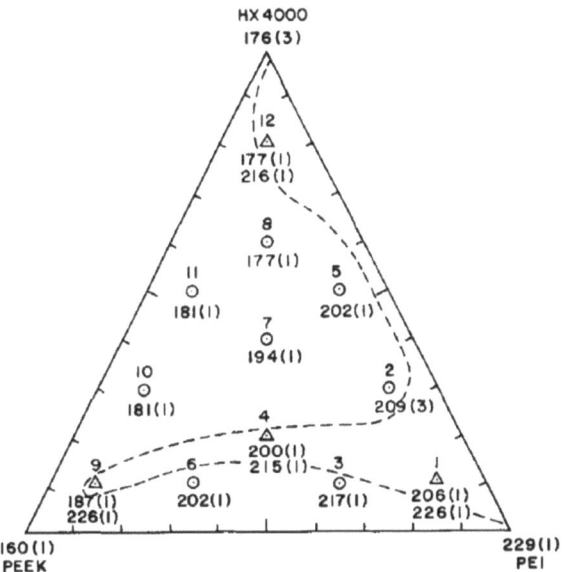

Figure 24. Phase diagram of T_g obtained by DSC of PEEK/PEI/LCP(HX400) blends after annealing, where ○ is composition with one T_g, △ is composition with two T_g, --- is approximate boundary between composition with one and two T_g, standard deviations are given in brackets, with permission from Elsevier [163].

11. Conclusions

Like conventional polymer blends, those of high-performance polymers are seldom naturally miscible. Among the blends reviewed, PEEK/PEI is the only one that gives miscible blends when the blends are simply elaborated through melt-mixing. Other PAEK are partially miscible with PEI; their degree of miscibility depends on the composition. The degree of miscibility of PAEK/PAI and PAEK/TPI depends on the ketone linkages of the PAEK, while PAEK/PBI and PAEK/PTFE are immiscible. PAEK/PES blends are miscible if the processing temperature is below a given threshold.

Another way of preparation is solution blending, meaning dissolving both polymers in a suitable solvent. Those PAEK/HPT blends with the HPT aforementioned enable a higher degree of miscibility. In particular, solution blending of sulfonated PAEK enabled high miscibility with PEI, PAI, PBI, TPI and PES thanks to the specific interactions between the sulfonate groups of the PAEK and mainly the imide functions (respectively, the sulfonate group of the PES).

The mechanical properties of the blends are not systematically reported in the literature reviewed, but instead, the properties relevant to the specific end-use envisioned are. The mechanical properties were reported for PAEK/PEI and PAEK/PES blends, as the primary function of the amorphous polymers was to enhance PAEK processability and to enable its use above its T_g. Thus, the mechanical properties of those blends were of utmost importance. Reports found that in the amorphous state, the overall properties decreased concerning those of PAEK, while semicrystalline blends had properties similar to that of neat PAEK. Those properties are not reported in the case of PAEK/PAI, PAEK/PBI and PAEK/TPI. The primary role of the PAEK in such studies is to reduce water uptake and increase proton conductivity. In this scope, sulfonation of PAEK was required, and miscible SPAEK/PBI and SPAEK/TPI are mentioned as serious candidates for PEMFC membranes.

Similarly, few reports have focused on the morphology of the obtained blends, mainly because their aimed properties were achieved, especially for the blends used in membranes.

Compatibilization of polymer blends is widely used for low-performance thermoplastic polymers, such as anhydride maleic acid being used to compatibilize PET/PP blends. This route remains to be explored for HPT blends, and the importance of doing so will inevitably increase. Indeed, most of the partial blends overviewed are obtained through solution blending, using highly concentrated strong acids, producing large quantities of waste with a high environmental impact. Indeed, solution blending does not fall under the latest environmental standards such as the REACH regulation, which aims to improve the protection of human health and the environment from the risks of chemicals. Producing compatible HPT blends respecting those standards will be one of the next challenges to overcome shortly. The first answer would be the study of ternary- or multicomponent blends. The most interest in the multiphasic system would be the fine control of the morphology, especially targeting the selective location of the components. This kind of system has already been extensively studied for low-performance thermoplastic alloys but seldom reported for HPT blends. Control of the morphology may also be realized by controlling the processing; it has been demonstrated that some immiscible blends, such as PEEK/PES, may be mechanically compatible. This path deserves a thorough and systematic investigation, the control of morphology being little studied in the literature reviewed.

The future trends will deal with the characterization of the spatial distribution of each phase in PAEK-based blends. More progress is necessary to give a fine mapping of the spatial distribution of the morphology at the surface and inside the volume of blends. The crystallization of polymer chains confined in nanodomains has not been resolved for PAEK blends until now. Also, the synthesis of tailored copolymers will be useful to improve the miscibility of PAEK-based polymer blends. Another future challenge will be the manufacturing of composites with PAEK-based blends as matrices in order to benefit from the remarkable properties of these blends.

Interest in HPT blends continues to grow as they represent an alternative to metals to face the upcoming environmental and energy transition. This interest will only grow when the above challenges are met.

Author Contributions: Conceptualization, A.K. and F.C. (France Chabert); methodology, A.K., O.T. and F.C. (France Chabert).; validation, C.G., T.D. and F.C. (France Chabert).; formal analysis, A.K., F.C. (Fabrice Carassus), O.T., C.G., T.D. and F.C. (France Chabert); investigation, A.K., F.C. (Fabrice Carassus) and O.T.; writing—original draft preparation, A.K., F.C. (Fabrice Carassus) and O.T.; writing—review and editing, A.K., F.C. (Fabrice Carassus), O.T., C.G., T.D. and F.C. (France Chabert); supervision, C.G., T.D. and F.C. (France Chabert); funding acquisition, C.G., T.D. and F.C. (France Chabert). All authors have read and agreed to the published version of the manuscript.

Funding: This work is supported by France Relance program "Préservation des emplois de recherche et développement" of the French National Research Agency (ANR).

Institutional Review Board Statement: Not applicable.

Data Availability Statement: Not applicable.

Conflicts of Interest: The authors declare that the research was conducted in the absence of any commercial or financial relationships that could be construed as a potential conflict of interest.

Nomenclature of Polymers

PAEK	Polyaryletherketone	HPT	High-performance thermoplastic
PEKK	Polyetherketoneketone	PEI	Polyetherimide
PEKEKK	Polyetherketonetherketoneketone	PAI	Polyamideimide
PEK	Polyetherketone	PBI	Polybenzimidazole
PEEKK	Polyetheretherketoneketone	PES	Polyethersulfone
PEEKEK	Polyetheretherketonetherketone	TPI	Thermoplastic polyimide
PEEK	Polyetheretherketone	PTFE	Polyterafluoroethylene
PEEEK	Polyetheretheretherketone	LCP	Liquid crystal polymer
SPEEK	Sulfonated polyetheretherketone	TLCP	Thermoplastic liquid crystal polymer

References

1. Lu, S.X.; Cebe, P.; Capel, M. Thermal stability and thermal expansion studies of PEEK and related polyimides. *Polymer* **1996**, *37*, 2999–3009. [CrossRef]
2. Blundell, D.J.; Osborn, B.N. The morphology of poly(aryl-ether-ether-ketone). *Polymer* **1983**, *24*, 953–958. [CrossRef]
3. Rigby, R.B. Polyetheretherketone PEEK. *Polym. News* **1984**, *9*, 325.
4. Velisaris, C.N.; Seferis, J.C. Crystallization kinetics of polyetheretherketone (PEEK) matrices. *Polym. Eng. Sci.* **1986**, *26*, 1574–1581. [CrossRef]
5. Parameswaranpillai, J.; Thomas, S.; Grohens, Y. Polymer blends: State of the art, new challenges, and opportunities. In *Characterization of Polymer Blends*; Thomas, S., Grohens, Y., Jyotishkumar, P., Eds.; Wiley: Weinheim, Germany, 2014; pp. 1–6. ISBN 978-3-527-64560-2.
6. Stuart, B.H.; Briscoe, B.J. Scratch hardness studies of poly(ether ether ketone). *Polymer* **1996**, *37*, 3819–3824. [CrossRef]
7. Dosière, M.; Villers, D.; Zolotukhin, M.G.; Koch, M.H.J. Comparison of the structure and thermal properties of a poly(aryl ether ketone ether ketone naphthyl ketone) with those of poly(aryl ether ketone ether ketone ketone). *e-Polymers* **2007**, *7*, 130. [CrossRef]
8. Harsha, A.P.; Tewari, U.S. Tribo performance of polyaryletherketone composites. *Polym. Test.* **2002**, *21*, 697–709. [CrossRef]
9. Schmidt, M.; Pohle, D.; Rechtenwald, T. Selective laser sintering of PEEK. *CIRP Ann.* **2007**, *56*, 205–208. [CrossRef]
10. Technical data—Kepstan 6000. Arkema Innovative Chemistry: Colombes, France, 2013.
11. Technical data—Kepstan 7000. Arkema Innovative Chemistry: Colombes, France, 2012.
12. Technical data—Kepstan 8000. Arkema Innovative Chemistry: Colombes, France, 2013.
13. Villar Montoya, M. Procédé de soudage laser de polymères haute performance: Établissement des relations entre les paramètres du procédé, la structure et la morphologie du polymère et les propriétés mécaniques de l'assemblage. Ph.D. Thesis, Université de Toulouse, Toulouse, France, 2018.
14. Technical data—Victrex HT G45. Victrex High Performance Polymers: Lancashire, UK, 2014.
15. Team Xomertry Polyether Ether Ketone (PEEK): Characteristics, Features, and Process. Available online: https://www.xometry.com/resources/materials/polyether-ether-ketone/ (accessed on 27 July 2023).
16. Friedrich, K.; Lu, Z.; Hager, A.M. Recent advances in polymer composites' tribology. *Wear* **1995**, *190*, 139–144. [CrossRef]
17. Heath, D.R.; Pittsfield; Mass; Wirth, J.G. Schenectady Polyetherimides. Patent US003847867, 1974.
18. Technical data—Sabic ULTEM™ Resin 1000—Europe. Sabic: Sittard, The Netherlands, 2018.
19. Frigione, M.; Naddeo, C.; Acierno, D. The rheological behavior of polyetheretherketone (PEEK)/polyetherimide (PEI) blends. *J. Polym. Eng.* **1996**, *16*, 14. [CrossRef]
20. Ramani, R.; Alam, S. Composition optimization of PEEK/PEI blend using model-free kinetics analysis. *Thermochim. Acta* **2010**, *511*, 179–188. [CrossRef]
21. Briscoe, B.J.; Stuart, B.H. A fourier transform raman spectroscopy study of the crystallization behaviour of poly (ether ether ketone)/poly (ether imide) blends. *Spectrochim. Acta* **1993**, *49A*, 753–758. [CrossRef]
22. Meakin, P.J.; Cogswell, F.N.; Halbritter, A.J.; Smiley, A.J.; Staniland, P.A. Thermoplastic interlayer bonding of aromatic polymer composites—Methods for using semi-crystallized polymers. *Compos. Manuf.* **1991**, *2*, 86–91. [CrossRef]
23. Korycki, A.; Garnier, C.; Bonmatin, M.; Laurent, E.; Chabert, F. Assembling of carbon fibre/PEEK composites: Comparison of ultrasonic, induction, and transmission laser welding. *Materials* **2022**, *15*(18), 6365. [CrossRef] [PubMed]
24. Yoo, J.H.; Eiss, N.S., Jr. Tribological behavior of blends of polyether ether ketone and polyether imide. *Wear* **1993**, *162–164*, 418–425.
25. Díez Pascual, A.M.; Díez Vicente, A.L. Nano-TiO2 reinforced PEEK/PEI blends as biomaterials for load-bearing implant applications. *ACS Appl. Mater. Interfaces* **2015**, *7*, 5561–5573. [CrossRef]
26. Cafiero, L.; Alfano, O.; Iannone, M.; Esposito, F.; Iannace, S.; Sorrentino, L. Microcellular foams from PEEK/PEI miscible blends. In Proceedings of the Regional Conference Graz, Graz, Austria, 6–10 September 2015; AIP Publishing: Graz, Austria, 2016; Volume 1779, p. 090009_1-5.
27. Harris, J.E.; Robeson, L.M. Miscible blends of poly(aryl ether ketone)s and polyetherimides. *J. Appl. Polym. Sci.* **1988**, *35*, 1877–1891. [CrossRef]
28. Hudson, S.D.; Davis, D.D.; Lovinger, A.J. Semicrystalline morphology of poly(aryl ether ketone)/poly(ether imide) blends. *Macromolecules* **1992**, *25*, 1759–1765. [CrossRef]
29. Hsiao, B.S.; Sauer, B.B. Glass transition, crystallization, and morphology relationships in miscible poly(aryl ether ketones) and poly(ether imide) blends. *J. Polym. Sci. Part B Polym. Phys.* **1993**, *31*, 901–915. [CrossRef]
30. Sauer, B.B.; Hsiao, B.S. Miscibility of three different poly(aryl ether ketones) with a high melting thermoplastic polyimide. *Polymer* **1993**, *34*, 3315–3318. [CrossRef]
31. Nemoto, T.; Takagi, J.; Ohshima, M. Nanocellular foamscell structure difference between immiscible and miscible PEEK/PEI polymer blends. *Polym. Eng. Sci.* **2010**, *50*, 2408–2416. [CrossRef]
32. Mark, J.E. *Polymer Data Handbook*; Oxford University Press: New York, NY, USA, 1999.
33. Wu, H.L.; Ma, C.C.M.; Li, C.H.; Lee, T.M.; Chen, C.Y.; Chiang, C.L.; Wu, C. Sulfonated poly(ether ether ketone)/poly(amide imide) polymer blends for proton conducting membrane. *J. Membr. Sci.* **2006**, *280*, 501–508. [CrossRef]
34. Wang, Y.; Goh, S.H.; Chung, T.S. Miscibility study of Torlon® polyamide-imide with Matrimid® 5218 polyimide and polybenzimidazole. *Polymer* **2007**, *48*, 2901–2909. [CrossRef]

35. Jones, D.J.; Rozière, J. Recent advances in the functionalisation of polybenzimidazole and polyetherketone for fuel cell applications. *J. Membr. Sci.* **2001**, *185*, 41–58. [CrossRef]
36. Liu, P.; Mullins, M.; Bremner, T.; Browne, J.A.; Sue, H.J. Hygrothermal behavior of polybenzimidazole. *Polymer* **2016**, *93*, 88–98. [CrossRef]
37. Brooks, N.W.; Duckett, R.A.; Rose, J.; Ward, I.M.; Clements, J. An n.m.r. study of absorbed water in polybenzimidazole. *Polymer* **1993**, *34*, 4038–4042. [CrossRef]
38. Chung, T.S. A critical review of polybenzimidazoles. *Polym. Revs.* **1997**, *37*, 277–301. [CrossRef]
39. Zhang, H.; Li, X.; Zhao, C.; Fu, T.; Shi, Y.; Na, H. Composite membranes based on highly sulfonated PEEK and PBI: Morphology characteristics and performance. *J. Membr. Sci.* **2008**, *308*, 66–74. [CrossRef]
40. Linkous, C. Development of solid polymer electrolytes for water electrolysis at intermediate temperatures. *Int. J. Hydrog. Energy* **1993**, *18*, 641–646. [CrossRef]
41. Yu, X.; Zheng, Y.; Wu, Z.; Tang, X.; Jiang, B. Study on the compatibility of the blend of poly(aryl ether ether ketone) with poly(aryl ether sulfone). *J. Appl. Polym. Sci.* **1990**, *41*, 2649–2654. [CrossRef]
42. Liu, J.; Li, J.; Wang, T.; Huang, D.; Li, Z.; Zhong, A.; Liu, W.; Sui, Y.; Liu, Q.; Niu, F.; et al. Organosoluble thermoplastic polyimide with improved thermal stability and UV absorption for temporary bonding and debonding in ultra-thin chip package. *Polymer* **2022**, *244*, 124660. [CrossRef]
43. Tsutsumi, T.; Nakakura, T.; Takahashi, T.; Morita, A.; Gotoh, Y.; Oochi, H. Polyimide based resin composition. Patent US005516837A, 1996.
44. Hou, T.H.; Reddy, R.M. *Thermoplastic Polyimide New-TPI*; NASA: Hampton, VA, USA, 1990.
45. Yang, H.; Liu, J.; Ji, M.; Yang, S. Novel thermoplastic polyimide composite materials. In *Thermoplastic—Composite Materials*; Intech Open Science: London, UK, 2012; pp. 1–11.
46. Tsutsumi, T.; Nakakura, T.; Morikawa, S.; Shimamura, K.; Takahashi, T.; Morita, A.; Koga, N.; Yamaguchi, A.; Ohta, M.; Gotoh, Y.; et al. Polyimide based resin composition. Patent US5312866A, 1994.
47. Kemmish, D.J. *High performance engineering plastics*; Report 86; Smithers Rapra Publishing: Shawbury, UK, 1995.
48. Kiss, G. In situ composites: Blends of isotropic polymers and thermotropic liquid crystalline polymers. *Polym. Eng. Sci.* **1987**, *27*, 410–423. [CrossRef]
49. Sariban, A.; Binder, K. Citical properties of the Flory–Huggins lattice model of polymer mixtures. *J. Chem. Phys.* **1987**, *86*, 5859–5873. [CrossRef]
50. Patterson, D.; Robard, A. Thermodynamics of polymer compatibility. *Macromolecules* **1978**, *11*, 690–695. [CrossRef]
51. Helfand, E.; Tagami, Y. Theory of the interface between immiscible polymers. II. *J. Chem. Phys.* **1972**, *56*, 3592–3601. [CrossRef]
52. Olabisi, O. Process for the moulding of plastic structural web and the resulting articles. Patent US004136220, 1979.
53. DeMeuse, M.T. High temperature polymer blends: An overview of the literature. *Polym. Adv. Technol.* **1995**, *6*, 76–82. [CrossRef]
54. Higgins, J.S.; Lipson, J.E.G.; White, R.P. A simple approach to polymer mixture miscibility. *Philos. Trans. R. Soc. A* **2010**, *368*, 1009–1025. [CrossRef]
55. Scott, C.E.; Macosko, C.W. Morphology development during the initial stages of polymer-polymer blending. *Polymer* **1995**, *36*, 461–470. [CrossRef]
56. Sundararaj, U.; Macosko, C.W. Drop breakup and coalescence in polymer blends: The effects of concentration and compatibilization. *Macromolecules* **1995**, *28*, 2647–2657. [CrossRef]
57. Faria, J.; Ruiz, M.P.; Resasco, D.E. Phase-selective catalysis in emulsions stabilized by janus silica-nanoparticles. *Adv. Synth. Catal.* **2010**, *352*, 2359–2364. [CrossRef]
58. Raphaël, E. Equilibre d'un "grain janus" à une interface eau/huil. *CR Acad. Sci. Paris Série II* **1988**, *307*, 9–12.
59. Gao, C.; Zhang, S.; Li, X.; Zhu, S.; Jiang, Z. Synthesis of poly(ether ether ketone)-block-polyimide copolymer and its compatibilization for poly(ether ether ketone)/thermoplastic polyimide blends. *Polymer* **2014**, *55*, 119–125. [CrossRef]
60. Pötschke, P.; Paul, D.R. Formation of co-continuous structures in melt-mixed immiscible polymer blends. *J. Macromol. Sci. Part C Polym. Rev.* **2003**, *43*, 87–141. [CrossRef]
61. Chen, H.L.; Porter, R.S. Phase and crystallization behavior of solution-blended poly(ether ether ketone) and poly(ether imide). *Polym. Eng. Sci.* **1992**, *32*, 1870–1875. [CrossRef]
62. Crevecoeur, G.; Groeninckx, G. Binary blends of poly(ether ether ketone) and poly(ether imide). Miscibility, crystallization behavior and semicrystalline morphology. *Macromolecules* **1991**, *24*, 1190–1195. [CrossRef]
63. MacKnight, W.J.; Karasz, F.E.; Fried, J.R. Solid state transition behavior of blends. In *Polymer Blends*; Academic Press: Cambridge, MA, USA, 1978; pp. 185–242. [CrossRef]
64. Couchman, P.R. Compositional variation of glass-transition temperatures. II. Application of the thermodynamic theory to compatible polymer blends. *Macromolecules* **1978**, *11*, 1156–1161. [CrossRef]
65. Fox, T.G. Influence of diluent and of copolymer composition on the glass temperature of a polymer system. *Bull. Am. Phys. Soc.* **1956**, *1*, 123.
66. Gordon, M.; Taylor, J.S. Ideal copolymers and the second-order transitions of synthetic rubbers. I. Non-crystalline copolymers. *J. Appl. Chem.* **1952**, *2*, 493–500. [CrossRef]
67. Wood, L.A. Glass transition temperatures of copolymers. *J. Polym. Sci.* **1958**, *28*, 319–330. [CrossRef]
68. Utracki, L.A.; Jukes, J.A. Dielectric studies of poly(vinyl chloride). *J. Vinyl Addit. Technol.* **1984**, *6*, 85–94. [CrossRef]

69. Aubin, M.; Prud'homme, R.E. Analysis of the glass transition temperature of miscible polymer blends. *Macromolecules* **1988**, *21*, 2945–2949. [CrossRef]
70. Brostow, W.; Chiu, R.; Kalogeras, I.M.; Vassilikou Dova, A. Prediction of glass transition temperatures: Binary blends and copolymers. *Mater. Lett.* **2008**, *62*, 3152–3155. [CrossRef]
71. Wang, C.B.; Cooper, S.L. Morphology and properties of segmented polyether polyurethaneureas. *Macromolecules* **1983**, *16*, 775–786. [CrossRef]
72. Di Lorenzo, M.L.; Frigione, M. Compatibilization criteria and procedures for binary blends: A review. *J. Polym. Eng.* **1997**, *17*, 429–459. [CrossRef]
73. Utracki, L.A.; Favis, B.D. Handbook of polymer science and technology. In *Polymer Alloys and Blends*; Marcel Dekker, INC.: New York, NY, USA, 1989; Volume 4, pp. 121–185.
74. Mamat, A.; Vu Khanh, T.; Cigana, P.; Favis, B.D. Impact fracture behavior of nylon-6 / ABS blends. *J. Polym. Sci. Part B Polym. Phys.* **1997**, *35*, 2583–2592. [CrossRef]
75. Zhumad, B. The European Lubricants Industry. *Lube Magazin*, 2014; Volume 121, 93, 22–27.
76. Larson, R.G. *The Structure and Rheology of Complex Fluids*; Oxford University Press: New York, NY, USA, 1990; Volume 150.
77. Bakrani Balani, S.; Chabert, F.; Nassiet, V.; Cantarel, A.; Garnier, C. Toward improvement of the properties of parts manufactured by FFF (fused filament fabrication) through understanding the influence of temperature and rheological behaviour on the coalescence phenomenon. In Proceedings of the 20th International Conference on Material Forming, Dublin, Ireland, 26–28 April 2017.
78. Bakrani Balani, S.; Mokhtarian, H.; Coatanéa, E.; Chabert, F.; Nassiet, V.; Cantarel, A. Integrated modeling of heat transfer, shear rate, and viscosity for simulation-based characterization of polymer coalescence during material extrusion. *J. Manuf. Process.* **2023**, *90*, 443–459. [CrossRef]
79. Korycki, A. Study of the Selective Laser Sintering Process: Materials Properties and Effect of Process Parameters. Ph.D. Thesis, Université de Toulouse, Tarbes, France, 2020.
80. Yuan, M.; Galloway, J.A.; Hoffman, R.J.; Bhatt, S. Influence of molecular weight on rheological, thermal, and mechanical properties of PEEK. *Polym. Eng. Sci.* **2011**, *51*, 94–102. [CrossRef]
81. Bonmatin, M.; Chabert, F.; Bernhart, G.; Djilali, T. Rheological and crystallization behaviors of low processing temperature poly(aryl ether ketone). *J. Appl. Polym. Sci.* **2021**, *138*, 51402. [CrossRef]
82. Chen, J.; Liu, X.; Yang, D. Rheological properties in blends of poly(aryl ether ether ketone) and liquid crystalline poly(aryl ether ketone). *J. Appl. Polym. Sci.* **2006**, *102*, 4040–4044. [CrossRef]
83. Nandan, B.; Kandpal, L.D.; Mathur, G.N. Poly(ether ether ketone)/poly(aryl ether sulfone) blends: Melt rheological behavior. *J. Polym. Sci. Part B Polym. Phys.* **2004**, *42*, 1548–1563. [CrossRef]
84. Ma, C.C.M.; Hsia, H.C.; Liu, W.L.; Hu, J.T. Thermal and rheological properties of poly(phenylene sulfide) and poly(ether etherketone) resins and composites. *Polym. Compos.* **1987**, *8*, 256–264. [CrossRef]
85. Bretas, R.E.S.; Collias, D.; Baird, D.G. Dynamic rheological properties of polyetherimide/polyetheretherketone/liquid crystalline polymer ternary blends. *Polym. Eng. Sci.* **1994**, *34*, 1492–1496. [CrossRef]
86. Basseri, G.; Mehrabi Mazidi, M.; Hosseini, F.; Razavi Aghjeh, M.K. Relationship among microstructure, linear viscoelastic behavior and mechanical properties of SBS triblock copolymer-compatibilized PP/SAN blend. *Polym. Bull.* **2014**, *71*, 465–486. [CrossRef]
87. Rosa, M.; Grassia, L.; D'Amore, A.; Carotenuto, C.; Minale, M. Rheology and mechanics of polyether(ether)ketone—Polyetherimide blends for composites in aeronautics. In *AIP Conference Proceedings, Proceedings of the VIII International conference on "Times of Polymers and Composites", Naples, Italy, 19–23 June 2016*; AIP Publishing: Long Island, NY, USA, 2016; p. 020177. [CrossRef]
88. Day, M.; Sally, D.; Wiles, D.M. Thermal degradation of poly(aryl-ether-ether-ketone): Experimental evaluation of crosslinking reactions. *J. Appl. Polym. Sci.* **1990**, *40*, 1615–1625. [CrossRef]
89. Day, M.; Cooney, J.D.; Wiles, D.M. The kinetics of the oxidative degradation of poly(aryl-ether-ether-ketone) (PEEK). *Thermochim. Acta* **1989**, *147*, 189–197. [CrossRef]
90. Zhang, H. Fire-Safe Polymers and Polymer Composites. Ph.D. Thesis, University of Massachusetts Amherst, Amherst, MA, USA, 2014.
91. Perng, L.H.; Tsai, C.J.; Ling, Y.C. Mechanism and kinetic modelling of PEEK pyrolysis by TG/MS. *Polymer* **1999**, *40*, 7321–7329. [CrossRef]
92. Lee, J.; Marvel, C.S. Polyaromatic ether-ketone sulfonamides prepared from polydiphenyl ether-ketones by chlorosulfonation and treatment with secondary amines. *J. Polym. Sci. Polym. Chem. Ed.* **1984**, *22*, 295–301. [CrossRef]
93. Fathima, N.N.; Aravindhan, R.; Lawrence, D.; Yugandhar, U.; Moorthy, T.S.R.; Nair, B.U. SPEEK polymeric membranes for fuel cell application and their characterization: A review. *J. Sci. Ind. Res.* **2007**, *66*, 209–219.
94. Bishop, M.T.; Karasz, F.E.; Russo, P.S.; Langley, K.H. Solubility and properties of a poly(aryl ether ketone) in strong acids. *Macromolecules* **1985**, *18*, 86–93. [CrossRef]
95. Shibuya, N.; Porter, R.S. Kinetics of PEEK sulfonation in concentrated sulfuric acid. *Macromolecules* **1992**, *25*, 6495–6499. [CrossRef]
96. Arigonda, M.; Deshpande, A.P.; Varughese, S. Effect of PES on the morphology and properties of proton conducting blends with sulfonated poly(ether ether ketone). *J. Appl. Polym. Sci.* **2013**, *127*, 5100–5110. [CrossRef]
97. Jin, X.; Bishop, M.T.; Ellis, T.S.; Karasz, F.E. A sulphonated poly(aryl ether ketone). *Br. Polym. J.* **1985**, *17*, 4–10. [CrossRef]

98. Ogawa, T.; Marvel, C.S. Polyaromatic ether-ketones and ether-keto-sulfones having various hydrophilic groups. *J. Polym. Sci. Polym. Chem. Ed.* **1985**, *23*, 1231–1241. [CrossRef]
99. Kerres, J.; Ullrich, A.; Meier, F.; Häring, T. Synthesis and characterization of novel acid–base polymer blends for application in membrane fuel cells. *Solid State Ion.* **1999**, *125*, 243–249. [CrossRef]
100. Goodwin, A.A.; Hay, J.N.; Mouledous, G.A.C.; Biddlestone, F. A compatible blend of poly(ether ether ketonex) (PEEK) and poly(ether imide) (Ultem 1000). In *Integration of Fundamental Polymer Science and Technology—5*; Springer: Limburg, The Netherlands, 1991; pp. 44–50. ISBN 978-1-85166-587-7.
101. Torre, L.; Kenny, J.M. Blends of semicrystalline and amorphous polymeric matrices for high performance composites. *Polym. Compos.* **1992**, *13*, 380–385. [CrossRef]
102. Ramani, R.; Alam, S. Influence of poly(ether imide) on the free volume hole size and distributions in poly(ether ether ketone). *J. Appl. Polym. Sci.* **2012**, *125*, 3200–3210. [CrossRef]
103. Ramani, R.; Alam, S. Free volume study on the miscibility of PEEK/PEI blend using positron annihilation and dynamic mechanical thermal analysis. *J. Phys. Conf. Ser.* **2015**, *618*, 012035. [CrossRef]
104. Chen, J.; Guo, W.; Li, Z.; Tian, L. Crystalline and thermal properties in miscible blends of PEEK, PPS and PEI obtained by melt compounding. *Funct. Mater.* **2016**, *23*, 55–62. [CrossRef]
105. Dingemans, T.J.; Mendes, E.; Hinkley, J.J.; Weiser, E.S.; StClair, T.L. Poly(ether imide)s from diamines with para-, meta-, and ortho-arylene substitutions: Synthesis, characterization, and liquid crystalline properties. *Macromolecules* **2008**, *41*, 2474–2483. [CrossRef]
106. Kong, X.; Tang, H.; Dong, L.; Teng, F.; Feng, Z. Miscibility and crystallization behavior of solution-blended PEEK/PI blends. *J. Polym. Sci. Part B Polym. Phys.* **1998**, *36*, 2267–2274. [CrossRef]
107. Dominguez, S.; Derail, C.; Léonardi, F.; Pascal, J.; Brulé, B. Study of the thermal properties of miscible blends between poly(ether ketone ketone) (PEKK) and polyimide. *Eur. Polym. J.* **2015**, *62*, 179–185. [CrossRef]
108. Zimmermann, H.J.; Könnecke, K. Crystallization of poly(aryl ether ketones): 3. The crystal structure of poly(ether ether ketone ketone) (PEEKK). *Polymer* **1991**, *32*, 3162–3169. [CrossRef]
109. Shibata, M.; Fang, Z.; Yosomiya, R. Miscibility and crystallization behavior of poly(ether ether ketone ketone)/poly(ether imide) blends. *J. Appl. Polym. Sci.* **2001**, *80*, 769–775. [CrossRef]
110. Lee, H.S.; Kim, W.N. Glass transition temperatures and rigid amorphous fraction of poly(ether ether ketone) and poly(ether imide) blends. *Polymer* **1997**, *38*, 2657–2663. [CrossRef]
111. Frigione, M.; Naddeo, C.; Acierno, D. Crystallization behavior and mechanical properties of poly(aryl ether ether ketone)/poly(ether imide) blends. *Polym. Eng. Sci.* **1996**, *36*, 2119–2128. [CrossRef]
112. Sauer, B.B.; Hsiao, B.S. Broadening of the glass transition in blends of poly(aryl ether ketones) and a poly(ether imide) as studied by thermally stimulated currents. *J. Polym. Sci. Part B Polym. Phys.* **1993**, *31*, 917–932. [CrossRef]
113. Hsiao, B.S.; Verma, R.K.; Sauer, B.B. Crystallization study of poly(ether ether ketone)/poly(ether imide) blends by real-time small-angle x-ray scattering. *J. Macromol. Sci. Part B* **1998**, *37*, 365–374. [CrossRef]
114. Ding, Y.; Bikson, B. Preparation and characterization of semi-crystalline poly(ether ether ketone) hollow fiber membranes. *J. Membr. Sci.* **2010**, *357*, 192–198. [CrossRef]
115. Huang, T.; Chen, G.; He, Z.; Xu, J.; Liu, P. Pore structure and properties of poly(ether ether ketone) hollow fiber membranes: Influence of solvent-induced crystallization during extraction. *Polym. Int.* **2019**, *68*, 1874–1880. [CrossRef]
116. Wang, J.; Cao, J.; Chen, Y.; Ke, Y.; Wu, Z.; Mo, Z. Crystallization behavior of poly(ether ether ketone ketone). *J. Appl. Polym. Sci.* **1996**, *61*, 1999–2007. [CrossRef]
117. Karcha, R.J.; Porter, R.S. Miscible blends of modified poly(aryl ether ketones) with aromatic polyimides. *J. Polym. Sci. Part B Polym. Phys.* **1993**, *31*, 821–830. [CrossRef]
118. Smyser, G.L.; Brooks, G.T. Injection moldable blends of poly(etherketones) and polyamide-imides. Patent US4963627A, 1990.
119. Berr, C.E. Polyketone copolymers. Patent US3516966, 1970.
120. Thornton, R.L. Boron trifluoride-hydrogen fluoride catalyzed synthesis of poly(aromatic sulfone) and poly(aromatic ketone) polymers. Patent US3442857, 1966.
121. Arzak, A.; Eguiazabal, J.I.; Nazabal, J. Phase behaviour and mechanical properties of poly(ether ether ketone)-poly(ether sulphone) blends. *J. Mater. Sci.* **1991**, *26*, 5939–5944. [CrossRef]
122. Nandan, B.; Kandpal, L.D.; Mathur, G.N. Glass transition behaviour of poly(ether ether ketone)/poly(aryl ether sulphone) blends: Dynamic mechanical and dielectric relaxation studies. *Polymer* **2003**, *44*, 1267–1279. [CrossRef]
123. Nandan, B.; Kandpal, L.D.; Mathur, G.N. Poly(ether ether ketone)/poly(aryl ether sulfone) blends: Relationships between morphology and mechanical properties. *J. Appl. Polym. Sci.* **2003**, *90*, 2887–2905. [CrossRef]
124. Ni, Z. The preparation, compatibility and structure of PEEK–PES blends. *Polym. Adv. Technol.* **1994**, *5*, 612–614. [CrossRef]
125. Korycki, A.; Garnier, C.; Abadie, A.; Nassiet, V.; Sultan, C.T.; Chabert, F. Poly(etheretherketone)/poly(ethersulfone) blends with phenolphthalein: Miscibility, thermomechanical properties, crystallization and morphology. *Polymers* **2021**, *13*, 1466. [CrossRef]
126. Malik, T.M. Thermal and mechanical characterization of partially miscible blends of poly(ether ether ketone) and polyethersulfone. *J. Appl. Polym. Sci.* **1992**, *46*, 303–310. [CrossRef]
127. Nandan, B.; Kandpal, L.D.; Mathur, G.N. Polyetherether ketone/polyarylethersulfone blends: Thermal and compatibility aspects. *J. Polym. Sci. Part B Polym. Phys.* **2002**, *40*, 1407–1424. [CrossRef]

128. Arzak, A.; Eguiazabal, J.I.; Nazabal, J. Compatibility in immiscible poly(ether ether ketone)/poly(ether sulfone) blends. *J. Appl. Polym. Sci.* **1995**, *58*, 653–661. [CrossRef]
129. Sung Chun, Y.; Weiss, R.A. Thermal behavior of poly(ether ketone ketone)/thermoplastic polyimide blends. *J. Appl. Polym. Sci.* **2004**, *94*, 1227–1235. [CrossRef]
130. Sauer, B.B.; Hsiao, B.S.; Faron, K.L. Miscibility and phase properties of poly(aryl ether ketone)s with three high temperature all-aromatic thermoplastic polyimides. *Polymer* **1996**, *37*, 445–453. [CrossRef]
131. Kong, X.; Teng, F.; Tang, H.; Dong, L.; Feng, Z. Miscibility and crystallization behaviour of poly(ether ether ketone)/polyimide blends. *Polymer* **1996**, *37*, 1751–1755. [CrossRef]
132. Mehta, A.; Isayev, A.I. The dynamic properties, temperature transitions, and thermal stability of poly (etherether ketone)-thermotropic liquid crystalline polymer blends. *Polym. Eng. Sci.* **1991**, *31*, 963–970. [CrossRef]
133. De Carvalho, B.; Bretas, R.E.S. Crystallization kinetics of a PEEK/LCP blend. *J. Appl. Polym. Sci.* **1995**, *55*, 233–246. [CrossRef]
134. Jung, H.C.; Lee, H.S.; Chun, Y.S.; Kim, S.-B.; Kim, W.N. Blends of a thermotropic liquid crystalline polymer and some flexible chain polymers and the determination of the polymer-polymer interaction parameter of the two polymers. *Polym. Bull.* **1998**, *41*, 387–394. [CrossRef]
135. Liu, P.; Mullins, M.; Bremner, T.; Benner, N.; Sue, H.J. Interfacial phenomena and mechanical behavior of polyetheretherketone/polybenzimidazole blend under hygrothermal environment. *J. Phys. Chem. B* **2017**, *121*, 5396–5406. [CrossRef]
136. Stuart, B.H.; Briscoe, B.J. A fourier transform raman spectroscopy study of poly (ether ether ketone)/polytetrafluoroethylene (PEEK/PTFE) blends. *Spectrochim. Acta Part A Mol. Spectrosc.* **1994**, *50*, 2005–2009. [CrossRef]
137. Frick, A.; Sich, D.; Heinrich, G.; Lehmann, D.; Gohs, U.; Stern, C. Properties of melt processable PTFE/PEEK blends: The effect of reactive compatibilization using electron beam irradiated melt processable PTFE. *J. Appl. Polym. Sci.* **2013**, *128*, 1815–1827. [CrossRef]
138. Thomas, P.S.; Stuart, B.H. DSC characterisation of compression moulded PEEK-PTFE plaques. *J. Therm. Anal. Calorim.* **2003**, *72*, 675–679. [CrossRef]
139. Karcha, R.J.; Porter, R.S. Miscible blends of a sulfonated poly(aryl ether ketone) and aromatic polyimides. *J. Polym. Sci. Part B Polym. Phys.* **1989**, *27*, 2153–2155. [CrossRef]
140. Karcha, R.J. Blends of Polyimides with poly(ether ether ketone) and PEEK Derivatives. Ph.D. Thesis, University of Massachusetts Amherst, Amherst, MA, USA, 1990.
141. Wu, H.-L.; Ma, C.-C.M.; Liu, F.-Y.; Chen, C.-Y.; Lee, S.-J.; Chiang, C.-L. Preparation and characterization of poly(ether sulfone)/sulfonated poly(ether ether ketone) blend membranes. *Eur. Polym. J.* **2006**, *42*, 1688–1695. [CrossRef]
142. Zaidi, S.M.J. Preparation and characterization of composite membranes using blends of SPEEK/PBI with boron phosphate. *Electrochim. Acta* **2005**, *50*, 4771–4777. [CrossRef]
143. Kerres, J.; Ullrich, A.; Häring, T.; Baldauf, M.; Gebhardt, U.; Preidel, W. Preparation, characterization and fuel cell application of new acid-base blend membranes. *J. New Mater. Electrochem. Syst.* **2000**, *3*, 229–239.
144. Silva, V.S.; Weisshaar, S.; Reissner, R.; Ruffmann, B.; Vetter, S.; Mendes, A.; Madeira, L.M.; Nunes, S. Performance and efficiency of a DMFC using non-fluorinated composite membranes operating at low/medium temperatures. *J. Power Sources* **2005**, *145*, 485–494. [CrossRef]
145. Silva, V.S.; Ruffmann, B.; Vetter, S.; Mendes, A.; Madeira, L.M.; Nunes, S.P. Characterization and application of composite membranes in DMFC. *Catal. Today* **2005**, *104*, 205–212. [CrossRef]
146. Pasupathi, S.; Ji, S.; Bladergroen, B.J.; Linkov, V. High DMFC performance output using modified acid–base polymer blend. *Int. J. Hydrog. Energy* **2008**, *33*, 3132–3136. [CrossRef]
147. Arzak, A.; Eguiazábal, J.I.; Nazábal, J. Mechanical performance of directly injection-molded PEEK/PEI blends at room and high temperature. *J. Macromol. Sci. Part B* **1997**, *36*, 233–246. [CrossRef]
148. Miao, Z.C.; Wang, D.; Xing, Y.; Gao, H.; Zhang, J. In situ composites based on blends of PEEK and thermotropic liquid crystalline polymer. *Mol. Cryst. Liq. Cryst.* **2016**, *630*, 139–143. [CrossRef]
149. Alvarez, E.; DiSano, L.P. Molded polybenzimidiazole/-polyaryleneketone articles and method of manufacture. Patent US005070153, 1991.
150. Bijwe, J.; Sen, S.; Ghosh, A. Influence of PTFE content in PEEK–PTFE blends on mechanical properties and tribo-performance in various wear modes. *Wear* **2005**, *258*, 1536–1542. [CrossRef]
151. Harris, J.E.; Robeson, L.M.; Gavula, J.P. Blends of poly(aryl ketone) and a polyetherimide. Patent US004609714, 1992.
152. Gensler, R.; Plummer, C.J.G.; Kausch, H.H.; Münstedt, H. Thin film and bulk deformation behaviour of poly(ether ether ketone)/poly(ether imide) blends. *J. Mater. Sci.* **1997**, *32*, 3037–3042. [CrossRef]
153. Browne, M.M.; Forsyth, M.; Goodwin, A.A. The effect of solvent uptake on the relaxation behaviour, morphology and mechanical properties of a poly(ether ether ketone)/poly(etherimide) blend. *Polymer* **1997**, *38*, 1285–1290. [CrossRef]
154. Briscoe, B.J.; Yao, L.H.; Stolarski, T.A. The fraction and wear of poly(tetrafluoroethylene)-poly(etheretherketone) composites: An initial appraisal of the optimum composition. *Wear* **1986**, *108*, 357–374. [CrossRef]
155. Lu, Z.P.; Friedrich, K. On sliding friction and wear of PEEK and its composites. *Wear* **1995**, *181–183*, 624–631. [CrossRef]
156. Zhang, R.; Häger, A.M.; Friedrich, K.; Song, Q.; Dong, Q. Study on tribological behaviour of plasma-treated PEEK and its composites. *Wear* **1995**, *181–183*, 613–623. [CrossRef]

157. Daud, W.R.W.; Ghasemi, M.; Chong, P.S.; Jahim, J.M.; Lim, S.S.; Ismail, M. SPEEK/PES composite membranes as an alternative for proton exchange membrane in microbial fuel cell (MFC). In Proceedings of the Conference on Clean Energy and Technology, Kuala Lumpur, Malaysia, 27–29 June 2011; IEEE: Kuala Lumpur, Malaysia, 2011; pp. 400–403.
158. Ismail, A.F.; Lau, W.J. Theoretical studies on structural and electrical properties of PES/SPEEK blend nanofiltration membrane. *AIChE J.* **2009**, *55*, 2081–2093. [CrossRef]
159. Harris, J.E.; Robeson, M.; Station, W. Blends of poly(aryl ketones). Patent US004609714, 1986.
160. Dawkins, B.G.; Gruender, M.; Copeland, G.S.; Zucker, J. Tri-blended resin of PBI, PAEK, and PEI. Patent US007629420B2, 2009.
161. Chen, J.; Guo, Q.; Zhao, Z.; Wang, X.; Duan, C. Structures and mechanical properties of PEEK/PEI/PES plastics alloys blent by extrusion molding used for cable insulating jacketing. *Procedia Eng.* **2012**, *36*, 96–104. [CrossRef]
162. Karcha, R.J.; Porter, R.S. Ternary blends of sulphonated PEEK and two aromatic polyimides. *Polymer* **1992**, *33*, 4866–4867. [CrossRef]
163. Bretas, R.E.S.; Baird, D.G. Miscibility and mechanical properties of poly(ether imide)/poly(ether ether ketone)/liquid crystalline polymer ternary blends. *Polymer* **1992**, *33*, 5233–5244. [CrossRef]
164. Morales, A.R.; Bretas, R.E.S. Polyetherimide/poly(ether ether ketone)/liquid crystalline polymer ternary blends-II. Approximited interaction parameters and phase diagrams. *Eur. Polym. J.* **1996**, *32*, 365–373. [CrossRef]

Disclaimer/Publisher's Note: The statements, opinions and data contained in all publications are solely those of the individual author(s) and contributor(s) and not of MDPI and/or the editor(s). MDPI and/or the editor(s) disclaim responsibility for any injury to people or property resulting from any ideas, methods, instructions or products referred to in the content.

Article

Improved Energy Storage Performance of Composite Films Based on Linear/Ferroelectric Polarization Characteristics

Chen Chen [1,2,*], Lifang Shen [2], Guang Liu [1,2,*], Yang Cui [1,2] and Shubin Yan [2]

[1] Nanxun Innovation Institute, Zhejiang University of Water Resources and Electric Power, Hangzhou 310018, China; cuiy@zjweu.edu.cn

[2] School of Electrical Engineering, Zhejiang University of Water Resources and Electric Power, Hangzhou 310018, China; shenlf@zjweu.edu.cn (L.S.); yanshb@zjweu.edu.cn (S.Y.)

* Correspondence: cc@zjweu.edu.cn (C.C.); lg@zjweu.edu.cn (G.L.)

Abstract: The development and integration of high-performance electronic devices are critical in advancing energy storage with dielectric capacitors. Poly(vinylidene fluoride-trifluoroethylene-chlorofluoroethylene) (PVTC), as an energy storage polymer, exhibits high-intensity polarization in low electric strength fields. However, a hysteresis effect can result in significant residual polarization, leading to a severe energy loss, which impacts the resultant energy storage density and charge/discharge efficiency. In order to modify the polarization properties of the polymer, a biaxially oriented polypropylene (BOPP) film with linear characteristics has been selected as an insulating layer and combined with the PVTC ferroelectric polarization layer to construct PVTC/BOPP bilayer films. The hetero-structure and polarization characteristics of the bilayer film have been systematically studied. Adjusting the BOPP volume content to 67% resulted in a discharge energy density of 10.1 J/cm^3 and an energy storage efficiency of 80.9%. The results of this study have established the mechanism for a composite structure regulation of macroscopic energy storage performance. These findings can provide a basis for the effective application of ferroelectric polymer-based composites in dielectric energy storage.

Keywords: polarization characteristics; PVTC/BOPP bilayer films; interfaces; energy storage properties

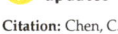

Citation: Chen, C.; Shen, L.; Liu, G.; Cui, Y.; Yan, S. Improved Energy Storage Performance of Composite Films Based on Linear/Ferroelectric Polarization Characteristics. *Polymers* 2024, *16*, 1058. https://doi.org/10.3390/polym16081058

Academic Editors: Phuong Nguyen-Tri, Ana Pilipović and Mustafa Özcanli

Received: 8 March 2024
Revised: 3 April 2024
Accepted: 9 April 2024
Published: 11 April 2024

Copyright: © 2024 by the authors. Licensee MDPI, Basel, Switzerland. This article is an open access article distributed under the terms and conditions of the Creative Commons Attribution (CC BY) license (https:// creativecommons.org/licenses/by/ 4.0/).

1. Introduction

Dielectric capacitors exhibit high power density and fast charge/discharge rates [1,2] and are widely used in pulse power devices in electrical/electronic engineering [3]. When compared with inorganic dielectric capacitors, polymer-based systems offer distinct advantages in terms of high breakdown strength, low dielectric loss, and low density. Moreover, the associated flexibility and processability offer many options for the micro-structure design and macroscopic performance control of thin film capacitors and embedded capacitors. Polymer-based dielectric capacitors have significant scope for application in high-energy weapons, electronic communication, and new energy vehicles [4,5]. The energy storage density of pure polymers is generally low and cannot meet the practical requirements of many compact pulse power devices. Current research in engineering dielectrics has focused on improvements in the energy storage characteristics of polymers. The latter depends on the polymer breakdown and polarization characteristics and the interrelationship of these factors. Improvements in the energy storage density of polymers require the targeted development of polymer-based composite systems [6].

Doping polymers with high dielectric constant inorganic fillers is the most common modification strategy used to enhance the energy storage performance of the composite due to the dielectric properties and interface polarization effect of the fillers. Tang et al. increased the relative dielectric constant of the composite to 17.5 at room temperature (ca. 25 °C) by doping a polyvinylidene fluoride (PVDF) matrix with 7.5 vol.% Ba$_{0.2}$Sr$_{0.8}$TiO$_3$ paraelectric nanofibers. When an electric field of 450 kV/mm was applied, the discharge energy density reached 14.86 J/cm^3, which was 42.9% higher than pure PVDF. However, a hysteresis effect due to polarization/depolarization resulted in a ca. 40% energy loss and a resultant energy storage efficiency of 60% [7]. Zhang and co-workers enhanced the interfacial polarization effect significantly by doping PVDF with BaTiO$_3$@TiO$_2$ fibers, generating a maximum polarization intensity of 9 µC/cm^2. The application of a 650 kV/mm electric field delivered a discharge energy density of 21.2 J/cm^3, where the energy loss was 25% [8]. The introduction of high dielectric fillers can improve the dielectric constant of the polymer to a certain extent, but it ultimately leads to a decrease in breakdown strength. This may be attributed to the large dielectric difference between the polymer matrix and inorganic fillers in addition to the local electric field distortion caused by interface incompatibility. An accumulation of ceramic filler aggregates results in conductive pathways along the electric field direction in the polymer matrix, causing a decrease in the breakdown strength of the composite material and an inability to achieve a high-energy storage density.

Doping the polymer matrix with the varying content, type, and morphology of the functional filler can achieve a high breakdown, high polarization, and a high thermal conductivity layer. By utilizing the dielectric mismatch between adjacent film layers to regulate the spatial electric field distribution, an interface barrier effect is formed that effectively blocks the breakdown path and improves the energy storage performance of the composite film. Introducing nanoscale fillers into the polymer generates a hard layer and soft layer, representing film layers with high breakdown strength and high polarization strength, respectively. The hard layer helps to suppress the formation of conductive channels, while the soft layer provides a high polarization intensity in high-electric-strength fields [9,10]. The combination of different hard and soft layers and the interfacial potential barriers between the film layers have a significant impact on the dielectric properties of composite films. The soft layer typically involves barium titanate (BaTiO$_3$) and titanium dioxide (TiO$_2$) doping [11,12], whereas high-insulation fillers (two-dimensional nano sheets) are associated with hard layers [13,14]. Research has shown that the formation of a "hard–soft" bilayer composite film can effectively reduce the leakage current density and improve overall dielectric properties. Liu et al. incorporated BaTiO$_3$ particles into poly(vinylidene fluoride-co-hexafluoropropene) (P(VDF-HFP)) and constructed a bilayer composite film with poly(tetrafluoroethylene vinylidene fluoride hexafluoropropene) (THV). At a field strength of 575.6 kV/mm, a discharge energy density (U_d) of 22.7 J/cm^3 and a discharge efficiency (η) of 79% were obtained [15]. In addition to the hard–soft double-layer structure, sandwich composite films with a "soft–hard–soft" arrangement have also been studied. Li et al. utilized 2% MgO nw/P(VDF-HFP) as the hard layer and 20% BaTiO$_3$ nw/P(VDF-HFP) as the soft layer [16]; at an applied field strength of 416 MV/m, the U_d was 15.5 J/cm^3. While the soft–hard–soft structure can increase the energy storage density, there are certain negative effects. Choosing a nano-filler polarization layer as the outer layer facilitates electrode charge injection under high-electric fields, and the consequent improvement of energy storage is limited.

In multi-layer polymer-based composite films, the polarization effect at the interlayer interface perpendicular to the direction of the applied electric field is beneficial for improving dielectric properties. The inorganic filler and polymer film are combined to form a multi-layer composite film structure. As the number of film layers increases, the interlayer interface component increases, hindering the expansion of the electrical tree along the direction of the electric field. Jiang et al. investigated the influence of the number of layers on the dielectric properties, preparing multi-layer composite films by alternately stacking P(VDF-HFP) and BT/P(VDF-HFP) composite fiber layers [16]. As the number of interlayer interfaces increased, the dielectric constant of the multi-layer structure showed a slight increase, while the dielectric loss gradually decreased. The application of the phase field simulation also established that the interlayer interface suppressed the movement of charge carriers, reducing the leakage current density and improving overall energy storage performance. Differences in the relative dielectric constant and conductivity of adjacent film layers in the alternating multi-layer thin film result in a severe distortion of the electric field at the interface. Gradient structured composite films can regulate the distribution of electric field gradients by continuously changing the electrical parameters. Taking a three-layer gradient structure thin film as an example, the relative dielectric constant of each film layer gradually increases from top to bottom. Under an external electric field, the electric field distributed by each film layer gradually decreases. Assuming that the electrical breakdown is caused by the migration of charge carriers from electrode A to electrode B, the electrical tree will grow downwards from the upper film with a higher electric field strength. When the electrical tree grows to reach interface A, the electric field strength of the middle film layer is lower than the upper layer, which limits the growth of the electrical tree to a certain extent. A minor component of electrical trees will continue to grow in the middle film layer. When the electrical tree grows to interface B, the lower film bears a lower electric field strength, and it is possible to terminate the growth of the electrical tree. Assuming that electrical breakdown is caused by the migration of charge carriers from electrode B to electrode A, the electric field strength of the lower film is lower, and it is difficult to induce electrical tree branches. The penetration of electrical tree branches through the lower film requires a higher electric field strength to improve the overall breakdown field strength of the composite film [17].

Multi-layer nano-composite films not only integrate the high dielectric constant of inorganic nano-fillers and the high breakdown strength of polymer films, but the multi-layer structure also enables an adjustment of filler content, circumventing filler aggregation with a consequent decline in performance caused by high doping content. Due to the difference in the dielectric constant between film layers in nanoscale multi-layer structures, the spatial electric field is redistributed, which can effectively suppress the development of electrical tree branches and improve the dielectric properties. The introduction of nano-fillers not only serves to optimize polymer energy storage performance but also leads to an electric field distortion near the filler due to the large dielectric differential between the filler and matrix, which is not conducive to improved energy storage performance. Some polymers require biaxial stretching treatment during production, where the incorporation of fillers can generate many defects during the stretching process that inhibit flexibility. The complexity of the preparation process limits the feasibility of large-scale production. The utilization of linear polymers with good insulation properties as the insulation layer and ferroelectric polymers, as the polarization layer enables the construction of a fully organic multi-layer thin film by stacking growth. Adjusting the volume ratio or stacking method of the insulation and polarization layers facilitates the formation of a composite thin film that combines high-energy density and high discharge efficiency. As a new generation of high-energy density dielectrics, PVDF-derived copolymers exhibit high discharge energy density, but the discharge efficiency sharply decreases (<70%) under a high field strength. Researchers have employed PVDF as the I-layer and P(VDF-TrFE-CTFE) as the P-layer. Due to the difference in the dielectric constant between the film layers, the middle P-layer bears a lower electric field and is less prone to premature breakdown, which enhances the breakdown field strength of the composite film. When a field strength of 660 MV/m was applied, U_d reached 20.86 J/cm^3, with a η of 62%. The use of polymers such as polyimide (PI) and polymethyl methacrylate (PMMA) with high breakdown strength and high discharge efficiency (>90%) as the I-layer to construct linear/ferroelectric hetero-structure composite films represents an effective strategy for achieving a high density and high rate [18,19].

In this study, we have set out to address PVTC polymer polarization, preparing a bilayer film with a linear/ferroelectric hetero-structure by incorporating BOPP. The difference in the dielectric properties of the linear and ferroelectric polymers is utilized to redistribute the electric field within the bilayer film. At a BOPP volume content of 67%, the PVTC/BOPP bilayer film exhibited excellent energy storage characteristics. At an electric field strength of 550 kV/mm, the energy storage density and charge/discharge efficiency reached 10.1 J/cm^3 and 80.9%, respectively. The organic multi-layer composite structure utilizes the performance characteristics of the constituent polymers, avoiding mismatches between inorganic fillers and polymer matrices, and simplifying materials preparation. Moreover, the hetero-structure interface enhances the dielectric properties of the multi-layer composite film. When compared with nano-composite multi-layer structures, all organic multi-layer films have shown significant advantages in terms of charge/discharge cycle stability and mechanical bending.

2. Experimental Section

2.1. Raw Materials

BOPP films with different thicknesses (5 µm, 7.5 µm, 10 µm, and 15 µm) and poly (vinylidene fluoride-trifluoroethylene-chlorofluoroethylene) (PVTC) powder were purchased from PolyK Technologies (Philipsburg, PA, USA); the ratios of vinylidene fluoride (VDF), trifluoroethylene (TrFE), and chlorofluoroethylene (CFE) monomers were 62%, 31%, and 7%, respectively. The N,N-dimethylformamide (DMF) reagent was obtained from the Sinopharm Chemical Reagent Co., Ltd., Shanghai, China.

2.2. Preparation of the Bilayer Films

BOPP films of different thicknesses were placed on a glass substrate for fixation. A PVTC/DMF solution was then cast on the BOPP film after vacuum degassing. A bilayer wet organic film was obtained in a stepwise thermal treatment from 80 °C to 200 °C in a vacuum oven. The sample was then immediately quenched in ice water and dried to give the required bilayer films. The volume ratios of BOPP in the bilayer films were 33%, 50%, and 67%.

2.3. Structural Characterization and Performance Testing

The crystal structure of the polymer film was determined using a PANalytical Empire (PANARKO, Alemlo, The Netherlands) unit at an operating voltage of 40 kV and 40 mA current. Characterization of the molecular structure and chemical bonding in the polymer films was performed using JASCO 6100 equipment (EQUINOX55 from Bruker, Karlsruhe, Germany). The film morphology and structure were assessed using FESEM and Hitachi SU8020 UHR equipment (Hitachi, Tokyo, Japan). The energy level structure and bandgap widths of the polymer films were obtained with ultraviolet and visible spectrophotometry (UV-Vis) (UV-3600i Plus from Shimadzu, Kyoto, Japan) the lowest and highest energy levels were determined with ultraviolet photoelectron spectroscopy (UPS) (Thermo Fischer, ESCALAB 250Xi, Waltham, MA, USA). The dielectric properties (dielectric constant and dielectric loss) of the polymer films under different frequency conditions were assessed using a broadband impedance analyzer (GmbH Novo control Alpha-A, Montabaur, Germany) at room temperature; the sample was 30 mm in length and 30 mm in width. The DC breakdown characteristics of the composite films were tested using the DC breakdown module of the dielectric ferroelectric integrated test system (PolyK Technologies, Philipsburg, PA, USA), with a boosting rate of 200 V/s. The polarization strength was evaluated using the Radiant Premier II (Albuquerque, NM, USA) ferroelectric testing system to generate polarization curves at different electric field strengths. The energy storage parameters (discharge energy density and charge/discharge efficiency) were obtained from an integration of the polarization curves.

3. Results and Discussion

In order to establish the successful preparation of bilayer thin films, XRD and FTIR techniques were employed to analyze the phase composition of the composite films, and scanning electron microscopy was used to assess the microstructure. As shown in Figure 1a, the diffraction peaks at 17.9°, 18.5°, and 20.1° correspond to PVTC [20–22], and the peaks at 14.04°, 16.85°, and 18.48° may be attributed to BOPP [23,24]. These characteristic peaks were observed in both pristine films and the prepared bilayer films. The FTIR spectra of the polymer films are presented in Figure 1b, where the peaks at 976 cm^{-1}, 812 cm^{-1}, 532 cm^{-1}, and 776 cm^{-1} are due to the α- and γ-phases of PVTC [25–27]. The absorption peak at 1720 cm^{-1} may be attributed to (C=O) stretching, which is the result of oxidative degradation [28–30]. All the characteristic peaks were present in the bilayer films. The cross-sectional SEM images are given in Figure 1c–e, where the thickness of composite films is ca. 15 μm; there is no clearly discernible interface separation in the bilayer films.

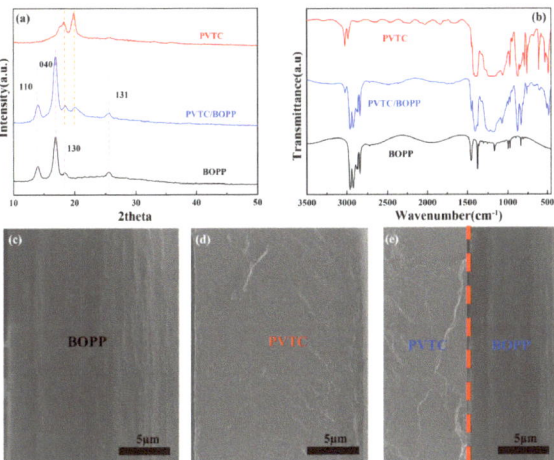

Figure 1. Micro-structure analyses of polymer films. (**a**) XRD patterns. (**b**) FTIR spectra. (**c**–**e**) SEM images.

The first key factor in determining energy storage performance is the dielectric property of the polymer films. The dielectric constants (Figure 2a) of the synthesized PVTC/BOPP bilayer films lie between the dielectric constants of pristine BOPP (ca. 2.37 at 1 kHz) and pristine PVTC (ca. 10.64 at 1 kHz). As shown in Figure 2b, the bilayer dielectric constants films at a frequency of 1 kHz are 7.14, 5.27, and 3.33, respectively. The dielectric constants decreased with increasing BOPP volume fraction in accordance with the theoretical dielectric constant law [31–33]. The dielectric loss associated with the PVTC/BOPP bilayer film is lower than that of the PVTC film in a given frequency range, which results from the lower dielectric loss of the incorporated BOPP film with linear polarization characteristics. As shown in Figure 2b, the dielectric loss of the PVTC/BOPP bilayer films at a frequency of 1 kHz is 0.010, 0.006, and 0.004, respectively. This dielectric loss includes polarization loss and conductivity loss. The dielectric loss of the bilayer films decreases with increasing frequency in the low frequency range but gradually increases in the high frequency range. This response demonstrates the relaxation characteristics of these dielectric materials, notably Maxwell–Wagner (M-W) relaxation at low frequencies and the α-relaxation phenomenon at high frequencies. In order to further investigate the factors that influence energy storage, the breakdown performance of the bilayer film was tested. Following a linear fit of the data, the Weibull distribution curve was generated [34,35], as shown in Figure 2c. The data analysis has revealed that the introduction of linear polymers results in an increase in the breakdown field strength (E_b) of the PVTC/BOPP bilayer film relative to the pristine PVTC film (ca. 350 kV/mm). When the breakdown probability is 63.2%, the breakdown field strengths (E_b) of the PVTC/BOPP bilayer films are 450.8 kV/mm, 487.9 kV/mm, and 528.9 kV/mm, respectively; the associated slopes of the breakdown data (β) are 7.6, 8.5, and 9.1. The larger the E_b value, the higher the fitting slope of the breakdown data is. The increase in breakdown field strength means that the dielectric difference between linear BOPP and ferroelectric PVTC results in a redistribution of the electric field. The BOPP layer with a smaller dielectric constant distributes a higher local electric field, and the breakdown field strength borne by PVTC in the bilayer film decreases. Simulation using the method in COMSOL Multiphysics (Figure 2e) has established that space charges tend to accumulate at the interface between PVTC and BOPP layers in the bilayer films due to the large mismatch in the electrical resistivity or dielectric constant of adjacent layers. The interfacial charges give rise to an increase in the average dielectric constant of the bilayer films [36,37]. The data presented in Figure 2f illustrate that the introduction of a linear layer is accompanied by a redistribution of the electric field inside the bilayer film.

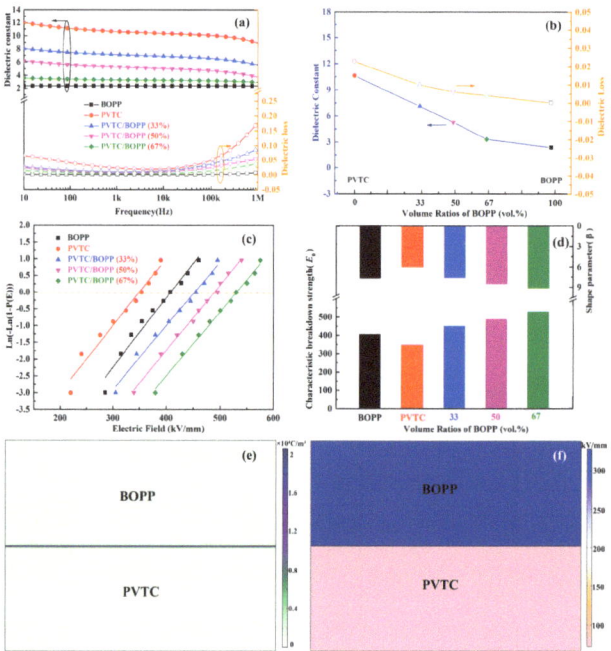

Figure 2. (a) Frequency-dependent changes of the dielectric constant and dielectric loss of the bilayer films. (b) The relationship between dielectric properties and BOPP volume ratios in bilayer films measured at 1 kHz. (c) Weibull distribution of the bilayer films. (d) Characteristic breakdown strength (E_b) and shape parameter (β). (e) The space charge density simulation for the bilayer film. (f) The distribution of electric field simulated for the bilayer film.

In order to gain a better understanding of the performance mechanism, the energy band structures of the PVTC ferroelectric and BOPP linear layers were analyzed. The UPS spectra of the polymer films are presented in Figure 3a,b, where the minimum binding energy (E_{HOMO}) associated with the right-hand side and the maximum binding energy (E_{SEE}) in the left-hand side can be obtained and the ionization potential (IP) calculated using Equation (1).

$$IP_{HOMO} = h\nu - (E_{SEE} - E_{HOMO}) \quad (1)$$

The calculated IP_{HOMO} allows for a determination of the positions of HOMO energy levels in the inorganic layer and polymer thin film; the positions of LUMO energy levels can be obtained from the bandgap width. Given that the work function of Au is 5.2 eV, combined with the HOMO and LUMO energy levels, the electron (ϕ_e) and hole(ϕ_h) potential barriers for each material can be calculated [38,39].

The UV-Vis analysis of the PVTC and BOPP films was undertaken to calculate the bandgap widths, as shown in Figure 3c,d. The UV absorption intensity of the material is presented as a function of the photoelectron wavelength. The relationship between the absorption rate and photon energy in the bandgap can be expressed by T_{auc} Equation (2):

$$(\alpha h\nu)^2 = B(h\nu - E_g) \qquad (2)$$

where α represents the absorption coefficient, $h\nu$ is the photon energy, h is the Planck constant ($4.13567 10^{-15}$ eV·s), ν is the frequency of incident photons ($\nu = c/\lambda$, where c is the speed of light (3×10^8 m/s) and λ is the wavelength of the incident light), B is the proportional constant, and E_g is the bandgap width of the material. The calculated E_g values for BOPP and PVTC are 5.09 eV and 4.81 eV, respectively [40,41]. Based on the calculated bandgap width, electron barrier height, and hole barrier height, the band structures were generated and are shown in Figure 3e,f. The barrier height associated with hole and electron injection in the case of BOPP and PVTC are 3.81 eV, 3.40 eV and 1.28 eV, and 1.80 eV, respectively. Given the variation of the electron barrier height at the electrode dielectric interface, the introduction of a higher insulation layer can suppress the charge injection at the interface between the electrode and dielectric, resulting in an improved breakdown resistance for the bilayer films [42–45].

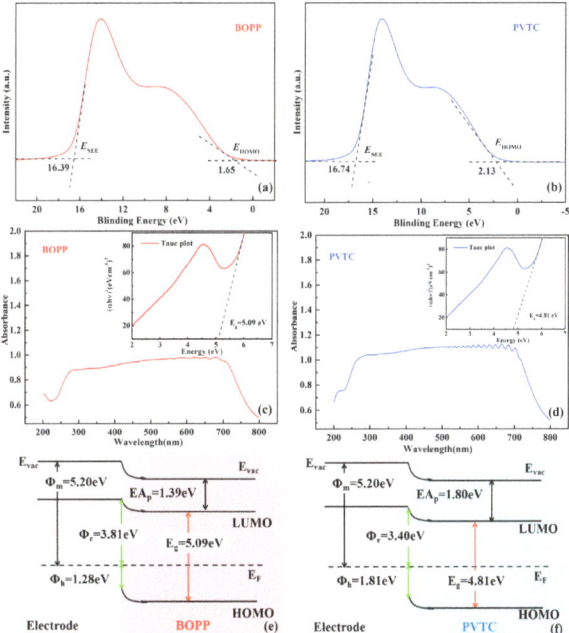

Figure 3. UPS spectra of (a) single-layer BOPP films and (b) single-layer PVTC films. UV-Vis absorption spectra (with inset showing the T_{auc} analyses) of (c) single-layer BOPP films and (d) single-layer PVTC films. Energy band structure diagram of (e) Au/BOPP and (f) Au/PVTC.

The incorporation of linear BOPP in constructing bilayer films serves to improve energy storage performance to a certain extent. In order to determine the relevant energy storage parameters, the charge/discharge curves of the polymer film were measured and are shown in Figure 4a,b. It can be seen that at the same electric field intensity, the PVTC/BOPP bilayer film exhibits a smaller curve area relative to the PVTC film. The maximum electric displacements (D_m) at the same electric field intensity for the PVTC/BOPP(33%), PVTC/BOPP(50%), and PVTC/BOPP(67%) bilayer films are 5.38 µC/cm², 4.35 µC/cm², and 3.96 µC/cm², respectively. The remanent electric displacements (P_r) at the same electric field intensity for PVTC/BOPP(33%), PVTC/BOPP(50%), and PVTC/BOPP(67%) bilayer films are 1.32 µC/cm², 0.92 µC/cm², and 0.43 µC/cm², respectively. These results indicate that BOPP, a linear dielectric with a smaller dielectric constant, can improve to some extent the large residual polarization associated with PVTC.

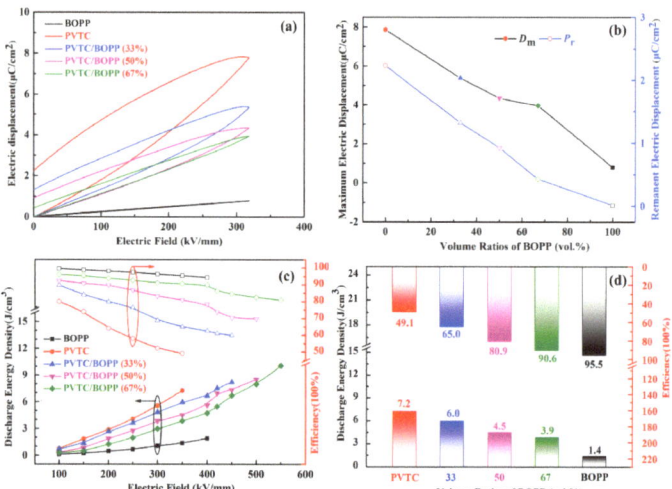

Figure 4. (**a**) The polarization characteristics of PVTC/BOPP bilayer films at the same electric field strength. (**b**) Maximum electric displacements (D_m) and Remanent electric displacements (P_r) of PVTC/BOPP bilayer films at the same electric field strength. (**c**) Relationship between the applied electric field and energy storage performances of PVTC/BOPP bilayer films. (**d**) Relationship between BOPP volume ratios and energy storage performances at the same electric field strength.

As shown in Figure 4c, at the maximum applied electric field (400 kV/mm), the efficiency of the pristine BOPP film remains at 94%, but the maximum discharge energy density is only 1.9 J/cm³. In the case of the pristine PVTC film, when an electric field of 350 kV/mm was applied, the maximum discharge energy density was 7.2 J/cm³, with an efficiency of 49%. In the bilayer films, as the volume ratio of BOPP increases, the energy storage density gradually increases; the PVTC/BOPP(33%), PVTC/BOPP(50%), and PVTC/BOPP(67%) bilayer films achieved a maximum energy storage density of 8.2 J/cm³, 8.5 J/cm³, and 10.1 J/cm³ at electric field strengths of 450 kV/mm, 500 kV/mm, and 550 kV/mm, respectively. The energy storage efficiency of all the bilayer films gradually decreases with increasing electric field strength. The charge/discharge efficiency of the three bilayer films at the maximum electric field strength is 60.1%, 69.8%, and 80.9%, respectively. We have compared the discharge energy density and charge/discharge efficiency of PVTC/BOPP bilayer films with different BOPP volume fractions and the same thickness under an electric field intensity of 350 kV/mm; the results are presented in Figure 4d. It can be seen that an increase in BOPP volume fraction is accompanied by a decrease in energy storage density and an increase in charge/discharge efficiency. The PVTC/BOPP bilayer films with thicker PVTC layers have a larger relative dielectric constant, which is beneficial for achieving greater potential shift polarization and a high charging energy density. However, the PVTC/BOPP bilayer films with thicker PVTC layers also have a larger residual polarization, which can cause energy loss and result in a low charge/discharge efficiency. This is consistent with the electrical properties discussed above. Therefore, the thicker the PVTC layer, the greater the residual polarization of the bilayer films is, with a consequent lower charge/discharge efficiency. Overall, the bilayer films combine the advantages of BOPP and PVTC, exhibiting a higher potential shift polarization compared with BOPP alone and a higher charge/discharge efficiency than PVTC.

The discharge energy density (U_d) and charge/discharge efficiency (η) recorded in this work are compared with recently reported polymer composites [46–54] in Figure 5. In all cases, the same matrix (PVDF co-polymer) and testing temperature (room temperature) were used. The majority of the reported η values are lower than the efficiency achieved in this study. A low discharge efficiency for a capacitor results in significant Joule heat loss that is detrimental to stable performance. The PVTC/BOPP bilayer films in this work exhibit a good balance of U_d (10.1 J/cm^3) and η (80.9%) and can effectively address the dilemma that "high U_d is always accompanied by low η".

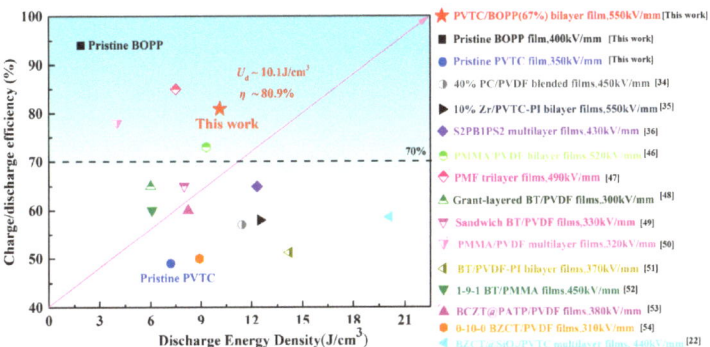

Figure 5. Comparison of the energy density U_d and charge/discharge efficiency η of this work's PVTC/BOPP(67%) bilayer film and previously reported results [22,34–36,46–54].

4. Conclusions

This study has thoroughly examined the impact of BOPP composition on the electrical and energy storage characteristics of PVTC/BOPP bilayer films with heterostructures. At a BOPP content of 67 vol%, the PVTC/BOPP bilayer film exhibited an exceptional energy storage capacity (U_d = ca. 10.1 J/cm^3, η = ca. 80.9% at 550 kV/mm). When compared with the energy storage performance of pristine PVTC at a maximum field strength of 350 kV/mm (U_d = ca. 7.2 J/cm^3, η = ca. 49.1%), the U_d is improved by a factor of 1.4, η by a factor of 1.7, and E_b by a factor of 1.6. By utilizing the dielectric properties of linear/ferroelectric layers, the electric field distribution of the composite film can be controlled, achieving a balance between relative dielectric constant and breakdown field strength, enhancing the energy storage density and charge/discharge efficiency. The concept of polymer-based composites with linear/ferroelectric heterostructures offers a new design paradigm for developing high-performance dielectric materials for flexible energy storage applications, and extends our understanding of dielectric breakdown and polarization mechanisms.

Author Contributions: Writing—original draft preparation, C.C.; writing—review and editing, G.L., Y.C., L.S. and S.Y. All authors have read and agreed to the published version of the manuscript.

Funding: This research was funded by Nanxun Scholars Program for Young Scholars of ZJWEU (NO. RC2022021089) and Scientific Research Project of Zhejiang Provincial Department of Education (No. Y202352960).

Institutional Review Board Statement: Not applicable.

Data Availability Statement: Data are contained within the article.

Conflicts of Interest: The authors declare no conflicts of interest.

References

1. Pei, J.; Yin, L.; Zhong, S.; Dang, Z. Suppressing the loss of polymer-based dielectrics for high power energy storage. *Adv. Mater.* **2023**, *35*, 2203623. [CrossRef] [PubMed]

2. Chen, J.; Huang, X. Dielectric polymers for emerging energy applications. *Sci. Bull.* **2023**, *68*, 1478–1483. [CrossRef]
3. Li, Q.; Chen, L.; Matthew, R.; Zhang, S.; Zhang, G.; Li, H.; Aman, H.; Chen, L.; Jackson, T.; Wang, Q. Flexible high-temperature dielectric materials from polymer nanocomposites. *Nature* **2015**, *523*, 576–579. [CrossRef] [PubMed]
4. Zha, J.; Tian, Y.; Zheng, M.; Wan, B.; Yang, X.; Chen, G. High-temperature energy storage polyimide dielectric materials: Polymer multiple-structure design. *Mater. Today Energy* **2023**, *31*, 101217. [CrossRef]
5. Dehghani-Sanij, A.; Tharumalingam, E.; Dusseault, M.; Fraser, R. Study of energy storage systems and environmental challenges of batteries. *Renew. Sustain. Energy Rev.* **2019**, *104*, 192–208. [CrossRef]
6. Zhang, Y.; He, X.; Cong, X.; Wang, Q.; Yi, H.; Li, S.; Zhang, C.; Zhang, T.; Wang, X.; Chi, Q. Enhanced energy storage performance of polyethersulfone-based dielectric composite via regulating heat treatment and filling phase. *J. Alloy. Compd.* **2023**, *960*, 170539. [CrossRef]
7. Tang, H.; Sodano, H. Ultra-high energy density nanocomposite capacitors with fast discharge using $Ba_{0.2}Sr_{0.8}TiO_3$ nanowires. *Nano Lett.* **2013**, *13*, 1373–1379. [CrossRef] [PubMed]
8. Zhang, X.; Shen, Y.; Zhang, Q.; Gu, L.; Hu, Y.; Du, J.; Lin, Y.; Nan, C. Ultrahigh energy density of polymer nanocomposites containing $BaTiO_3@TiO_2$ nanofibers by atomic-scale interface engineering. *Adv. Mater.* **2015**, *27*, 819–824. [CrossRef]
9. Zhang, T.; Chen, X.; Thakur, Y.; Lu, B.; Zhang, Q.; Runt, J.; Zhang, Q. A highly scalable dielectric metamaterial with superior capacitor performance over a broad temperature. *Sci. Adv.* **2020**, *6*, eaax6622. [CrossRef]
10. Wang, Y.; Cui, J.; Yuan, Q.; Niu, Y.; Bai, Y.; Wang, H. Significantly enhanced breakdown strength and energy density in sandwich-structured barium titanate/poly (vinylidene fluoride) nanocomposites. *Adv. Mater.* **2015**, *27*, 6658–6663. [CrossRef] [PubMed]
11. Liu, X.; Hu, P.; Yu, J.; Fan, M.; Ji, X.; Sun, B.; Shen, Y. Topologically distributed one-dimensional TiO_2 nanofillers maximize the dielectric energy density in a P(VDF-HFP) nanocomposite. *J. Mater. Chem. A* **2020**, *8*, 18244–18253. [CrossRef]
12. Jiang, Y.; Wang, J.; Zhang, Q.; Yang, H.; Shen, D.; Zhou, F. Enhanced dielectric performance of P(VDF-HFP) composites filled with Ni@polydopamine@$BaTiO_3$ nanowires. *Colloids Surf. A Physicochem. Eng. Asp.* **2019**, *576*, 55–62. [CrossRef]
13. Li, H.; Yao, B.; Zhou, Y.; Xu, W.; Ren, L.; Ai, D.; Wang, Q. Bilayer-structured polymer nanocomposites exhibiting high breakdown strength and energy density via interfacial barrier design. *ACS Appl. Energy Mater.* **2020**, *3*, 8055–8063. [CrossRef]
14. Bai, H.; Zhu, K.; Wang, Z.; Shen, B.; Zhai, J. 2D fillers highly boost the discharge energy density of polymer-based nanocomposites with trilayered architecture. *Adv. Funct. Mater.* **2021**, *31*, 2102646. [CrossRef]
15. Liu, T.; Hou, Y.; Ji, Q.; Wei, S.; Du, P.; Luo, L.; Li, W. Significant enhancement of energy storage performances by regulating the dielectric contrast between adjacent layers in the heterostructural composites. *ACS Appl. Energy Mater.* **2020**, *3*, 3015–3023. [CrossRef]
16. Li, Z.; Liu, F.; Li, H.; Ren, L.; Dong, L.; Xiong, C.; Wang, Q. Largely enhanced energy storage performance of sandwich-structured polymer nanocomposites with synergistic inorganic nanowires. *Ceram. Int.* **2019**, *45*, 8216–8221. [CrossRef]
17. Feng, Y.; Wu, Q.; Deng, Q.; Xu, Z. High-energy density in Si-based layered nanoceramic/polymer composites based on gradient design of ceramic bandgaps. *Ceram. Int.* **2019**, *45*, 16600–16607. [CrossRef]
18. Feng, Y.; Xue, J.; Zhang, T.; Chi, Q.; Li, J.; Chen, Q.; Wang, J.; Chen, L. Double-gradients design of polymer nanocomposites with high energy density. *Energy Storage Mater.* **2022**, *44*, 73–81. [CrossRef]
19. Wang, L.; Luo, H.; Zhou, X.; Yuan, X.; Zhou, K.; Zhang, D. Sandwich-structured all-organic composites with high breakdown strength and high dielectric constant for film capacitor. *Compos. Part A Appl. Sci. Manuf.* **2019**, *117*, 369–376. [CrossRef]
20. Daljeet, K.; Naveen, K.; Gagan, A.; Ranvir, S.; Charu, M. Enhancement of energy storage in nanocomposite thin films: Investigating PVDF-ZnO and PVDF-TZO for improved dielectric and ferroelectric characteristics. *Phys. Scr.* **2024**, *99*, 036101.
21. Uwa, O.; Abimbola, P.; Olawale, M. Carbon nanotubes enhancement of tribological and nanomechanical properties of PVDF-BN nanocomposites. *Polym. Bull.* **2024**. [CrossRef]
22. Zhang, Y.; Zhang, C.; Feng, Y.; Zhang, T.; Chen, Q.; Chi, Q.; Liu, L.; Wang, X.; Lei, Q. Energy storage enhancement of P(VDF-TrFE-CFE)-based composites with double shell structured BZCT nanofibers of parallel and orthogonal configurations. *Nano Energy* **2019**, *66*, 104195. [CrossRef]
23. Wang, X.; An, Z.; Zhuo, M. Energy storage performance and dielectric properties of surface fluorinated BOPP films. *IEEE Trans. Dielectr. Electr. Insul.* **2023**, *30*, 1950–1957. [CrossRef]
24. Chi, Q.; Wang, T.; Zhang, C.; Yu, H.; Zhao, X.; Yang, X.; Lei, Q.; Zhao, H.; Zhang, T. Significantly improved high-temperature energy storage performance of commercial BOPP films by utilizing ultraviolet grafting modification. *iEnergy* **2022**, *1*, 374–382. [CrossRef]
25. Yu, Y.; Li, J.; Xie, Z.; Gong, X.; Gao, G.; Wang, J.; Li, J. Characterizing piezoelectric properties of PVDF film under extreme loadings. *Smart Mater. Struct.* **2024**, *33*, 015026. [CrossRef]
26. Wang, X.; Lu, Y.; Jiang, J.; Lv, C.; Fu, H.; Xie, M. A flexible piezoelectric PVDF/MXene pressure sensor for roughness discrimination. *IEEE Sens. J.* **2024**, *24*, 7176–7184. [CrossRef]
27. Adel, M.; Abdulrahman, I.; Hosam, H.; Penchal, R.; Mohamed, M.; Essam, A.; Waleed, H.; Teresa, D. Enhanced corrosion resistance and surface wettability of PVDF/ZnO and PVDF/TiO_2 composite coatings: A comparative study. *Coatings* **2023**, *13*, 1729. [CrossRef]
28. Zhang, T.; Liang, S.; Yu, H.; Zhang, C.; Tang, C.; Li, H.; Chi, Q. Improved high-temperature energy storage density at low-electric field in BOPP/PVDF multilayer films. *J. Appl. Polym. Sci.* **2023**, *140*, e54729. [CrossRef]

29. Wang, D.; Yang, J.; Li, L.; Wang, X.; Yang, X.; Zhao, X.; Zhao, H.; Li, L.; Liu, H. Effect of shear action on structural and electrical insulation properties of polypropylene. *Energies* **2023**, *16*, 1421. [CrossRef]
30. Gao, M.; Yang, J.; Zhao, H.; He, H.; Hu, M.; Xie, S. Preparation methods of polypropylene/nano-silica/styrene-ethylene-butylene-styrene composite and its effect on electrical properties. *Polymers* **2019**, *11*, 797. [CrossRef] [PubMed]
31. Tang, Y.; Gao, Y.; Yu, C.; Gao, C.; Zhang, Y.; Li, E. Experimental study of the dielectric properties of energy-containing materials at variable temperatures. *Rev. Sci. Instrum.* **2023**, *94*, 035113. [CrossRef] [PubMed]
32. Zhang, C.; Yan, W.; Zhang, Y.; Cui, Y.; Zhang, T.; Tang, C.; Liu, X.; Chi, Q. Enhanced energy storage performance of doped modified PC/PVDF coblended flexible composite films. *ACS Appl. Electron. Mater.* **2023**, *5*, 3817–3829. [CrossRef]
33. Li, Y.; Zhou, Y.; Cheng, S.; Hu, J.; He, J.; Li, Q. Polymer nanocomposites with high energy density utilizing oriented nanosheets and high-dielectric-constant nanoparticles. *Materials* **2021**, *14*, 4780. [CrossRef] [PubMed]
34. Chen, C.; Zhang, T.; Zhang, C.; Feng, Y.; Zhang, Y.; Zhang, Y.; Chi, Q.; Wang, X.; Lei, Q. Improved energy storage performance of P(VDF-TrFE-CFE) multilayer films by utilizing inorganic functional layers. *ACS Appl. Energy Mater.* **2021**, *4*, 11726–11734. [CrossRef]
35. Zhang, C.; Tong, X.; Liu, Z.; Zhang, Y.; Zhang, T.; Tang, C.; Liu, X.; Chi, Q. Enhancement of energy storage performance of PMMA/PVDF composites by changing the crystalline phase through heat treatment. *Polymers* **2023**, *15*, 2486. [CrossRef] [PubMed]
36. Zhang, C.; Wang, H.; Zhang, T.; Zhang, Y.; Zhang, Y.; Tang, C. Significantly enhanced energy storage density and efficiency of sandwich polymer-based composite via doped MgO and TiO_2 nanofillers. *J. Mater. Sci.* **2023**, *58*, 12724–12735. [CrossRef]
37. Chen, C.; Zhang, C.; Zhang, T.; Feng, Y.; Zhang, Y.; Chi, Q.; Wang, X.; Lei, Q. Improved energy storage performances of solution-processable ferroelectric polymer by modulating of microscopic and mesoscopic structure. *Compos. Part B* **2020**, *199*, 108312. [CrossRef]
38. Zhang, T.; Yang, L.; Zhang, C.; Feng, Y.; Wang, J.; Shen, Z.; Chen, Q.; Lei, Q.; Chi, Q. Polymer dielectric films exhibiting superior high-temperature capacitive performance by utilizing an inorganic insulation interlayer. *Mater. Horiz.* **2022**, *9*, 1273–1282. [CrossRef]
39. Zhou, Y.; Li, Q.; Dang, B.; Yang, Y.; Shao, T.; Li, H.; Hu, J.; Zeng, R.; He, J.; Wang, Q. A scalable, high-throughput, and environmentally benign approach to polymer dielectrics exhibiting significantly improved capacitive performance at high temperatures. *Adv. Mater.* **2018**, *30*, 1805672. [CrossRef]
40. Tong, J.; Song, Z.; Kim, D.; Chen, X.; Chen, C.; Axel, F.; Paul, K.; Matthew, O.; Sean, P.; Obadiah, G.; et al. Carrier lifetimes of >1 ms in Sn-Pb perovskites enable efficient all-perovskite tandem solar cells. *Science* **2019**, *364*, 475–479. [CrossRef]
41. Tong, J.; Jiang, Q.; Andrew, J.; Axel, F.; Wang, X.; Hao, J.; Sean, P.; Amy, E.; Steven, P.; Li, C.; et al. Carrier control in Sn-Pb perovskites via 2D cation engineering for all-perovskite tandem solar cells with improved efficiency and stability. *Nat. Energy* **2022**, *7*, 642–651. [CrossRef]
42. Zhang, T.; Yang, L.; Ruan, J.; Zhang, C.; Chi, Q. Improved high-temperature energy storage performance of PEI dielectric films by introducing an SiO_2 insulating layer. *Macromol. Mater. Eng.* **2021**, *306*, 2100514. [CrossRef]
43. Li, L.; Dong, J.; Hu, R.; Chen, X.; Niu, Y.; Wang, H. Wide-bandgap fluorides/polyimide composites with enhanced energy storage properties at high temperatures. *Chem. Eng. J.* **2022**, *435*, 135059. [CrossRef]
44. Zhang, T.; Yu, H.; Young, J.; Zhang, C.; Feng, Y.; Chen, Q.; Lee, K.; Chi, Q. Significantly improved high-temperature energy storage performance of BOPP films by coating nanoscale inorganic layer. *Energy Environ. Mater.* **2022**, *7*, e12549. [CrossRef]
45. Liu, G.; Lei, Q.; Feng, Y.; Zhang, C.; Zhang, T.; Chen, Q.; Chi, Q. High-temperature energy storage dielectric with inhibition of carrier injection/migration based on band structure regulation. *InfoMat* **2023**, *5*, e12368. [CrossRef]
46. Zhu, Y.; Jiang, P.; Huang, X. Poly(vinylidene fluoride) terpolymer and poly(methyl methacrylate) composite films with superior energy storage performance for electrostatic capacitor application. *Compos. Sci. Technol.* **2019**, *179*, 115–124. [CrossRef]
47. Sun, S.; Shi, Z.; Sun, L.; Liang, L.; Dastan, D.; He, B.; Wang, H.; Huang, M.; Fan, R. Achieving concurrent high energy density and efficiency in all polymer layered paraelectric/ferroelectric composites via introducing a moderate layer. *ACS Appl. Mater. Interfaces* **2021**, *13*, 27522–27532. [CrossRef] [PubMed]
48. Wang, Y.; Li, Y.; Wang, L.; Yuan, Q.; Chen, J.; Niu, Y.; Xu, X.; Wang, Q.; Wang, H. Gradient-layered polymer nanocomposites with significantly improved insulation performance for dielectric energy storage. *Energy Storage Mater.* **2020**, *24*, 626–634. [CrossRef]
49. Wang, Y.; Cui, J.; Wang, L.; Yuan, Q.; Niu, Y.; Chen, J.; Wang, Q.; Wang, H. Compositional tailoring effect on electric field distribution for significantly enhanced breakdown strength and restrained conductive loss in sandwich-structured ceramic/polymer nanocomposites. *J. Mater. Chem. A* **2017**, *5*, 4710–4718. [CrossRef]
50. Cui, Y.; Zhang, T.; Feng, Y.; Zhang, C.; Chi, Q.; Zhang, Y.; Chen, Q.; Wang, X.; Lei, Q. Excellent energy storage density and efficiency in blend polymer-based composites by design of coreshell structured inorganic fibers and sandwich structured films. *Composites Part B* **2019**, *177*, 107429. [CrossRef]
51. Xie, B.; Zhang, Q.; Zhang, L.; Zhu, Y.; Guo, X.; Fan, P.; Zhang, H. Ultrahigh discharged energy density in polymer nanocomposites by designing linear/ferroelectric bilayer heterostructure. *Nano Energy* **2018**, *54*, 437–446. [CrossRef]
52. Marwat, M.; Xie, B.; Zhu, Y.; Fan, P.; Ma, W.; Liu, H.; Ashtar, M.; Xiao, J.; Salamon, D.; Samater, C.; et al. Largely enhanced discharge energy density in linear polymer nanocomposites by designing a sandwich structure. *Compos. Part A* **2019**, *121*, 115–122. [CrossRef]

53. Pan, Z.; Yao, L.; Zhai, J.; Wang, H.; Shen, B. Ultrafast discharge and enhanced energy density of polymer nanocomposites loaded with $0.5(Ba_{0.7}Ca_{0.3})TiO_3$-$0.5Ba(Zr_{0.2}Ti_{0.8})O_3$ one-dimensional nanofibers. *ACS Appl. Mater. Interfaces* **2017**, *9*, 14337–14346. [CrossRef]
54. Zhang, Y.; Zhang, T.; Liu, L.; Chi, Q.; Zhang, C.; Chen, Q.; Cui, Y.; Wang, X.; Lei, Q. Sandwich-structured PVDF-based composite incorporated with Hybrid Fe_3O_4@BN nanosheets for excellent dielectric properties and energy storage performance. *J. Phys. Chem. C* **2018**, *122*, 1500–1512. [CrossRef]

Disclaimer/Publisher's Note: The statements, opinions and data contained in all publications are solely those of the individual author(s) and contributor(s) and not of MDPI and/or the editor(s). MDPI and/or the editor(s) disclaim responsibility for any injury to people or property resulting from any ideas, methods, instructions or products referred to in the content.

Communication

A Sandwich Structural Filter Paper–AgNWs/MXene Composite for Superior Electromagnetic Interference Shielding

Xiaoshuai Han [1,2], Hongyu Feng [2], Wei Tian [2], Kai Zhang [1], Lei Zhang [1], Jiangbo Wang [3] and Shaohua Jiang [2,*]

1 State Key Laboratory of Biobased Material and Green Papermaking, Qilu University of Technology, Shandong Academy of Sciences, Jinan 250353, China; xiaoshuai.han@njfu.edu.cn (X.H.)
2 Jiangsu Co-Innovation Center of Efficient Processing and Utilization of Forest Resources, International Innovation Center for Forest Chemicals and Materials, College of Materials Science and Engineering, Nanjing Forestry University, Nanjing 210037, China
3 School of Materials and Chemical Engineering, Ningbo University of Technology, Ningbo 315211, China
* Correspondence: shaohua.jiang@njfu.edu.cn; Tel.: +86-156-2496-0675

Abstract: A thin, lightweight and flexible electromagnetic interference (EMI) shielding paper composite is an urgent need for modern military confrontations. Herein, a sandwich-structured EMI shielding paper composite with an easy pavement consisting of a filter paper layer, middle AgNWs/MXene layer, and polyvinyl butyral (PVB) layer was constructed by vacuum-assisted filtration, spraying and air-drying. The middle AgNWs/MXene compound endowed the filter paper with excellent electrical conductivity (166 S cm^{-1}) and the fabricated filter paper–AgNWs/MXene–PVB composite exhibits superior EMI shielding (30 dB) with a 141 μm thickness. Remarkably, the specific EMI shielding effectiveness (SSE/t) of the filter paper–AgNWs/MXene–PVB composite reached 13,000 dB cm^2 g^{-1} within the X-band frequency range. This value represents one of the highest reported for cellulose-based EMI shielding materials. Therefore, our sandwich-structured filter paper composite with superior EMI shielding performance can be used in the medical and military fields.

Keywords: sandwich structure; electromagnetic interference (EMI) shielding; silver nanowires (AgNWs); MXene

1. Introduction

With the advancement of contemporary society, daily life is increasingly dominated by a plethora of electronic devices, which, while facilitating routine activities, also contribute to a significant amount of electromagnetic pollution. Many studies have indicated that long-time exposure to electromagnetic waves may adversely affect human health. Therefore, investigations of electromagnetic interference (EMI) shielding materials have emerged as a prominent area of research in recent years [1–6].

Recently, biomass materials have been used to fabricate EMI shielding materials using a top-down method or bottom-up strategy [7]. Wood, characterized as a sustainable biomass resource, stands out as an excellent prospect for electromagnetic interference (EMI) shielding materials due to its cost-effectiveness, lightweight nature, and inherently porous and stratified structural properties [8–11]. A diverse array of carbon composites derived from wood have been engineered to yield high-performance EMI shielding materials, demonstrating considerable shielding effectiveness [12,13]. However, the above-fabricated materials are bulky and hard to use in practical applications. A bottom-up strategy can transform wood-derived biomass materials (mainly cellulose and lignin) into flexible and mechanically strong films/papers with superior EMI shielding properties [14]. However, this fabrication process is time-consuming, toxic, expensive, and hard to adapt to large-scale production. Another example is cellulose-based carbon aerogel [15–17]. Although these aerogels are lightweight, insulating and have good EMI shielding properties, their construction also comes with the aforementioned drawbacks. Moreover, these aerogels are commonly brittle and demonstrate poor physical and mechanical performance.

The paper making industry has a well-established history of using cellulose materials, including wood and recycled paper, to fabricate new paper via a sequence of processing and manufacturing steps. With the arrival of the Industrial Revolution, the paper industry grew to a larger scale and implemented greater efficiency. Traditional papermaking methods were gradually replaced by mechanization and automation, resulting in significantly increased production capacity. Currently, a multitude of paper products find utility across diverse sectors, encompassing decoration, automotives, and aerospace, among others. Therefore, the direct employment of paper substrates for producing EMI shielding materials will be practical and favorable. More recently, MXene and silver nanowires (AgNWs) have proved particularly suitable for EMI shielding applications, owing to their remarkable electrical conductivity and solution compatibility [16,18–20]. They can be coated on the surface of the matrix to fabricate superior EMI shielding materials. Unfortunately, the coating can easily be destroyed by physical friction, causing a low EMI SE. Polyvinyl butyral (PVB) is a prevalent thermoplastic resin frequently used to improve mechanical properties, thermal stability, and water- and oil-proofing performance [21].

In this study, we present an innovative and promising technique for fabricating large-format electromagnetic interference (EMI) shielding paper composites utilizing a vacuum impregnation, spray deposition, and evaporation drying strategy. Readily available filter paper serves as the substrate, with MXene/AgNWs acting as the intermediate functional layer, and an external layer of polyvinyl butyral (PVB) providing protection. This layered configuration bestows the filter paper–AgNWs/MXene–PVB composite with superior EMI shielding effectiveness (SE) along with robust physical and mechanical characteristics.

2. Materials and Methods

2.1. Materials and Chemicals

The filter paper (diameter: 60 mm; pore size: 20~25 μm) and hydrophilic PTFE microporous membrane were purchased from Cytiva (Shanghai, China). Sulfuric acid (H_2SO_4, 72%), ethanol (99.7%), acetone (99.5%), iron chloride ($FeCl_3$, 99.9%), silver nitrate ($AgNO_3$), ethylene glycol (EG), polyvinyl pyrrolidone (PVP), and polyvinyl butyral (PVB) were purchased from Aladdin (Shanghai, China). A delaminated solution of titanium carbide (Ti_3C_2 MXene solution, 5 mg/mL) was purchased from Jilin 11 Technology Co., Ltd. (Changchun, China).

2.2. Synthesis of Silver Nanowires (AgNWs)

AgNWs were prepared using a modified polyol method based on previous reports [12]. Specifically, 0.2 g PVP was absolutely dissolved in 25 mL EG under magnetic stirring at room temperature. Then, 0.25 g $AgNO_3$ was added into the PVP/EG solution and stirred to form a transparent, uniform solution. Afterward, 3.5 g 0.6 mmol/L $FeCl_3$ salt solution was dripped into the above mixture with stirring for 10 min to produce a uniform solution. Finally, the mixed solution was transferred into oil bath reactor (180 °C) for 45 min to grow AgNWs at a slow stirring speed. After the end of reaction, generated AgNWs were purified five times using a solvent exchange method with acetone and ethanol with the aid of centrifugation (5000 rpm, 5 min for each time and type of instrument). Finally, the AgNW precipitate was redispersed in ethanol with a concentration of 2.35 mg/mL for use. The microstructure of the AgNWs resembles long rods (Figure S1).

2.3. Preparation of Sandwich-Structured Filter Paper–AgNWs/MXene–PVB Composite

MXene solution was homogeneously dispersed in deionized water (5 mg/mL) and AgNWs was dispersed in ethanol (2.35 mg/mL) for after use. The filter paper was placed on the PTFE microporous membrane (0.22 μm pore size), and then 1.82 mL MXene solution and 0.39 mL AgNWs solution were applied and filtered through vacuum-assisted filtration, resulting in the formation of an MXene/AgNWs composite film. After that, the filter paper–MXene/AgNWs was peeled off from the filter, followed by air-drying. In the final stage, the air-dried filter paper–MXene/AgNWs composite film underwent a spraying process with 2 mL of a 2 wt% PVB solution, culminating in the creation of a functional filter paper–MXene/AgNWs–PVB (FMAP) composite for EMI shielding applications. The density of the fabricated FMAP is 0.73 g cm^{-3}.

2.4. Characterizations

The lignin contents (Klason lignin) of the filter paper were determined by following a standard TAPPI T 222 om^{-2} method [22]. The Fourier transform infrared (FTIR) spectra of filter paper, filter paper–MXene/AgNWs, and filter paper–MXene/AgNWs–PVB were obtained using a Fourier transform infrared spectrometer (VERTEX 80 V, Bruker, Bremen, Germany) from 4000 to 400 cm^{-1} at a spectral resolution of 6 cm^{-1} and a total of 32 scans. The morphologies and microstructures of AgNWs, filter paper, filter paper–MXene/AgNWs and filter paper–MXene/AgNWs–PVB were observed by Phenom scanning electron microscopy (SEM) (Phenom XL G2, Phenom-World BV, Eindhoven, The Netherlands). A Four-Point-Probe instrument (Guangzhou Four-Point-Probe Technology, SDY-4, Guangzhou, China) was used to test the conductivity of samples. At least four parts were tested for all samples, and the average and standard deviation were reported. A Vector Network Analyzer (Agilent Technologies N5063A, Palo Alto, CA, USA) was used to measure the EMI shielding effectiveness in the frequency range of 8.2–12.4 GHz (X-band) (Figure S2), according to our previous study [23].

2.5. EMI Shielding Parameters

According to Schelkunoff theory, the total EMI shielding effectiveness (SE$_T$) consists of absorption (SE$_A$), reflection (SE$_R$), and multi-reflection (SE$_M$), where SE$_M$ is often ignored when the SE$_T$ is over 15 dB. The S-parameter is derived from the wave quantities a and b of the incident and reflected waves in the vector analysis tester.

$$S_{11} = \frac{b_1}{a_1}|a_2 = 0 \tag{1}$$

$$S_{21} = \frac{b_2}{a_1}|a_2 = 0 \tag{2}$$

$$S_{12} = \frac{b_1}{a_2}|a_1 = 0 \tag{3}$$

$$S_{11} = \frac{b_1}{a_2}|a_1 = 0 \tag{4}$$

The reflection coefficient R, the absorption coefficient A, and the transmission coefficient T are calculated from the S-parameters S11 and S21.

$$R = |S_{11}|^2 \tag{5}$$

$$T = |S_{21}|^2 \tag{6}$$

$$A = 1 - T - R \tag{7}$$

Reflection is the main mechanism of shielding, which occurs at the interface of two different media with different refractive index or impedance characteristics. The reflection loss is given by the Frensel equation, as shown in the following equation:

$$SE_R = -10\lg(1 - R) = -10\lg\left(1 - |S_{11}|^2\right) \tag{8}$$

Absorption attenuation is given by the following equation:

$$SE_A = -10\lg\left(\frac{T}{1-R}\right) = -10\lg\left(\frac{|S_{21}|^2}{1-|S_{11}|^2}\right) \tag{9}$$

3. Results

The schematic for fabrication of the sandwich structural filter paper–AgNWs/MXene–PVB composite for electromagnetic shielding is depicted in Figure 1. Specifically, the filter paper was positioned beneath the Buchner flask, and then MXene solution and AgNWs solution were vacuum-filtered to fabricate the filter paper–MXene/AgNWs composite (FMA). Next, the FMA was peeled off from the filter flask followed by spraying with PVB and evaporation–drying to obtain the filter paper–MXene/AgNWs–PVB composite (FMAP). In this work, SEM was applied to investigate the surfaces of materials [24–29]. Figure 2 shows the morphologies of the filter paper, FMA, and FMAP. Figure 2a–d shows the microstructure of the pure filter paper. The filter paper was fabricated using smooth fibers through the papermaking process. After vacuum filtration of the MXene/AgNWs solution, the top surface of the filter paper was covered by the MXene layer and AgNWs rods. More importantly, the MXene and AgNWs were inserted into the inner of the filter paper through the cutting surface morphologies of FMA (Figure 2e–h). After PVB coating, the filter paper became denser, and we also observed that the PVB formed a membrane-like layer on the top surface of FMA. Meanwhile, the MXene and AgNWs were not influenced by the PVB membrane, having a good intrinsic morphology, which plays an important role in EMI SE.

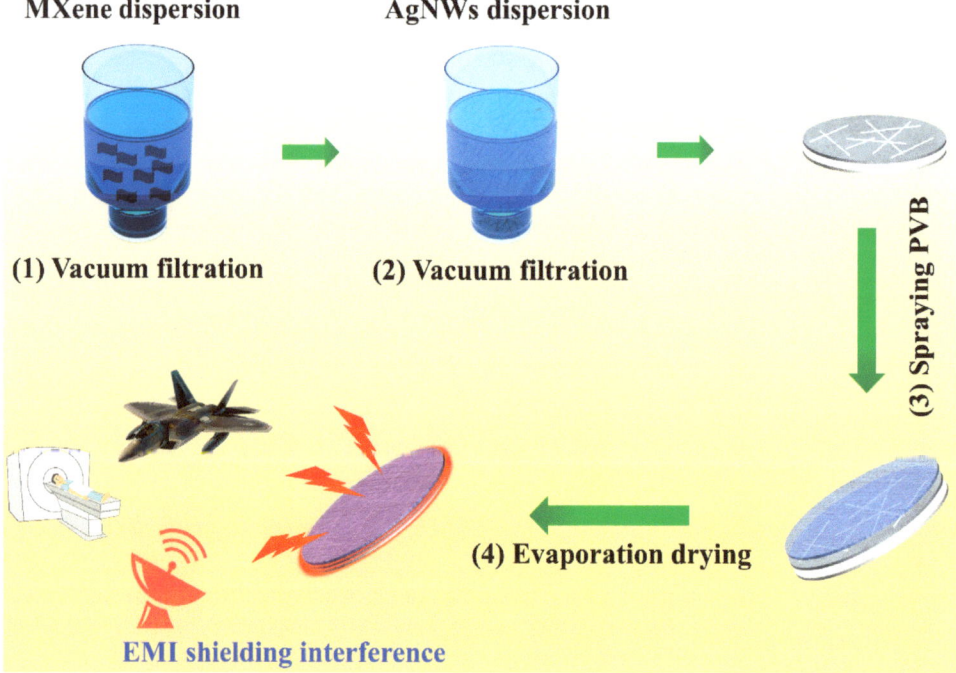

Figure 1. Schematic illustration for fabrication of the sandwich-structured filter paper–AgNWs/MXene–PVB composite.

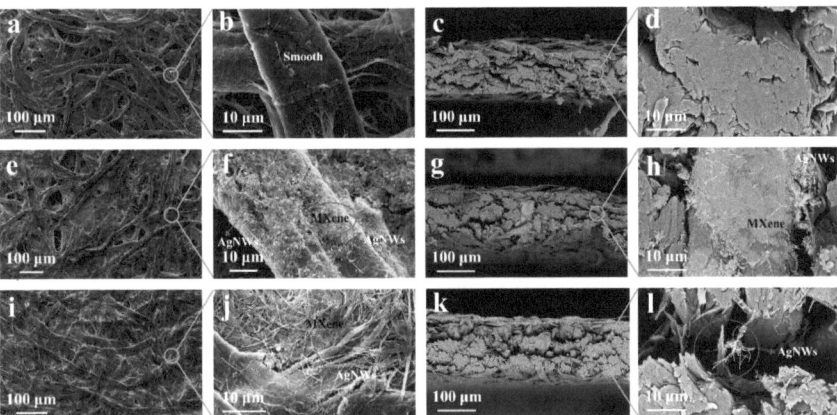

Figure 2. (a) Surface microstructure of filter paper and (b) its magnified SEM image. (c) Cutting surface morphologies of filter paper and (d) corresponding magnified SEM image. (e) Surface microstructure of filter paper–MXene/AgNWs sample and (f) its magnified SEM image. (g) Cutting surface morphologies of filter paper–MXene/AgNWs sample and (h) corresponding magnified SEM image. (i) Surface microstructure of filter paper–MXene/AgNWs–PVB sample and (j) its magnified SEM image. (k) Cutting surface morphologies of filter paper–MXene/AgNWs–PVB sample and (l) corresponding magnified SEM image.

The chemical composition of the filter paper was analyzed using the standard Technical Association of the Pulp and Paper Industry (TAPPI)T 222 om^{-2} method. The result showed the filter paper is composed of 95.65% cellulose, 1.87% hemicellulose, and 0.50% lignin (Figure 3a), suggesting that the cellulose is the most dominated by matrix, which facilitates its mechanical flexibility. FTIR was used to obtain an infrared spectrum of the absorption or emission of the solid, liquid, or gas, which is helpful in determining the chemical structures of materials [30–36]. The FTIR spectra of the filter paper, FMA sample, and FMAP sample are displayed and compared in Figure 3b. Obvious characteristic peaks appear at 3329 cm^{-1} (O–H stretching vibration), 2897 cm^{-1} (C–H stretching vibration), and 1030 cm^{-1} (C–O stretching vibration) of cellulose, respectively. In addition, there are small peaks at 1733 cm^{-1}, which correspond to the unconjugated carbonyl C=O of hemicellulose. The above analysis shows that the filter paper is mainly composed of cellulose, which is in accordance with the result of composition data. Commonly, silver itself does not contain organic elements, leading to no obvious absorption peak in the FTIR spectra. Yet, the surface of MXene has several characteristic peaks at 1395 cm^{-1}, 557 cm^{-1}, 1621 cm^{-1}, 1234 cm^{-1}, and 1030 cm^{-1} belonging to the C–F, –OH, C=O of alkone, C=C–O, and C–O–C stretching vibration functional groups. These typical peaks actually appeared in the FTIR spectra of FMA sample and FMAP sample, which indicated that there is a good connection between the filter paper and the AgNWs/MXene composite.

Conductivity is a critical factor influencing the electromagnetic interference (EMI) shielding performance of materials. The unmodified filter paper is dielectric (=0 S cm^{-1}) (Figure 4a), having no EMI shielding effectiveness (~0 dB) (Figure 4b). After depositing conductive MXene and AgNWs on the surface of filter paper, the fabricated FMA sample shows superior conductivity with 166 S cm^{-1} (Figure 4a), further exhibiting good EMI shielding effectiveness. The total shielding effectiveness (SE$_T$) of FMA achieves 21 dB at a thickness of 141 μm in the 8.2–12.4 GHz frequency range (Figure 4b). To provide a more comprehensive assessment of material performance, the specific EMI shielding effectiveness (SSE/t), calculated by dividing SE by density and thickness, is utilized. As depicted in Figure 4c, the EMI SSE/t values of FMA sample are above 8000 dB cm^2 g^{-1}. It is well known that the surface coatings of MXene and AgNWs are susceptible to damage, potentially leading to a loss of EMI shielding effectiveness. To address this issue, the FMA sample was coated with a layer of polyvinyl butyral (PVB), known for its high physical and mechanical strength, thus forming the FMAP composite MXene. Then, the conductivity and EMI shielding effectiveness of FMAP were evaluated, and the results showed that the FMA was covered by dielectric PVB (the conductivity of FMAP = 0 S cm^{-1}) (Figure 4a). Surprisingly, the EMI shielding effectiveness did not decrease; conversely, it increased to up to 30 dB (Figure 4b), and the specific EMI SE of FMAP reached a very high value of 13,000 dB cm^2 g^{-1} (Figure 4c). In addition, there was greater absorption efficiency (SE$_A$) than reflection efficiency (SE$_R$) in the FMA and FMAP samples, indicating SE$_A$ plays a more important role than SE$_R$ in EMI SE. The result also shows that MXene/AgNWs are more effective in terms of EMI SE through enhanced SE$_A$ due to the PVB protection effect (Figure 4d). The reason is that when PVB is sprayed on paper, ethanol will loosen AgNWs and Mxene at the interface, and after drying again, more bonding points will appear at the interface, which is beneficial for improving absorption efficiency (SE$_A$). In order to make a good comparison, we also tested the EMI SE of the carbon pencil and explained our reasoning (Figure S3).

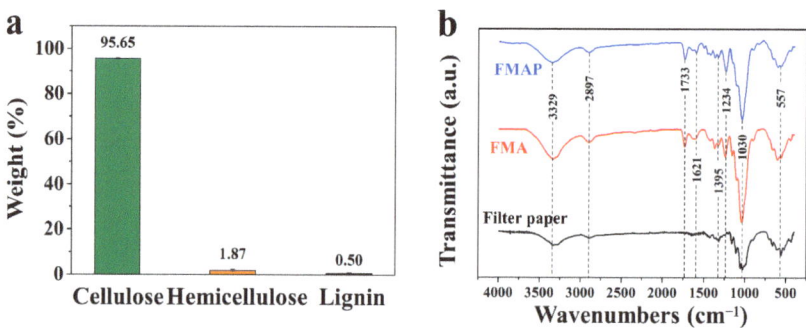

Figure 3. (a) Chemical composition of filter paper. (b) FTIR spectra of filter paper, FMA, and FMAP samples.

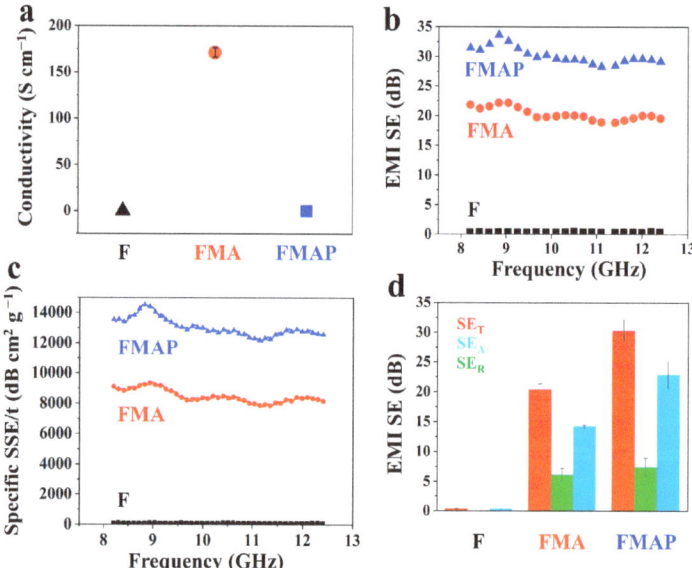

Figure 4. (a) Electrical conductivities of the F (filter paper), FMA, and FMAP samples. (b) EMI SE$_T$ and (c) the corresponding specific SST/t of the F, FMA and FMAP samples. (d) Comparison of the SE$_T$, SE$_A$, and SE$_R$ of the aforementioned specimens.

4. Conclusions

In summary, this work showcases the fabrication of a sandwich-structured filter paper–MXene/AgNWs–PVB composite for high-performance electromagnetic interference shielding. This was achieved through a two-step vacuum-assisted filtration process, followed by a methodical spraying and evaporation drying technique. An exhaustive examination of the filter paper–MXene/AgNWs–PVB composite's chemical structure, microstructure, electrical conductivity, and EMI shielding effectiveness was undertaken. The middle MXene/AgNWs layer showed a high electrical conductivity of 166 S cm^{-1}, achieving good EMI SE (30 dB) and a high SSE/t value (13,000 dB cm^2 g^{-1}) in the filter paper–MXene/AgNWs–PVB composite. The composite's comprehensive properties make it an ideal candidate for future applications in smart homes and the aerospace, military, and artificial intelligence domains.

Supplementary Materials: The following supporting information can be downloaded at: https://www.mdpi.com/article/10.3390/polym16060760/s1, Figure S1: (1) SEM image of AgNWs; (2) TEM micrograph of MXene nanosheets. Figure S2: Vector Network Analyzer (Agilent Technologies N5063A, Palo Alto, State of California, USA) was used to measure the EMI shielding effectiveness in the frequency range of 8.2−12.4 GHz (X-band). The cross-section size is 22.86 mm × 10.16 mm. Figure S3: Comparison of SET, SEA, and SER of the F, FMA, FMAP and FC samples.

Author Contributions: X.H.: investigation, conceptualization, methodology, writing—original draft preparation; H.F.: data curation; W.T.: resources and investigation; K.Z.: methodology, validation; L.Z.: conceptualization, reviewing, and editing; J.W.: reviewing and supervision; S.J.: writing—reviewing, editing, and supervision. All authors have read and agreed to the published version of the manuscript.

Funding: This research was funded by the Foundation (GZKF202129) of State Key Laboratory of Biobased Material and Green Papermaking, Qilu University of Technology, Shandong Academy of Sciences; The National Natural Science Foundation of China (32301518); The Natural Science Foundation of the Jiangsu Higher Education Institutions of China (23KJB220003); the Chunhui Research Grant (HZKY20220168, 202201269), Department of International Cooperation and Exchanges, Ministry of Education of the People's Republic of China.

Institutional Review Board Statement: Not applicable.

Data Availability Statement: The data presented in this study are available upon request from the corresponding author. The data are not publicly available due to privacy restrictions.

Conflicts of Interest: The authors declare that they have no known competing financial interests or personal relationships that could have appeared to influence the work reported in this paper.

References

1. Mao, F.Z.; Fan, X.K.; Long, L.; Li, Y.; Chen, H.; Zhou, W. Constructing 3D hierarchical CNTs/VO$_2$ composite micro-spheres with superior electromagnetic absorption performance. *Ceram. Int.* **2023**, *49 Pt A*, 16924–16931. [CrossRef]
2. Zhu, H.H.; Qin, G.; Zhou, W.; Li, Y.; Zhou, X.B. Constructing flake-like ternary rare earth Pr3Si2C2 ceramic on SiC whiskers to enhance electromagnetic wave absorption properties. *Ceram. Int.* **2024**, *50 Pt A*, 134–142. [CrossRef]
3. Mao, F.Z.; Long, L.; Pi, W.Q.; Li, Y.; Zhou, W. X-band electromagnetic absorption and mechanical properties of mul-lite/Ti$_3$AlC$_2$ composites. *Mater. Chem. Phys.* **2022**, *292*, 126819. [CrossRef]
4. Liu, Y.; Wang, Y.D.; Wu, N.; Han, M.R.; Liu, W.; Liu, J.R.; Zeng, Z.H. Diverse Structural Design Strategies of MXene-Based Macrostructure for High-Performance Electromagnetic Interference Shielding. *Nano-Micro Lett.* **2023**, *15*, 240. [CrossRef]
5. Li, X.L.; Li, M.H.; Li, X.; Fan, X.M.; Zhi, C.Y. Low Infrared Emissivity and Strong Stealth of Ti-Based MXenes. *Research* **2022**, *2022*, 9892628. [CrossRef]
6. Verma, R.; Thakur, P.; Chauhan, A.; Jasrotia, R.; Thakur, A. A review on MXene and its? composites for electromagnetic interference (EMI) shielding applications. *Carbon* **2023**, *208*, 170–190. [CrossRef]
7. Mao, F.Z.; Long, L.; Zeng, G.L.; Chen, H.; Li, Y.; Zhou, W. Achieving excellent electromagnetic wave absorption property by constructing VO$_2$ coated biomass carbon heterostructures. *Diam. Relat. Mater.* **2022**, *130*, 109422. [CrossRef]
8. Chen, Y.Y.; Zhang, Q.T.; Chi, M.C.; Guo, C.Y.; Wang, S.F.; Min, D.Y. Preparation and performance of different carbon-ized wood electrodes. *J. For. Eng.* **2022**, *7*, 127–135. [CrossRef]
9. Zhang, Y.Y.; Xiao, H.N.; Xiong, R.H.; Huang, C.B. Xylan-based ratiometric fluorescence carbon dots composite with delignified wood for highly efficient water purification and photothermal conversion. *Sep. Purif. Technol.* **2023**, *324*, 124513. [CrossRef]

10. Gan, W.T.; Wang, Y.X.; Xiao, K.; Zhai, M.K.; Wang, H.G.; Xie, Y.J. Research review of energy storage and conversion materials based on wood cell wall functional modification. *J. For. Eng.* **2022**, *7*, 1–12. [CrossRef]
11. Lu, Y.; Liang, Z.X.; Fu, Z.Y.; Zhang, S.F. Research advances and prospect of wood cell wall nanotechnology. *J. For. Eng.* **2022**, *7*, 1–11. [CrossRef]
12. Chen, Y.M.; Pang, L.; Li, Y.; Luo, H.; Duan, G.G.; Mei, C.T.; Xu, W.H.; Zhou, W.; Liu, K.M.; Jiang, S.H. Ultra-thin and highly flexible cellulose nanofiber/silver nanowire conductive paper for effective electromagnetic interference shielding. *Compos. Part A Appl. Sci. Manuf.* **2020**, *135*, 105960. [CrossRef]
13. Ma, X.F.; Guo, H.T.; Zhang, C.M.; Chen, D.H.; Tian, Z.W.; Wang, Y.F.; Chen, Y.M.; Wang, S.W.; Han, J.Q.; Lou, Z.C.; et al. ZIF-67/wood derived self-supported carbon composites for electromagnetic interference shielding and sound and heat insulation. *Inorg. Chem. Front.* **2022**, *9*, 6305–6316. [CrossRef]
14. Zhou, B.; Li, Q.T.; Xu, P.H.; Feng, Y.Z.; Ma, J.M.; Liu, C.T.; Shen, C.Y. An asymmetric sandwich structural cellulose-based film with self-supported MXene and AgNW layers for flexible electromagnetic interference shielding and thermal management. *Nanoscale* **2021**, *13*, 2378–2388. [CrossRef] [PubMed]
15. Wang, H.R.; Jiang, Y.; Ma, Z.W.; Shi, Y.Q.; Zhu, Y.J.; Huang, R.Z.; Feng, Y.Z.; Wang, Z.B.; Hong, M.; Gao, J.F.; et al. Hyperelastic, Robust, Fire-Safe Multifunctional MXene Aerogels with Unprecedented Electromagnetic Interference Shielding Efficiency. *Adv. Funct. Mater.* **2023**, *33*, 2306884. [CrossRef]
16. Chen, Y.M.; Luo, H.; Guo, H.T.; Liu, K.M.; Mei, C.T.; Li, Y.; Duan, G.G.; He, S.J.; Han, J.Q.; Zheng, J.J.; et al. Anisotropic cellulose nanofibril composite sponges for electromagnetic interference shielding with low reflection loss. *Carbohydr. Polym.* **2022**, *276*, 118799. [CrossRef] [PubMed]
17. Xin, W.; Ma, M.G.; Chen, F. Silicone-Coated MXene/Cellulose Nanofiber Aerogel Films with Photothermal and Joule Heating Performances for Electromagnetic Interference Shielding. *Acs Appl. Nano Mater.* **2021**, *4*, 7234–7243. [CrossRef]
18. Weng, C.X.; Xing, T.L.; Jin, H.; Wang, G.R.; Dai, Z.H.; Pei, Y.M.; Liu, L.Q.; Zhang, Z. Mechanically robust ANF/MXene composite films with tunable electromagnetic interference shielding performance. *Compos. Part A Appl. Sci. Manuf.* **2020**, *135*, 105927. [CrossRef]
19. Liu, L.; Chen, W.; Zhang, H.; Wang, Q.; Guan, F.; Yu, Z. Flexible and multifunctional silk textiles with biomimetic leaf-like MXene/silver nanowire nanostructures for electromagnetic interference shielding, humidity monitoring, and self-derived hydrophobicity. *Adv. Funct. Mater.* **2019**, *29*, 1905197. [CrossRef]
20. Zhang, H.; Chen, J.; Ji, H.; Wang, N.; Feng, S.; Xiao, H. Electromagnetic interference shielding with absorption-dominant performance of Ti_3C_2TX MXene/non-woven laminated fabrics. *Text. Res. J.* **2021**, *91*, 2448–2458. [CrossRef]
21. Harandi, D.; Moradienayat, M. Multifunctional PVB nanocomposite wood coating by cellulose nanocrystal/ZnO nanofiller: Hydrophobic, water uptake, and UV-resistance properties. *Prog. Org. Coat.* **2023**, *179*, 107546. [CrossRef]
22. Han, X.S.; Ye, Y.H.; Lam, F.; Pu, J.W.; Jiang, F. Hydrogen-bonding-induced assembly of aligned cellulose nanofibers into ultrastrong and tough bulk materials. *J. Mater. Chem. A* **2019**, *7*, 27023–27031. [CrossRef]
23. Ma, X.F.; Liu, S.Y.; Luo, H.; Guo, H.T.; Jiang, S.H.; Duan, G.G.; Zhang, G.Y.; Han, J.Q.; He, S.J.; Lu, W.; et al. MOF@wood Derived Ultrathin Carbon Composite Film for Electromagnetic Interference Shielding with Effective Absorption and Electrothermal Management. *Adv. Funct. Mater.* **2023**, *34*, 202310126. [CrossRef]
24. Zhu, L.Y.; Li, Y.C.; Zhao, J.Y.; Liu, J.; Lei, J.D.; Wang, L.Y.; Huang, C.B. A novel green lignosulfonic acid/Nafion composite membrane with reduced cost and enhanced thermal stability. *Chem. Commun.* **2021**, *57*, 9288–9291. [CrossRef]
25. Deng, W.N.; Xu, Y.X.; Zhang, X.C.; Li, C.Y.; Liu, Y.X.; Xiang, K.X.; Chen, H. $(NH_4)_2Co_2V_{10}O_{28} \cdot 16H_2O/(NH_4)_2V_{10}O_{25} \cdot 8H_2O$ heterostructure as cathode for high-performance aqueous Zn-ion batteries. *J. Alloys Compd.* **2022**, *903*, 163824. [CrossRef]
26. Qu, Q.L.; Zhang, X.L.; Yang, A.Q.; Wang, J.; Cheng, W.X.; Zhou, A.Y.; Deng, Y.K.; Xiong, R.H.; Huang, C.B. Spatial confinement of multi-enzyme for cascade catalysis in cell-inspired all-aqueous multicompartmental microcapsules. *J. Colloid Interface Sci.* **2022**, *626*, 768–774. [CrossRef] [PubMed]
27. Deng, W.N.; Li, Y.H.; Xu, D.F.; Zhou, W.; Xiang, K.X.; Chen, H. Three-dimensional hierarchically porous nitro-gen-doped carbon from water hyacinth as selenium host for high-performance lithium–selenium batteries. *Rare Met.* **2022**, *41*, 3432–3445. [CrossRef]
28. Zhou, W.; Niu, Z.B.; Chen, X.; Xiao, P.; Li, Y. Synergistic effect of water vapour on the thermal corrosion of CFAS melt to $Yb_2Si_2O_7$ environmental barrier coating material. *Corros. Sci.* **2023**, *225*, 111625. [CrossRef]
29. Wen, X.Y.; Luo, J.H.; Xiang, K.X.; Zhou, W.; Zhang, C.F.; Chen, H. High-performance monoclinic WO_3 nanospheres with the novel NH4+ diffusion behaviors for aqueous ammonium-ion batteries. *Chem. Eng. J.* **2023**, *458*, 141381. [CrossRef]
30. Deng, W.N.; Liu, W.M.; Zhu, H.; Chen, L.; Liao, H.Y.; Chen, H. Click-chemistry and ionic cross-linking induced double cross-linking ionogel electrolyte for flexible lithium-ion batteries. *J. Energy Storage* **2023**, *72*, 108509. [CrossRef]
31. Wu, D.D.; Wang, D.M.; Ye, X.M.; Yuan, K.R.; Xie, Y.L.; Li, B.H.; Huang, C.B.; Kuang, T.R.; Yu, Z.Q.; Chen, Z. Fluo-rescence detection of Escherichia coli on mannose modified ZnTe quantum dots. *Chin. Chem. Lett.* **2020**, *31*, 1504–1507. [CrossRef]
32. Cui, J.X.; Lu, T.; Li, F.H.; Wang, Y.L.; Lei, J.D.; Ma, W.J.; Zou, Y.; Huang, C.B. Flexible and transparent composite nanofibre membrane that was fabricated via a "green" electrospinning method for efficient particulate matter 2.5 capture. *J. Colloid Interface Sci.* **2021**, *582*, 506–514. [CrossRef] [PubMed]
33. Zeng, G.L.; Wang, Y.Q.; Lou, X.M.; Chen, H.; Jiang, S.H.; Zhou, W. Vanadium oxide/carbonized chestnut needle com-posites as cathode materials for advanced aqueous zinc-ion batteries. *J. Energy Storage* **2024**, *77*, 109859. [CrossRef]

34. Ma, W.J.; Ding, Y.C.; Li, Y.S.; Gao, S.T.; Jiang, Z.C.; Cui, J.X.; Huang, C.B.; Fu, G.D. Durable, self-healing superhy-drophobic nanofibrous membrane with self-cleaning ability for highly-efficient oily wastewater purification. *J. Membr. Sci.* **2021**, *634*, 119402. [CrossRef]
35. Deng, Y.K.; Lu, T.; Zhang, X.L.; Zeng, Z.Y.; Tao, R.P.; Qu, Q.L.; Zhang, Y.Y.; Zhu, M.M.; Xiong, R.H.; Huang, C.B. Multi-hierarchical nanofiber membrane with typical curved-ribbon structure fabricated by green electrospinning for effi-cient, breathable and sustainable air filtration. *J. Membr. Sci.* **2022**, *660*, 120857. [CrossRef]
36. Lu, T.; Liang, H.B.; Cao, W.X.; Deng, Y.K.; Qu, Q.L.; Ma, W.J.; Xiong, R.H.; Huang, C.B. Blow-spun nanofibrous composite Self-cleaning membrane for enhanced purification of oily wastewater. *J. Colloid Interface Sci.* **2022**, *608*, 2860–2869. [CrossRef]

Disclaimer/Publisher's Note: The statements, opinions and data contained in all publications are solely those of the individual author(s) and contributor(s) and not of MDPI and/or the editor(s). MDPI and/or the editor(s) disclaim responsibility for any injury to people or property resulting from any ideas, methods, instructions or products referred to in the content.

Article

Bioinspired Thermal Conductive Cellulose Nanofibers/Boron Nitride Coating Enabled by Co-Exfoliation and Interfacial Engineering

Xinyuan Wan [1], Xiaojian Xia [1], Yunxiang Chen [1], Deyuan Lin [1], Yi Zhou [2] and Rui Xiong [2,*]

[1] State Grid Fujian Electric Research Institute, Fuzhou 350007, China; wanxinyuan89103@163.com (X.W.); xia.xiaojian@gmail.com (X.X.); rogerchen614@163.com (Y.C.); lindeyuan_fj@126.com (D.L.)
[2] State Key Laboratory of Polymer Materials Engineering, Polymer Research Institute of Sichuan University, Chengdu 610065, China; yizhou@stu.scu.edu.cn
* Correspondence: rui.xiong@scu.edu.cn

Citation: Wan, X.; Xia, X.; Chen, Y.; Lin, D.; Zhou, Y.; Xiong, R. Bioinspired Thermal Conductive Cellulose Nanofibers/Boron Nitride Coating Enabled by Co-Exfoliation and Interfacial Engineering. *Polymers* **2024**, *16*, 805.
https://doi.org/10.3390/polym 16060805

Academic Editors: Ana Pilipović, Phuong Nguyen-Tri and Mustafa Özcanli

Received: 5 February 2024
Revised: 5 March 2024
Accepted: 12 March 2024
Published: 14 March 2024

Copyright: © 2024 by the authors. Licensee MDPI, Basel, Switzerland. This article is an open access article distributed under the terms and conditions of the Creative Commons Attribution (CC BY) license (https:// creativecommons.org/licenses/by/ 4.0/).

Abstract: Thermal conductive coating materials with combination of mechanical robustness, good adhesion and electrical insulation are in high demand in the electronics industry. However, very few progresses have been achieved in constructing a highly thermal conductive composites coating that can conformably coat on desired subjects for efficient thermal dissipation, due to their lack of materials design and structure control. Herein, we report a bioinspired thermal conductive coating material from cellulose nanofibers (CNFs), boron nitride (BN), and polydopamine (PDA) by mimicking the layered structure of nacre. Owing to the strong interfacial strength, mechanical robustness, and high thermal conductivity of CNFs, they do not only enhance the exfoliation and dispersion of BN nanoplates, but also bridge BN nanoplates to achieve superior thermal and mechanical performance. The resulting composites coating exhibits a high thermal conductivity of 13.8 W/(m·K) that surpasses most of the reported thermal conductive composites coating owing to the formation of an efficient thermal conductive pathway in the layered structure. Additionally, the coating material has good interface adhesion to conformably wrap around various substrates by scalable spray coating, combined with good mechanical robustness, sustainability, electrical insulation, low-cost, and easy processability, which makes our materials attractive for electronic packaging applications.

Keywords: cellulose nanofibers; boron nitride; layered structure; thermal conductive coating; mechanical properties

1. Introduction

The urgent requirement of miniaturized, densified, and multi-functional electronics significantly increases the power density of electronics, leading to fast heat accumulation in a limited or confined space [1,2]. This rapidly increased temperature inevitably affects the service life, safety, reliability, and speed of the electronics, even causing severe equipment damage and major fires. To address this issue, progress has been made to develop thermally conductive materials that can effectively dissipate the accumulated heat. Metallic and nanocarbon materials, including aluminum, MXene, graphene, and carbon nanotube have been intensively incorporated into polymeric matrix for constructing flexible, light weight composites with high-performance thermal management capabilities [3–5]. Although these resulting composites exhibit outstanding thermal conductivity, they usually also possess high electrical conductivity [6–8], which easily cause undesired short-circuit problem in the application of electronics. On the other hand, to achieve efficient thermal management, a seamless interface between electronics and thermal conductive composites is strictly needed. However, the conformal integration of these pre-formed thermal conductive composites onto the electronics is difficult due to the poor interface adhesion, especially for some irregular surfaces. The addition of glue could solve the adhesion problem but introduces additional thermal resistance and cost. Therefore, it is desirable to develop a thermal conductive but electrical insulating composites coating that can be seamlessly assembled on various irregular objects.

Boron nitride (BN) not only has excellent thermal conductivity and electrical insulation property, but also has good chemical stability and oxidation resistance, making it a promising candidate as a nanofiller for constructing thermal conductive composite coating [9,10]. However, most BN-based composites coating still face the issues of low thermal conductivity and weak mechanical properties, due to the poor dispersion of conductive filler and lack of structure control, largely limiting their practical applications [11,12]. To realize high thermal conductivity and mechanical robustness, bioinspired structural hierarchy design is one of the most promising approaches to engineer the composites coating [13]. The most spectacular examples are nacre-like composites, using aligned micro/nanoplates in polymer matrix [14,15]. For instance, Han et al. report a strong nacre-mimetic BN/epoxy composite by using a bidirectional freezing technique, which can realize high thermal conductivity of 6.07 $Wm^{-1}K^{-1}$ combined with good electrical insulation performance [15]. Pan et al., took advantage of hot-pressing technique to construct brick and mortar-structured Ag-Al_2O_3 platelets/epoxy composites with a thermal conductivity of 6.71 $Wm^{-1}K^{-1}$ [16]. However, these techniques usually need complex processes, which are difficult to be apply in composite coating and constructing nanostructured composites coating with a combination of high thermal conductivity, mechanical robustness, and good adhesion is still a big challenge.

In nature, nacres utilize chitin nanofibers which are wrapped by proteins to glue $CaCO_3$ microplates to form hierarchical layered structures [17]. The highly ordered architecture and favorable interfacial strength enable nacre to have an amazing combination of mechanical strength and toughness, with added brilliant iridescence [18]. In this study, we report a bioinspired high-performance thermal conductive coating by exfoliation/dispersion of BN nanoplate and cellulose nanofibers (CNFs) and multiple interfacial interaction engineering. We take advantage of sustainable 1D CNF and dopamine as the building blocks to mimic the combination of chitin and proteins. CNFs have a similar structure to chitin, but it is more widely available and stronger [19–21], while polydopamine (PDA) is a well-known mussel-inspired adhesive that mimics the adhesive properties of mussels, which can adhere to a variety of surfaces under wet and dry conditions [22]. CNFs and BN nanoplates are integrated together to form viscose ink through the simultaneous exfoliation/dispersion induced by the strong shearing force of pan mill. The subsequent incorporation of dopamine not only significantly enhanced the interfacial strength of the composites, but also enabled strong adhesion to various substrates as conformal coating. The resulting coating demonstrates thermal conductivity of 13.8 W/(m·K) combined with good mechanical properties because the uniform layered structure provides prolonged phonon pathways. Thus, such an outstanding combination of thermal conductivity, mechanical properties, electrical insulation, adhesion, sustainability, and scalable process make our composites promising candidates for advanced electronic packaging technology applications.

2. Materials and Methods
2.1. Materials

Bleached wood pulp was purchased from Dalian Yangrun Trading Co., Ltd., Dalian, China. 2,2,6,6-Tetramethylpiperidine-1-oxyl (TEMPO), sodium hydroxide (NaOH), and dopamine hydrochloride were purchased from Aladdin Industrial Co., Shanghai, China. Sulfuric acid, sodium bromide (NaBr), sodium hypochlorite (NaClO), and hydrochloric acid were purchased from Xilong Scientific Co., Ltd., Guangdong, China. BN powder was purchased from Zhengzhou Jiajie Chemical Products Co., Ltd., Zhengzhou, China. Tris-HCl buffer was purchased from Fuzhou Feijin Biotechnology Co., Ltd., Fuzhou, China.

2.2. TEMPO Oxidation of Cellulose Fibers

Briefly, cellulose bleached wood pulp (50 g) was added to distilled water (5000 mL) to obtain homogenous dispersion with vigorous stirring. Then, a pH meter was placed in the suspension to monitor its pH value. TEMPO (0.78 g), NaBr (5.113 g), and 11 wt % NaClO (322.7 g) were separately added in the above suspension. Under constant stirring, the 0.1 M HCl was used to adjust the pH of the system to maintain it at around 10. As the reaction proceeds, the pH of the solution decreases. To maintain the pH of the system, 0.1 M NaOH solution was added slowly. The reaction was terminated when there is no change of the pH in the system for at least 30 min. The as-prepared TEMPO oxidized cellulose slurry was thoroughly washed until neutral and isolated by sieve. After measuring the mass of fibers, the cellulose slurry was diluted to 2 wt % cellulose fiber suspension.

2.3. Co-Exfoliation/Dispersion of BN Nanoplate and Cellulose Nanofibers (CNFs)

Co-exfoliation/dispersion of BN nanoplate and cellulose nanofibers (CNFs) were carried out according to the previous study [23]. Under continuous stirring, different amounts of BN powder were suspended in 2000 mL cellulose fiber suspension. Then, the suspensions were milled by an ultra-fine friction grinder Supermass colloider (MKCA6-2, Masuko Sangyo Co., Ltd., Kawaguchi, Japan) at 1500 rpm to obtain homogeneous CNFs/BN nanoplates mixture. The initial pan gap distance was set as 0 μm which was the point of slight contact between the two pans to avoid blocking. The pan gap distance was set from large to small. The distance between the two pans was reduced from 0 μm to −10 μm, −50 μm, −100 μm, and −150 μm, respectively. The suspensions were treated 10 times at every position to obtain 5 wt % homogeneous CNFs/BN nanoplates mixture (BCNF). The mixture with different BN contents related to the total mass of the suspension were coded to be BCNF10, BCNF30, BCNF50, and BCNF70 for 10 wt %, 30 wt %, 50 wt %, and 70 wt %, respectively. For a better comparison, the solid content of all BCNF samples (10–70 wt %) are the same 5 wt %.

2.4. Self-Polymerisation of Dopamine

The self-polymerization of dopamine was carried out according to the reported works [24–26]. Typically, dopamine hydrochloride (0.5 g) was added to 100 mL Tris-HCl buffer (10 mM, pH 8.5) under vigorous stirring for 24 h at room temperature. Then, the mixtures were centrifuged (10,000 rpm, 10 min) and washed with distilled water 3 times. After drying under vacuum, the PDA solution was obtained through re-dispersion in water at the concentration of 5 mg/mL.

2.5. Preparation of Thermal Conductive Film and Coating

In a typical process, 10 mL PDA solution (5 mg/mL) was added to 19 g BCNF (5 wt %) under stirring to prepare thermal conductive composite slurries (CNFs/BN/PDA). The composites films were fabricated by vacuum filtration of the slurries on a cellulose filter membrane with a 0.45 μm pore size. The as-prepared films were dried at room temperature and peeled off from the filter membrane. The thickness of membrane could be controlled by adjusting the volume of CNFs/BN/PDA slurries. Here, the thickness was adjusted to 200 ± 20 μm. The thermal conductive coating was fabricated by spraying CNFs/BN/PDA slurries on different substrates. In the typical spraying process, the substrates were fixed on fixtures and sprayed with different substrates using a spray gun fitted with CNFs/BN/PDA slurries. Then, they were air dried at room temperature.

2.6. Characterization

The Micro/nano structures of BN powder, cellulose bleached wood pulp, CNFs/BN/PDA slurries, and films were characterized by field emission scanning electron microscope (SEM, Thermo Fisher Scientific, Shanghai, China (FEI, Apreo S HiVoc)). The morphologies of BCNF were characterized using a high-resolution transmission electron microscopy (TEM, Tecnai G2 F20 S-TWIN, Hillsboro, OR, USA) at 200 Kv. To study the rheological behavior of

BCNF with different contents of BN, the rheological tests were performed by an advanced rotational rheometer (MCR302, Anton Paar, Graz, Austria). The stress–strain curves of the CNFs/BN/PDA films were obtained on the Instron 5966 universal testing machine (Instron, Boston, MA, USA) with a 1000 N load cell. The stress–strain curves of three identical spline patterns were repeated for samples with different BCNF ratios, and the error bars were obtained from the three sets of data. The thermal conductivity values (λ, W/m·K) of CNFs/BN/PDA films were characterized by a thermal constant analyzer (Hot Disk, 2500S, Goteborg, Switzerland). A pull-off method adhesion tester (BJZJ-M, Zhongjiaojianyi Testing Equipment Co., Beijing, China) was used to test the adhesion strength of CNFs/BN/PDA coatings on different substrates. The water contact angle of the coatings was characterized by an optical contact angle measuring instrument (KRUSS, DSA30, Hamburg, Germany).

3. Results and Discussion

The schematic representation shown in Figure 1a–d illustrates the fabrication of the CNFs/BN/PDA thermal conductive composites ink. To facilitate the exfoliation and dispersion of CNFs, soft wood pulp has been pretreated using TEMPO oxidation, which selectively oxidizes the primary hydroxyl groups (C6) on the surface of cellulose fibers to negatively charged carboxyl groups [27]. The pretreated cellulose fibers were mixed with BN and subjected to pan mill to co- exfoliate and disperse CNFs/BN nanoplates mixture. During the process of co-exfoliation and dispersion, stacked large-size BN particles are peeled off into nanosheets, using the strong shear force of two pans. Meanwhile, the pretreated cellulose fibers also become nano fibers (CNFs). CNFs could inhibit the re-stacking and agglomeration of BN nanosheets through hydrogen bonding, hydrophobic interactions, and spatial site resistance, resulting in stable CNFs/BN slurries (Figure S1) [23]. However, the resulting slurries are difficult to apply onto various substrates due to poor adhesion. To achieve a wide range of adaptability to different substrates, 5 wt % PDA were added into the above slurries to regulate the interactions. The strong multiple interactions, including hydrogen bonding, hydrophobic interaction, and π-π stacking interactions, would greatly enhance both adhesion with substrates and internal interactions [28]. As shown in Figure S2, the Fourier Transform Infrared Spectrometer (FTIR) peak at 3350 cm^{-1} assigned to the hydrogen bonded –OH stretching vibrations increased after the addition of PDA, indicating the formation of additional hydrogen bonding by PDA. Taking advantage of evaporation induced self-assembly strategy, conformal dense coating with highly organized layered structure can form from the viscous CNFs/BN/PDA slurry. The slurry with different BN contents is coded to be BCNF10, BCNF30, BCNF50, and BCNF70 for 10 wt %, 30 wt %, 50 wt %, and 70 wt % BN, respectively.

We investigate the morphological evolution of building blocks during the co-exfoliation and PDA formation. Pristine BN powder consists of nanoplates with a lateral size of 1–10 μm and a thickness of around 200 nm (Figure 2a), while original cellulose fibers from wood pulp have a diameter of around 10–15 μm (Figure 2d). Due to their large size and hydrophobic character, BN powders are difficult to disperse in water. After the pan mill treatment, the BN mixture and cellulose fibers are co-exfoliated into small sized particles to form a stable slurry (Figures 2b,e and 3a), where BN nanoplates are conformal wrapped by an elementary CNFs network with diameter of 3 nm (Figure 2c). We suggest that this co-exfoliation of CNFs/BN dispersion is facilitated by the amphiphilic property of CNFs, which consists of a hydrophobic and hydrophobic crystalline plane [29], while the –OH/-COOH groups of hydrophilic planes assist the dispersion of BN in water [30]. After the addition of PDA, numerous PDA nanoparticles are bonded on the CNFs and BN nanoplates' surface due to strong adhesion (Figure 2f). This good exfoliation and good dispersion of building blocks is favorable for constructing thermal conductive coating materials.

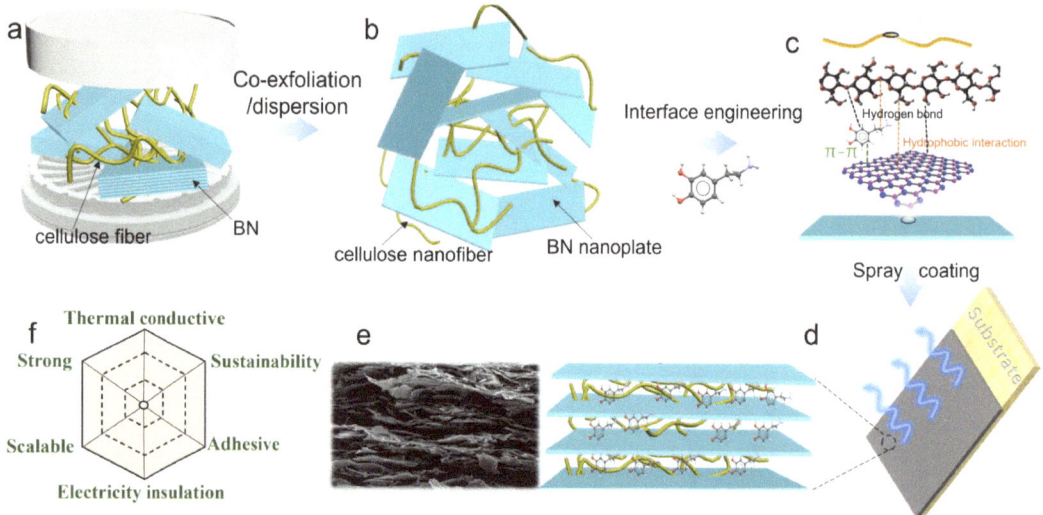

Figure 1. (**a–d**) The schematic illustration of the preparation of bioinspired thermal conductive composite slurries from BN, CNFs, and dopamine. (**e**) SEM image of cross section of composites coating. (**f**) The properties of the composites coating.

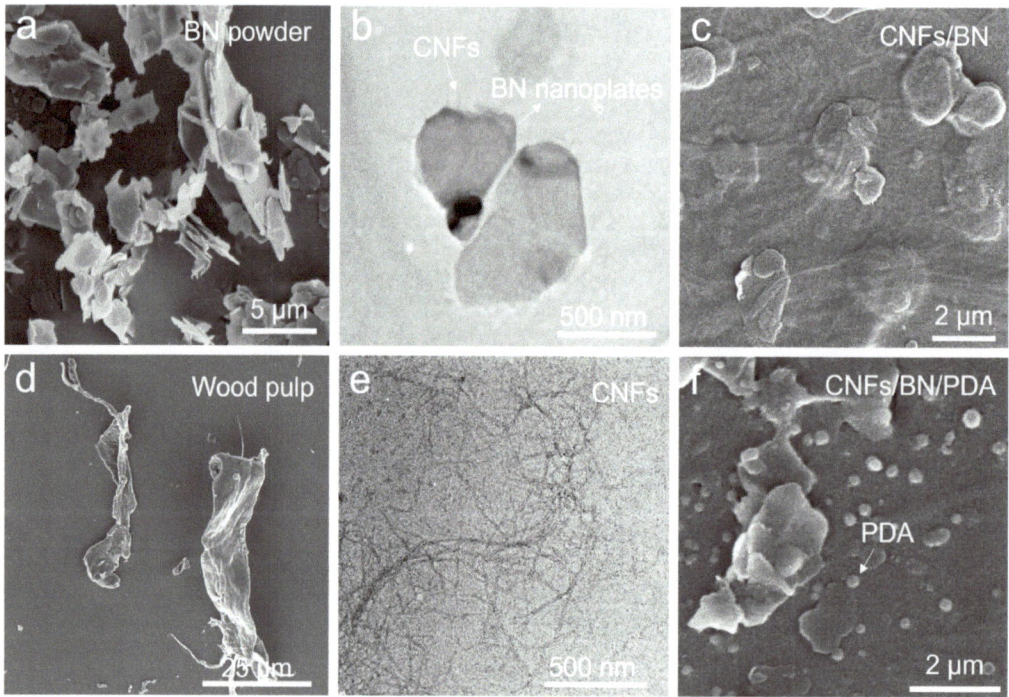

Figure 2. SEM images of pristine BN (**a**) and cellulose fibers (**d**); TEM images of BN (**b**) and CNFs (**e**) after co-exfoliation; SEM images of CNFs/BN mixture before (**c**) and after PDA decoration (**f**).

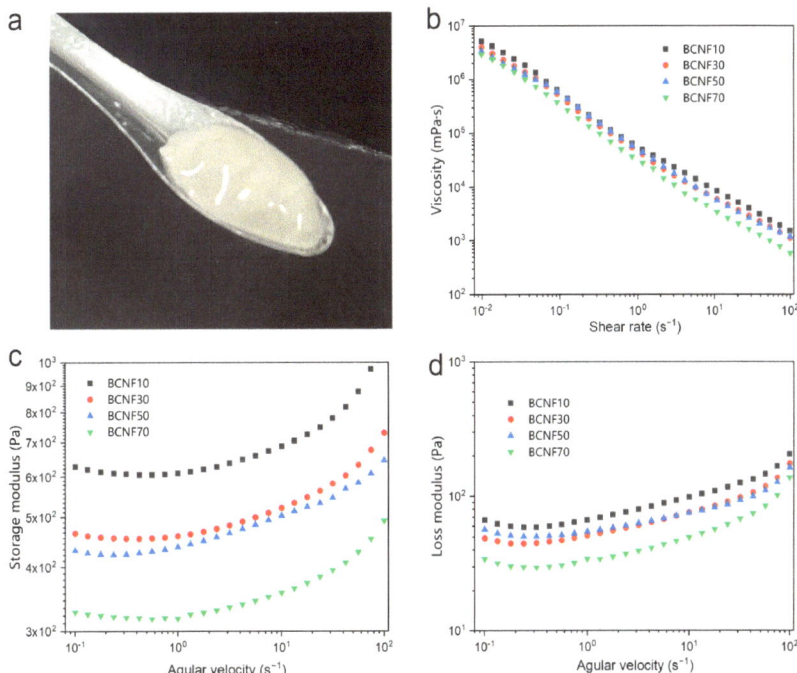

Figure 3. (**a**) The photo of obtained CNFs/BN/PDA slurry (3.5 wt %). (**b**) Viscosities of CNFs/BN/PDA slurries with different BN contents as a function of shear rate. Storage (**c**) and loss (**d**) modulus of CNFs/BN/PDA slurries with different BN contents as functions of frequency.

As shown in Figure 3a, the obtained CNFs/BN/PDA slurry exhibits a homogeneous morphology with good stability up to several months (Figure S3). This is because the amphiphilic character of CNFs allows them to firmly bond on the BN surface, while the high aspect ratio of CNFs forms a physical entanglement network to prevent aggregation through electrostatic repulsion [31]. As is already known, the rheological properties of dispersion play an important role in the coating processability, thus we systematically studied the rheological behavior of the resulting CNFs/BN/PDA slurries with different BN content. In the shear rate–viscosity curves, all the viscosity of slurries shows a similar downward trend with the increase in shear rate, indicating typical shear-thinning non-Newtonian behavior that is favorable for improving processability. This gradual decreased viscosity is caused by the orientation of CNFs/BN along the shear direction under low shear force [32]. Furthermore, with the increase in BN content, the viscoelastic characteristics of the suspensions decrease. The reason for this is that the BN dispersed in the slurry reduces entanglement between CNFs, so the higher the BN content, the less entanglement between the CNFs and the lower viscosity of the slurry [33]. Figure 3c,d show viscoelastic storage modulus (G′) and loss modulus (G″) behavior of CNFs/BN/PDA slurries as a function of angular frequency (ω). For all the slurries, the G′ was much higher than G″ in all of the investigated angular frequency ranges, indicating the formation of a gel-like structure in the slurries. The G′ increases continuously with the decreasing addition of BN content, mainly due to the formation of a more entangled CNFs network with more CNF contents. Additionally, the G′ of slurries has a frequency dependent behavior, where G′ experiences continuously decrease when the frequency is unchanged and increase with the decreasing frequency. This behavior can be explained by the dynamic nanostructure change of slurries. As a high frequency is applied, the initial percolating network of slurries would be broken, leading to the decreased modulus. At relatively low frequency, the destruction and reconstruction of CNFs/BN entanglements can reach a balance at the plateau. Further lowering the frequency provides enough time to significantly reconstruct the percolating network, leading to an increase in elastic and viscous moduli [34].

Owing to the good processability and adhesion of the obtained slurry, conformal thermal conductive coating can be applied to various substrate materials. As shown in Figure 4a, the BCNF70 slurries can uniformly coat the smooth steel, polytetrafluoroethylene (PTFE), and rough wood using a simple spray coating technique. The resulting coating exhibits a good smooth surface as illustrated in the 3D topographic images (Figure S4). Also, the aligned texture on the wood is still maintained after the coating treatment, indicating that the coating tightly wraps around the substrate surface. The interface adhesion of the coating was measured to be 0.81, 0.61, and 0.38 MPa for steel, wood and PTFE, respectively, which is good enough to enable the practical applications (Figure 4c). Then, we investigated the structure of the coating with different BN loading. All of the surface has a dense and smooth morphology with uniform BN nanoplates distribution. Increasing BN loading leads to more BN nanoplates being exposed on the coating surface, but they are still wrapped by CNF network to prevent the leakage in the practical applications (Figure S5). For the cross-section, all the coating has a nacre-like layered structure, where BN nanoplates are highly aligned along the substrate surface due to the strong capillary and gravity effect during drying. This uniform layered structure is not only favorable to connect BN nanoplates to form good thermal conductive networks for enhancing thermal conductivity, but also possesses multiple reinforcing mechanisms for mechanical enhancement. Additionally, the CNFs/BN/PDA coating is hydrophilic with a water contact angle of 82.4°. (Figure S6).

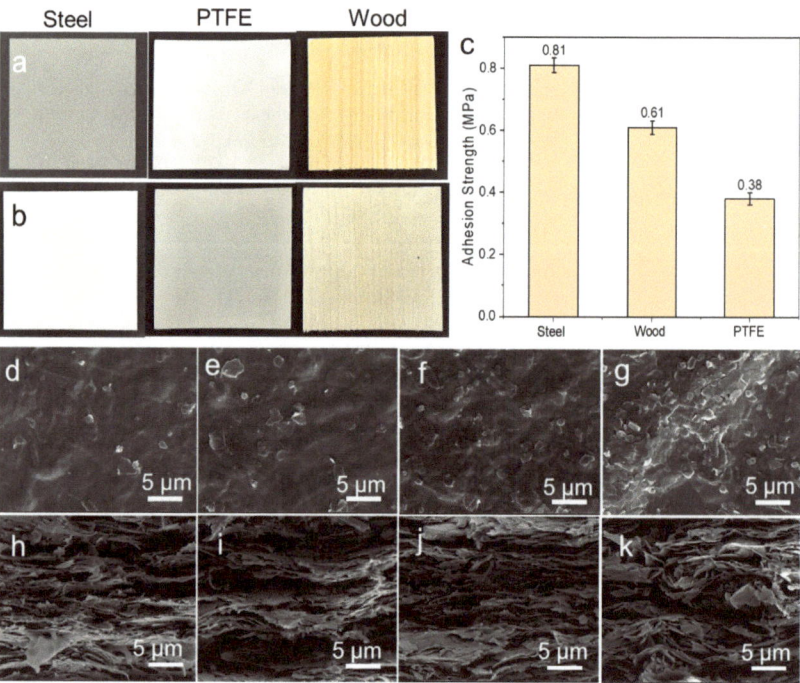

Figure 4. Photos of steel, PTFE and wood before (**a**) and after coating (**b**). (**c**) The adhesion strength of coating on different substrates. SEM images of CNFs/BN/PDA coatings' morphology. (**d–g**) The surfaces of BCNF10, BCNF30, BCNF50, and BCNF70, respectively. (**h–k**) Cross-sections of BCNF10, BCNF30, BCNF50, and BCNF70, respectively.

Next, the mechanical properties of the composites were investigated since it is critical for service life in practical applications. As illustrated in Figure 5a–d, composites with 10 wt % BN can achieve good mechanical robustness with a strength of 100 ± 4 MPa, modulus of 17 ± 1 GPa and toughness of 1.3 ± 0.2 MJ/cm^3. The increasing addition of BN nanoplates leads to the decrease in the mechanical properties, but the composites still maintain reasonable strength and toughness, even with 70 wt % BN loading. BN composites usually have limited strength and toughness because of the weak interfacial interactions between BN nanoparticles, while the superstrong CNFs can connect BN together to significantly enhance the interfacial strength [35]. To gain insight into the fracture mechanism of the composites, we investigated the fracture cross-section of the composites (Figure 5c). The SEM image of the cross-section exhibits hierarchical rough layered morphology where many nanofibers and nanoplates are pulled-out. We propose a multiple toughening mechanism response to the mechanical enhancement (Figure 5f). When applying force, the dynamic bonds are broken, followed by the stretching and slipping of CNFs and BN. With further loading, CNFs and BN reorient and align along the loading direction, which will further experience pull-out, delamination, and fracture. This process would largely dissipate fracture energy for achieving good mechanical properties [36].

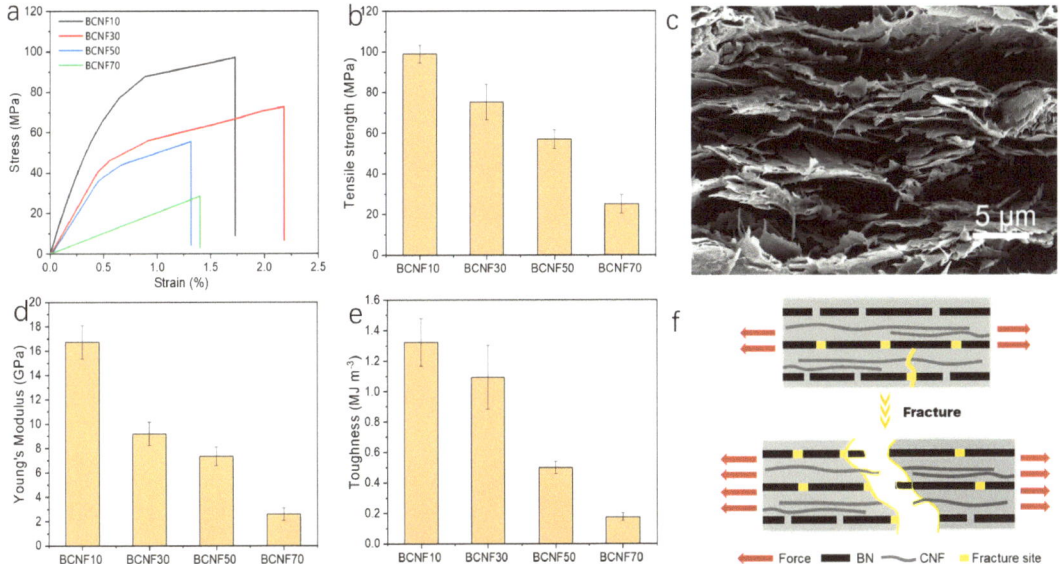

Figure 5. The stress-strain curves (**a**), strength (**b**), Young's modulus (**d**) and toughness (**e**) of CNFs/BN/PDA composites films with different BN loading (the error bars were obtained from the three sets of data). (**c**) The fractured cross-section SEM image of CNFs/BN/PDA composites. (**f**) The schematic illustration of the fracture mechanism.

The thermal conductivity of the composites coatings is illustrated in Figure 6b. The thermal conductivity of composites with 10 wt % BN loading is around 3.1 W/(m·K), which is superior to most plastics. The thermal conductivity further increases linearly with the increase in BN loading and reaches a high value up to 13.8 W/(m·K) with 70 wt % BN incorporation. We suggest that this high thermal conductivity is owed to the highly ordered thermal conducting network of the composites. These well-exfoliated CNFs and BN nanoplates are closely packed together to form a dense layered structure, where CNFs can work as a thermal conductive bridge to connect BN nanoplates because of its high intrinsic thermal conductivity [37]. Additionally, the smooth and homogenous surface could also reduce the phonon scattering during thermal conduction [23]. The possible mechanism for the high thermal conductivity was schematically illustrated in Figure 6a. The exfoliated BN nanosheets overlapped each other, and the CNFs attached around the BN nanosheets due to hydrophobic–hydrophobic interactions, forming a dense and oriented thermal conductivity network, which minimized the gaps between the BNs, reduced the thermal resistance between the composite interfaces of the BN and CNFs, and improved the thermal properties. When the CNFs/BN were heated, the heat flow diffused rapidly along the network of BN nanosheets and CNFs to the whole, due to the inherent high thermal conductivity, dense thermal network, and low thermal resistance of BN and cellulose, thus exhibiting excellent thermal conductivity [38]. To visually evaluate the heat transfer capability of our coating, we placed the coating films on a hotplate with a temperature of 80 °C. The heat dissipation performance is monitored by the temperature variations through infrared thermal imaging technology (Figure 6d). Obviously, the temperature of composites with higher BN loading increases faster, and can reach 75 °C within just 5 s due to their high thermal conductivity, exhibiting outstanding heat dissipation performance. Additionally, to highlight the heat dissipation performance, we compared the thermal conductivity of our coating materials with other reported thermal conductive coating materials (Table S1) [39–44]. As shown in Figure 6b, the thermal conductivity exhibited in this work is far beyond other thermal conductive coating materials. For instance, Xu et al., reported that the addition of 40 wt % BN into epoxy (EP) lead to conductivity of 1.5 W/(m·K), which can further increase to 2.4 W/(m·K) with the addition of graphene. Ligati et al. prepared graphene-loaded paint with a thermal conductivity of 1.6 W/(m·K). Clearly, the thermal conductivity of our work is far higher than these reported materials, even with the same BN loading (Figure S7).

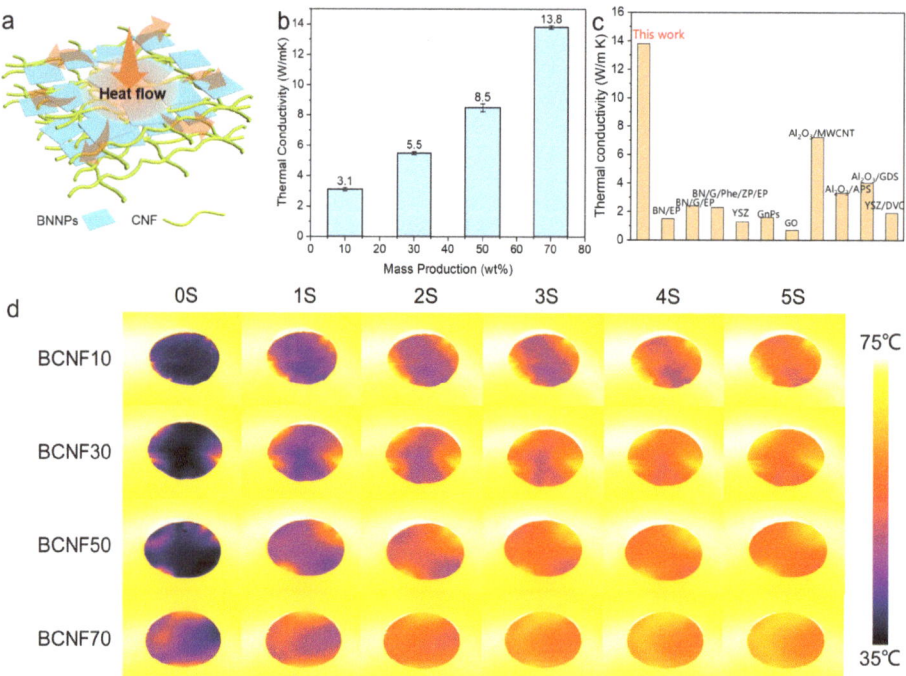

Figure 6. (**a**) Schematic of thermal conduction mechanism. (**b**) Thermal conductivity of CNFs/BN/PDA composites films with different BN loading. (**c**) Comparison of reported thermal conductive composite coatings. (**d**) Infrared thermal images of surface temperature changes of CNFs/BN/PDA composites films placed on an 80 °C hotplate.

4. Conclusions

In summary, we have realized the scalable production of highly thermal conductive CNF/BN/PDA coating by mimicking the layered nanostructure of nacre. Taking advantage of the strong interfacial interaction and the strong mechanical shearing of a pan mill, efficient co-exfoliation and dispersion of CNF and BN can be achieved to produce processible slurry. The addition of PDA can largely increase the adhesion of the slurry to various substrate surfaces by simple spray coating. The resulting coating exhibits nacre-like layered structure with horizontally aligned BN nanoplates that are connected by CNFs. These interconnected networks guarantee ultrafast thermal conductive pathways for phonon transport, as well as multiple reinforcement mechanisms for energy dissipation. As a result, the composites coatings exhibit a high thermal conductivity of 13.8 W/(m·K), which is well beyond most of the previously reported thermal conductive coating, combined with good mechanical properties, low-cost, good adhesion and sustainability. We expect this material will find many real-world applications in the electronic, auto, and aerospace industries.

Supplementary Materials: The following supporting information can be downloaded at: https://www.mdpi.com/article/10.3390/polym16060805/s1, Figure S1. The mechanism of Co-exfoliation/dispersion of BN nanoplate and CNFs. Figure S2. FTIR spectra of BCNF70 and BCNF70 without PDA. Figure S3. (a) The photos of CNFs/BN/PDA slurry. (b) Viscosities of CNFs/BN/PDA slurries. Figure S4. 3D topographic images of BCNF70 on steel, PTFE and wood. Figure S5. SEM image of BCNF70 coating where CNFs (arrows) wrap around BN. Figure S6. Water contact angle of the CNFs/BN/PDA coating. Figure S7. Thermal conductivity comparison as a function of the mass fraction of thermal conductive nanofillers. Table S1. Comparison of thermal conductivity of reported composites coating. Refs. [39–44] are cited in the supplementary materials.

Author Contributions: X.W.: conceptualization, funding acquisition, investigation, project administration, supervision, and writing—review and editing; X.X.: conceptualization, funding acquisition, investigation; Y.C.: funding acquisition, investigation; D.L.: funding acquisition, investigation; Y.Z.: data curation, methodology, writing—original draft; R.X.: conceptualization, formal analysis, investigation, project administration, supervision, writing—original draft, writing—review and editing. All authors have read and agreed to the published version of the manuscript.

Funding: This work is supported by the science and technology project of State Grid Fujian Electric Power Co., Ltd. (Grand No. 521304230012).

Institutional Review Board Statement: Not applicable.

Data Availability Statement: Data are available in the main article and Supplementary Materials.

Conflicts of Interest: Authors Xinyuan Wan, Xiaojian Xia, Yunxiang Chen, Deyuan Lin were employed by the company State Grid Fujian Electric Research Institute. The remaining authors declare that the research was conducted in the absence of any commercial or financial relationships that could be construed as a potential conflict of interest.

References

1. Burger, N.; Laachachi, A.; Ferriol, M.; Lutz, M.; Toniazzo, V.; Ruch, D. Review of Thermal Conductivity in Composites: Mechanisms, Parameters and Theory. *Prog. Polym. Sci.* **2016**, *61*, 1–28. [CrossRef]
2. Toberer, E.S.; Baranowski, L.L.; Dames, C. Advances in Thermal Conductivity. *Annu. Rev. Mater. Res.* **2012**, *42*, 179–209. [CrossRef]
3. Han, Z.; Fina, A. Thermal Conductivity of Carbon Nanotubes and Their Polymer Nanocomposites: A Review. *Prog. Polym. Sci.* **2011**, *36*, 914–944. [CrossRef]
4. Zhang, H.; Zhang, X.; Fang, Z.; Huang, Y.; Xu, H.; Liu, Y.; Wu, D.; Zhuang, J.; Sun, J. Recent Advances in Preparation, Mechanisms, and Applications of Thermally Conductive Polymer Composites: A Review. *J. Compos. Sci.* **2020**, *4*, 180. [CrossRef]
5. Chen, H.; Ginzburg, V.V.; Yang, J.; Yang, Y.; Liu, W.; Huang, Y.; Du, L.; Chen, B. Thermal Conductivity of Polymer-Based Composites: Fundamentals and Applications. *Prog. Polym. Sci.* **2016**, *59*, 41–85. [CrossRef]
6. Peng, L.; Xu, Z.; Liu, Z.; Guo, Y.; Li, P.; Gao, C. Ultrahigh Thermal Conductive yet Superflexible Graphene Films. *Adv. Mater.* **2017**, *29*, 1700589. [CrossRef]
7. Chen, X.; Cheng, P.; Tang, Z.; Xu, X.; Gao, H.; Wang, G. Carbon-Based Composite Phase Change Materials for Thermal Energy Storage, Transfer, and Conversion. *Adv. Sci.* **2021**, *8*, 2001274. [CrossRef]

8. Zhang, J.; Kong, N.; Uzun, S.; Levitt, A.; Seyedin, S.; Lynch, P.A.; Qin, S.; Han, M.; Yang, W.; Liu, J.; et al. Scalable Manufacturing of Free-Standing, Strong Ti3C2Tx MXene Films with Outstanding Conductivity. *Adv. Mater.* **2020**, *32*, 2001093. [CrossRef]
9. Cai, Q.; Scullion, D.; Gan, W.; Falin, A.; Zhang, S.; Watanabe, K.; Taniguchi, T.; Chen, Y.; Santos, E.J.G.; Li, L.H. High Thermal Conductivity of High-Quality Monolayer Boron Nitride and Its Thermal Expansion. *Sci. Adv.* **2019**, *5*, eaav0129. [CrossRef]
10. Guerra, V.; Wan, C.; McNally, T. Thermal Conductivity of 2D Nano-Structured Boron Nitride (BN) and Its Composites with Polymers. *Prog. Mater. Sci.* **2019**, *100*, 170–186. [CrossRef]
11. Kim, K.; Ryu, S.; Kim, J. Melt-Processable Aggregated Boron Nitride Particle via Polysilazane Coating for Thermal Conductive Composite. *Ceram. Int.* **2017**, *43*, 2441–2447. [CrossRef]
12. Abbas, A.; Zhao, Y.; Zhou, J.; Wang, X.; Lin, T. Improving Thermal Conductivity of Cotton Fabrics Using Composite Coatings Containing Graphene, Multiwall Carbon Nanotube or Boron Nitride Fine Particles. *Fibers Polym.* **2013**, *14*, 1641–1649. [CrossRef]
13. Tao, P.; Shang, W.; Song, C.; Shen, Q.; Zhang, F.; Luo, Z.; Yi, N.; Zhang, D.; Deng, T. Bioinspired Engineering of Thermal Materials. *Adv. Mater.* **2015**, *27*, 428–463. [CrossRef] [PubMed]
14. Wang, H.; Lu, R.; Li, L.; Liang, C.; Yan, J.; Liang, R.; Sun, G.; Jiang, L.; Cheng, Q. Strong, Tough, and Thermally Conductive Nacre-Inspired Boron Nitride Nanosheet/Epoxy Layered Nanocomposites. *Nano Res.* **2023**, *17*, 820–828. [CrossRef]
15. Han, J.; Du, G.; Gao, W.; Bai, H. An Anisotropically High Thermal Conductive Boron Nitride/Epoxy Composite Based on Nacre-Mimetic 3D Network. *Adv. Funct. Mater.* **2019**, *29*, 1900412. [CrossRef]
16. Pan, G.; Yao, Y.; Zeng, X.; Sun, J.; Hu, J.; Sun, R.; Xu, J.-B.; Wong, C.-P. Learning from Natural Nacre: Constructing Layered Polymer Composites with High Thermal Conductivity. *ACS Appl. Mater. Interfaces* **2017**, *9*, 33001–33010. [CrossRef] [PubMed]
17. Wegst, U.G.K.; Bai, H.; Saiz, E.; Tomsia, A.P.; Ritchie, R.O. Bioinspired Structural Materials. *Nat. Mater.* **2015**, *14*, 23–36. [CrossRef] [PubMed]
18. Sun, Z.; Liao, T.; Li, W.; Qiao, Y.; Ostrikov, K. (Ken) Beyond Seashells: Bioinspired 2D Photonic and Photoelectronic Devices. *Adv. Funct. Mater.* **2019**, *29*, 1901460. [CrossRef]
19. Ling, S.; Chen, W.; Fan, Y.; Zheng, K.; Jin, K.; Yu, H.; Buehler, M.J.; Kaplan, D.L. Biopolymer Nanofibrils: Structure, Modeling, Preparation, and Applications. *Prog. Polym. Sci.* **2018**, *85*, 1–56. [CrossRef]
20. Iioka, M.; Kawanabe, W.; Tsujimura, S.; Kobayashi, T.; Shohji, I. An Evaluation of the Wear Resistance of Electroplated Nickel Coatings Composited with 2,2,6,6-Tetramethylpiperidine 1-Oxyl-Oxidized Cellulose Nanofibers. *Polymers* **2024**, *16*, 224. [CrossRef]
21. Lisuzzo, L.; Cavallaro, G.; Lazzara, G.; Milioto, S. Supramolecular Systems Based on Chitosan and Chemically Functionalized Nanocelluloses as Protective and Reinforcing Fillers of Paper Structure. *Carbohydr. Polym. Technol. Appl.* **2023**, *6*, 100380. [CrossRef]
22. Liu, Y.; Ai, K.; Lu, L. Polydopamine and Its Derivative Materials: Synthesis and Promising Applications in Energy, Environmental, and Biomedical Fields. *Chem. Rev.* **2014**, *114*, 5057–5115. [CrossRef] [PubMed]
23. Li, Q.; Xue, Z.; Zhao, J.; Ao, C.; Jia, X.; Xia, T.; Wang, Q.; Deng, X.; Zhang, W.; Lu, C. Mass Production of High Thermal Conductive Boron Nitride/Nanofibrillated Cellulose Composite Membranes. *Chem. Eng. J.* **2020**, *383*, 123101. [CrossRef]
24. Lee, H.; Dellatore, S.M.; Miller, W.M.; Messersmith, P.B. Mussel-Inspired Surface Chemistry for Multifunctional Coatings. *Science* **2007**, *318*, 426–430. [CrossRef]
25. Ruppel, S.S.; Liang, J. Tunable Properties of Polydopamine Nanoparticles and Coated Surfaces. *Langmuir* **2022**, *38*, 5020–5029. [CrossRef] [PubMed]
26. Zeng, Y.; Du, X.; Hou, W.; Liu, X.; Zhu, C.; Gao, B.; Sun, L.; Li, Q.; Liao, J.; Levkin, P.A.; et al. UV-Triggered Polydopamine Secondary Modification: Fast Deposition and Removal of Metal Nanoparticles. *Adv. Funct. Mater.* **2019**, *29*, 1901875. [CrossRef]
27. Isogai, A.; Saito, T.; Fukuzumi, H. TEMPO-Oxidized Cellulose Nanofibers. *Nanoscale* **2011**, *3*, 71–85. [CrossRef]
28. Yang, K.-P.; Chen, H.; Han, Z.-M.; Yin, C.-H.; Liu, H.-C.; Li, D.-H.; Yang, H.-B.; Sun, W.-B.; Guan, Q.-F.; Yu, S.-H. Bioinspired Multifunctional High-Performance Electromagnetic Shielding Coatings Resistant to Extreme Space Environments. *Innov. Mater.* **2023**, *1*, 100010–100025. [CrossRef]
29. Xiong, R.; Kim, H.S.; Zhang, L.; Korolovych, V.F.; Zhang, S.; Yingling, Y.G.; Tsukruk, V.V. Wrapping Nanocellulose Nets around Graphene Oxide Sheets. *Angew. Chem. Int. Ed.* **2018**, *57*, 8508–8513. [CrossRef]
30. Li, Y.; Zhu, H.; Shen, F.; Wan, J.; Lacey, S.; Fang, Z.; Dai, H.; Hu, L. Nanocellulose as Green Dispersant for Two-Dimensional Energy Materials. *Nano Energy* **2015**, *13*, 346–354. [CrossRef]
31. Mattos, B.D.; Tardy, B.L.; Greca, L.G.; Kämäräinen, T.; Xiang, W.; Cusola, O.; Magalhães, W.L.E.; Rojas, O.J. Nanofibrillar Networks Enable Universal Assembly of Superstructured Particle Constructs. *Sci. Adv.* **2020**, *6*, eaaz7328. [CrossRef] [PubMed]
32. Li, M.-C.; Wu, Q.; Song, K.; Lee, S.; Qing, Y.; Wu, Y. Cellulose Nanoparticles: Structure–Morphology–Rheology Relationships. *ACS Sustain. Chem. Eng.* **2015**, *3*, 821–832. [CrossRef]
33. Chen, K.; Peng, L.; Fang, Z.; Lin, X.; Sun, C.; Qiu, X. Dispersing Boron Nitride Nanosheets with Carboxymethylated Cellulose Nanofibrils for Strong and Thermally Conductive Nanocomposite Films with Improved Water-Resistance. *Carbohydr. Polym.* **2023**, *321*, 121250. [CrossRef] [PubMed]
34. Feng, S.; Yi, Y.; Chen, B.; Deng, P.; Zhou, Z.; Lu, C. Rheology-Guided Assembly of a Highly Aligned MXene/Cellulose Nanofiber Composite Film for High-Performance Electromagnetic Interference Shielding and Infrared Stealth. *ACS Appl. Mater. Interfaces* **2022**, *14*, 36060–36070. [CrossRef]

35. Wu, K.; Fang, J.; Ma, J.; Huang, R.; Chai, S.; Chen, F.; Fu, Q. Achieving a Collapsible, Strong, and Highly Thermally Conductive Film Based on Oriented Functionalized Boron Nitride Nanosheets and Cellulose Nanofiber. *ACS Appl. Mater. Interfaces* **2017**, *9*, 30035–30045. [CrossRef]
36. Wang, J.; Cheng, Q.; Lin, L.; Jiang, L. Synergistic Toughening of Bioinspired Poly(Vinyl Alcohol)–Clay–Nanofibrillar Cellulose Artificial Nacre. *ACS Nano* **2014**, *8*, 2739–2745. [CrossRef]
37. Uetani, K.; Okada, T.; Oyama, H.T. Crystallite Size Effect on Thermal Conductive Properties of Nonwoven Nanocellulose Sheets. *Biomacromolecules* **2015**, *16*, 2220–2227. [CrossRef]
38. Nie, S.; Hao, N.; Zhang, K.; Xing, C.; Wang, S. Cellulose Nanofibrils-Based Thermally Conductive Composites for Flexible Electronics: A Mini Review. *Cellulose* **2020**, *27*, 4173–4187. [CrossRef]
39. Xu, F.; Zhang, M.; Cui, Y.; Bao, D.; Peng, J.; Gao, Y.; Lin, D.; Geng, H.; Zhu, Y.; Wang, H. A Novel Polymer Composite Coating with High Thermal Conductivity and Unique Anti-Corrosion Performance. *Chem. Eng. J.* **2022**, *439*, 135660. [CrossRef]
40. Chi, W.; Sampath, S.; Wang, H. Ambient and High-Temperature Thermal Conductivity of Thermal Sprayed Coatings. *J. Therm. Spray Technol.* **2006**, *15*, 773–778. [CrossRef]
41. Ligati, S.; Ohayon-Lavi, A.; Keyes, J.; Ziskind, G.; Regev, O. Enhancing Thermal Conductivity in Graphene-Loaded Paint: Effects of Phase Change, Rheology and Filler Size. *Int. J. Therm. Sci.* **2020**, *153*, 106381. [CrossRef]
42. Bakshi, S.R.; Balani, K.; Agarwal, A. Thermal Conductivity of Plasma-Sprayed Aluminum Oxide—Multiwalled Carbon Nanotube Composites. *J. Am. Ceram. Soc.* **2008**, *91*, 942–947. [CrossRef]
43. Shakhova, I.; Mironov, E.; Azarmi, F.; Safonov, A. Thermo-Electrical Properties of the Alumina Coatings Deposited by Different Thermal Spraying Technologies. *Ceram. Int.* **2017**, *43*, 15392–15401. [CrossRef]
44. Curry, N.; VanEvery, K.; Snyder, T.; Markocsan, N. Thermal Conductivity Analysis and Lifetime Testing of Suspension Plasma-Sprayed Thermal Barrier Coatings. *Coatings* **2014**, *4*, 630–650. [CrossRef]

Disclaimer/Publisher's Note: The statements, opinions and data contained in all publications are solely those of the individual author(s) and contributor(s) and not of MDPI and/or the editor(s). MDPI and/or the editor(s) disclaim responsibility for any injury to people or property resulting from any ideas, methods, instructions or products referred to in the content.

Article

Optimization of Thermal Conductivity and Tensile Properties of High-Density Polyethylene by Addition of Expanded Graphite and Boron Nitride

Lovro Travaš *, Maja Rujnić Havstad and Ana Pilipović

Faculty of Mechanical Engineering and Naval Architecture, University of Zagreb, Ivana Lucica 5, 10000 Zagreb, Croatia; mrujnic@fsb.hr (M.R.H.); apilipovic@fsb.hr (A.P.)
* Correspondence: ltravas@fsb.hr

Citation: Travaš, L.; Rujnić Havstad, M.; Pilipović, A. Optimization of Thermal Conductivity and Tensile Properties of High-Density Polyethylene by Addition of Expanded Graphite and Boron Nitride. *Polymers* 2023, 15, 3645. https://doi.org/10.3390/polym15173645

Academic Editor: Markus Gahleitner

Received: 8 August 2023
Revised: 27 August 2023
Accepted: 30 August 2023
Published: 4 September 2023

Copyright: © 2023 by the authors. Licensee MDPI, Basel, Switzerland. This article is an open access article distributed under the terms and conditions of the Creative Commons Attribution (CC BY) license (https://creativecommons.org/licenses/by/4.0/).

Abstract: Due to its mechanical, rheological, and chemical properties, high-density polyethylene (HDPE) is commonly used as a material for producing the pipes for transport of various media. Low thermal conductivity (0.4 W/mK) narrows down the usage of HDPE in the heat exchanger systems. The main goal of the work is to reduce the vertical depth of the HDPE pipe buried in the borehole by increasing the thermal conductivity of the material. This property can be improved by adding certain additives to the pure HDPE matrix. Composites made of HDPE with metallic and non-metallic additives show increased thermal conductivity several times compared to the thermal conductivity of pure HDPE. Those additives affect the mechanical properties too, by enhancing or degrading them. In this research, the thermal conductivity and tensile properties of composite made of HDPE matrix and two types of additives, expanded graphite (EG) and boron nitride (BN), were tested. Micro-sized particles of EG and two different sizes of BN particles, micro and nano, were used to produce composite. The objective behind utilizing composite materials featuring dual additives is twofold: firstly, to enhance thermal properties, and secondly, to improve mechanical properties when compared with the pure HDPE. As anticipated, the thermal conductivity of the composites exhibited an eightfold rise in comparison to the pure HDPE. The tensile modulus experienced augmentation across all variations of additive ratios within the composites, albeit with a marginal reduction in tensile strength. This implies that the composite retains a value similar to pure HDPE in terms of tensile strength. Apart from the enhancement observed in all the aforementioned properties, the most significant downside of these composites pertains to their strain at yield, which experienced a reduction, declining from the initial 8.5% found in pure HDPE to a range spanning from 6.6% to 1.8%, dependent upon the specific additive ratios and the size of the BN particles.

Keywords: boron nitride; composite; expanded graphite; HDPE pipe; tensile properties; thermal conductivity

1. Introduction

Nanocomposites are composites made of particles up to 100 nm large, while micro-composites contain particles sized from 0.1 to 100 μm. Generally, there is a significant effect on the mechanical and rheological properties of the composite caused by micro- and nano-constituents compared to the matrix material. Nanocomposites based on polymer matrix and non-polymer additives are the subject of various studies aiming for the enhancement of electric conductivity, antistatic features, tensile strength, flexural strength, water absorption, abrasion resistivity, etc. [1–6]. Due to the significantly higher thermal conductivity (λ) of carbon-based additives compared to polymers, the presence of these additives results in the increase of thermal conductivity of carbon–polymer composites [7]. The enhanced thermal properties of graphite are related to its structure, in which atoms of carbon build a hexagonal one-layered structure. Expanded graphite (EG) is one of the many modifications of graphite, which is produced by intercalation and can be exfoliated several hundred times compared to the original volume when exposed to heat. Three-dimensional wormlike structures of EG are the basis for achieving increased values of electro and thermal conductivity in EG–polymer composite [8]. Compared to polymers, most metals have a hundred times higher λ values. Some metal additives used in polymer matrices, such as nickel–copper alloy and titanium, are characterized by excellent chemical resistivity and can therefore be used as heat exchangers in media such as sea water or chemicals. On the other hand, polymer processing demands significantly lower temperatures (<300 °C) compared to metal processing [9]. Polyethylene, especially high-density polyethylene (HDPE), due to its low price, recyclability, nontoxicity, corrosion resistivity, and good processing properties, has a wide range of applications. Features of composites with HDPE as matrix are based on the interphase compatibility between matrix and additive, polarity between the contact surfaces of the matrix and additive, etc. The dispersion of additives in the matrix depends on size, shape, dispersion technique, equipment, and on processing parameters (time, temperature, etc.). To increase the dispersion of additive in matrix, and thus reduce the surface tension between components, various methods are applied. Some of them include the addition of maleic anhydride (MAH), resulting in an increase of strength, toughness, and ductility. Other methods include high speed shearing during the mixing of components (such as poly(methyl-methacrylate) and EG), resulting in similar improvements compared to MAH [10].

1.1. Composites with HDPE and EG

Sobaliček et al. exposed HDPE and untreated EG mixture to a temperature above the melting point of HDPE, which resulted in multiple times higher value of thermal conductivity compared to HDPE [11]. EG treated with poly(vinyl-alcohol) shows an increase of thermal conductivity in the polymer, even at low percentages of additives, which was researched by Yin et al. [12]. According to Panagiotis et al., the polystyrene (PS) matrix, compared to the HDPE matrix with EG additive, results in a multiple-fold increase in thermal conductivity [13]. A similar procedure and parameters were applied (temperature of kneader chamber of 185 °C, at 60 rpm/min, and for 10 min) for mixing HDPE/EG/carbon nano tubes (CNT) composite, with EG up to 20 wt.% and CNT up to 3 wt.%. It was then pressed at a temperature of 185 °C and pressure of 10 MPa. For the highest concentration of both additives, thermal conductivity reached a value of ~ 3 W/mK [14]. In their research, Sanchez et al. produced a composite with an ultra-high molecular weight polyethylene (UHMWPE) matrix and graphite as an additive by ultrasonic injection molding. Even at low mass percentage (7 wt.%), the tensile modulus of the composite increased 96% when compared to the tensile modulus of the matrix [15]. Abdelrazeq et al. researched the properties of a phase-change material made of HDPE matrix, paraffin wax, and 15 wt.% EG. The composite was exposed to UV radiation, temperature, and moisture. The highest value of thermal conductivity, 1.64 W/mK, was measured at the highest mass percentage of EG [16].

1.2. Composites with HDPE and BN

Muratov et al. used HDPE and hexagonal BN (hBN) with two particle sizes, 10 μm in mass proportions of 25% and 50% and 150 nm in 25 wt.%. The highest value of thermal conductivity, 2.08 W/mK, was achieved in the composite with 50 wt.% micro-sized hBN particles and without a compatibilizer. The yield strength for all composites reached between 22 MPa and 24.9 MPa, while the highest tensile modulus, 3829 MPa, was measured for the composite with 50 wt.% mass percentages of hBN micro particles and compatibilizer based on titanate (KR TTS) [17]. A composite with LDPE matrix and hexagonal boron nitride nanosheets (hBNNs) was studied by Ali et al., where hBNNs were added in a volume percentage up to 30%. A value of thermal conductivity of 1.46 W/mK was reached for the highest volume proportion of additive. The elasticity modulus reached 2.2 GPa for 25 vol.% and the tensile strength was 18.7 MPa for the same volume percentage of the additive [18].

A layered structure of LDPE/HDPE and BN composite was researched by Shang et al. BN was added in composites in 15 wt.%, which resulted in a measured thermal conductivity of 3.54 W/mK for the layered structure and 3.13 W/mK for the randomly oriented composite with the same mass percentage of BN [19]. A significant increase in thermal conductivity was achieved by Zhang et al. by stretching composite foil with HDPE matrix and boron nitride nanoplates (BNNPs) as additives. The measured thermal conductivity reached 3.1 W/mK for unstretched foil with 15 wt.% BNNP, while for stretched foil (with stretching ratio $\Lambda = 5$), thermal conductivity reached 106.2 W/mK [20]. A recycled PE matrix with high ratio of aluminum oxide and zinc oxide as matrix and BN as additive was used by Rasul et al. to produce the composite, where silane was used as a compatibilizer. The thermal conductivity of the matrix was 0.72 W/mK; for the composite with 5 wt.% of BN, it increased to 0.84 W/mK and for the composite with 5 wt.% of BN and 3 wt.% of silane, it reached 0.96 W/mK [21]. An increase of a composite's thermal conductivity was achieved by Shi et al. by solid-state extrusion. Using UHMWPE as the matrix and BN as an additive with 50 wt.%, it reached a thermal conductivity of 23.03 W/mK [22]. In his research, Lebedev compared the thermal conductivity of two composites produced by injection molding. The thermal conductivity of the composite made of polyacrylic acid (PLA) and BN reached 0.67 W/mK, while the composite made of LDPE and BN reached a thermal conductivity of 0.7 W/mK for the same BN ratio of 40 wt.% [23]. Güzdemir et al. researched the increase of thermal conductivity for LDPE/BN composite in the shape of extruded film. The initial thermal conductivity of the LDPE matrix of 0.4 W/mK increased to 1.8 W/mK for the composite with 30 vol% of BN additive [24]. Other research, such as the study conducted by Yang et al., included a composite with HDPE matrix with BN (25 wt.%) and coconut shell carbon (3 wt.%), reaching a thermal conductivity of 0.943 W/mK [25].

Considering the superior thermal conductivity inherent in both BN and EG compared to PE, it is foreseeable that the EG/BN/HDPE composite will exhibit significantly higher thermal conductivity in comparison to HDPE. While this particular composite has not yet been explored, the hypothesis suggests a substantial multiple-fold increase in thermal conductivity. The potential application of such a material would primarily target heat exchangers, an area where mechanical properties carry equal significance alongside thermal attributes. Consequently, the composite material will undergo testing to assess its tensile properties, providing valuable insights into its suitability for this purpose.

While existing research predominantly investigated either EG/PE or BN/PE composites with regards to mechanical properties, there is a lack of examination of their synergistic effects. According to the literature of previous authors, it is necessary to add up to 50% EG and 50% BN individually. The point of this work is to reduce the percentages of EG and BN and to obtain better thermal conductivity and tensile properties. Therefore, it is necessary to see how both additives together affect the properties. Additionally, the study investigated the interaction between the particle size of boron nitride and the incorporation of expanded graphite within the HDPE matrix. Based on a literature review, it is anticipated that BN will exert a more pronounced influence on tensile properties compared to EG. Furthermore, the work optimized the results according to the input and output parameters for the actual application of that material.

2. Materials and Methods

2.1. Materials

HDPE 6060R (*Sabic*, Riyadh; Saudi Arabia) was chosen as a matrix material in a granulate form. The type of material is PE-100, with granulate density of 0.959 g/cm^3 and melting point at 124 °C. HDPE was chosen because of its easy processability, low cost, light weight, and low melting point. This material is a classic material for the production of pipes to be installed in a borehole for the transfer of media. SABIC® Vestolen A 6060R 10000 (black) is a grade which has a high density and a bimodal distribution of the molecular mass. Due to its profile of properties, this material is typically used for gas, drinking water, and wastewater piping. This material meets (inter)national standards for use in gas, drinking water, and wastewater piping MRS class ISO 12162 MPa = 10.0 (PE 100). Expanded graphite *Sigratherm GFG75* was acquired from the producer *SGL Carbon* (Austria) in a powdered form. It has a density of 2.25 g/cm^3, with an average particle size of 75 μm. The manufacturer specifies > 98% carbon in the material. Boron nitride with micro-sized particles (35 μm) named *HeBoFill* from *Henze* (Lauben; Germany) has a density of 2.25 g/cm^3, while boron nitride with nano-sized particles (70 nm) produced by *IoLiTec* (Heilbronn; Germany) has a density of 2.3 g/cm^3. Both micro- and nano-sized BN additives have a hexagonal crystal structure. Compatibilizer PE-g-MAH was acquired from *BOC Sciences* and was added into each composite in 5 wt.%. All additional information can be seen in the data sheets from the manufacturer. The designations for the composite with expanded graphite and micro-particles of BN are HDPE/EG/mBN, and with expanded graphite and nanoparticles of BN HDPE/EG/nBN.

2.2. Processing

A laboratory twin screw kneader (*MetaStation 4E*, manufacturer *Brabender*; Duisburg; Germany) was used for mixing the components. A chamber sized 50 cm^3 was heated up to 200 °C, with a rotation speed of 60 min^{-1}, and for a duration of 20 min. Firstly, HDPE was added into the mixing chamber, followed by the compatibilizer, and finally EG and BN. After mixing was done, the composite was cooled down to room temperature and milled in a mechanical mill (*SM100*, manufacturer *Retsch*; Haan; Germany) into average particles sizes of 4 mm. The size of the particles is controlled by selecting a sieve that is replaceable. Ultimately, the composite was formed through compression molding, involving a hot pressing process at 15 MPa and 180 °C for a duration of 10 min to create the test specimens. Subsequently, the molded specimens were cooled to room temperature within a closed mold, utilizing water as the cooling medium. The composites are marked according to the mass percentage of additives, i.e., HDPE/EG5/mBN15 means that 5 wt.% of expanded graphite and 15 wt.% micro-sized particles of boron nitride were added into the HDPE matrix. Combinations of additive mass proportions in the matrix were determined according to the central composite design in the *Design Expert* software. Eleven experimental runs were determined for both micro- and nano-sized BN particles and are shown in Tables 1 and 2. The thermal conductivity of composites was measured according to ISO 22007-2:2022 by the transient hot bridge method. The thermal conductivity was determined on the *Linesis THB Advanced* device. A THBN1 sensor was used for the measurement with a measurement error of 2%. The measurements were carried out at a current of 0.056 A and at a temperature of 23 °C, and the measurement time was 50 s. Tensile properties were tested on a universal testing machine *Shimadzu AGS-X* with a maximum force of 10 kN according to HRN EN ISO 527-2 at a speed of 1 mm/min. The test was carried out on type 1BA test specimens with dimension 110 × 10 × 2 mm. The measurement error on the universal testing machine is 0.5% in the measurement range from 10 to 10,000 N. For each composite, six specimens were produced by the described procedure for tensile properties and two for thermal properties. Thermal conductivity was measured at 10 places on the test specimens in contact. The mean value and standard deviation were calculated and are shown in Tables 1 and 2. In the table, after the ± sign, the standard deviation, that is, the dispersion of data in the conducted testing of six test specimens, is shown. The effect of additives on the thermal and tensile properties of pure HDPE matrix was tested according to the design of the experiment generated by software *Design Expert*, using response surface methodology. Data acquired by testing were processed by ANOVA (analysis of variance) linear modeling method with three center points. Values crossed out in Tables 1 and 2 are excluded from further data analysis because of a high result deviation from the center point in the model calculated by *Design Expert*. Along with the results for different ratios of EG and BN in the HDPE matrix, the results of pure HDPE are also presented. The maximum and minimum limits of the percentage of BN for the HDPE/EG/mBN and HDPE/EG/nBN composites are different because the point is to see if the same or similar properties can be obtained with a smaller amount of nano particles of BN as with the use of cheaper micro particles of BN. Furthermore, only one particle size was chosen for the EG additive because the selected material proved to be the best with preliminary experiments. In the table, crossed values refer to values that were excluded from further analysis because they deviate from the model.

Table 1. Mass proportion of additives and values of measured properties of HDPE/EG/mBN composite.

Run	HDPE wt.%	Factor A: EG wt.%	Factor B: mBN wt.%	Thermal Conductivity λ (W/mK)	Tensile Strength σ_m (MPa)	Tensile Modulus E (MPa)	Strain at Yield ε_y (%)
1	80	5	15	1.21 ± 0.01	18.84 ± 1.04	1072 ± 98	4.90 ± 1.13
2	70	8	22	1.96 ± 0.14	19.65 ± 0.75	1601 ± 192	2.91 ± 0.40
3	70	15	15	1.96 ± 0.13	23.45 ± 0.48	1723 ± 244	4.94 ± 0.34
4	60	25	15	1.26 ± 0.04	23.85 ± 0.99	1915 ± 324	2.65 ± 0.51
5	20	15	5	1.69 ± 0.10	20.84 ± 1.01	1149 ± 181	5.92 ± 1.2
6	84	8	8	1.21 ± 0.06	19.58 ± 1.95	1164 ± 219	5.71 ± 1.51
7	70	15	15	2.39 ± 0.17	20.13 ± 2.65	1548 ± 152	2.86 ± 0.95
8	60	15	25	0.24 ± 0.00	24.37 ± 3.76	2153 ± 404	2.65 ± 0.28
9	56	22	22	3.01 ± 0.08	23.06 ± 1.77	2087 ± 306	1.85 ± 0.23
10	70	15	15	2.03 ± 0.06	21.86 ± 1.16	1593 ± 154	3.63 ± 0.75
11	70	22	8	1.91 ± 0.07	22.55 ± 1.58	1687 ± 233	3.34 ± 0.4
/	100	/	/	0.37 ± 0.003	25.58 ± 0.99	1104 ± 84	8.52 ± 0.83

The underlined values apply only to the implementation of the analysis of these individual properties, and not for all analyzed properties.

Table 2. Mass proportion of additives and values of measured properties of HDPE/EG/nBN composite.

Run	HDPE wt.%	Factor A: EG wt.%	Factor B: nBN wt.%	Thermal Conductivity λ (W/mK)	Tensile Strength σ_m (MPa)	Tensile Modulus E (MPa)	Strain at Yield ε_y (%)
1	64.5	22	13.5	2.79 ± 0.09	23.31 ± 1.85	1899 ± 189	2.32 ± 0.51
2	71.5	22	6.5	2.09 ± 0.07	23.06 ± 1.27	1985 ± 236	3.33 ± 2.99
3	70	15	15	1.95 ± 0.08	24.03 ± 2.86	2209 ± 293	3.66 ± 1.28
4	75	15	10	1.88 ± 0.10	22.80 ± 2.26	1733 ± 101	3.94 ± 0.94
5	75	15	10	1.61 ± 0.07	22.84 ± 2.32	1814 ± 212	4.01 ± 1.26
6	75	15	10	1.44 ± 0.02	21.23 ± 2.07	1477 ± 101	3.86 ± 1.5
7	80	15	5	1.55 ± 0.06	20.83 ± 2.06	1432 ± 227	4.43 ± 4.57
8	85.5	8	6.5	1.11 ± 0.08	20.94 ± 1.17	1328 ± 310	6.61 ± 0.72
9	65	25	10	1.48 ± 0.02	22.99 ± 1.93	2006 ± 241	2.24 ± 0.51
10	85	5	10	0.91 ± 0.04	21.59 ± 1.49	1144 ± 167	6.23 ± 1.23
11	78.5	8	13.5	1.61 ± 0.21	21.95 ± 2.02	1381 ± 168	5.21 ± 1.39
/	100	/	/	0.37 ± 0.003	25.58 ± 0.99	1104 ± 84	8.52 ± 0.83

The underlined values apply only to the implementation of the analysis of these individual properties, and not for all analyzed properties.

3. Results

Composites were divided into two groups, one with nano-sized particles of BN and the other with micro-sized particles of BN, to compare thermal and tensile properties depending on additive proportions.

3.1. Thermal Properties

Even at the lowest proportions, the inclusion of additives led to an increase in the thermal conductivity of the composites, reaching its peak value at the greatest mass proportions of additives. From a starting value of 0.37 W/mK for pure HDPE, the thermal conductivity increased up to 3.00 W/mK for the HDPE/EG22/mBN22 composite and to 2.09 W/mK for the HDPE/EG22/nBN6.5 composite. In percentages, this means an increase of 710% and 465%, respectively, compared to the thermal conductivity of HDPE. For the same proportion of 15 wt.% of micro or nano BN additive, the value of the thermal conductivity is greater for mBN (2.13 W/mK), while for the same mass percentage of nBN, the thermal conductivity is 1.95 W/mK. Higher thermal conductivity values for the HDPE/EG/mBN composite indicate that a better thermal pathway is achieved through larger particles. The contribution of additives to the thermal conductivity of composites is comparable for both EG and BN. This similarity can be attributed to the notable thermal conductivity inherent in both EG and BN, stemming from their hexagonal crystal lattice structure (Tables 3 and 5).

Table 3. Analysis of variance—influence of EG and mBN on thermal conductivity.

Source	Sum of Squares	df	Mean Square	F-Value	p-Value (Risk of Rejection of H_0[1])
Model	2.26	3	0.7524	17.43	0.0044 significant
A	1.41	1	1.41	32.57	0.0023
B	0.9039	1	0.9039	20.94	0.006
AB	0.032	1	0.032	0.7415	0.4285
Residual	0.2158	5	0.0432		
Lack of Fit	0.1077	3	0.0359	0.6637	0.6476 not significant
Pure Error	0.1081	2	0.0541		
Cor Total	2.47	8			

[1] H_0—null hypothesis: there are no factor effects.

ANOVA analysis of variance indicates that the 2-factor interaction (2FI) model best fits the influence of additives EG and mBN on the thermal conductivity. The details of analysis are shown in Table 3. In the table, df stands for degrees of freedom.

The model F-value (variation between sample means) of 17.43 implies that the model is significant. There is only a 0.44% chance that an F-value this large could occur due to noise. p-values less than 0.05 indicate that model terms are significant. In this case, both factor A and B (percentage of EG and mBN) are significant model terms. Values greater than 0.1 indicate that the model terms are not significant. If there are many insignificant model terms (not counting those required to support hierarchy), model reduction may improve the model. The lack of fit F-value of 0.6637 implies that the lack of fit is not significant relative to the pure error. There is a 64.76% chance that a lack of fit F-value this large could occur due to noise. Non-significant lack of fit is good because it means that the model fits.

The statistical data (mean value, standard deviation, and R^2) about the model are given in Table 4. The coefficient of determination R^2 is a measure of deviation from the arithmetic mean which is explained by the model. The closer R^2 is to 1, the better the model follows the data, that is, the phenomenon is better explained.

Table 4. Summary statistics about the model for thermal conductivity of HDPE/EG/mBN composite.

Standard Deviation	0.2077	R^2	0.9127
Mean	1.93	Adjusted R^2	0.8604
C.V. %	10.76	Predicted R^2	0.683
		Adeq Precision	12.8045

From Table 4, it can be concluded that the model followed the data very well since the coefficient of determination is $R^2 = 0.9127$. The predicted R^2 of 0.683 is in reasonable agreement with the adjusted R^2 of 0.8604; i.e., the difference is less than 0.2. An adequate precision measures the signal to noise ratio. A ratio greater than 4 is desirable. The ratio of 12.805 indicates an adequate signal.

Predicted thermal conductivity λ for HDPE/EG/mBN composite can be described by Equation (1) in actual parameters:

$$\lambda = 0.582 + 0.043 \times EG + 0.029 \times mBN + 0.002 \times EG \times mBN \tag{1}$$

where: λ (W/mK)—thermal conductivity, EG and mBN are the mass percentages of expanded graphite and micro boron nitride in %.

In the case of the HDPE/EG/nBN composite with nano BN, ANOVA analysis implies that the linear model best describes the influence of additives on thermal conductivity. Analysis of variance is shown in Table 5.

Table 5. Analysis of variance—influence of EG and nBN on thermal conductivity.

Source	Sum of Squares	df	Mean Square	F-Value	p-Value (Risk of Rejection of H_0)
Model	1.01	2	0.5046	16.82	0.0035 significant
A	0.9653	1	0.9653	32.18	0.0013
B	0.1846	1	0.1846	6.15	0.0477
Residual	0.18	6	0.03		
Lack of Fit	0.0838	4	0.0209	0.4354	0.7833 not significant
Pure Error	0.0962	2	0.0481		
Cor Total	1.19	8			

The model F-value (variation between sample means) of 16.82 implies that the model is significant. There is only a 0.35 % chance that an F-value this large could occur due to noise. P-values less than 0.05 indicate that the model terms are significant. Both Factor A and B (EG and nBN additives) are significant model terms. Lack of fit F-value of 0.44 implies that the lack of fit is not significant relative to the pure error. There is a 78.33% chance that a lack of fit F-value this large could occur due to noise.

Table 6 shows basic statistical data about the model.

Table 6. Summary statistics for the model for thermal conductivity of HDPE/EG/nBN composite.

Standard Deviation	0.1732	R^2	0.8486
Mean	1.57	Adjusted R^2	0.7982
C.V. %	11.01	Predicted R^2	0.7249
		Adeq Precision	10.0348

The predicted R^2 of 0.7249 is in reasonable agreement with the adjusted R^2 of 0.7982. The ratio of 10.0348 indicates an adequate signal. From Table 6, it can be concluded that the model followed the data very well since the coefficient of determination is R^2 = 0.8486.

The predicted thermal conductivity λ for HDPE/EG/nBN composite can be described by Equation (2) in actual parameters:

$$\lambda = 0.210348 + 0.068865 \times EG + 0.047779 \times nBN \tag{2}$$

where: λ (W/mK)—thermal conductivity, EG and nBN are the mass percentages of expanded graphite and nano boron nitride in %.

The contribution of both additives in HDPE/EG/mBN and HDPE/EG/nBN composites proves the assumption that the structure of these materials will lead to an increase of thermal conductivity of the composites. Two-factor interaction and a linear model that describe the contribution of additives for EG/mBN and EG/nBN can be applied to predict thermal conductivity λ accurately. Figure 1 shows the thermal conductivity for each run of the experiment (except for those excluded from the analysis, in this case, run 4 and 8 for composites HDPE/EG/mBN and run 1 and 9 for composites HDPE/EG/nBN) and pure HDPE to compare the obtained results, while Figure 2 shows the thermal conductivity dependence on the mass proportions of additives. In addition to the values, error bars, i.e., deviations from the results, have been added to the diagram. The diagram indicates a significant impact of boron nitride particle size on thermal conductivity. Notably, the addition of BN nanoparticles to pure HDPE results in a 1 W/mK lower thermal conductivity value than with larger BN particles. It can be concluded that larger BN particles will yield a more substantial enhancement in thermal conductivity.

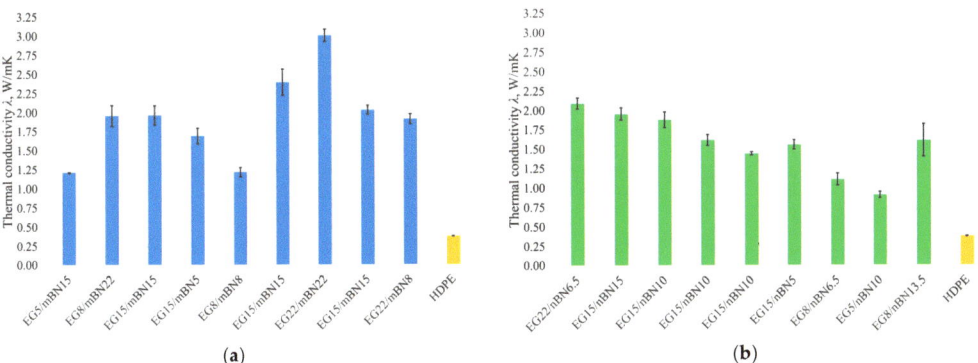

Figure 1. Thermal conductivity of all tests run: (**a**) HDPE/EG/mBN composites; (**b**) HDPE/EG/nBN composites.

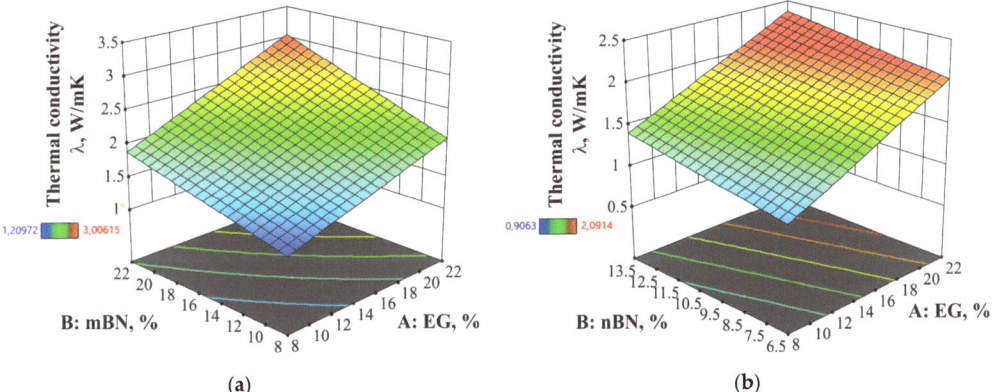

Figure 2. Dependence of proportion of EG and BN additives on the thermal conductivity of: (**a**) HDPE/EG/mBN composite; (**b**) HDPE/EG/nBN composite.

3.2. Tensile Properties
3.2.1. Tensile Strength

The highest value of tensile strength for the composite with micro-particles reached 24.37 MPa (EG15/mBN25) and 24.03 MPa for the composite with nano-particles of BN (EG15/nBN15). For both groups of composites (micro- and nano-sized BN particles), this was the highest ratio of BN. Accordingly, the lowest values of tensile strength were achieved for EG5/mBN15 and EG15/nBN5. With various percentages of both additives, the approximate tensile strength value of the pure HDPE was maintained (25.58 MPa).

Based on the analysis of variance (ANOVA), the linear model provides the most suitable fit for assessing the impact of additives on the tensile strength of the composite containing micro-sized BN. Analysis of variance is shown in Table 7.

The model F-value (variation between sample means) of 9.78 implies that the model is significant. There is only a 0.71% chance that an F-value this large could occur due to noise. In this case, factor A–EG is a significant model term. A lack of fit F-value of 0.32 implies that the lack of fit is not significant relative to the pure error. There is an 88.01% chance that a lack of fit F-value this large could occur due to noise.

Table 7. Analysis of variance—influence of EG and mBN on tensile strength.

Source	Sum of Squares	df	Mean Square	F-Value	p-Value (Risk of Rejection of H_0)
Model	26.61	2	13.3	9.78	0.0071 significant
A	22.67	1	22.67	16.67	0.0035
B	3.93	1	3.93	2.89	0.1275
Residual	10.88	8	1.36		
Lack of Fit	5.37	6	0.8945	0.3243	0.8801 not significant
Pure Error	5.52	2	2.76		
Cor Total	37.49	10			

Furthermore, statistical data about the model for tensile strength is shown in Table 8.

Table 8. Summary statistics for the model for tensile strength of HDPE/EG/mBN composite.

Standard Deviation	1.17	R^2	0.7097
Mean	21.65	Adjusted R^2	0.6371
C.V. %	5.39	Predicted R^2	0.4942
		Adeq Precision	7.8567

The predicted R^2 of 0.4942 is in reasonable agreement with the adjusted R^2 of 0.6371; i.e., the difference is less than 0.2. An adequate precision ratio of 7.857 indicates an adequate signal. The tensile strength σ_m for HDPE/EG/mBN composite can be described by Equation (3):

$$\sigma_m = 16.5691 + 0.2393 \times EG + 0.0997 \times mBN \qquad (3)$$

where: σ_m (N/mm^2 or MPa)—tensile strength, EG and mBN are the mass percentages of expanded graphite and micro boron nitride in %.

As for composites HDPE/EG/nBN, ANOVA implies that the linear model best describes the influence of additives on tensile strength. Analysis of variance is shown in Table 9.

Table 9. Analysis of variance—influence of EG and nBN on tensile strength.

Source	Sum of Squares	df	Mean Square	F-Value	p-Value (Risk of Rejection of H_0)
Model	7.94	2	3.97	9.21	0.0084 significant
A	3.71	1	3.71	8.61	0.0189
B	4.23	1	4.23	9.8	0.014
Residual	3.45	8	0.4314		
Lack of Fit	1.77	6	0.2955	0.3521	0.8644 not significant
Pure Error	1.68	2	0.839		
Cor Total	11.39	10			

In this case, both additives EG and nBN are significant model parameters. The lack of fit F-value of 0.35 implies that the lack of fit is not significant relative to the pure error. There is an 86.44% chance that a lack of lit F-value this large could occur due to noise, which indicates a well-chosen model for tensile strength.

The statistical data about model for tensile strength for HDPE/EG/nBN composite are shown in Table 10.

The predicted R^2 of 0.4641 is in reasonable agreement with the adjusted R^2 of 0.6214. The adequate precision ratio of 8.1705 indicates an adequate signal. R^2 is only 0.6971, which is a slightly lower value, but in accordance with the other presented parameters of the model, both additives are significant for the display of tensile strength. When comparing the R^2 values for the HDPE/EG composite with micro- or nano-sized BN particles, the values are the same.

Table 10. Summary statistics for the model for tensile strength of HDPE/EG/nBN composite.

Standard Deviation	0.6568	R^2	0.6971
Mean	22.32	Adjusted R^2	0.6214
C.V. %	2.94	Predicted R^2	0.4641
		Adeq Precision	8.1705

The tensile strength σ_m for HDPE/EG/nBN composite can be described by Equation (4):

$$\sigma_m = 18.80401 + 0.096851 \times EG + 0.206639 \times nBN \tag{4}$$

where: σ_m (N/mm^2 or MPa)—tensile strength, EG and nBN are the mass percentages of expanded graphite and nano boron nitride in %.

In the case of the HDPE/EG/nBN composite, both additives exhibit a comparable effect on tensile strength. However, in the HDPE/EG/mBN composite, it is evident that EG holds a greater influence over the tensile strength. Figure 3 displays tensile strength for each run of the experiment, while Figure 4 shows the dependence of tensile strength on the amount of EG and BN.

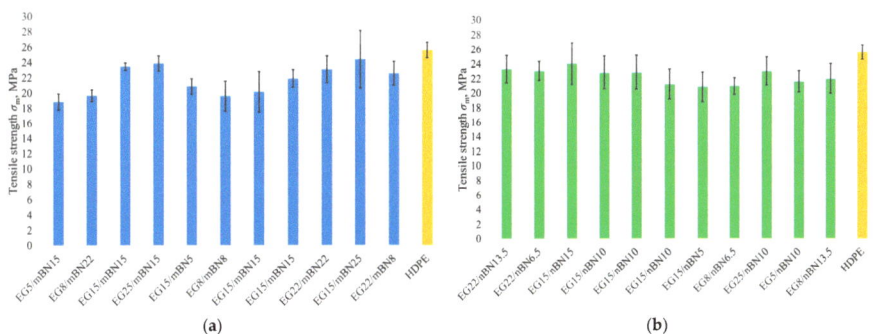

Figure 3. Tensile strength of all tests run: (a) HDPE/EG/mBN composites; (b) HDPE/EG/nBN composites.

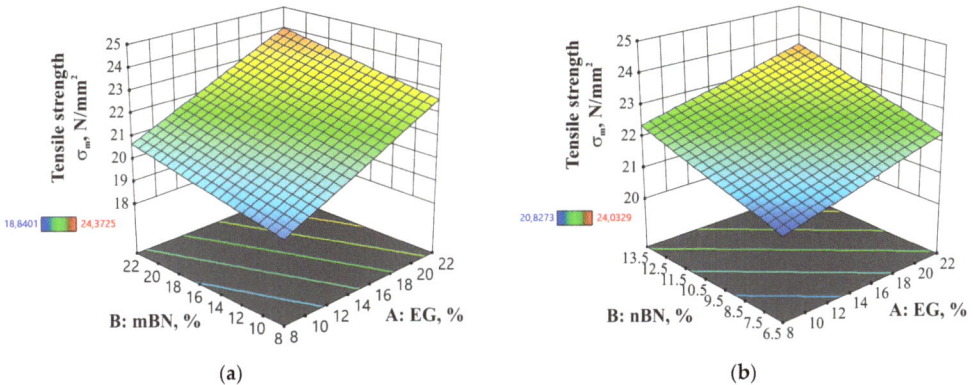

Figure 4. Dependence of percentage of EG and BN additives on the tensile strength of: (a) HDPE/EG/mBN; (b) HDPE/EG/nBN composite.

3.2.2. Tensile Modulus

In addition to tensile strength, the values of the tensile modulus for both composites are dispersed in the range from 1072 MPa (HDPE/EG5/mBN15) to 2153 MPa

(HDPE/EG15/mBN25) and from 1144 MPa (HDPE/EG5/nBN10) to 2006 MPa (HDPE/EG25/nBN10). In contrast to the pure HDPE value of 1104 MPa, the highest value observed for EG15/mBN25 showcased a 95% increase. As expected, the values of the tensile modulus increased with the addition of additives to the composite. For HDPE/EG/mBN, the ANOVA implies that the reduced quadratic model fits best to describe the impact of additives on tensile modulus, as shown in Table 11.

Table 11. Analysis of variance—influence of EG and mBN on tensile modulus.

Source	Sum of Squares	df	Mean Square	F-Value	p-Value (Risk of Rejection of H_0)
Model	1.09×10^6	4	2.734×10^5	48.73	0.0003 significant
A	6.15×10^5	1	6.15×10^5	109.63	0.0001
B	6.427×10^5	1	6.427×10^5	114.57	0.0001
AB	21,640.4	1	21,640.4	3.86	0.1067
A^2	31,280.84	1	31,280.84	5.58	0.0646
Residual	28,049.55	5	5609.91		
Lack of Fit	11,677.13	3	3892.38	0.4755	0.7314 not significant
Pure Error	16,372.42	2	8186.21		
Cor Total	1.122×10^6	9			

Both factor A and B (EG and mBN) are significant model terms (p-value of 0.0001). The value of 0.48 implies that the lack of fit F-value is not significant relative to pure error—there is a 73.14% chance that a lack of fit F-value this large could occur due to noise, which indicates that the model is well chosen.

The statistical data about the model for tensile modulus for HDPE/EG/mBN composite are shown in Table 12.

Table 12. Summary statistics for the model for tensile modulus of HDPE/EG/mBN composite.

Standard Deviation	74.9	R^2	0.975
Mean	1652.81	Adjusted R^2	0.955
C.V. %	4.53	Predicted R^2	0.8494
		Adeq Precision	20.1876

From Table 12, it can be concluded that the model followed the data very well since the coefficient of determination is $R^2 = 0.975$.

The tensile modulus E for HDPE/EG/mBN composite can be described by Equation (5):

$$E = -493.85333 + 117.1943 \times EG + 74.78531 \times mBN - 1.91297 \times EG \times mBN - 1.44711 \times EG^2 \quad (5)$$

where: E (MPa)—tensile modulus, EG and mBN are the mass percentages of expanded graphite and micro boron nitride in %.

Based on the analysis of variance, the most suitable model to describe the interaction of additives with the tensile modulus of the HDPE/EG/nBN composite is the two-factor interaction model. The analysis of measured values is shown in Table 13.

Table 13. Analysis of variance—influence of EG and nBN on tensile modulus.

Source	Sum of Squares	df	Mean Square	F-Value	p-Value (Risk of Rejection of H_0)
Model	7.31×10^5	3	2.437×10^5	14	0.0041 significant
A	7.166×10^5	1	7.166×10^5	41.17	0.0007
B	9491.21	1	9491.21	0.5453	0.4881
AB	4872.04	1	4872.04	0.2799	0.6158
Residual	1.044×10^5	6	17,405.55		
Lack of Fit	42,647.43	4	10,661.86	0.3451	0.8332 not significant
Pure Error	61,785.88	2	30,892.94		
Cor Total	8.354×10^5	9			

In this case, only factor A (EG) is a significant model term (*p*-value of 0.0007). The value of 0.35 implies that the lack of fit *F*-value is not significant relative to pure error—there is an 83.32% chance that a lack of fit *F*-value this large could occur due to noise.

The statistical data about the model for tensile modulus for HDPE/EG/nBN composite are shown in Table 14.

Table 14. Summary statistics for the model for tensile modulus of HDPE/EG/nBN composite.

Standard Deviation	131.93	R^2	0.875
Mean	1619.93	Adjusted R^2	0.8125
C.V. %	8.14	Predicted R^2	0.7292
		Adeq Precision	10.1967

R^2 in the case of HDPE/EG/mBN is closer to the number one than in the case of the application of nano particles boron nitride in composite, but the value of 0.875 and the values of other statistical data indicate that the model is well chosen.

The tensile modulus *E* for HDPE/EG/nBN composite can be described by Equation (6):

$$E = 658.69725 + 56.78545 \times EG + 32.88882 \times nBN - 1.42449 \times EG \times nBN \quad (6)$$

where: *E* (MPa)—tensile modulus, EG and nBN are the mass percentages of expanded graphite and nano boron nitride in %.

In the case of the HDPE/EG/mBN composite, both additives contribute to the impact on the tensile modulus. Conversely, in the HDPE/EG/nBN composite, only EG demonstrates an influential effect. Figure 5 illustrates the tensile modulus for each experimental run and pure HDPE, whereas Figure 6 illustrates the relationship between the mass proportion of additives and the tensile modulus.

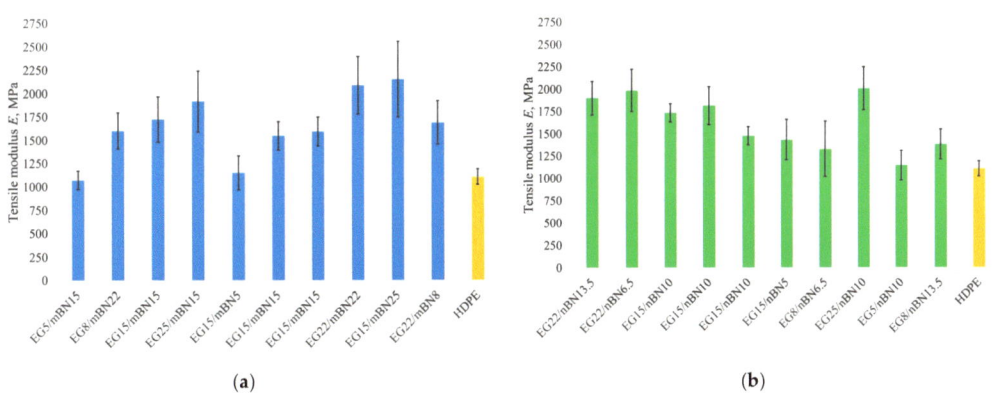

Figure 5. Tensile modulus of all tests run: (**a**) HDPE/EG/mBN composites; (**b**) HDPE/EG/nBN composites.

3.2.3. Strain at Yield

Table 15 gives the ANOVA results for the strain at yield for HDPE/EG/mBN composites and Table 16 shows the basic statistical data for the model.

Table 15 reveals that both factors A and B (EG and mBN) significantly impact the strain at yield. In contrast, the lack of fit is not statistically significant, suggesting that the selected linear model adequately followed the strain at yield data.

From Table 16, it can be concluded that the model followed the data very well since the coefficient of determination is $R^2 = 0.8103$.

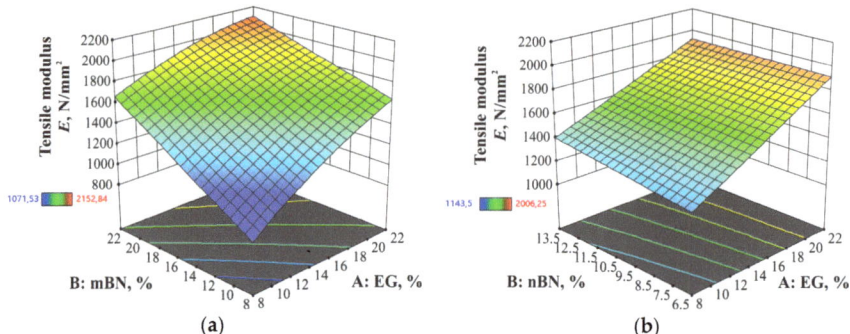

Figure 6. Dependence of percentage of EG and BN additives on the tensile modulus of: (**a**) HDPE/EG/mBN; (**b**) HDPE/EG/nBN composite.

Table 15. Analysis of variance—influence of EG and mBN on the strain at yield.

Source	Sum of Squares	df	Mean Square	F-Value	p-Value (Risk of Rejection of H_0[1])
Model	15.39	2	7.69	17.08	0.0013 significant
A	5.47	1	5.47	12.14	0.0083
B	9.92	1	9.92	22.02	0.0016
Residual	3.60	8	0.4504		
Lack of Fit	1.39	6	0.2322	0.2101	0.9422 not significant
Pure Error	2.21	2	1.10		
Cor Total	18.99	10			

[1] H_0—null hypothesis: there are no factor effects.

Table 16. Summary statistics for the model for strain at yield of HDPE/EG/mBN composite.

Standard Deviation	0.6711	R^2	0.8103
Mean	3.76	Adjusted R^2	0.7628
C.V. %	17.84	Predicted R^2	0.6909
		Adeq Precision	11.0156

The strain at yield ε_y for HDPE/EG/mBN composite can be described by Equation (7):

$$\varepsilon_y = 7.89820 - 0.117499 \times EG - 0.158257 \times mBN \quad (7)$$

where: ε_y (%)—strain at yield, EG and mBN are the mass ratio of expanded graphite and micro boron nitride in %.

Table 17 gives the analysis results for the strain at yield for HDPE/EG/nBN composites and Table 18 shows the basic statistical data for the model. Table 17 indicates that factors A, B, and A^2 exert a significant influence on the strain at yield of HDPE/EG/nBN composites.

Table 17. Analysis of variance—influence of EG and nBN on the strain at yield.

Source	Sum of Squares	df	Mean Square	F-Value	p-Value (Risk of Rejection of H_0[1])
Model	13.49	5	2.7	475.96	< 0.0001 significant
A	12.02	1	12.02	2120.07	< 0.0001
B	0.5102	1	0.5102	89.98	0.0007
AB	0.0017	1	0.0017	0.3080	0.6085
A^2	0.0947	1	0.0947	16.70	0.0150
B^2	0.0073	1	0.0073	1.29	0.3189
Residual	0.0227	4	0.0057		
Lack of Fit	0.0104	2	0.0052	0.8501	0.5405 not significant
Pure Error	0.0123	2	0.0061		
Cor Total	13.52	9			

[1] H_0—null hypothesis: there are no factor effects.

Table 18. Summary statistics for the model for strain at yield of HDPE/EG/nBN composite.

Standard Deviation	0.0753	R^2	0.9983
Mean	3.89	Adjusted R^2	0.9962
C.V. %	1.94	Predicted R^2	0.9907
		Adeq Precision	68.7844

From the given data in Table 18, it can be concluded that the model followed the data excellently. The coefficient of determination is $R^2 = 0.9983$, which is higher than the determination coefficient for the composite HDPE/EG/mBN. This means that the presented quadratic model follows the data very well.

The strain at yield ε_y for HDPE/EG/nBN composite can be described by Equation (8):

$$\varepsilon_y = 8.50832 - 0.269906 \times EG - 0.125754 \times nBN - 0.001098 \times EG \times nBN + 0.002676 \times EG^2 + 0.002979 \times nBN^2 \quad (8)$$

where: ε_y (%)—strain at yield, EG and nBN are the mass percentages of expanded graphite and nano boron nitride in %.

Figure 7 displays the strain at yield for each run of the experiment and pure HDPE, while the dependence of the mass percentage of additives on the strain at yield is shown in Figure 8.

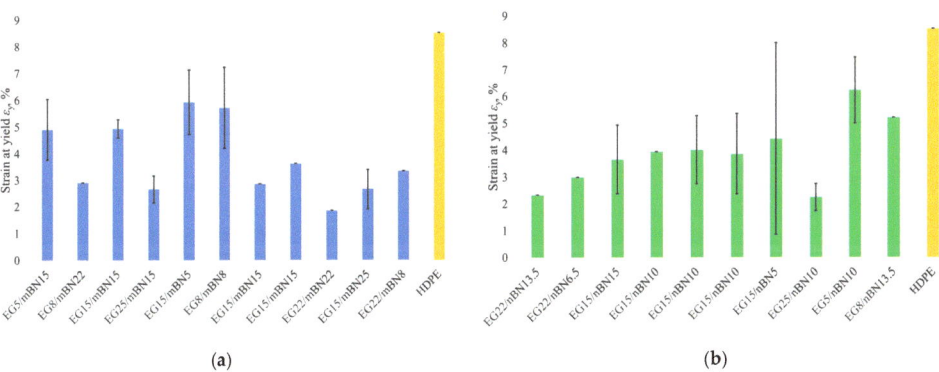

Figure 7. Strain at yield of all tests run: (**a**) HDPE/EG/mBN composites; (**b**) HDPE/EG/nBN composites.

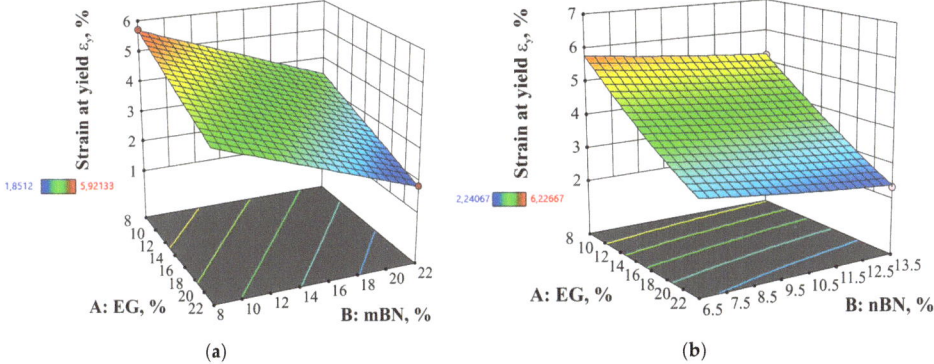

Figure 8. Dependence of percentage of EG and BN additives on the strain at yield of: (**a**) HDPE/EG/mBN; (**b**) HDPE/EG/nBN composite.

3.3. Optimization

Software package *Design Expert* also includes an optimization module, in which optimization can be performed based on the desirability functions. The optimization process searches for a combination of factor values that simultaneously satisfy the criteria (wishes and priorities) placed on each of the responses and factors.

In this research, the optimization criteria (for optimal solution) were maximum thermal conductivity and strain at yield, while the tensile modulus and tensile strength with input percentage of EG and BN were within the chosen limits of the experiment.

The optimization was carried out in accordance with the basic goal of achieving the highest possible thermal conductivity within the selected limits of the input data (percentage of EG and BN additives in the HDPE matrix). Regarding the tensile properties, given that the tensile strength results for all combinations of percentages of additives in the HDPE matrix approximately maintained the value of pure HDPE, and the modulus in all combinations is greater than pure HDPE regardless of the size of the boron nitride particles, the values of the entire obtained spectrum results were considered for optimization. However, for easier processing, the strain at yield value should be as high as possible. Such a combination of the selected values for optimization of HDPE/EG/mBN composite gives the result presented in Table 19.

Table 19. Constraints and optimization solution for HDPE/EG/mBN composite—version 1.

Name	Goal	Lower Limit	Upper Limit	Solution
Factor A: EG, %	is in range	5	25	5
Factor B: mBN, %	is in range	5	25	20.99
Thermal conductivity, W/mK	max	1.21	3.0	1.9
Tensile strength, N/mm^2	is in range	18.84	24.37	19.86
Tensile modulus, N/mm^2	is in range	1072	2153	1458
Strain at yield, %	max	1.85	5.92	3.99

For the given optimization conditions, eleven solutions were found with desirability $d = 0.449$, which did not completely satisfy the set criteria/goal (thermal conductivity is only 1.9 W/mK). Such a low desirability is the consequence of trying to achieve the highest value of strain at yield. However, when this value is constrained to fall within the obtained limits, the desirability rises to $d = 0.727$.

The optimization constraints and criteria for version 2 for HDPE/EG/mBN composite are given in Table 20.

Table 20. Constraints and optimization solution for HDPE/EG/mBN composite—version 2.

Name	Goal	Lower Limit	Upper Limit	Solution
Factor A: EG, %	is in range	5	25	17.79
Factor B: mBN, %	is in range	5	25	25
Thermal conductivity, W/mK	max	1.21	3.0	2.52
Tensile strength, N/mm^2	is in range	18.84	24.37	23.32
Tensile modulus, N/mm^2	is in range	1072	2153	2119
Strain at yield, %	is in range	1.85	5.92	1.85

The desirability curve for the percentages of EG and mBN within the optimization constraints is shown in Figure 9.

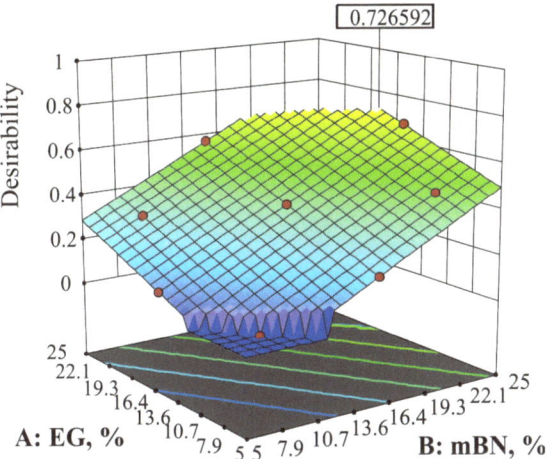

Figure 9. Desirability curve for the solution of HDPE/EG/mBN composite according to Table 18.

The optimization of HDPE/EG/nBN composite gives the results presented in Table 21.

Table 21. Constraints and optimization solution for HDPE/EG/nBN composite—version 1.

Name	Goal	Lower Limit	Upper Limit	Solution
Factor A: EG, %	is in range	5	22	7.86
Factor B: mBN, %	is in range	5	15	15
Thermal conductivity, W/mK	max	0.91	2.09	1.61
Tensile strength, N/mm^2	is in range	20.83	24.03	22.67
Tensile modulus, N/mm^2	is in range	1144	2006	1160
Strain at yield, %	max	2.24	6.23	5.11

Under the specified optimization conditions for the HDPE/EG/nBN composite, the desirability stands at $d = 0.655$. This is unquestionably an improvement compared to the scenario involving composites with micro-sized BN particles. However, by adhering to the same assumption, it becomes feasible to elevate the desirability to its highest value of $d = 1$ (as shown in Figure 10 and Table 22).

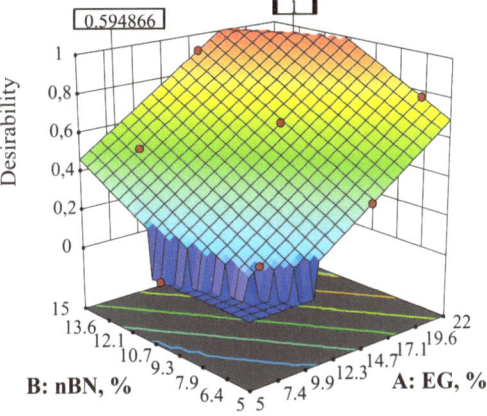

Figure 10. Desirability curve for the solution of HDPE/EG/nBN composite according to Table 18.

Table 22. Constraints and optimization solution for HDPE/EG/nBN composite—version 2.

Name	Goal	Lower Limit	Upper Limit	Solution
Factor A: EG, %	is in range	5	22	18.6
Factor B: mBN, %	is in range	5	15	14.14
Thermal conductivity, W/mK	max	0.91	2.09	2.12
Tensile strength, N/mm^2	is in range	20.83	24.03	23.53
Tensile modulus, N/mm^2	is in range	1144	2006	1984
Strain at yield, %	is in range	2.24	6.23	2.9

The desirability curve for the percentages of EG and mBN within the optimization constraints is shown in Figure 10.

4. Discussion

In comparison with research carried out by Sever et al., which included an EG/HDPE composite with up to 40 wt%. of EG, the achieved values of tensile strength are similar.

In line with the findings of Sever et al., the inclusion of an additive led to a rise in tensile strength, increasing from 26.93 MPa for HDPE to 31.97 MPa at a 40 wt.% content of EG [26]. Considering the starting value of HDPE 25.58 MPa, a tensile strength of 24.37 MPa for HDPE/EG15/mBN25 shows a slight decrease compared to the 18% increase measured by Sever et al. Moreover, it is worth noting that both this study and the research by Sever et al. employed an identical total additive proportion ratio of 40 wt.%. However, it is important to highlight that the study conducted by Sever et al. focused solely on a single component. Li et al. achieved a tensile strength of 26.90 MPa for EG additive of 50 wt.%, at a mixing temperature of 190 °C and for a duration of 30 min [27]. Soboličak et al. reported an increase in tensile modulus up to 1440 MPa at an EG of 50 wt.%, but also a decrease of tensile strength by 40% compared to pure HDPE [11], which is comparable to the tensile modulus results obtained for some combination of HDPE/EG/mBN and HDPE/EG/nBN composites. Zhang et al. blended BN and HDPE composite with the addition of a PE-g-MAH compatibilizer, achieving a tensile strength of ~28 MPa with a BN of 20 wt.% [28]. With the same percentage of BN, Zhang et al. reached the highest value of tensile strength (~28.5 MPa), while with the further addition of BN, the value of tensile strength decreased [29]. This stands in opposition to the findings presented in this study, as the introduction of any amount of additive percentage consistently resulted in a decrease in tensile strength. In the research mentioned above, the size of the EG particles was 5 μm, while the sizes of the BN particles were 4 μm and 5 μm [26,28,29].

Compared to [26], where a tensile modulus of 3390 MPa was achieved for 40 wt.% of EG, and [11], where the tensile modulus reached 1440 MPa at 50 wt.% of EG, in our tests, only 15% EG and 25% mBN were enough to achieve 2153 MPa, which is 1.5 times higher than in the research in [11], i.e., 1.6 times less than in the research in [26]. The question arises as to why, despite the substantial 50 wt.% proportion utilized in the investigation [11], a comparatively small tensile modulus value was attained. The tensile property values typically arise from the bonding facilitated by compatibilizers between the additive and matrix, resulting in improved interfacial connections among these three phases. The increase of tensile modulus, while retaining consistent tensile strength, indicates the superior mechanical properties of additives compared to the matrix, particularly as the mass proportion of additives increases. Sofiti and Berto have reported that the tensile modulus of EG as an individual separate constituent falls within the range of 0.5 GPa to 3 GPa, whereas a single crystal of graphite exhibits a modulus of 36.5 MPa [30]. Compared to EG, a much higher tensile modulus of BN monolayer structure was reported by Falin et al., 0.865 TPa, while the tensile strength of bilayer structured BN reported by Song et al. was 0.334 TPa [31,32]. The findings of Cheurasia et al. from an investigation into the mechanical characteristics of composites featuring a PE matrix and hBN as an additive revealed that, even at a modest wt.% of BN, there was a significant 64% enhancement in the tensile modulus and a 27% increase in the tensile strength of the composite containing 5 wt.%, in contrast to the pure

HDPE material [33]. Lee et al. reported a rise in tensile modulus by 127% and an increase in yield strength by 69% when utilizing BNNP as a 5 wt.% additive, as compared to the pure PE [34]. While experiencing notable improvements in strength and modulus, the introduction of any proportion of additives to the HDPE matrix results in a reduction in the strain at yield.

Compared to our results, Wieme et al. reported a higher value of thermal conductivity for in-plane measurement (4.03 W/mK). For through plane conductivity, which was used in this research too, Wieme reported a thermal conductivity of 0.69 W/mK [7]. With a high percentage of EG (50 wt.%) in HDPE, Soboliček reported a thermal conductivity of 2.18 W/mK, which is an increment of 372% compared to the HDPE matrix [11]. Our tests reveal that the HDPE/EG22/mBN22 composite exhibits a thermal conductivity of 3 W/mK. This value showcases a significant 137% increase when compared to the study conducted by Soboliček et al., even when employing only half the quantity of EG. A slightly higher percentage of EG (55 wt.%) in the HDPE matrix resulted in an increase in the thermal conductivity of composites (1.97 W/mK), as evidenced in a study by Klonos et al. [13], which can be compared with HDPE/EG15/BN15 regardless of the size of BN particles. Research on composites with BN as an additive show similar thermal properties to EG at the same percentage. Muratov et al. reported an increase of thermal conductivity to 2.08 W/mK for the HDPE/BN composite with 50 wt.% of a micro-sized hexagonal BN additive. For the same additive ratio, which was treated with titanate, thermal conductivity reached 0.9 W/mK [17]. Better results were reported by Rasul et al., who applied silane as a coupling agent. With the inclusion of 5 wt.% BN nanosheets, the thermal conductivity rose to 0.96 W/mK, whereas the matrix itself exhibited a thermal conductivity of 0.72 W/mK. The extraordinarily high value of matrix thermal conductivity is explained by the high mass proportion of aluminum oxide and zinc oxide, totaling 37% [21]. The importance of a coupling agent was researched by Zhang et al., where BN was treated with PE-g-MAH. The difference between treated and non-treated additive was obvious, 2.6 W/mK compared to 2.2 W/mK, respectively [28]. The literature surveyed from various authors reveals considerable disparities in thermal conductivity outcomes. It is noteworthy that achieving a thermal conductivity of 3 W/mK often demands substantial quantities of expanded graphite. However, our experiments demonstrate that this value can be attained using just half the EG amount when incorporating a specific proportion of BN. Notably, the overall additive quantity required to achieve this desired thermal conductivity is smaller in comparison to other investigations.

Salavagione et al. [35], to increase thermal conductivity, applied two different covalent functionalization approaches on boron nitride nanotubes (BNNTs) with short polyethylene (PE) chains, and from that work, it can be concluded that with a rather large amount of up to 40% boron nitride nanotubes (BNNTs), a 250% higher thermal conductivity than the pure HDPE matrix can be achieved, which is compared to composites with micro-particles of BN HDPE/EG8/mBN8, while for nano-particles of BN, composites HDPE/EG5/nBN10 or HDPE/EG8/nBN6.5. Furthermore, authors Huang and Qian et al. [36], to improve thermal conductivity, used ultra-high molecular weight polyethylene composites, to which they added a hybrid filler network of boron nitride sheets (BNs) and carbon nanotubes (CNTs) in the matrix. They concluded that it was necessary to add 40% BN sheet and 7 wt.% CNTs to obtain 2.38 W/mK. These results can be compared with the composite HDPE/EG15/mBN15 and HDPE/EG22/nBN6.5, i.e., with a smaller amount of BN in both cases and with the addition of EG, the same value was obtained. The development of the material has reached such a level that the properties of the composite can be significantly improved by adding new additives and with a smaller amount.

5. Conclusions

The synergetic effect of both additives EG and BN (micro- and nanoparticles) has a significant impact on the mechanical and thermal properties of polyethylene composite. The addition of PE-g-MAH compatibilizer results in better bonds between the polymer matrix and additives and equal dispersion of particles in the matrix. The differences in the values of tensile strength between the ones of the matrix (25.58 MPa) and the ones with the highest value of additives (24.37 MPa for HDPE/EG15/mBN25 and 24.03 MPa for HDPE/EG15/nBN15) show a negligible drop in strength and a drastic improvement of tensile modulus. At a high percentage of additives, the tensile modulus increased by 105% and 91% for composites with micro- and nano-sized BN particles, respectively.

The impact of particle size is not directly correlated with this characteristic, as the variance in average values is merely 4.5%. The contribution of EG to the increase of tensile strength is dominant compared to BN, which is especially noticeable in the EG/mBN composite group.

Furthermore, the addition of both additives led to the anticipated elevation in the thermal conductivity of the composites. In the case of the HDPE/EG/mBN composite, the thermal conductivity achieved a notable 3.0 W/mK, whereas for the HDPE/EG/nBN composite, the thermal conductivity reached 2.09 W/mK. This outcome provides evidence of established heat transfer pathways at the interface of surfaces. Even at a relatively low ratio of additive (i.e., EG8/mBN8), the thermal conductivity increased by 227% compared to the pure HDPE matrix.

Considering all the findings, composites combining EG and BN could serve as efficient heat conductors in scenarios demanding chemical resistance within the temperature range of 0 to 60 °C (because of the polymer chain degradation at higher temperatures). Further research should prioritize the study of rheological characteristics to determine the optimal parameters for large-scale manufacturing. Based on the optimization outcomes (as presented in Tables 18 and 20) and the obtained values, the next step involves the fabrication of a pipe to assess the impact of the low strain at yield value on the production process.

Author Contributions: Conceptualization, A.P., M.R.H. and L.T.; methodology, A.P., M.R.H. and L.T.; software, A.P. and L.T.; validation, A.P. and L.T.; formal analysis, A.P. and L.T.; investigation, L.T.; writing—original draft preparation, L.T.; writing—review and editing, A.P. and M.R.H.; visualization, A.P.; supervision, A.P.; project administration, A.P.; funding acquisition, A.P. All authors have read and agreed to the published version of the manuscript.

Funding: This research was funded from the Operational Program "Competitiveness and Cohesion 2014-2020" from the European Fund for Regional Development as part of the call KK.01.2.1.02—Increasing the development of new products and services resulting from research and development activities (IRI), phase II.

Institutional Review Board Statement: Not applicable.

Informed Consent Statement: Not applicable.

Data Availability Statement: The data presented in this study are available on request from the corresponding author.

Acknowledgments: The work was created as part of the project Development of innovative systems for the use of geothermal energy sources and energy from biological waste—RazInoGeoBio (Project code: KK.01.2.1.02.0314) financed from the Operational Program "Competitiveness and Cohesion 2014-2020" from the European Fund for Regional Development as part of the call KK.01.2.1.02—Increasing the development of new products and services resulting from research and development activities (IRI), phase II. The authors would like to thank the EU for funding the project.

Conflicts of Interest: The authors declare no conflict of interest.

References

1. Alshammari, B.A.; Al-Mubaddel, F.S.; Karim, M.R.; Hossain, M.; Al-Mutairi, A.S.; Wilkinson, A.N. Addition of Graphite Filler to Enhance Electrical, Morphological, Thermal, and Mechanical Properties in Poly (Ethylene Terephthalate): Experimental Characterization and Material Modeling. *Polymers* **2019**, *11*, 1411. [CrossRef]
2. Imiolek, P.; Kasprowicz, K.; Laska, J. Antistatic polyethylene free-standing films modified with expanded graphite—Technological aspects. *Polimery* **2020**, *65*, 275–279. [CrossRef]

3. Bazan, P.; Mierzwiński, D.; Bogucki, R.; Kuciel, S. Bio-Based Polyethylene Composites with Natural Fiber: Mechanical, Thermal, and Ageing Properties. *Materials* **2020**, *13*, 2595. [CrossRef] [PubMed]
4. Makhlouf, A.; Belaadi, A.; Amroune, S.; Bourchak, M.; Satha, H. Elaboration and Characterization of Flax Fiber Reinforced High Density Polyethylene Biocomposite: Effect of the Heating Rate on Thermo-mechanical Properties. *J. Nat. Fibers* **2022**, *19*, 3928–3941. [CrossRef]
5. Zhang, Q.; Zhang, D.; Xu, H.; Lu, W.; Ren, X.; Cai, H.; Lei, H.; Huo, E.; Zhao, Y.; Qian, M.; et al. Biochar filled high-density polyethylene composites with excellent properties: Towards maximizing the utilization of agricultural wastes. *Ind. Crop. Prod.* **2020**, *146*, 112185. [CrossRef]
6. Eze, I.O.; Igwe, I.O.; Ogbobe, O.; Obasi, H.C.; Ezeamaku, U.L.; Nwanonenyi, S.C.; Anyanwu, E.E.; Nwachukwu, I. Effects of Compatibilization on Mechanical Properties of Pineapple Leaf Powder Filled High Density Polyethylene. *Int. J. Eng. Technol.* **2017**, *10*, 22–28. [CrossRef]
7. Wieme, T.; Duan, L.; Mys, N.; Cardon, L.; D'hooge, D.R. Effect of Matrix and Graphite Filler on Thermal Conductivity of Industrially Feasible Injection Molded Thermoplastic Composites. *Polymers* **2019**, *11*, 87. [CrossRef]
8. Wei, B.; Zhang, L.; Yang, S. Polymer composites with expanded graphite network with superior thermal conductivity and electromagnetic interference shielding performance. *Chem. Eng. J.* **2021**, *404*, 126437. [CrossRef]
9. Chen, X.; Su, Y.; Reay, D.; Riffat, S. Recent research developments in polymer heat exchangers—A review. *Renew. Sustain. Energy Rev.* **2016**, *60*, 1367–1386. [CrossRef]
10. Wang, Y.; Lu, L.; Hao, Y.; Wu, Y.; Li, Y. Mechanical and Processing Enhancement of a Recycled HDPE/PPR-Based Double-Wall Corrugated Pipe via a POE-g-MAH/CaCO3/HDPE Polymer Composite. *ACS Omega* **2021**, *6*, 19705–19716. [CrossRef]
11. Sobolčiak, P.; Abdulgader, A.; Mrlik, M.; Popelka, A.; Abdala, A.A.; Aboukhlewa, A.A.; Karkri, M.; Kiepfer, H.; Bart, H.-J.; Krupa, I. Thermally Conductive Polyethylene/Expanded Graphite Composites as Heat Transfer Surface: Mechanical, Thermo-Physical and Surface Behavior. *Polymers* **2020**, *12*, 2863. [CrossRef] [PubMed]
12. Yin, X.; Jie, X.; Wei, K.; He, G.; Feng, Y. In-situ exfoliation and thermal conductivity in phase-transition-assisted melt blending fabrication of low-density polyethylene/expanded graphite nanocomposite. *Polym. Eng. Sci.* **2022**, *62*, 3487–3497. [CrossRef]
13. Klonos, P.A.; Papadopoulos, L.; Kourtidou, D.; Chrissafis, K.; Peoglos, V.; Kyritsis, A.; Bikiaris, D.N. Effects of Expandable Graphite at Moderate and Heavy Loadings on the Thermal and Electrical Conductivity of Amorphous Polystyrene and Semicrystalline High-Density Polyethylene. *Appl. Nano* **2021**, *2*, 31–45. [CrossRef]
14. Che, J.; Wu, K.; Lin, Y.; Wang, K.; Fu, Q. Largely improved thermal conductivity of HDPE/expanded graphite/carbon nanotubes ternary composites via additive network-network synergy. *Compos. Part A Appl. Sci. Manuf.* **2017**, *99*, 32–40. [CrossRef]
15. Sánchez-Sánchez, X.; Elias-Zuñiga, A.; Hernández-Avila, M. Processing of ultra-high molecular weight polyethylene/graphite composites by ultrasonic injection moulding: Taguchi optimization. *Ultrason. Sonochemistry* **2018**, *44*, 350–358. [CrossRef]
16. Abdelrazeq, H.; Sobolčiak, P.; Al-Maadeed, M.A.-A.; Ouederni, M.; Krupa, I. Recycled Polyethylene/Paraffin Wax/Expanded Graphite Based Heat Absorbers for Thermal Energy Storage: An Artificial Aging Study. *Molecules* **2019**, *24*, 1217. [CrossRef]
17. Muratov, D.S.; Stepashkin, A.A.; Anshin, S.M.; Kuznetsov, D.V. Controlling thermal conductivity of high density polyethylene filled with modified hexagonal boron nitride (hBN). *J. Alloys Compd.* **2018**, *735*, 1200–1205. [CrossRef]
18. Ali, M.; Abdala, A. Large scale synthesis of hexagonal boron nitride nanosheets and their use in thermally conductive polyethylene nanocomposites. *Int. J. Energy Res.* **2022**, *46*, 10143–10156. [CrossRef]
19. Shang, Y.; Yang, G.; Su, F.; Feng, Y.; Ji, Y.; Liu, D.; Yin, R.; Liu, C.; Shen, C. Multilayer polyethylene/ hexagonal boron nitride composites showing high neutron shielding efficiency and thermal conductivity. *Compos. Commun.* **2020**, *19*, 147–153. [CrossRef]
20. Zhang, R.-C.; Huang, Z.; Huang, Z.; Zhong, M.; Zang, D.; Lu, A.; Lin, Y.; Millar, B.; Garet, G.; Turner, J.; et al. Uniaxially stretched polyethylene/boron nitride nanocomposite films with metal-like thermal conductivity. *Compos. Sci. Technol.* **2020**, *196*, 108154. [CrossRef]
21. Rasul, G.; Kiziltas, A.; Bin Hoque, S.; Banik, A.; Hopkins, P.E.; Tan, K.-T.; Arfaei, B.; Shahbazian-Yassar, R. Improvement of the thermal conductivity and tribological properties of polyethylene by incorporating functionalized boron nitride nanosheets. *Tribol. Int.* **2022**, *165*, 107277. [CrossRef]
22. Shi, A.; Li, Y.; Liu, W.; Xu, J.-Z.; Yan, D.-X.; Lei, J.; Li, Z.-M. Highly thermally conductive and mechanically robust composite of linear ultrahigh molecular weight polyethylene and boron nitride via constructing nacre-like structure. *Compos. Sci. Technol.* **2019**, *184*, 107858. [CrossRef]
23. Lebedev, S.M. A comparative study on thermal conductivity and permittivity of composites based on linear low-density polyethylene and poly(lactic acid) filled with hexagonal boron nitride. *Polym. Compos.* **2021**, *43*, 111–117. [CrossRef]
24. Güzdemir; Kanhere, S.; Bermudez, V.; Ogale, A.A. Boron Nitride-Filled Linear Low-Density Polyethylene for Enhanced Thermal Transport: Continuous Extrusion of Micro-Textured Films. *Polymers* **2021**, *13*, 3393. [CrossRef]
25. Yang, X.; Song, X.; Hu, Z.; Li, C.; Guo, T. Improving comprehensive properties of high-density polyethylene matrix composite by boron nitride/coconut shell carbon reinforcement. *Polym. Test.* **2022**, *115*, 107728. [CrossRef]
26. Sever, K.; Tavman, I.H.; Seki, Y.; Turgut, A.; Omastova, M.; Ozdemir, I. Electrical and mechanical properties of expanded graphite/high density polyethylene nanocomposites. *Compos. Part B Eng.* **2013**, *53*, 226–233. [CrossRef]
27. Li, Y.; Wu, D.; Chen, G. Preparation and characterization of high-density polyethylene/expanded graphite conducting masterbatch. *J. Appl. Polym. Sci.* **2007**, *106*, 3119–3124. [CrossRef]

28. Zhang, X.; Wu, H.; Guo, S. Effect of Interfacial Interaction on Morphology and Properties of Polyethylene/Boron Nitride Thermally Conductive Composites. *Polym. Technol. Eng.* **2015**, *54*, 1097–1105. [CrossRef]
29. Zhang, X.; Shen, L.; Wu, H.; Guo, S. Enhanced thermally conductivity and mechanical properties of polyethylene (PE)/boron nitride (BN) composites through multistage stretching extrusion. *Compos. Sci. Technol.* **2013**, *89*, 24–28. [CrossRef]
30. Solfiti, E.; Berto, F. Mechanical properties of flexible graphite: Review. *Procedia Struct. Integr.* **2020**, *25*, 420–429. [CrossRef]
31. Falin, A.; Cai, Q.; Santos, E.J.G.; Scullion, D.; Qian, D.; Zhang, R.; Yang, Z.; Huang, S.; Watanabe, K.; Taniguchi, T.; et al. Mechanical properties of atomically thin boron nitride and the role of interlayer interactions. *Nat. Commun.* **2017**, *8*, 15815. [CrossRef]
32. Song, L.; Ci, L.; Lu, H.; Sorokin, P.B.; Jin, C.; Ni, J.; Kvashnin, A.G.; Kvashnin, D.G.; Lou, J.; Yakobson, B.I.; et al. Large Scale Growth and Characterization of Atomic Hexagonal Boron Nitride Layers. *Nano Lett.* **2010**, *10*, 3209–3215. [CrossRef]
33. Chaurasia, A.; Verma, A.; Parashar, A.; Mulik, R.S. Experimental and Computational Studies to Analyze the Effect of h-BN Nanosheets on Mechanical Behavior of h-BN/Polyethylene Nanocomposites. *J. Phys. Chem. C* **2019**, *123*, 20059–20070. [CrossRef]
34. Lee, D.; Lee, B.; Park, K.H.; Ryu, H.J.; Jeon, S.; Hong, S.H. Scalable Exfoliation Process for Highly Soluble Boron Nitride Nanoplatelets by Hydroxide-Assisted Ball Milling. *Nano Lett.* **2015**, *15*, 1238–1244. [CrossRef]
35. Quiles-Díaz, S.; Martínez-Rubí, Y.; Guan, J.; Kim, K.S.; Couillard, M.; Salavagione, H.J.; Gómez-Fatou, M.A.; Simard, B. Enhanced Thermal Conductivity in Polymer Nanocomposites via Covalent Functionalization of Boron Nitride Nanotubes with Short Polyethylene Chains for Heat-Transfer Applications. *ACS Appl. Nano Mater.* **2019**, *2*, 440–451. [CrossRef]
36. Guo, Y.; Cao, C.; Luo, F.; Huang, B.; Xiao, L.; Qian, Q.; Chen, Q. Largely enhanced thermal conductivity and thermal stability of ultra high molecular weight polyethylene composites via BN/CNT synergy. *RSC Adv.* **2019**, *9*, 40800–40809. [CrossRef]

Disclaimer/Publisher's Note: The statements, opinions and data contained in all publications are solely those of the individual author(s) and contributor(s) and not of MDPI and/or the editor(s). MDPI and/or the editor(s) disclaim responsibility for any injury to people or property resulting from any ideas, methods, instructions or products referred to in the content.

Article

Effect of Resin Bleed Out on Compaction Behavior of the Fiber Tow Gap Region during Automated Fiber Placement Manufacturing

Von Clyde Jamora [1,*], Virginia Rauch [1], Sergii G. Kravchenko [2] and Oleksandr G. Kravchenko [1]

1. Department of Aerospace and Mechanical Engineering, Old Dominion University, Norfolk, VA 23529, USA; vrauch.vmr@gmail.com (V.R.); okravche@odu.edu (O.G.K.)
2. Department of Materials Engineering, The University of British Columbia, Vancouver, BC V6T 1Z4, Canada; sergey.kravchenko@ubc.ca
* Correspondence: vonjamora@gmail.com

Citation: Jamora, V.C.; Rauch, V.; Kravchenko, S.G.; Kravchenko, O.G. Effect of Resin Bleed Out on Compaction Behavior of the Fiber Tow Gap Region during Automated Fiber Placement Manufacturing. *Polymers* 2024, *16*, 31. https://doi.org/10.3390/polym16010031

Academic Editors: Phuong Nguyen-Tri, Ana Pilipović and Mustafa Özcanli

Received: 11 November 2023
Revised: 1 December 2023
Accepted: 7 December 2023
Published: 21 December 2023

Copyright: © 2023 by the authors. Licensee MDPI, Basel, Switzerland. This article is an open access article distributed under the terms and conditions of the Creative Commons Attribution (CC BY) license (https:// creativecommons.org/licenses/by/ 4.0/).

Abstract: Automated fiber placement is a state-of-the-art manufacturing method which allows for precise control over layup design. However, AFP results in irregular morphology due to fiber tow deposition induced features such as tow gaps and overlaps. Factors such as the squeeze flow and resin bleed out, combined with large non-linear deformation, lead to morphological variability. To understand these complex interacting phenomena, a coupled multiphysics finite element framework was developed to simulate the compaction behavior around fiber tow gap regions, which consists of coupled chemo-rheological and flow-compaction analysis. The compaction analysis incorporated a visco-hyperelastic constitutive model with anisotropic tensorial prepreg viscosity, which depends on the resin degree of cure and local fiber orientation and volume fraction. The proposed methodology was validated using the compaction of unidirectional tows and layup with a fiber tow gap. The proposed approach considered the effect of resin bleed out into the gap region, leading to the formation of a resin-rich pocket with a complex non-uniform morphology.

Keywords: process modeling; compaction; finite element analysis; automated fiber placement

1. Introduction

High-throughput manufacturing processes for fiber-reinforced composites allow for complex layups with a lower processing cycle time [1,2]. Particularly, automated fiber placement (AFP) is a robotic system that deposits pre-impregnated (prepreg) fiber tow strips at specified locations and enables tailored orientations [3–7]. For example, Wu et al. created a variable stiffness structure with a cut-and-restart functionality of an AFP machine in order to manufacture a prototype of an airplane fuselage [3]. However, deposition induced AFP tow features, especially fiber tow gaps, inevitably lead to irregular, local non-uniform morphology upon cure and compaction [8]. Manufacturing features, such as voids and resin-rich regions, develop when plies neighboring to the gap tow sink into the gap and experience large non-linear deformation caused by the flow of liquid resin into the empty space (Figure 1). Previous reports show that AFP manufacturing defects created during the early stages of cure before resin gelation can significantly affect the mechanical properties of the composites [8–10]. These regions with irregular morphology can serve as weak points and reduce the strength [11–13].

Fiber tow gaps and overlaps can be restricted during manufacturing or allowed within a specific tolerance. However, imperfections in the cured composites propagate from these AFP deposition features within the preform and produce local non-uniform morphology, such as ply thickness variations, waviness, and resin-rich pockets (Figure 1). These irregularities produced from AFP deposition features cannot be entirely avoided and should be considered as a part of AFP-manufactured composites to properly predict the strength of composite structures [14,15]. Furthermore, microscopic features, such as fiber volume fraction redistribution, and mesoscopic features (ply thickness, waviness) are not formed in isolation from each other and should be analyzed within a comprehensive framework that considers their coupling [7,16,17].

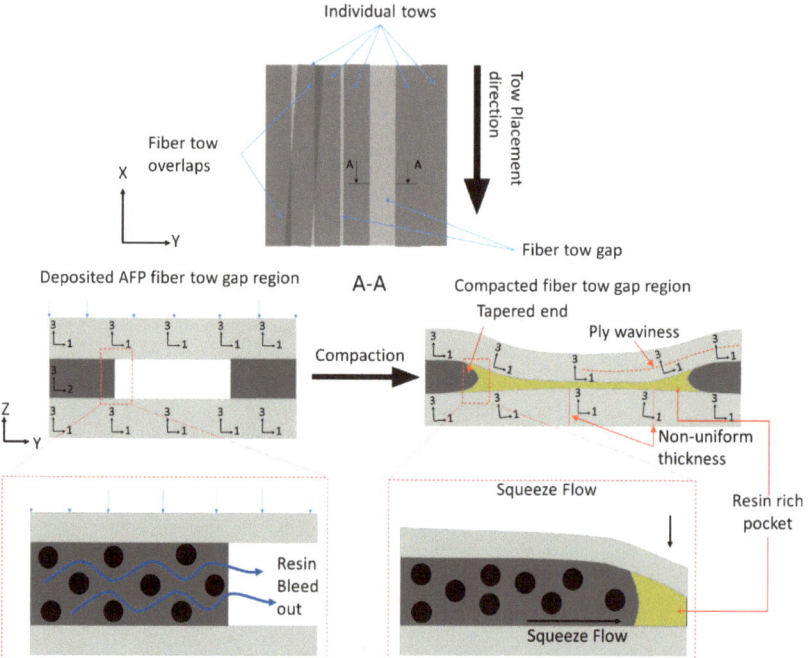

Figure 1. Schematics of gaps within a laminate and the effects of resin bleed out on the non-linear morphology.

Prior to vitrification, resin changes from an incompressible viscoelastic liquid to a viscoelastic solid upon gelation. The compaction of AFP-deposited preform develops prior to composite gelation, wherein resin viscosity depends on the degree of cure and temperature [18,19]. At this stage, the deformation of the prepreg acts as a homogeneous medium, which is treated as a squeeze flow. In the case of AFP preform compaction, the squeeze flow develops when the fiber tow above the gap region is pushed inside of the void space during the compaction. However, resin bleed out through the fiber bed also occurs simultaneously (Figure 2a), causing the formation of resin-rich zones in the tow gap void and local fiber volume fraction variation in the composite [20]. Conventionally, the effect of resin bleed out in manufacturing of prepregs is considered insignificant, but in the case of AFP preforms with fiber tow gaps, it was shown to affect the waviness of the ply [21].

Recently, a hyper-viscoelastic material model was applied by Belnoue et al. to simulate the squeeze flow deformation during compaction [7,22–25]. However, the research did not consider the anisotropic viscosity of prepreg tape and its dependence on the fiber volume fraction. Furthermore, earlier studies did not consider the effects of the resin bleed out from the compacted prepreg into the gap region. The present study considers the effect of resin bleed out on the formation of non-uniform morphology around the region of the fiber tow gap. To model prepreg behavior around the internal void, like in the case of debulking, a negative internal pressure gradient can be considered inside of the void using porous-cohesive elements [26]. Another approach is to consider the flow front filling the gap via a one-dimensional numerical model [27]. These studies represent a two-phase modeling scheme where a region other than the composite is represented [21,27]. In the present work, a single-phase model was used and the effects of the resin bleed out into the gap was shown to have a larger impact on the morphology of the composite prior to resin gelation.

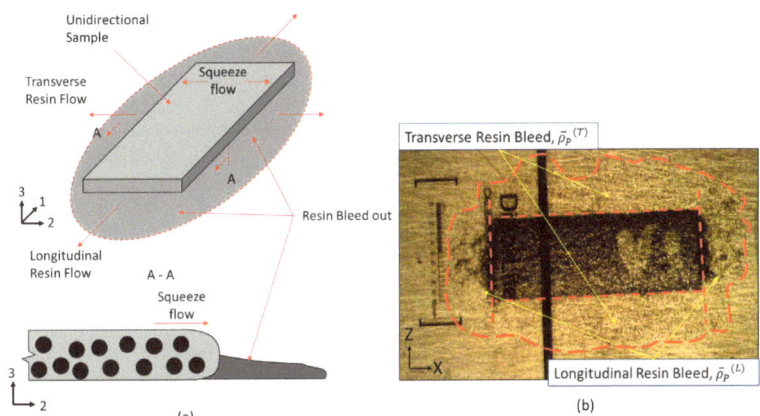

Figure 2. (**a**) Schematics of the bleed out in a unidirectional composite sample. (**b**) A unidirectional sample which shows a dashed red line dividing the transverse resin bleed, longitudinal resin bleed and the area of the unidirectional tow.

The present treatise extends the use of a hyper-viscoelastic model by introducing the fiber volume-dependent anisotropic tensorial viscosity of the prepreg into the formulation of the viscous component [22,28,29]. This approach introduces the local fiber volume fraction as a coupling between the squeeze flow and resin bleed out. Resin bleed-out measurements were obtained from the unidirectional tape compaction experiments and used to predict the amount of resin bleed out around the tow gap region. In the present case, a single-phase model was developed based on sequential compaction flow analysis, which adds the influence of resin flow into fiber tow gaps without explicitly modeling the resin. First, the hyper viscoelastic finite element analysis (FEA) squeeze flow was performed and the tow gap volume decrease was predicted. Once the tow gap volume reached the resin bleed-out volume, the time of filling was predicted. To capture the presence of resin inside of the tow gap, internal pressure was introduced in the tow gap based on the predicted time of void filling. The interaction between the squeeze flow and resin bleed out was investigated and the predicted resin-rich pocket shape agreed closely with the experimental observations. The proposed sequential compaction flow methodology can be extended to modeling various combinations of fiber tow gaps in AFP manufacturing to predict fiber waviness, resin-rich regions, as well as ply thickness variation in the region around the fiber tow gaps.

2. Sample Fabrication and Experimental Measurement

2.1. Unidirectional Tow Compaction Experiments

Experiments were performed on unidirectional samples with different geometries designed to measure the tow spreading and to predict resin bleed out from the prepreg during compaction. The material used was tows of IM7/8552 (Hexcel, Stamford, CT, USA) with low tack behavior with a width of 6.57 mm and a thickness of 0.14 mm. To measure the compaction of unidirectional tows in the early stages of manufacturing before resin gelation, the first hold stage of the manufacturer's cure cycle was used. This included a hold temperature of 110 °C with that of an initial ramp of 3 °C/min. A pressure of 0.551 MPa (80 psi) was applied to the composite at a ramp of 0.034 MPa/min. Six configurations were made for the unidirectional samples. The tow width and length were varied between one, two, and four plies. This measured to be 0.14 mm, 0.28 mm, and 0.56 mm for the thickness and the widths were 6.46 mm, 12.92 mm, and 25.84 mm, respectively. The length of the samples was a constant 25.84 mm. Six samples were made for each of the configurations and the experimentally measured results were averaged for each configuration.

To analyze the tow spreading during cure, the thickness and width of the unidirectional samples were measured at 5, 15, 30, and 60 min into the hold stage by taking images of the top x-y plane of the unidirectional samples using a digital microscope. At the end of the first isothermal stage of the cure cycle, the bleed-out resin was observed on the outside of the ply (Figure 2b). This resin was cut with a blade and its mass was measured. The images taken throughout the cure cycles were used to estimate the resin bleed out in fiber as well as transverse to the fiber direction throughout the cure process.

2.2. Compaction Experiments with an Embedded Fiber Tow Gap

A cross-ply configuration with an embedded tow gap was laid up and used to study the compaction behavior around the region of the tow gap. The fiber tow stacking sequence [0°/90°/0°] was used with the centrally located 90° tow gap oriented in the y direction. The width of the sample consisted of four fiber tows with 0° and two tows with 90° layers (Figure 3). The overall dimensions of the fabricated sample before cure were 25.15 mm × 25.15 mm × 0.40 mm. A hand roller was used to apply slight uniform contact force to ensure proper contact in the layup. Once the top 0° layers were deposited, no additional fiber spreading was observed and the 0° tows sank into the tow gap region in the central location of the gap.

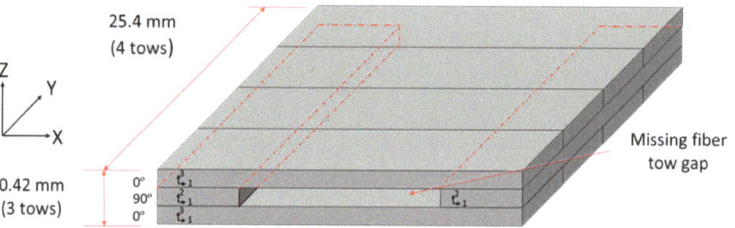

Figure 3. Schematics of the manufactured fiber tow gap dimensions and stacking sequence. The dashed line indicates the 90° ply between the 0° plies in the laminate.

The same cure cycle was used to cure the tow gap sample, which was post-cured in the oven at 180 °C for two hours. To observe the cross section of the fiber tow gap, the samples were cut and polished. To determine the effects of the compaction on the cured morphology, the x-z plane of the samples was analyzed using a Leica DM6 upright microscope, and different ply features were measured, including the fiber angles in the top 0° tow, size, and shape of the resin-rich region, as well as ply thickness variations in the bottom 0° tow.

3. Multiphysics FEA Simulation Framework for AFP Fiber Tow Preform Compaction

The interconnected nature of the cure and compaction process was studied via a unified process modeling approach in the form of a physics-based simulation that captured the chemo-rheological transformation of the resin, squeeze flow, and resin bleed out. Figure 4 shows a highly non-linear multivariable model in which chemo-rheological resin behavior was used to determine the effective flow-compaction characteristics of the prepreg. Therefore, the proposed model was implemented in a commercially available FEA software ABAQUS/Explicit 2021 (Dassualt Systemes, Vélizy-Villacoublay, France) using custom-built subroutines. The rheological and physical state properties, such as the degree of cure, viscosity, fiber volume fraction, and fiber orientation, updated the in situ effective anisotropic viscosities of the fiber tow. The elements of the present model may be calibrated for different elastic and rheological properties for different material systems. With a large number of factors involved in the model, future algorithms, such as smoothed-FEA and machine learning, can be used to predict the deformation response of composites during manufacturing [30,31]. However, the present methodology is a more accessible approach for the process modeling of composites [32,33].

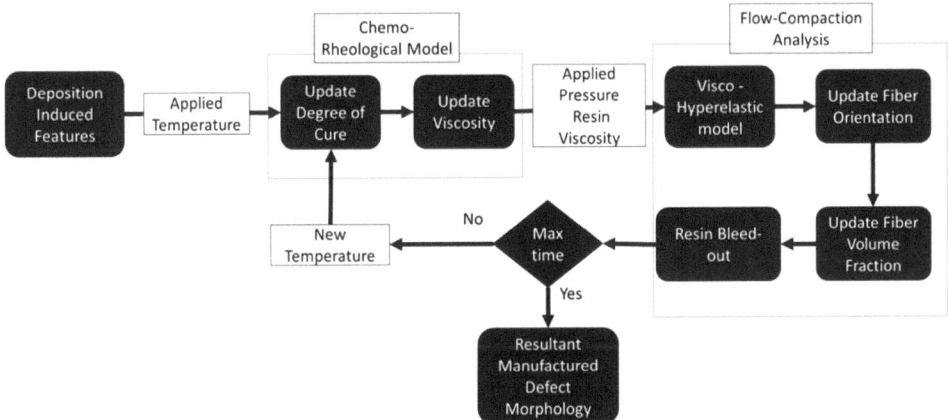

Figure 4. Coupled chemo-rheological and compaction model for predicting defect morphology in AFP-manufactured composites.

3.1. Chemo-Rheological Model of Thermosetting Prepreg

The degree of cure and viscosity of resin govern the squeeze flow behavior of the prepreg during the compaction process [34]. No significant exothermic reaction was observed due to the small thickness [16]. The chemo-rheological model of resin provided the degree of cure, α, and the resin viscosity, μ. The modified autocatalytic resin kinetics model were used in the present work [35]:

$$\frac{d\alpha}{dT} = \frac{Ae^{-\frac{\Delta E}{RT}} \alpha^{m_\alpha} (1-\alpha)^{n_\alpha}}{1 + e^{C_\alpha \{\alpha - (\alpha_{c0} + \alpha_{CT} T)\}}} \quad (1)$$

where $\Delta E = 66,500 \frac{J}{mol \cdot K}$ is the activation energy, $A = 153,000 \frac{1}{s}$ is a pre-exponential cure rate coefficient, and R is the gas constant. $m_\alpha = 0.813$ is the first exponential coefficient, $n_\alpha = 2.74$ is a second exponential coefficient, $C_\alpha = 4.31$ is a diffusion constant, $\alpha_{C0} = -1.684$ is the critical degree of cure at $T = 0$ K, and finally, $\alpha_{CT} = 0.005473$ is a constant that governs the increase in the critical resin degree of cure with temperature, where α is the degree of cure and T is the imposed temperature. The initial degree of cure of the composite samples was measured using digital scanning calorimetry (DSC), which determined it to be 0.145.

As the resin cures, the viscosity increases exponentially. The function for the viscosity uses the degree of cure and temperature as shown in Equation (2):

$$\mu = \mu_\infty e^{k_i \alpha} e^{\frac{U}{RT}} \quad (2)$$

where $\mu_\infty = 83 \times 10^{-8}$ MPa·s is a viscosity constant, $U = 49,100$ kJ is the activation energy of the viscosity, and $k_1 = 25.49 \frac{kJ}{mol}$ and $k_2 = -1.32 \frac{kJ}{mol}$ are temperature-independent constants determined via the experimental data [36]. The applied temperature, the evolution of the degree of cure, and the viscosity is shown in Figure 5. The degree of cure was seen only reaching a 17% cure at the end of the hold stage, which means the resin remains in the viscous state during the experiment [37,38]. Upon gelation, the viscosity increases exponentially and the material develops solid viscoelastic properties. Therefore, the flow-compaction analysis was restricted to the first stage of the cure cycle, since the large non-linear deformation and resin flow will develop prior to gelation. Also, the chemical and thermally induced shrinkages that develop prior to gelation will be relaxed due to the liquid resin state [39,40].

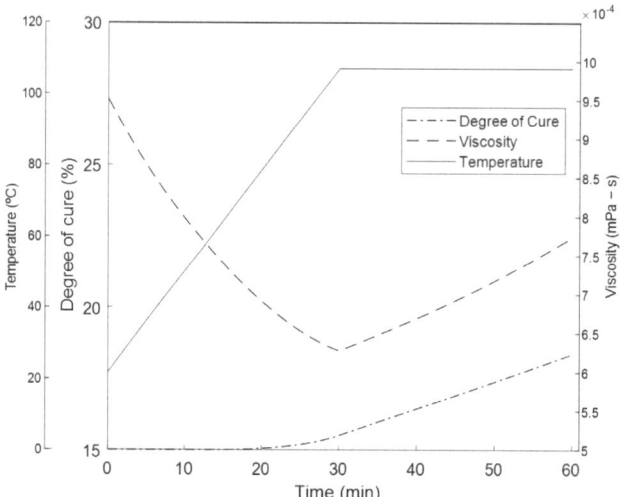

Figure 5. The applied temperature and the evolution of the degree of cure and viscosity using the thermo-rheological model.

3.2. Constitutive Squeeze Flow Model during Prepreg Fiber Tow Compaction

When heated above the instantaneous glass transition temperature, the resin becomes a viscoelsatic fluid, while rigid continuous fibers result in anisotropic viscous behavior of the uncured prepreg. Specifically, the inextensible fibers lead to the axial viscosity being six orders of magnitude higher than transverse and shear viscosity [29]. Therefore, the constitutive model simulates the unidirectional material with transversely isotropic properties. The flow-compaction model incorporated visco-hyperelastic behavior to capture the contribution of the liquid resin and included an elastic part and a viscous part [41]. The elastic part was calculated from a hyperelastic potential energy equation, ψ, using the five invariants of the Cauchy–Green deformation tensor, C.

$$\hat{\sigma} = \hat{\sigma}^e + \hat{\sigma}^v = 2\left(\frac{\partial \psi^e}{\partial \hat{C}} + \frac{\partial \psi^v}{\partial \dot{\hat{C}}}\right) = 2\sum_{\alpha=1}^{5}\left(\frac{\partial \psi^e}{\partial I_\alpha}\frac{\partial I_\alpha}{\partial \hat{C}}\right) + \hat{\sigma}^v = \frac{2}{J}\left(\hat{F}\cdot\frac{\partial \psi^e}{\partial \hat{C}}\cdot\hat{F}^T\right) + \hat{\sigma}^v \quad (3)$$

Through a push-forward operation, the stress becomes a function of the deformation gradient, $\hat{F}$, and its Jacobian, J. The fiber direction was solved using the linear-elastic behavior from the fiber instead of the homogenized composite since it is governed by inextensible fibers. The constitutive equation for the stress is given by Equation (4) [42].

$$\sigma_{11}^e = 2\mu_f \varepsilon_{11} + \lambda_f tr(\varepsilon) \quad (4)$$

where ε_{11} is the strain in the fiber direction and $tr(\varepsilon)$ is the trace of the strain. The constants μ and λ are related to the fiber material properties as

$$\mu_f = \frac{E_f}{2(1+\nu_f)} \quad (5)$$

$$\lambda_f = \frac{\mu_f \nu_f}{(1-2\nu_f)} \quad (6)$$

where E_f = 168 GPa and $v_f = 0.027$ are the modulus of elasticity and Poisson's ratio of the IM7 fiber, respectively.

For the transverse and shear properties, a transversely isotropic neo-Hookean model was used to calculate the stress of the current configuration. The strain energy potential used for the 2–3 plane for the stress is shown in Equation (7).

$$\psi^e = \frac{1}{2}\mu(I_1 - 3) - \mu\ln(J) + \frac{1}{2}\lambda(J-1)^2 + [\beta + \zeta\ln(J) + \kappa(I_4 - 1)](I_4 - 1) - \frac{1}{2}\beta(I_5 - 1) \tag{7}$$

I_1–I_5 are the invariants of the Cauchy–Green strain tensor. The material constants, λ, μ, β, ζ, and κ, are material constants derived from the material constants of the homogenized composites using Equations (8)–(11) [43]. The model uses the invariants of the Cauchy–Green strain tensor.

$$\lambda = \frac{E_{22}(v + nv^2)}{m(1+v)} \tag{8}$$

$$\mu = \frac{E_{22}}{2(1+v)} \tag{9}$$

$$\beta = \mu - G_{23} \tag{10}$$

$$\zeta = \frac{E_{22}v(1-n)}{4m(1+v)} \tag{11}$$

$$\kappa = \frac{E_{11}(1-v)}{8m} - \frac{\lambda + 2\mu}{8} + \frac{\beta}{2} - \zeta \tag{12}$$

$$m = 1 - v - 2nv^2 \tag{13}$$

$$n = \frac{E_{11}}{E_{22}} \tag{14}$$

where E_{11} = 25 MPa and E_{22} = 1 Pa are the Young's modulus in the fiber and transverse to the fiber directions of the homogenized uncured composite to ensure the correct bending stiffness and the stability of the FEA, respectively. $v = 0.01$, and $G_{23} = 1$ MPa are the Poisson's ratio and the shear modulus of the homogenized composite, respectively [41]. The Cauchy stress tensor can then be derived by differentiating the strain energy potential and pushing forward with a double multiplication of the deformation gradient provided by Abaqus [43]. Since the flow and the hyperelasticity constitutive equations are dependent on the fiber direction, the fiber orientation is updated every step. A model assumed that the fiber orientation follows an affine motion based on the deformation gradient characterized by the following equation [44]:

$$\vec{P} = \frac{\hat{F} \cdot \vec{p_0}}{\left\| \hat{F} \cdot \vec{p_0} \right\|} \tag{15}$$

where $\vec{p}$ is the fiber direction vector and $\vec{p_0}$ is the initial fiber direction vector. The fiber direction is used to inform the anisotropic component of the hyperelastic model, where it becomes the variable for the function of stress based on the strain energy potential [24].

The microscopic fibers suspended in resin show time-dependent behavior based on the strain rate effects and resin viscous behavior. Research from Kelly [28] used the multiplicative decomposition of apparent viscosity to describe the squeeze flow. To quantify the homogenized prepreg tape response, the proposed approach for the viscous term uses multiplicative components that consist of strain- and strain rate-dependent terms [28,29,41]. The strain dependent terms take into account the ply geometry and the strain rate described as the resin behavior [41]. The viscous stress term followed the behavior as shown in Equation (16).

$$\sigma^v{}_{ij} = \eta_{rate_{ij}} \cdot \eta_{ply_{ij}} \cdot \eta_{micro_{ij}} \cdot \hat{\dot{\varepsilon}}_{ij} \tag{16}$$

where $\hat{\eta}_{rate}$ is a strain rate-dependent viscosity tensor component and $\hat{\dot{\varepsilon}}$ is the strain rate tensor. $\hat{\eta}_{ply}$ is the term responsible for ply geometry and $\hat{\eta}_{micro}$ is the phenomenological strain-dependent term derived from micromechanical considerations at the micro level.

During the curing, the fiber–resin system is assumed to act as a power-law viscous fluid. Previous work described the anisotropic viscosity of oriented discontinuous fibers suspended in Newtonian fluid [29]. The same principles may be applied to continuous fiber with a finite length. Therefore, a tensorial viscosity approach was used to model the $\hat{\eta}_{rate}$ term where the resin matrix follows a power-law behavior. Since the composite tow is transversely isotropic, the in-plane directions (transverse to the fiber) were assumed to have the same properties, while the fiber direction is assumed to be inextensible.

In Nixon's work [23], the strain rate term used phenomenological parameters that were determined experimentally. The proposed rate term for the current study uses effective properties of the composite derived from the anisotropic viscosity as a function of the strain rate, fiber volume fraction, and a power-law exponent using the following tensor components [29].

$$\eta_{rate_{11}} = \eta_{resin} 2^{-mf} \left[\frac{\sqrt{\bar{f}}}{1-\sqrt{\bar{f}}} \right]^m \left(\frac{L}{D}\right)^{m+1} \dot{\varepsilon}_{11}{}^{m-1} \tag{17}$$

$$\eta_{rate_{22}} = \eta_{rate_{33}} = \eta_{resin} 2^{m+1} \left[1 - \sqrt{\bar{f}}\right]^{-m} \dot{\varepsilon}_{22}{}^{m-1} \tag{18}$$

$$\eta_{rate_{23}} = \eta_{resin} \left[1 - \sqrt{\bar{f}}\right]^{-m} \dot{\gamma}_{23}{}^{m-1} \tag{19}$$

$$\eta_{rate_{12}} = \eta_{rate_{13}} = \eta_{resin} 2^{-m} \left[\frac{1-\sqrt{\bar{f}}}{1-\sqrt{\bar{f}}}\right]^m \dot{\gamma}_{12}{}^{m-1} \tag{20}$$

where $\bar{f}$ is the fiber volume fraction, L is the fiber length, D is the fiber diameter, and $m = 0.5$ is the power-law exponent.

The ply geometry term is derived from an analytical solution by Kelly [28] and describes the geometry and surface conditions of the ply. A frictionless boundary condition was assumed for the solution since the scope of the curing and compaction has the resin as a fluid. A solution to the work, rearranged by Belnoue et al. [41], was used in this study, as shown in the equations below.

$$\eta_{ply_{11}} = \hat{\eta}_{ply_{12}} = \hat{\eta}_{ply_{13}} = \hat{\eta}_{ply_{23}} = 1 \tag{21}$$

$$\eta_{ply_{22}} = 2\left(\frac{w_0}{h_0}\right)^2 e^{-4\varepsilon_{22}} \tag{22}$$

$$\eta_{ply_{33}} = 2\left(\frac{w_0}{h_0}\right)^2 e^{-4\varepsilon_{33}} \tag{23}$$

where w_0 is the initial width of the ply, and h_0, is the initial height of the ply.

The micro term depends on the phenomenological parameters that govern the flow regimes between an incompressible solid to a compressible solid due to resin bleed out in the fiber direction as a result of fiber locking during compaction [45].

$$\eta_{micro_{11}} = \eta_{micro_{12}} = \eta_{micro_{13}} = \eta_{micro_{23}} = 1 \tag{24}$$

$$\eta_{micro_{22}} = 2\sqrt{\chi_l}e^{\varepsilon_{22}}k\left(\left(\frac{k}{\sqrt{\chi_f}e^{\varepsilon_{22}} - k}\right)^2 + 3\right) \tag{25}$$

$$\eta_{micro_{33}} = 2\sqrt{\chi_l}e^{\varepsilon_{33}}k\left(\left(\frac{k}{\sqrt{\chi_f}e^{\varepsilon_{22}} - k}\right)^2 + 3\right) \tag{26}$$

where k is a stepwise function of the temperature, $\chi_l = 0.63$ is the aspect ratio during locking, and $\chi_f = 0.785$ is the final aspect ratio based on the maximum packing fraction of the composite [46]. The packing in this case is interpreted as the fiber volume fraction.

3.3. Finite Element Modeling of Compaction in Unidirectional Tows and Fiber Tow Gap Samples

The proposed flow-compaction constitutive behavior was validated using the experimental results from unidirectional fiber tows with varying widths and thicknesses (Figure 6a). The same geometry was modeled and simulated using FEA with the proposed visco-hyperelastic behavior (Figure 6b). There were 60 to 200 C3D8T elements depending on the width of the composite, with one element through the thickness for all configurations. The model was constrained in the z-direction to simulate the rigid tool used in the autoclave curing process. The strain was measured using the average deformation of the nodes on the top of the ply.

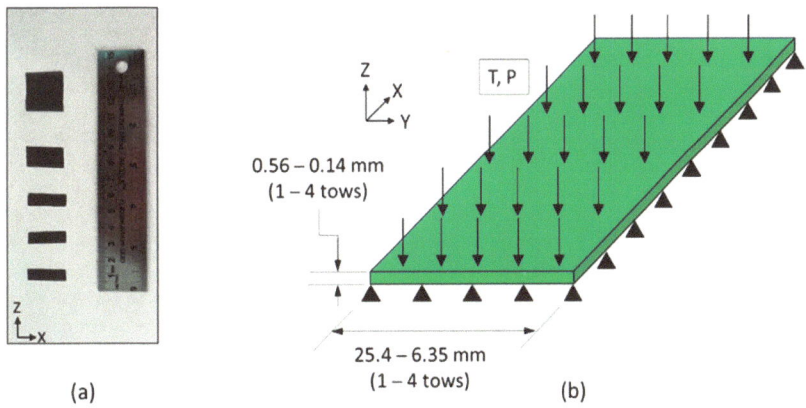

Figure 6. (a) The fully cured samples of the unidirectional experiments and the aspect ratio in order from left to right and labeled with the aspect ratio (ξ): one ply thick ($\xi = 1.0$), two ply thick ($\xi = 0.5$), four ply thick ($\xi = 0.25$), two ply wide ($\xi = 2.0$), and four ply thick ($\xi = 4.0$) (**b**) Schematics of the unidirectional FEA models, where it was varied by the same aspect ratios. (**b**) Schematics of the unidirectional experiments, where the width and thicknesses were varied.

The tow gap configuration was used for understanding the effect of resin bleed out on the cured morphology around the gap region and used the same stacking sequence and geometry as the experimentally manufactured composite (Figure 7). To incorporate the initial sinking of the top 0° ply created via a manually applied roller during the layup, the initial slope was incorporated by connecting the bottom and top 0° plies at the middle of the gap. Symmetrical boundary conditions were used on the x-z plane of the 0° plies towards the gap and one side towards of the *y-z* plane to reduce processing time. To simulate the tool, the bottom of the laminate model was constrained in the *z*-direction. The model consisted of 2088 C3D8RT elements to capture the non-linear deformation. Frictionless contact between the layers of 0° and 90° plies was used. Each ply had a nominal 0.14 mm in thickness with five elements through the thickness. The models used a density for the homogenized composite of 1590 kg/m^3 with a mass scaling of 10 to reduce the amount of time to complete the FEA simulations. No significant rise in kinetic energy was found with the addition of the mass scaling. The material properties for both the IM7 carbon fibers and the 8552 resin were derived from the literature and are utilized in the subroutines [19,35,36]. Geometric non-linearity was activated for both models to accommodate the large deformations during the compaction.

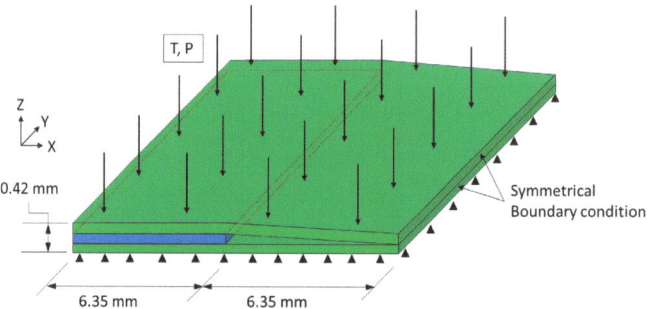

Figure 7. The schematics of the fiber tow gap model with boundary conditions, applied pressure The green plies indicate 0° and the blue ply indicates 90°. Internal resin pressure is applied to the gap in the middle of the 0° plies.

4. Results

4.1. Experimental Quantification of Resin Bleed Out in Unidirectional Prepreg Tows

Six configurations with six samples each of the unidirectional composites were compacted and cured in the first 90 min of the cure cycle (Figure 5) A tow aspect ratio in unidirectional tow was introduced based on the number of tows in the width and through the thickness:

$$\frac{N_w}{N_T} = \zeta \qquad (27)$$

where N_w is the number of plies in the tow width (transverse of the fiber direction), N_T is the number of tows through the thickness of the composite, and ζ is the geometric aspect ratio. For example, a sample with a thickness of four tows and a width of one tow had $\zeta = 0.25$. The experimental samples and corresponding simulations used the proposed fiber tow aspect notation.

Resin bleeds out of the porous fiber bed when heat and pressure are applied [47]. This resin bleed out is a complex process which depends on resin viscosity, local fiber volume fraction, applied autoclave pressure, and pressure distribution throughout the laminate. In the present study, a phenomenological approach was proposed which considered the amount of resin bleed out as a function of time for the given cure cycle. The experimental results of unidirectional tow compaction demonstrated that resin bleed out develops at the initial stages of the manufacturing process where the flow was considered separately in the longitudinal and transverse to the fiber directions. The amount of resin that came out in the respective directions were cut from the composite and weighed. The mass loss in each direction was calculated as a ratio of the measured resin mass.

The average bleed-out mass per unit of surface area, based on the initial dimensions, was measured. While the mass loss for the transverse resin weighs more, when the corresponding area is considered, the resin flowing longitudinally along the fiber direction was responsible for a larger fraction of mass loss per surface area. Longitudinal bleed out occurred through the surface as normal to the fibers, while transverse flow occurred on the side of the prepreg tape—no transverse flow occurred due the tool surface on the bottom and the release film on top. Greater resin loss in the longitudinal direction corresponds to the higher permeability in the fiber direction compared to permeability in the transverse direction [48]. The single tow configuration ($\xi = 1.0$) shows that the longitudinal direction bleeds out more resin mass per unit area, with a coefficient $\bar{\rho}_P{}^{(L)} = 0.400$ mg/mm^2, compared to the transverse direction, with a mass loss coefficient $\bar{\rho}_P{}^{(T)} = 0.0198$ mg/mm^2 at 90 min into the cure cycle. The area of resin around a single tow (Figure 2b) was measured transversely and longitudinally over time and used as a scale to estimate the mass loss over time per unit area over time (Figure 8). Functions representing the mass loss coefficient over time in longitudinal, $\bar{\rho}_P{}^{(L)}(t)$, and transverse, $\bar{\rho}_P{}^{(T)}(t)$, directions were based on the logistic curve functions in Equation (28). Note that only the transverse resin bleed out was present around the surface area of the gap (Figure 3).

$$\bar{\rho}_P{}^{(T)}(t) = \frac{A^{(T)}{}_1}{\left(1 + e^{\frac{A^{(T)}{}_2 - t}{A^{(T)}{}_3}}\right)} \tag{28a}$$

$$\bar{\rho}_P{}^{(L)}(t) = \frac{A^{(L)}{}_1}{\left(1 + e^{\frac{A^{(L)}{}_2 - t}{A^{(L)}{}_3}}\right)} \tag{28b}$$

where $A^{(T)}{}_1 = 0.0195$ mg, $A^{(T)}{}_2 = 33.542$ min, and $A^{(T)}{}_3 = 3.10$ min are transverse fitting parameters and $A^{(L)}{}_1 = 0.0293$ mg, $A^{(L)}{}_2 = 43.881$ min, and $A^{(L)}{}_3 = 11.78$ are the longitudinal parameters.

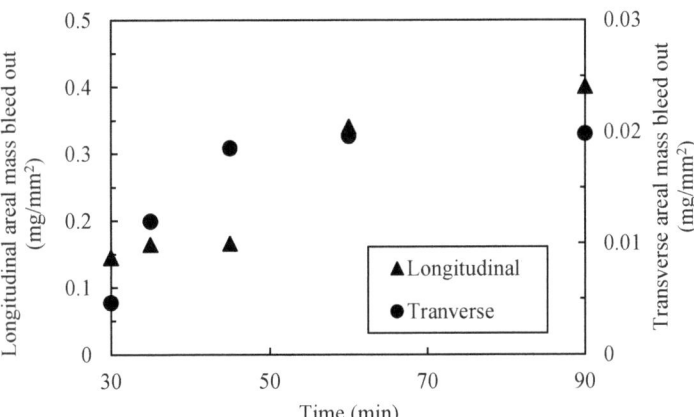

Figure 8. An estimated mass loss fraction over time of the fiber tow gap using experimental data from unidirectional experiments.

The experimental data shows that the resin begins flowing between 20 and 30 min into the cure cycle before plateauing due to decreased permeability and increased viscosity. The resin mass loss data was also used to quantify the fiber volume fraction as a function of the through-thickness strain. An average initial value of the fiber volume fraction of 0.55 was derived from thermogravimetric (TGA) analysis [49]. The initial weight of the resin was determined using the matrix volume fraction based on TGA analysis, and the known densities of the matrix and the fiber was determined via the rule of mixtures. The change in width and length and the known initial volume of a single unidirectional tow

(ξ = 1.0) was used to calculate the through-thickness strain, ε_{33} over time. This allowed to fit the fiber volume with respect to the compaction strain as shown in Figure 9 as an exponential plateau:

$$V_f(\varepsilon_{33}) = V_{f_0} + B_1 * e^{\left(\frac{-B_2}{\varepsilon_{33}}\right)} \tag{29}$$

where $B_1 = 0.0673$ and $B_2 = 0.0084$ are fitting parameters, ε_{33} is the strain in the compaction direction, and V_{f_0} is the initial fiber volume fraction. The fiber volume fraction rapidly increased at the early stages of compaction and plateaus as the permeability decreased and resin viscosity increased. Furthermore, the resin bleed out present at the beginning of the cure cycle influenced the squeeze flow via the evolution of the fiber volume fraction, which affects the anisotropic apparent viscosity in Equations (17)–(20). The fiber volume fraction formulation was measured for the prescribed cure cycle and material system. The same methods can be recreated with different temperature profiles and composite materials. With the resin bleed out and the development of the fiber volume fraction for the current cure cycle established, a single-phase model was made where the resin is not explicitly modeled. Modeling the flow of resin explicitly can be achieved with a higher resolution model. However, this would require computationally expensive representative volume elements to model the microscopic behavior. The present analytical equation allows for an adequate estimate of the evolution of the fiber volume fraction.

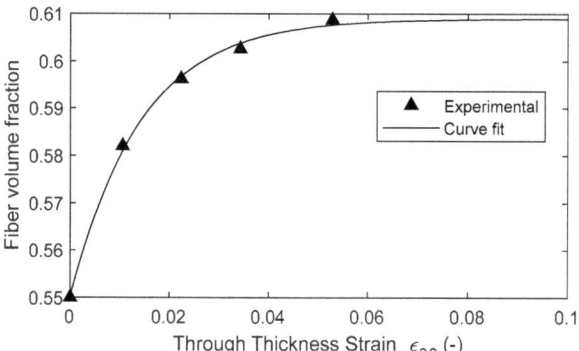

Figure 9. The fiber volume fraction plotted with respect to the strain.

4.2. Modeling and Validation of Compaction Behavior in Unidirectional Tows

The compaction results of the unidirectional tows showed that the thicker samples display more tow spreading in comparison to the tows with a higher width. For a configuration of a single tow (ξ = 1), the maximum compaction strain was shown to be 0.053 at 90 min into the cure cycle. The largest compaction strain, ε_{33} = 0.137, was found for the four tows arranged in the thickness direction (ξ = 0.25), while the smallest strain ε_{33} = 0.012 was found for the four tows arranged in the width direction (ξ = 4.0). The range in the observed compaction strain shows the importance of the tow geometry, and more specifically, the thickness of the ply, on the squeeze flow behavior.

The undeformed and deformed configuration of the unidirectional models are shown in Figure 10. The baseline configuration for the unidirectional models with one ply (ξ = 1.0) shows the thickness decreased and the tow spread at the Y-direction (transverse to the fibers). The modeling results in terms of the through-thickness compaction strains were compared with the experimental measurements to validate the proposed flow-compaction model. The same trend between the experimental and numerical results demonstrated that the transverse spreading increased, with more plies through the thickness, and decreased with more tows across the width.

The through-thickness strains, ε_{33}, in the FEA models of unidirectional tows were quantified using the average displacement on the top y-x surface and the known initial thicknesses. Figure 11a shows the rapid decay of the through-thickness strain, ε_{33}, at 30 min into the hold stage with respect to the aspect ratio. The tow thickness during the compaction simulations and the experimental measurements followed a non-linear visco-hyperelastic curve and decreased with the aspect ratio (Figure 11b). Both the experimental and modeling prediction of compaction agreed closely that up to 90 min of the hold stage was observed, and deviation was noticed at the end of the hold stage. The inclusion of updated fiber volume fraction as a function of the compaction strain (Equation (29)), due to resin bleed out, resulted in improved accuracy of the tow compaction strain (Figure 11a).

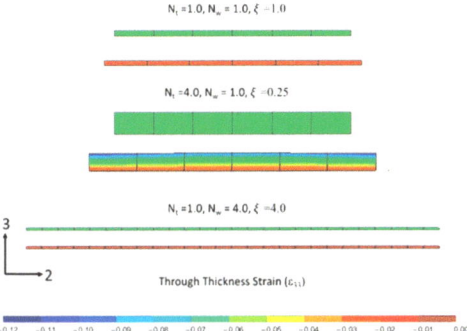

Figure 10. Undeformed and deformed configuration of tows from compaction simulation: a one-tow wide and thick composite (ξ = 1.0), a four-tow thick composite (ξ = 0.25), and a four-tow wide composite (ξ = 4.0).

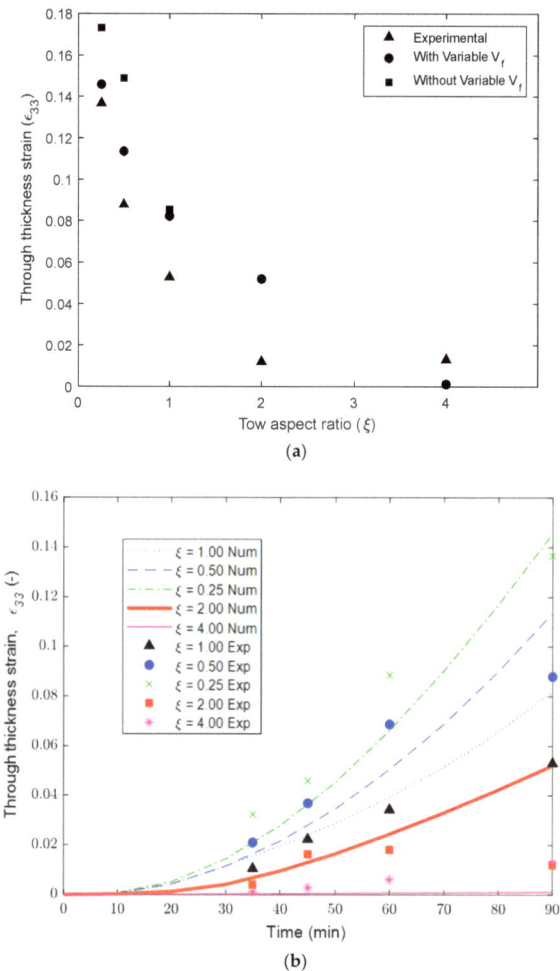

Figure 11. (**a**) Comparison of compaction strain in tows with different aspect ratios. (**b**) Compaction strain evolution during cure time in tows with different aspect ratios.

4.3. Squeeze Flow Simulation of Layup with Embedded Tow Gap

To further validate the proposed compaction constitutive model, the region around the fiber tow gap was analyzed. The initial undeformed configuration of the fiber tow gap model is shown in Figure 12a, which demonstrates the layup compaction during the simulation. The initial FEA simulation did not consider coupling between the squeeze flow and resin bleed out, which is discussed in the following section. The top ply sank into the fiber tow gap, creating a curvilinear triangular void near the edge of the 90° ply. Early in the cure cycle (before 13 min), the gap closed rapidly until the autoclave pressure reached the maximum 0.541 MPa (80 psi). After the initial gap closure, the gap volume decreased slowly until the end of the isothermal hold stage. Tow spreading was observed at the edge of the 90° ply.

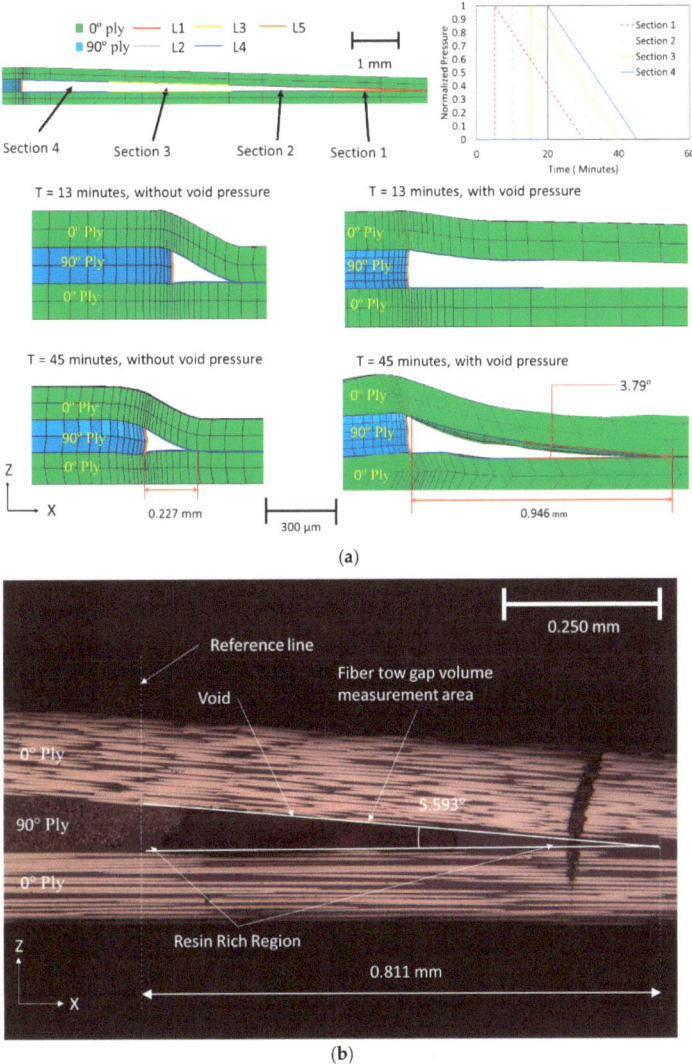

Figure 12. (**a**) The effect of including internal pressure on the size of the void region. (**b**) The geometry of the resin rich region around the initial tow gap in cured composite.

The simulation results with only squeeze flow were compared to the experimentally manufactured sample with the fiber tow gap. The micrograph of the area of interest showed the morphology due to the fiber tow gap in Figure 12b, in which the top 0° ply sank into the gap. The resin-rich region was formed because of resin bleed out from the adjacent plies. The experimental size of the resin-rich region was significantly larger (0.811 mm in length) compared to the one predicted from squeeze flow analysis (0.227 mm in the closed gap), as shown in Figure 12. A much steeper slope in the top 0° ply was more pronounced in the simulation than in the physical piece. The explanation for this discrepancy in the predicted and observed tow morphology is due to resin bleed out, which provides the support against the sinking to a 0° ply. This result illustrates that resin bleed out must be considered to capture the cured defect morphology.

4.4. Effect of Resin Bleed Out during Compaction of Layup with Embedded Tow Gap

The squeeze flow simulation of the fiber tow gap was used to track the changes in the gap volume. The size of the void in squeeze flow analysis was shown to rapidly decrease because of top ply sinking, where the void volume quickly dropped from 0 to 13 min and then plateaued. However, it is evident that resin bleed out occurs simultaneously with the squeeze flow in the early stages of the cure cycle (see Figure 8), which slows down the volume of the gap from decreasing. In order to capture the coupling between the two phenomena, the squeeze flow, and the resin bleed out, the following procedure was used. The volume of the bled-out resin was estimated using the following equation:

$$V_{Res}(t) = \frac{\overline{\rho}_P^{(T)}(t) \cdot SA_{Gap}}{\rho_r} \tag{30}$$

where the density of resin, ρ_r, was 1100 g/cm^3; SA_{Gap} is the initial surface of the fiber tow gap volume $V_{Gap}(t)$, which changes during the compaction; and $\overline{\rho}_P^{(T)}$ is the experimentally determined resin bleed-out areal density for the prescribed cure cycle, as given by Equation (28a).

Over time, the volume of the gap region, $V_{Gap}(t)$, is expected to close, as a result of ply sinking and squeeze flow, while at the same time, the volume of the bled-out resin is increasing, $V_{Res}(t)$. Once the volume of the bled-out resin reaches the volume of the tow gap void where the resin-rich region is formed. The time of gap filling with resin was predicted to occur from the squeeze flow simulation at t_{fill} = 26 min (Figure 13). On the other hand, the estimated experimental volume of the resin pocket in the micrograph is shown as a horizontal line in Figure 13. The volume of the resin-rich pocket was found from a micrograph in Figure 12b and corresponded to the volume of resin in the gap at a slightly later time of 29 min, which indicates the progressive filling and coupling between the squeeze flow and resin bleed out.

The following method was used to describe this flow coupling effect. A pressure, matching the autoclave pressure, was applied on the internal surfaces of the closing gap to capture the presence of bled-out resin as it fills the void. Furthermore, to simulate the progressive filling of the gap, the internal void pressure was activated gradually in partitioned regions 1–4, as shown in Figure 12a, until the predicted moment in time, t_{fill}, when the gap void is supported by the resin. The internal pressure was enabled, starting from the symmetrical end of the model, which was selected based on the proximity of the void surfaces in that region. The pressure was activated at 6, 12, 18, and 24 min, respectively. To capture the stress relaxation effects in the viscoelastic liquid resin, the applied pressure followed a linear decrease over 25 min (Figure 12a). The time was selected to correspond to the overall time of filling of the tow gap region. With the addition of the pressure, the initial sinking of the fiber tow gap was slowed. With the incorporation of the coupling between squeeze flow and resin bleed out, the void filling was predicted to occur at a later time than in squeeze flow analysis (28 min compared to 26 min), which was closer to the moment corresponding to the predicted time of filling (at 29 min), based on the experimental size of the resin-rich pocket (Figure 13). In conjunction with the change in the fiber volume fraction, the resin bleed out was calibrated to the specific material and cure history. The sequential analysis and modeling method presented was calibrated for the current configuration but can be used on other defect features.

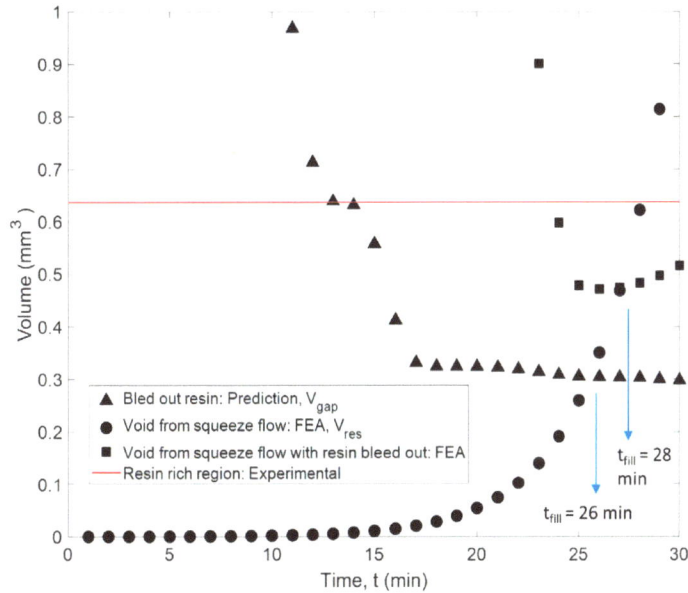

Figure 13. The measured volumes of the model, a theoretical mass flow volume, and the experimental gap volume measured from the micrograph.

4.5. Evaluation of Morphology and Fiber Volume Fraction in Cured Layup with Fiber Tow Gap

With the bled-out resin providing structural support against the sinking ply above the tow gap, the final morphology of the fiber tow gap model with the internal void pressure is shown in Figure 12a. When comparing the two FEA models (squeeze flow with and without the effect of resin bleed out) at 13 min, the void pressure applied prevented the ply from sinking too rapidly. The out-of-plane waviness for the top 0° ply, during the hold stage (45 min), was more gradual and developed over a larger distance, which matched the experimental observation. The out-of-plane angle measured in the top 0° ply is shown in Figure 14, where the simulation captured the evolution of waviness along the length of the gap; however, the experimental results were not as steep as the modeling prediction. The size of the predicted resin-rich region from the coupled simulation had a length of 0.946 mm (compared to the 0.227 mm without the internal void pressure), which agreed with the experimental size of 0.811 mm. The thickness variation in the bottom 0° ply is shown in Figure 15. The measured thickness in the micrograph was found to increase from the left of the reference line towards the center of the composite. In comparison, the numerical model followed a curve which had the thickest part at the reference line and decreased in both directions. The resulting morphology can be utilized in further analysis of the composite, such as through progressive damage analysis, to compare with fabricated laminates in the future [50].

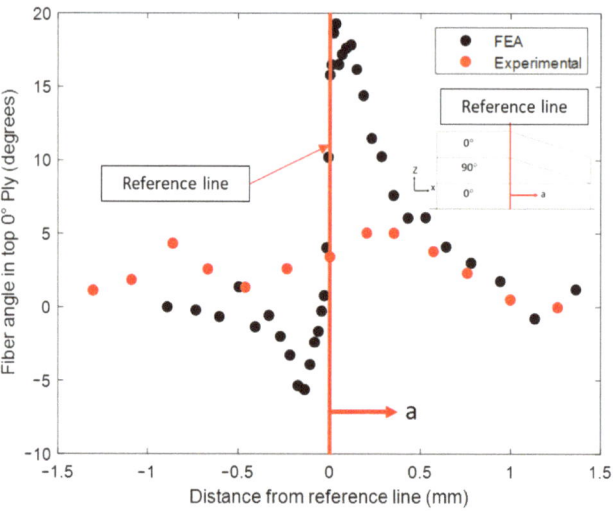

Figure 14. The out-of-plane fiber waviness in the top 0° ply.

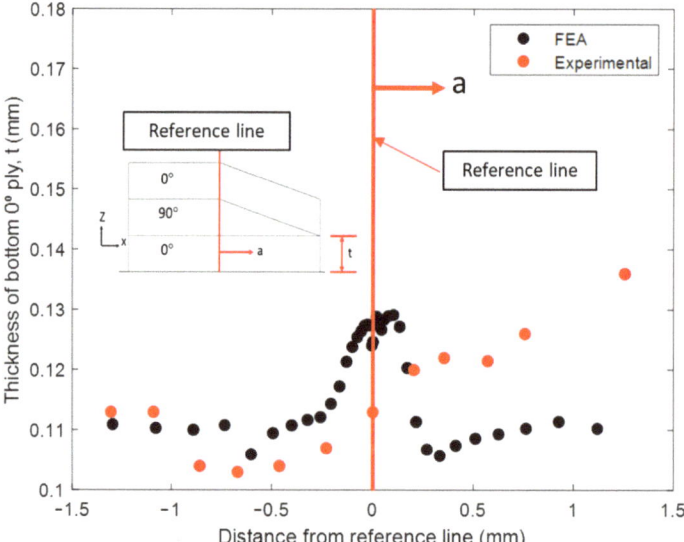

Figure 15. The bottom 0° ply thickness variation.

Predictions of the variable volume fraction of the 90° ply is shown in Figure 16a. A micrograph of the center 90° ply from the cured sample was discretized into rectangular sections where the fiber volume fraction, V_f, was quantified. The micrograph showed a V_f range of 0.40–0.57 with a resin-rich region in the top corner that had a V_f of about 0.10. FEA simulations predicted V_f to a range from 0.60 to 0.70. This study did not consider the initial distribution of the local fiber volume fraction that can account for some of this variation.

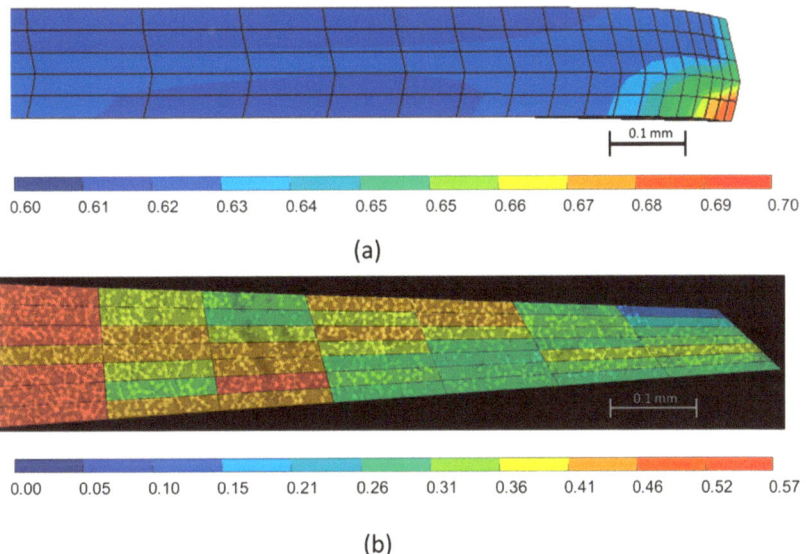

Figure 16. (**a**) The distribution of the fiber volume fraction in the fiber tow gap simulation of the 90° ply. (**b**) The micrograph based measurement of the fiber volume fraction of the 90° ply.

5. Conclusions

The interconnected multiphysics framework, which incorporated chemo-rheological and coupled flow-compaction analyses, was developed to simulate the compaction of AFP prepreg preform in order to predict complex morphology around the region of deposited fiber tow gap. Anisotropic tensorial prepreg viscosity was used to capture tow spreading during the cure cycle. The deformation in unidirectional and fiber tow gap samples was observed to be a result of the coupling between the squeeze flow and resin bleed out. Resin bleed out was quantified by measuring the volumetric flux of the resin from unidirectional prepreg tape. The bled-out resin was found to fill the void in the samples with a single fiber tow gap. To capture this phenomenon in simulation, sequential compaction-flow analysis was proposed and validated. The multi-physics modeling approach connects the squeeze flow simulation with the effect of resin bleed out through the evolution of void closure in the simulation. As the first step, squeeze flow simulation is used to predict the evolution of the gap and estimate the time at which the void is filled with resin. At the second step, the calculated time of void filling was used to activate the internal pressure inside of the void. Instead of modeling the resin explicitly, a one-phase model was proposed, which considers the effect of resin bleed out via the activated internal void pressure. The void pressure is activated progressively to simulate the void filling with resin based on measured resin bleed-out behavior in unidirectional tows. By restricting early sinking of the top ply into the gap, the proposed modeling approach resulted in the cured morphology that captured the non-uniform morphology more accurately around the initial tow gap region.

The modeling framework demonstrated that morphological features develop due to the coupled manufacturing phenomena and cannot be considered in isolation from each other. The phenomena include the chemo-rheology of the resin and visco-hyperelasticity of the laminate in the early stages of curing. The compaction strain was connected to the local fiber volume fraction via tensorial viscosity and it allowed us to consider the interaction between resin bleed out and squeeze flow. The initial deposition gap regions develop into the non-uniform morphological features in the composite, which includes the non-uniform ply thicknesses, fiber waviness, resin-rich regions, and the fiber volume fraction variation. Moreover, the proposed multi-physics modeling framework can be adopted to accommodate various material systems with different chemo-rheological properties.

The proposed sequential modeling approach has been used to predict cured morphology in more complex layups with spatially distributed, staggered fiber tow gaps [51]. The proposed modeling approach can be used to decouple the effect of various effects on the cure morphology in AFP composites, including the material uncertainty, such as initial fiber volume fraction and fiber orientation distributions.

Author Contributions: Conceptualization, S.G.K. and O.G.K.; Methodology, V.R., S.G.K. and O.G.K.; Software, O.G.K.; Validation, V.C.J., V.R. and O.G.K.; Formal analysis, V.C.J.; Investigation, V.C.J., V.R. and S.G.K.; Resources, S.G.K. and O.G.K.; Data curation, S.G.K. and O.G.K.; Writing—original draft, V.C.J. and V.R.; Writing—review & editing, V.C.J., S.G.K. and O.G.K.; Visualization, V.C.J.; Supervision, O.G.K.; Project administration, O.G.K.; Funding acquisition, O.G.K. All authors have read and agreed to the published version of the manuscript.

Funding: This researched has been funded by the NSF Scholarship in STEM (Award # 1833896) and the Rao V. Arimilli and V. Laskshmi Graduate Award in Mechanical and Aerospace Engineering.

Institutional Review Board Statement: Not applicable.

Data Availability Statement: The data presented in this study are available on request from the corresponding author. The data are not publicly available due to the proprietary model.

Acknowledgments: Von Clyde Jamora acknowledges Old Dominion University Composite Modeling and Manufacturing group for their support of the study.

Conflicts of Interest: The authors declare no conflict of interest.

References

1. Lukaszewicz, D.H.-J.A.; Ward, C.; Potter, K.D. The engineering aspects of automated prepreg layup: History, present and future. *Compos. Part B Eng.* **2012**, *43*, 997–1009. [CrossRef]
2. Peeters, D.M.; Lozano, G.G.; Abdalla, M.M. Effect of steering limit constraints on the performance of variable stiffness laminates. *Comput. Struct.* **2018**, *196*, 94–111. [CrossRef]
3. Wu, K.C. Design and analysis of tow-steered composite shells using fiber placement. In Proceedings of the American Society for Composites 23rd Annual Technical Conference, Memphis, TN, USA, 9–11 September 2008.
4. Denkena, B.; Schmidt, C.; Weber, P. Automated Fiber Placement Head for Manufacturing of Innovative Aerospace Stiffening Structures. *Procedia Manuf.* **2016**, *6*, 96–104. [CrossRef]
5. Gregory, E.D.; Juarez, P.D. In-situ thermography of automated fiber placement parts. *AIP Conf. Proc.* **2018**, *1949*, 060005. [CrossRef]
6. Liu, Y.-N.; Yuan, C.; Liu, C.; Pan, J.; Dong, Q. Study on the resin infusion process based on automated fiber placement fabricated dry fiber preform. *Sci. Rep.* **2019**, *9*, 7440. [CrossRef]
7. Jamora, V.C.C.; Rauch, V.M.; Kravchenko, S.; Kravchenko, O. Mechanics of Fiber Tow Compaction for Predicting Defect Morphology in AFP composites. In *AIAA SCITECH 2022 Forum*; American Institute of Aeronautics and Astronautics: Reston, VA, USA, 2022. [CrossRef]
8. Croft, K.; Lessard, L.; Pasini, D.; Hojjati, M.; Chen, J.; Yousefpour, A. Experimental study of the effect of automated fiber placement induced defects on performance of composite laminates. *Compos. Part Appl. Sci. Manuf.* **2011**, *42*, 484–491. [CrossRef]
9. Nguyen, M.H.; Vijayachandran, A.A.; Davidson, P.; Call, D.; Lee, D.; Waas, A.M. Effect of automated fiber placement (AFP) manufacturing signature on mechanical performance of composite structures. *Compos. Struct.* **2019**, *228*, 111335. [CrossRef]
10. Trochez, A.; Jamora, V.C.; Larson, R.; Wu, K.C.; Ghosh, D.; Kravchenko, O.G. Effects of automated fiber placement defects on high strain rate compressive response in advanced thermosetting composites. *J. Compos. Mater.* **2021**, *55*, 4549–4562. [CrossRef]
11. Greenhalgh, E.S. Defects and damage and their role in the failure of polymer composites. In *Failure Analysis and Fractography of Polymer Composites*; Woodhead Publishing Series in Composites Science and Engineering; Woodhead Publishing: Cambridge, UK, 2009; pp. 356–440. [CrossRef]
12. Cairns, D.S.; Mandell, J.F.; Scott, M.E.; Maccagnano, J.Z. Design and manufacturing considerations for ply drops in composite structures. *Compos. Part B Eng.* **1999**, *30*, 523–534. [CrossRef]
13. Wang, J.; Potter, K.D.; Etches, J. Experimental investigation and characterisation techniques of compressive fatigue failure of composites with fibre waviness at ply drops. *Compos. Struct.* **2013**, *100*, 398–403. [CrossRef]
14. Thor, M.; Mandel, U.; Nagler, M.; Maier, F.; Tauchner, J.; Sause, M.G.R.; Hinterhölzl, R.M. Numerical and experimental investigation of out-of-plane fiber waviness on the mechanical properties of composite materials. *Int. J. Mater. Form.* **2020**, *14*, 19–37. [CrossRef]
15. Díaz-Montiel, P.; Ayala, G.G.; Rivera, A.; Mauk, R.; Reiman, C.; Venkataraman, S. Quantification of Material and Geometric Defects Variability in Fiber-Reinforced Composites with Ply Waviness Defects. In *ASME Aerospace Structures, Structural Dynamics, and Materials Conference*; American Society of Mechanical Engineers: New York, NY, USA, 2023.

16. Jamora, V. Compaction and Residual Stress Modeling in Composite Manufactured with Automated Fiber Placement. Master's Thesis, Old Dominion University, Norfolk, VA, USA, 2020. [CrossRef]
17. Jamora, V.C.; Wu, K.C.; Kravchenko, O.G. Residual deformation analysis in composite shell structures manufactured using automated fiber placement. *Compos. Struct.* **2020**, *248*, 112482. [CrossRef]
18. Yenilmez, B.; Senan, M.; Sozer, E.M. Variation of part thickness and compaction pressure in vacuum infusion process. *Compos. Sci. Technol.* **2009**, *69*, 1710–1719. [CrossRef]
19. Lee, W.I.; Loos, A.C.; Springer, G.S. Heat of Reaction, Degree of Cure, and Viscosity of Hercules 3501-6 Resin. *J. Compos. Mater.* **1982**, *16*, 510–520. [CrossRef]
20. Costa, V.A.F.; Sousa, A.C.M. Modeling of flow and thermo-kinetics during the cure of thick laminated composites. *Int. J. Therm. Sci.* **2003**, *42*, 15–22. [CrossRef]
21. Kermani, N.N.; Gargitter, V.; Simacek, P.; Advani, S.G. Gap filling mechanisms during the thin ply Automated Tape Placement process. *Compos. Part Appl. Sci. Manuf.* **2021**, *147*, 106454. [CrossRef]
22. Belnoue, J.P.-H.; Mesogitis, T.; Nixon-Pearson, O.J.; Kratz, J.; Ivanov, D.S.; Partridge, I.K.; Potter, K.D.; Hallett, S.R. Understanding and predicting defect formation in automated fibre placement pre-preg laminates. *Compos. Part Appl. Sci. Manuf.* **2017**, *102*, 196–206. [CrossRef]
23. Nixon-Pearson, O.J.; Belnoue, J.P.H.; Ivanov, D.S.; Hallett, S.R. The Compaction Behaviour of Uncured Prepregs: 20th International Conference on Composite Materials, ICCM 2015. 2015. Available online: http://www.scopus.com/inward/record.url?scp=85044681739&partnerID=8YFLogxK (accessed on 30 May 2022).
24. Limbert, G.; Middleton, J. A transversely isotropic viscohyperelastic material: Application to the modeling of biological soft connective tissues. *Int. J. Solids Struct.* **2004**, *41*, 4237–4260. [CrossRef]
25. Rogers, T.G.; O'neill, J. Theoretical analysis of forming flows of fibre-reinforced composites. *Compos. Manuf.* **1991**, *2*, 153–160. [CrossRef]
26. Seon, G.; Nikishkov, Y.; Makeev, A. A numerical method based on pore-pressure cohesive zone modeling for simulation of debulking in resin-saturated composite prepregs. *Int. J. Numer. Methods Eng.* **2022**, *123*, 2791–2813. [CrossRef]
27. Simacek, P.; Kermani, N.N.; Gargitter, V.; Advani, S.G. Role of resin percolation in gap filling mechanisms during the thin ply thermosetting automated tape placement process. *Compos. Part Appl. Sci. Manuf.* **2022**, *152*, 106677. [CrossRef]
28. Kelly, P.A. A viscoelastic model for the compaction of fibrous materials. *J. Text. Inst.* **2011**, *102*, 689–699. [CrossRef]
29. Pipes, R.B. Anisotropic Viscosities of an Oriented Fiber Composite with a Power-Law Matrix. *J. Compos. Mater.* **1992**, *26*, 1536–1552. [CrossRef]
30. He, T.; Zhang, H.; Zhang, K. A smoothed finite element approach for computational fluid dynamics: Applications to incompressible flows and fluid–structure interaction. *Comput. Mech.* **2018**, *62*, 1037–1057. [CrossRef]
31. Saquib, M.N.; Larson, R.; Sattar, S.; Li, J.; Kravchenko, S.; Kravchenko, O. Experimental Validation of Reconstructed Microstructure via Deep Learning in Discontinuous Fiber Platelet Composite. *J. Appl. Mech.* **2023**, *91*, 041004. [CrossRef]
32. Larson, R.; Horque, R.; Jamora, V.C.; Li, J.; Kravchenko, S.G.; Kravchenko, O.G. Prediction of Local Fiber Orientation State in Prepreg Platelet Molded Composites via Deep Convolutional Neural Network. In *AIAA SCITECH 2022 Forum*; AIAA: Reston, VA, USA, 2022.
33. Larson, R.A.; Hoque, R.; Jamora, V.; Li, J.; Kravchenko, S.; Kravchenko, O. Hyperparameters Effect in Deep Convolutional Neural Network Model on Prediction of Fiber Orientation Distribution in Prepreg Platelet Molded Composites. In *AIAA SCITECH 2022 Forum*; American Institute of Aeronautics and Astronautics: Reston, VA, USA, 2021. [CrossRef]
34. Chapman, T.J.; Gillespie, J.W.; Pipes, R.B.; Manson, J.-A.E.; Seferis, J.C. Prediction of Process-Induced Residual Stresses in Thermoplastic Composites. *J. Compos. Mater.* **1990**, *24*, 616–643. [CrossRef]
35. Hubert, P.; Johnston, A.; Poursartip, A.; Nelson, K. Cure kinetics and viscosity models for Hexcel 8552 epoxy resin. In Proceedings of the International SAMPE Symposium and Exhibition, Long Beach, CA, USA, 6–10 May 2001; Volume 46, pp. 2341–2354.
36. Ng, S.J.; Boswell, R.; Claus, S.J.; Arnold, F.; Vizzini, A. Degree of Cure, Heat of Reaction, and Viscosity of 8552 and 977-3 HM Epoxy Resin. Naval Air Warfare Center Aircraft Div Patuxent River MD. 2000. Available online: https://apps.dtic.mil/sti/citations/ADA377439 (accessed on 1 April 2021).
37. Kravchenko, O.G.; Kravchenko, S.G.; Pipes, R.B. Chemical and thermal shrinkage in thermosetting prepreg. *Compos. Part Appl. Sci. Manuf.* **2016**, *80*, 72–81. [CrossRef]
38. Kravchenko, O.G.; Kravchenko, S.G.; Pipes, R.B. Cure history dependence of residual deformation in a thermosetting laminate. *Compos. Part Appl. Sci. Manuf.* **2017**, *99*, 186–197. [CrossRef]
39. Kravchenko, O.G.; Kravchenko, S.G.; Casares, A.; Pipes, R.B. Digital image correlation measurement of resin chemical and thermal shrinkage after gelation. *J. Mater. Sci.* **2015**, *50*, 5244–5252. [CrossRef]
40. Kravchenko, O.; Li, C.; Strachan, A.; Kravchenko, S.; Pipes, B. Prediction of the chemical and thermal shrinkage in a thermoset polymer. *Compos. Part Appl. Sci. Manuf.* **2014**, *66*, 35–43. [CrossRef]
41. Belnoue, J.P.-H.; Nixon-Pearson, O.J.; Ivanov, D.; Hallett, S.R. A novel hyper-viscoelastic model for consolidation of toughened prepregs under processing conditions. *Mech. Mater.* **2016**, *97*, 118–134. [CrossRef]
42. Pikey, W. Stress and Strain. In *Formulas for Stress, Strain, and Structural Matrices*; John Wiley & Sons, Ltd.: Hoboken, NJ, USA, 2004; pp. 89–147. [CrossRef]

43. Bonet, J.; Burton, A.J. A simple orthotropic, transversely isotropic hyperelastic constitutive equation for large strain computations. *Comput. Methods Appl. Mech. Eng.* **1998**, *162*, 151–164. [CrossRef]
44. Favaloro, A.J.; Sommer, D.E.; Denos, B.R.; Pipes, R.B. Simulation of prepreg platelet compression molding: Method and orientation validation. *J. Rheol.* **2018**, *62*, 1443–1455. [CrossRef]
45. Belnoue, J.P.-H.; Nixon-Pearson, O.; Ivanov, D.; Hallet, S.R. Numerical and Experimental Investigation of Prepreg Compaction for Defect Formation Mechanisms. *Des. Manuf. Appl. Compos.* **2014**. Available online: https://dpi-proceedings.com/index.php/dmac2014/article/view/21306 (accessed on 10 August 2023).
46. Messiry, M.E. Theoretical analysis of natural fiber volume fraction of reinforced composites. *Alex. Eng. J.* **2013**, *52*, 301–306. [CrossRef]
47. Hubert, P.; Poursartip, A. A Review of Flow and Compaction Modelling Relevant to Thermoset Matrix Laminate Processing. *J. Reinf. Plast. Compos.* **1998**, *17*, 286–318. [CrossRef]
48. Stadtfeld, H.C.; Erninger, M.; Bickerton, S.; Advani, S.G. An Experimental Method to Continuously Measure Permeability of Fiber Preforms as a Function of Fiber Volume Fraction. *J. Reinf. Plast. Compos.* **2002**, *21*, 879–899. [CrossRef]
49. Marlett, K.; Ng, Y.; Tomblin, J. *Hexcel 8552 IM7 Unidirectional Prepreg 190 Gsm & 35% RC Qualification Material Property Data Report*; NIAR: Wichita, KS, USA, 2011; pp. 1–238.
50. Larson, R.; Bergan, A.; Leone, F.; Kravchenko, O.G. Influence of stochastic adhesive porosity and material variability on failure behavior of adhesively bonded composite sandwich joints. *Compos. Struct.* **2023**, *306*, 116608. [CrossRef]
51. Ravangard, A.; Jamora, V.C.; Bhagatji, J.D.; Kravchenko, O. Origin and Significance of Non-Uniform Morphology in AFP Composites. In Proceedings of the American Society for Composites—38th Technical Conference, Greater Boston, MA, USA, 17–20 September 2023. [CrossRef]

Disclaimer/Publisher's Note: The statements, opinions and data contained in all publications are solely those of the individual author(s) and contributor(s) and not of MDPI and/or the editor(s). MDPI and/or the editor(s) disclaim responsibility for any injury to people or property resulting from any ideas, methods, instructions or products referred to in the content.

Article

Preparation Scheme Optimization of Thermosetting Polyurethane Modified Asphalt

Min Sun [1,*], Shuo Jing [1], Haibo Wu [2], Jun Zhong [3], Yongfu Yang [3,*], Ye Zhu [4] and Qingpeng Xu [5]

1. School of Transportation Engineering, Shandong Jianzhu University, Jinan 250101, China; jingshuo99@163.com
2. MCC Road & Bridge Construction Co., Ltd., Jinan 250031, China; 13561669518@163.com
3. Shandong Transportation Institute, Jinan 250031, China; zhongjun@sdjtky.cn
4. Shandong Provincial Communications Planning and Design Institute Co., Ltd., Jinan 250031, China; zhuye1223@163.com
5. Jinan Urban Construction Group Co., Ltd., Jinan 250031, China; xuqingpeng2005@163.com
* Correspondence: 15253170143@163.com or sunmin20@sdjzu.edu.cn (M.S.); yangyongfu001@163.com (Y.Y.)

Abstract: To solve the issue of the poor temperature stability of conventional modified asphalt, polyurethane (PU) was used as a modifier with its corresponding curing agent (CA) to prepare thermosetting PU asphalt. First, the modifying effects of the different types of PU modifiers were evaluated, and the optimal PU modifier was then selected. Second, a three-factor and three-level L9 (3^3) orthogonal experiment table was designed based on the preparation technology, PU dosage, and CA dosage to prepare the thermosetting PU asphalt and asphalt mixture. Further, the effect of PU dosage, CA dosage, and preparation technology on the 3d, 5d, and 7d splitting tensile strength, freeze-thaw splitting strength, and tensile strength ratio (TSR) of the PU asphalt mixture was analyzed, and a PU-modified asphalt preparation plan was recommended. Finally, a tension test was conducted on the PU-modified asphalt and a split tensile test was performed on the PU asphalt mixture to analyze their mechanical properties. The results show that the content of PU has a significant effect on the splitting tensile strength of PU asphalt mixtures. When the content of the PU modifier is 56.64% and the content of CA is 3.58%, the performance of the PU-modified asphalt and mixture is better when prepared by the prefabricated method. The PU-modified asphalt and mixture have high strength and plastic deformation ability. The modified asphalt mixture has excellent tensile performance, low-temperature performance, and water stability, which meets the requirements of epoxy asphalt and the mixture standards.

Keywords: road engineering; thermosetting polyurethane asphalt; orthogonal experiment; preparation plan

1. Introduction

As the main form of road surface in China, asphalt pavement has distinct advantages, such as a good driving comfort, excellent road performance, strong environmental adaptability, and significant economic benefits [1,2]. As a binder for asphalt pavement, asphalt has an excellent bonding performance. However, with the gradual improvements in road construction at all levels in China, the traffic volume and vehicle axle load have sharply increased, along with road disease also gradually increasing as a result, meaning that the mechanical properties of asphalt mixture can no longer meet the developmental needs for high-quality roads. Therefore, modified asphalt technologies have been proposed [3–5]. At present, modified asphalt technologies, such as SBS modification, rubber powder modification, graphene modification, and composite modification have been rapidly developed. Modified asphalt technology can improve the water damage resistance, as well as the high and low temperature performances of asphalt [6–8]; however, there are still issues regarding temperature stability and durability [9,10]. Epoxy asphalt has the characteristics of a thermosetting material, with excellent strength, stiffness, durability, and temperature stability, but there are limitations in terms of its high price, harsh construction conditions, and insufficient toughness [11–13]. Therefore, it is imperative to develop novel modified asphalt materials.

PU is a green modified asphalt material, and is a multi-block polymer that can adjust the type and proportion of isocyanate, polyol, and other additives to design different structures. With its high degree of designability, it can be utilized used across various industrial production fields, such as plastics, films, adhesives, synthetic leather, and elastomers [14–16]. Currently, the primary application of PU in the domain of road pavement mainly focuses on two research directions, with one of them involving using PU to replace asphalt as a binder. The prepared PU mixture has high strength, high-temperature performance, low-temperature performance, excellent water stability, and fatigue resistance. It can substantially decrease the layer thickness of the pavement structure, prolong the service life of the pavement, save energy, reduce emissions, and is environmentally friendly [17–19]. The second involves using PU as the modifier for the preparation of PU asphalt. Due to the flexible formulation, diverse product forms, and excellent performance of the PU materials, the types and characteristics of PU-modified asphalt differ. The compatibility, storage stability, and economy of PU as an asphalt modifier must therefore be considered [20–22].

To explore the compatibility between polyurethane and asphalt and its influence on asphalt performance, in recent years, researchers have explored the modification mechanisms and compatibility of PU asphalt through molecular simulation, infrared spectroscopy (FI-TR), and fluorescence microscopy. A study on molecular dynamics has shown that the PU modifier has the best effect on asphalt modification at 140 °C [23]. The PU modifier exhibits a positive direct effect on the swelling and dispersion between aromatics–aromatics and saturated–saturated components in asphalt, decreasing the congregation of the two molecules. At the same time, the PU content and shear temperature will also affect the compatibility of asphalt and PU. The characteristic peak of asphalt aging in FT-IR spectra was found to have weakened. It was considered that the isocyanate group reacted with active hydrogen in asphalt, leading to cross-linking, and the PU chain was subsequently broken down and degraded after aging, which further enhanced the compatibility of PU and asphalt [24]. The microstructure of PU composite modified asphalt can be investigated with scanning electron microscopy and FT-IR [25], which are both helpful to understand the modification mechanism and corresponding properties of this asphalt. The results of dynamic thermodynamic analysis and differential scanning calorimetry showed that the addition reaction between the polyisocyanate in the polyurethane prepolymer and the aromatic compounds in base asphalt improved the performance of asphalt [20].

At present, PU asphalt is mainly composed of thermoplastic PU (TPU) asphalt and composite modified polyurethane asphalt. In the field of thermoplastic PU (TPU) asphalt, Zhang et al. [26] studied a new type of PU thermoplastic (PUTE) as an asphalt modifier and conducted a series of experimental studies to explore the effect of PU content on asphalt performance. The results indicated that PUTE can significantly enhance the high-temperature, low-temperature, and mechanical performance of asphalt, and the optimum content was found to be 11%. Salas et al. [27] used PU foam to prepare PU-modified asphalt and found that, in comparison to conventional polymer-modified asphalt mixtures, PU-modified asphalt has improved stability and low-temperature deformability. Jin et al. [28] used TPU as a modifier to be incorporated into base asphalt and found that TPU-modified asphalt had better properties, such as ductility, softening point, needle penetration, and rotational viscosity after the experimental analysis of the chemical, microstructural, and rheological properties of the modified asphalt, and pointed out that polyester-based TPU modifiers have better high-temperature properties than polyether-based TPU modifiers.

In the field of composite modified polyurethane asphalt, Zhang et al. [29] prepared an epoxy/PU (EPU) modified asphalt using PU and epoxy resin. The mechanical properties and microstructure of EPU-modified asphalt were evaluated by FT-IR testing, tensile tests, and scanning electron microscopy. It was found that EPU has good mechanical strength, flexibility, and storage stability. Yan et al. [30] prepared bone glue/polyurethane composite modified asphalt (CMA). The relationship between the modifier content and CMA's conventional properties and rheological properties was determined using the response surface method. It was found that the low-temperature crack resistance, water stability, and dynamic stability of CMA mixtures were enhanced. Jin et al. [21] analyzed the rheological properties, chemical properties, and microstructure of PU asphalt by adding a mixture of PU and RA rock asphalt. The results showed that PU composite modified asphalt had a good performance in terms of permeability, ductility, softening point, and rotational viscosity. Moreover, the isocyanate reacted with phenol and carboxylic acid in modified asphalt to further improve the performance.

Most current research on PU asphalt is either on a certain modifier or composite modification, and there are a lack of comparisons among various modifiers. There are still issues that remain from its utilization, such as high construction temperatures and poor temperature stability [31]. Therefore, thermosetting polyurethane modified asphalt has been studied by several scholars. Zhang et al. [32] determined the appropriate formula for a thermosetting PU (TS-PU) modified binder through a series of laboratory tests, studied the comprehensive properties of asphalt modifiers containing various TS-PU contents, and determined the optimum content of modifiers. It has also been shown that a TS-PU modified modifier is superior to epoxy modifiers in terms of the high-temperature rutting resistance and tensile strength, flexibility, and cost. Jia [33] evaluated the performance of a thermosetting PU (TS-PU) asphalt mixture and a TPU asphalt mixture through conducting a series of tests. The results showed that the two kinds of PU asphalt mixture have better flexibility and low-temperature crack resistance than epoxy asphalt mixtures. The performance difference between these PU asphalt mixtures is also significant, indicating that the PU asphalt modifier is of great significance for obtaining PU asphalt pavement with customized performance. Yang et al. [22] also found that thermosetting PU can enhance the low-temperature flexibility through BBR tests. In addition, although the tensile strength of TSPU is lower than that of epoxy asphalt, when the TSPU content exceeds 40%, the fracture energy is comparable to that of epoxy asphalt through direct tensile tests. In summary, thermosetting PU-modified asphalt can improve the temperature stability, flexibility, and mechanical properties of asphalt, and has a greater development value.

Therefore, this study conducted a comparative analysis of the tests of each component material of thermosetting PU asphalt and optimized the best PU modifier. Based on the three-factor and three-level orthogonal test of the preparation process, PU content, and CA content, a PU asphalt mixture was prepared. Then, the 3d, 5d, and 7d splitting tensile strength, freeze-thaw splitting strength, and TSR of the PU asphalt mixture were all tested. The effects of various factor levels and combinations on the basic mechanical properties of PU asphalt mixture were analyzed, and the best preparation scheme of PU-modified asphalt was recommended, which provides a reference for future research on thermosetting PU asphalt.

2. Materials and Methods

2.1. Materials

The base asphalt was selected as the raw material for PU asphalt, and four kinds of PU and two kinds of CA were used to modify the original asphalt by single mixing. The technical indexes of the base asphalt are shown in Table 1.

Table 1. Base asphalt technical indicators.

Performance Indicators	Penetration (25 °C)/0.1 mm	Ductility (10 °C)/cm	Softening Point/°C	Flash Point/°C	Wax Content/%	Density (15 °C)/(g·cm^{-3})
Test results	65.3	56	48	330	203	1.036

The aggregate was provided by the Shandong Provincial Transportation Planning and Design Institute, composed of limestone gravel and manufactured sand, which was mainly divided into 0–4 mm, 4–6 mm, 6–12 mm, 12–16 mm, and mineral powder. The aggregate indexes of each grade are displayed in Table 2. The AC-13 PU asphalt mixture gradation composition is shown in Figure 1.

Table 2. Aggregate density summary.

Aggregate Size	Apparent Relative Gravity/γa	Relative Density of Bulk Volume/γb	Water Absorption (%)	Asphalt Absorption Coefficient	Effective Relative Density (g/cm³)
Limestone 12–16 mm	3.315	3.227	1.10	0.38	3.307
Limestone 6–12 mm	2.935	2.823	1.33	0.59	2.895
Limestone 4–6 mm	2.773	2.721	1.68	0.79	2.715
Limestone 0–4 mm	2.692	2.573	1.80	0.92	2.687
Mineral powder	2.636	2.636	-	-	2.636

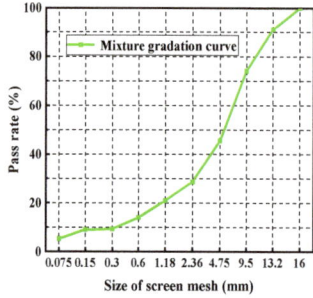

Figure 1. Mixture gradation curve.

Four kinds of PU modifiers produced by the Wanhua Group were selected to prepare PU-modified asphalt with CA. These modifiers were labelled as GXJ-1, GXJ-2, GXJ-3, and GXJ-4. The material composition characteristics of the modifier and its corresponding CA are shown in Table 3.

Table 3. PU modifier raw material description.

Numbering	Type	CA Type	Tensile Strength/MPa	Breaking Elongation/%
GXJ-1	PU prepolymer	Small molecule polyol (CA-1)	11.9	330
GXJ-2	PU prepolymer	Small molecule polyol (CA-1)	20.4	183
GXJ-3	Castor oil	Isocyanate (CA-2)	15.2	80
GXJ-4	Polyether polyol	Isocyanate (CA-2)	18.6	245

The volumetric properties of mixtures are shown in Table 4.

Table 4. AC-13 asphalt mixture component material.

Numbering	Gradation Type	PU Type	OAC/%	VV/%	VMA/%	VFA/%
PU asphalt	AC-13	GXJ-1	6.1	4.1	14.9	72.5
70#A asphalt	AC-13	-	4.5	4.3	15.3	71.89

2.2. Test Method

2.2.1. Splitting Strength and Height Test of PU Specimens

At room temperature, the PU was mixed with the CA, and the standard Marshall test specimen—with a diameter of 101.6 mm and a height of 63.5 mm—was made as according to the design grading requirements [34]. After the mixture was mixed, Marshall compaction was conducted. The number of compactions made was fifty times for each side, and the compaction temperature was 120 °C. After the specimens were cured for 3d, 5d, and 7d at room temperature, the splitting tensile test was conducted with reference to the test regulations [34]. The 3d, 5d, and 7d splitting tensile strengths of the different polyurethane modifier materials were assessed as performance indicators for evaluating the polyurethane materials.

The PU in the mixture reacts with the water molecules in the air to form CO_2 gas, which cannot be discharged due to the low porosity, causing the volumetric expansion of the specimen. After 7 days of curing, the height of the PU specimen was measured with a vernier calliper. In comparison to the height of the initial specimen, Figure 2 shows the measurement method for the PU specimen height. The change in height can reflect the expansion phenomenon of the specimen. When $H' = H$, there is no bulge in the specimen. When $H' > H$, the bulge of the specimen will increase its height.

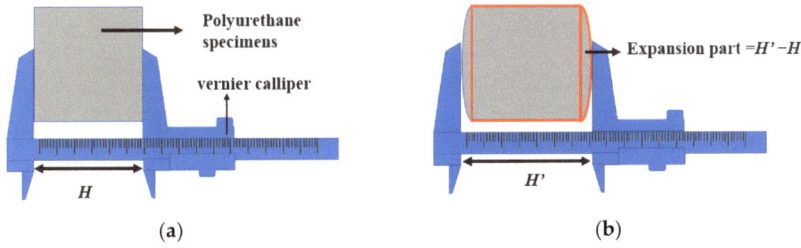

Figure 2. (a) Before specimen expansion; (b) After specimen expansion.

2.2.2. Optimization of PU

Four kinds of PU modifiers were modified through single blending to prepare the different modified asphalts. First, base asphalt and PU (GXJ-1, GXJ-2, GXJ-3, and GXJ-4) were preheated to 120 °C. Second, PU (mass fractions of 10%, 30%, and 50%) was added to the base asphalt, placed in a heating plate, and heated to 120 °C, and then sheared for 10 min with high-speed shear at 4000 rpm. After shearing was completed, CA-1 and CA-2 (mass fractions of 12% and 6%, respectively) were slowly added to continue shearing for 10 min to prepare the PU asphalt. After the preparation was completed, it was mixed with an aggregate to form Marshall specimens. The forming process is shown in Figure 3. The 3d, 5d, and 7d splitting tensile strengths of the different PU-modified asphalts were assessed; the splitting test is shown in Figure 4. A PU modifier was then selected.

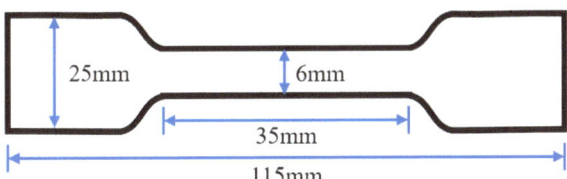

Figure 3. Marshall specimen forming process.

Figure 4. Marshall specimen splitting strength test.

2.2.3. Orthogonal Experiment of PU-Modified Asphalt

After optimizing a PU modifier, an L9 (3^3) orthogonal test table based on the preparation method, PU content, and CA content was designed to prepare the PU asphalt and mixture. The influences of PU content, CA content, and the preparation method on the 3d, 5d, and 7d splitting tensile strength, freeze-thaw splitting strength, and TSR of PU asphalt mixture were analyzed. The orthogonal factors and horizontal are listed in Table 5, the L9 (3^3) orthogonal test table is listed in Table 6, and the preparation process of PU asphalt is laid out in Table 7.

Table 5. Orthogonal factors and horizontal.

Orthogonal Factors	PU Content (A)	Preparation Method (B)	CA Content (C)
1	40%	Prefabrication method	3%
2	50%	Post-doping method	6%
3	60%	Pre-reaction method	12%

Table 6. Orthogonal design table.

Numbering	Orthogonal Composite	A	B	C
1	A1B1C3	40	1	12
2	A1B2C1	40	2	3
3	A1B3C2	40	3	6
4	A2B1C1	50	1	3
5	A2B2C2	50	2	6
6	A2B3C3	50	3	12
7	A3B1C2	60	1	6
8	A3B2C3	60	2	12
9	A3B3C1	60	3	3

Table 7. Introduction to the PU asphalt preparation process.

Preparation Method	Description
Prefabrication method	The base asphalt is sheared with polyurethane. After the shearing process, the CA is added and then mixed with the aggregate.
Post-doping method	The base asphalt is sheared with the CA; after the shearing process, it is mixed with the aggregate, and polyurethane is added during the mixing process.
Pre-reaction method	The polyurethane is sheared with the CA. Following the shearing process, the base asphalt is added and then mixed with the aggregate.

To further optimize the experimental conditions for the preparation of polyurethane asphalt, the effects of polyurethane content, curing agent content, and the preparation process on the mechanical properties of polyurethane asphalt were studied using the response surface optimization method. The three-factor three-level response surface test was designed in the Design-Expert.V8.0.6.1 software to obtain the best preparation scheme for polyurethane-modified asphalt.

2.2.4. Tensile Test

In accordance with the GB/T 528-1998 standard, a tensile performance test [35] was conducted. After preparing the polyurethane asphalt, it was molded into a tensile sample in a PTFE mold and cured at room temperature for one week. The tensile test of the specimen was conducted using a WDL-2000 electronic tensile tester. The tensile strength and the elongation at break of the polyurethane asphalt were assessed using a WDL-2000 machine at a tensile rate of 500 mm/min and a temperature of 23 °C. The dimensions of the tensile specimen are shown in Figure 5. The entire experimental process was conducted as shown in Figure 6.

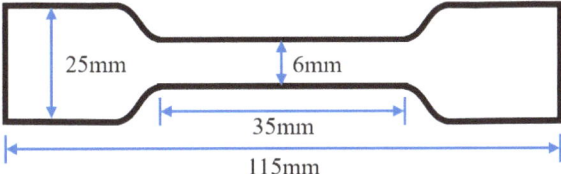

Figure 5. Tensile test specimen size.

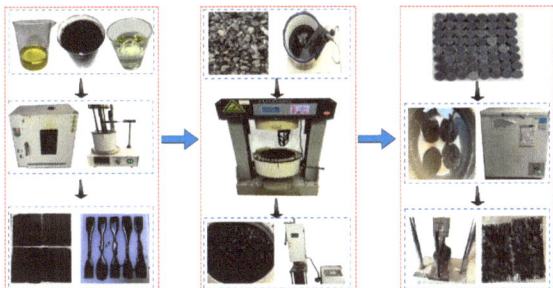

Figure 6. Test process flow chart.

3. Results and Discussion

3.1. PU-Modified Asphalt Optimization

The 3d, 5d, and 7d splitting tensile strength results of the different PU modifier materials are shown in Figure 7, and the changes in the expansion height of the specimen are indicated in Figure 8.

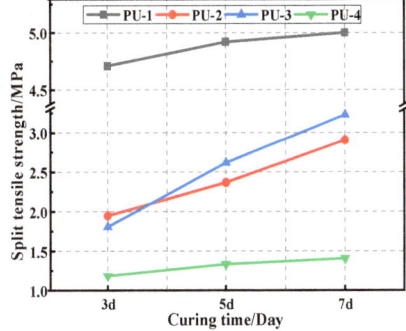

Figure 7. Splitting strength results.

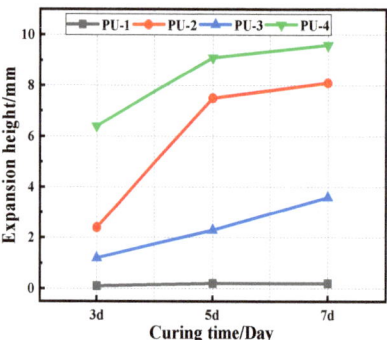

Figure 8. Expansion heights of the specimens.

It is clear from Figure 7 that the splitting strength of polyurethane mixtures increases accordingly with the increases in the curing time. This indicates that the longer the curing time of the polyurethane mixture, the more complete the curing reaction of the polyurethane, resulting in a higher strength of the mixture. The strength and formation rate of the four PU specimens are in the following order: PU-1 > PU-3 > PU-2 > PU-4, indicating that the PU-1 mixture has the highest strength and a rapid curing reaction. The height of expansion of the same polyurethane mixture with increasing curing time is shown in Figure 8. The expansion heights of the specimens are in the following order: PU-4 > PU-2 > PU-3 > PU-1, indicating that with the curing reaction, the polyurethane mixture continuously releases CO_2 gas. This makes the specimen expand, and with the volumetric expansion of PU-1 being the smallest, this was deemed as the best in comparison to other polyurethane mixtures.

The PU content was determined to be 10%. Under the condition of 12% CA, the splitting tensile strengths of PU asphalt with different modifier materials at 3d, 5d, and 7d were assessed. The results of this test are displayed in Figure 9.

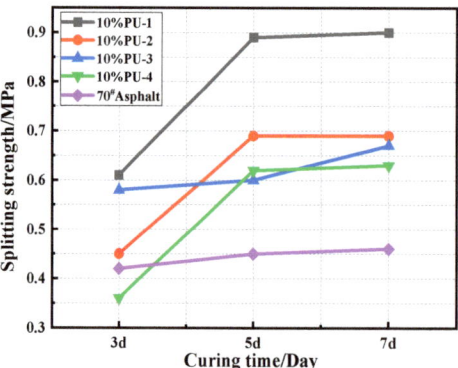

Figure 9. Splitting strengths of the different PU asphalt mixtures.

The data in Figure 9 indicates that the strength of the polyurethane asphalt specimens increases with the corresponding increases in the polyurethane content. The 7d splitting strength of the PU asphalt specimens is in the following order: PU-1 > PU-2 > PU-3 > PU-4. The 7d strength of PU-1, PU-2, PU-3, and PU-4 asphalt with 10% PU content was found to be 95%, 50%, 45%, and 37% higher than that of base asphalt, respectively, showing that PU can dramatically increase the strength of the asphalt specimens. PU-1 was found to have the highest strength and the greatest modification effectiveness.

The 3d, 5d, and 7d splitting tensile strengths of PU asphalt with different dosages and CA dosages were also evaluated. The test data are displayed in Figure 10.

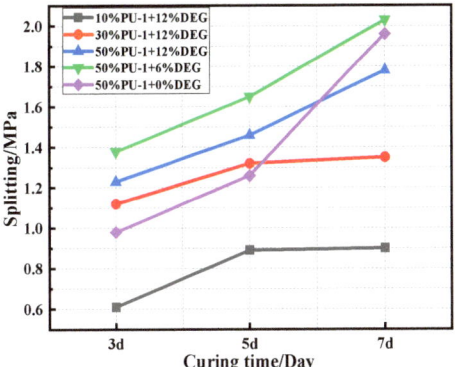

Figure 10. Splitting strength of the PU asphalt mixtures with various contents.

It is clear from Figure 10 that the splitting strength of 50% PU content is 31% and 97% higher than that of 30% and 10% PU, respectively, indicating that higher contents of PU lead to higher mechanical strengths. With improvements in the CA content, the splitting strength of the PU asphalt specimens exhibited a trend of initially increasing and then decreasing, indicating that the high content of CA may lead to the early curing of PU asphalt in the forming specimen along with a loss in strength.

In summary, the quantities of CA and PU affects the properties of PU asphalt, and the preparation method needs to be further optimized. Therefore, the optimal preparation scheme of PU asphalt was determined by designing the three-factor and three-level orthogonal test of the preparation process, PU dosage, and CA dosage.

3.2. Range Analysis of the Orthogonal Test

The test results of the 3d, 5d, and 7d splitting strength, freeze-thaw splitting strength, and TSR of the different orthogonal combinations are shown in Table 8. The impacts of the various factors on the splitting strength of the PU asphalt mixture are shown in Figure 11, and the influence of the freeze-thaw splitting strength and the TSR of the PU asphalt mixtures are displayed in Figure 12.

Table 8. Results of the orthogonal tests.

Number	Orthogonal Composite	3d Split Strength/MPa	5d Split Strength/Mpa	7d Split Strength/Mpa	Freeze-Thaw Splitting Strength/Mpa	TSR/%
1	A1B1C3	1.32	1.44	1.93	1.12	64.48
2	A1B2C1	1.23	1.74	1.96	1.4	79.37
3	A1B3C2	1.12	1.28	1.46	1.02	77.63
4	A2B1C1	1.57	1.97	2.26	1.98	87.82
5	A2B2C2	1.47	1.86	2.11	1.57	82.68
6	A2B3C3	1.13	1.14	1.64	1.15	77.91
7	A3B1C2	1.58	2.10	2.57	2.01	83.52
8	A3B2C3	1.77	2.08	2.37	1.8	75.01
9	A3B3C1	1.46	1.68	1.96	1.31	74.26

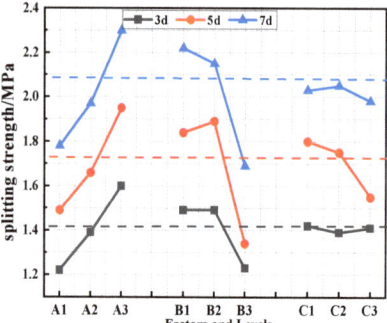

Figure 11. Effects of the various factors on the splitting strength of the specimens.

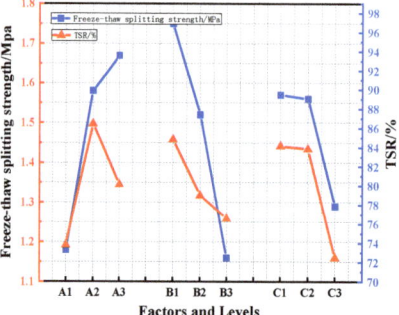

Figure 12. Effects of the various factors on the freeze-thaw splitting strength and TSR of the specimens.

It can be observed from Figure 11 that the splitting strength of polyurethane mixtures at all ages increases gradually with increases in the PU dosage. When the PU content is 60%, the splitting strength of 3d, 5d, and 7d is 1.31 times, 1.31 times, and 1.29 times higher than that of 40% PU, respectively. It can be seen that the PU content has a great influence on the splitting strength. On one hand, with increases in the PU content, the proportion of PU in the modified asphalt increases, and the mechanical properties of PU lead to an increase in the modified asphalt strength. On the other hand, a high content of PU can play a connecting role in the cross-linked network architecture developed by the PU asphalt, enhance the cohesion of the modified asphalt, and increase the splitting strength of the specimen. The 7d splitting strength of the PU asphalt specimen was found to be the highest, being 64.3% higher than that of the 3d curing condition. The analysis concluded that the strength of PU asphalt was affected by its gradual reaction with the moisture in the air. The early formation strength was low; however, with increases in the curing time, the curing reaction proceeds to gradually increase the strength.

It can be seen in Figure 12 that the freeze-thaw splitting strength increases with increases in the PU content. The freeze-thaw splitting strength at 50% and 60% PU is 34% and 41% higher than that at 40% PU, respectively, indicating that 50% and 60% PU can significantly improve the low-temperature performance and water stability of PU asphalt. With increases in the PU content, the TSR exhibited a trend of first increasing and then decreasing. When the PU content was 50%, the TSR reached its the maximum, indicating that the high content of PU will affect the water stability of the PU asphalt specimen. The best water stability and low-temperature properties of the polyurethane asphalt specimens were therefore obtained when the polyurethane content was 50%.

It can be also seen in Figure 11 how the preparation method has distinct effects on the splitting strengths at different ages. In the initial stages of the curing, there were no significant differences found between the precast method and the post-mixing method in terms of the strength. As the reaction progressed, the post-mixing method accelerated the reaction of the PU asphalt, and the strength of the specimen was found to be greater than that of the prefabricated method. After the reaction was completed, the 7d splitting strength of the specimens prepared by the prefabricated method were found to have increased by 3.2% and 31%, respectively, for the post-mixing method and the pre-reaction method. The pre-reaction method exhibited low strengths at each age; therefore, this method was excluded in subsequent testing. It can be concluded from Figure 12 that the prefabrication method has a significant impact on the freeze-thaw splitting strength and the TSR, indicating that the PU-modified asphalt prepared by the prefabrication method can provide a better water stability and low-temperature performance to the specimens.

It can be concluded from Figure 11 that the splitting strength of the specimens decreases with increases in the CA content. When the CA content changed from 3% to 6%, the strength of the specimen remained practically unchanged. When the content of CA increased to 12%, the splitting strength of the specimen decreased by 13%. Increases in the CA content led to the rapid formation of PU asphalt strength, and the specimen cannot therefore form strength in the subsequent curing process. Therefore, when the CA content was 3–6%, the reaction speed of the specimen was able to be improved without a loss in the strength. It can also be observed in Figure 12 how the freeze-thaw splitting strength and TSR of PU asphalt decreased as the CA content increased. Therefore, the water stability and low-temperature performance of PU asphalt can be enhanced when the CA content is less than 6%.

3.3. Orthogonal Experimental Difference Analysis

To further explain the effects of the various factors on PU asphalt, the results in Table 8 were used to perform ANOVA on the splitting strength, 7d freeze-thaw splitting strength, and TSR of the 3d, 5d, and 7d PU asphalt specimens. The results are displayed in Table 9. From the ANOVA results in Table 9, it can be concluded that the factors affecting the significance of the 5d and 7d splitting strength and freeze-thaw splitting strength are in the following order: preparation process > PU dosing > hardener dosing; the factors affecting the significance of 3d splitting strength are in the following order: PU dosing > preparation process > hardener dosing; and the factors affecting the significance of the TSR are in the following order: hardener dosing > PU dosing > preparation process.

Table 9. Variance analysis of the orthogonal experiment.

Response Value		PU Content	Preparation Method	Dosage of CA	Error Value
3d Splitting strength	SS	0.218	0.128	0.001	0.046
	d/f	2	2	2	2
	MS	0.109	0.064	0.001	0.023
	F	4.77	2.81	0.03	
5d Splitting strength	SS	0.335	0.501	0.099	0.027
	d/f	2	2	2	2
	MS	0.167	0.251	0.050	0.014
	F	12.29	18.42	3.64	
7d Splitting strength	SS	0.411	0.501	0.007	0.007
	d/f	2	2	2	2
	MS	0.205	0.251	0.003	0.003
	F	60.20	73.50	1.03	
Freeze-thaw splitting strength	SS	0.410	0.526	0.157	0.093
	d/f	2	2	2	2
	MS	0.205	0.263	0.078	0.046
	F	4.43	5.68	1.69	
TSR	SS	0.024	0.011	0.027	0.018
	d/f	2	2	2	2
	MS	0.012	0.005	0.013	0.009
	F	1.31	0.59	1.46	

Two factors, the PU content, and the preparation process, have significant effects on the splitting strength and freeze-thaw splitting strength, while the amount of curing agent has a significant effect on the TSR. Based on the range analysis and variance analysis methods, the better ratio was found to be A2B1C1 and A3B1C1, that is, the polyurethane contents were 50% and 60%, the prefabrication preparation method was used, and the curing agent content was 3%.

3.4. Response Surface Model

The experimental conditions of the three factors and the three levels were optimized using the Design-Expert software. Based on the measured data, the response surface fitting equations Y_1, Y_2, and Y_3 of the 7d splitting strength, freeze-thaw splitting strength, and splitting tensile strength were obtained through interpolation.

$$Y_1 = 2.02 + 0.29A - 0.22B - 0.15C - 0.037AB + 0.023AC + 0.084BC \quad (1)$$

$$Y_2 = 1.53 + 0.28A - 0.32B - 0.15C - 0.13AB + 0.070AC + 0.022BC \quad (2)$$

$$Y_3 = 80.60 + 1.05A - 1.26B - 3.57C + 0.16AB - 1.52AC - 0.26BC \quad (3)$$

Taking the response surface of the prefabricated method as an example, as shown in Figure 13, model variance analysis and reliability analysis were conducted, the results of which are shown in Table 10.

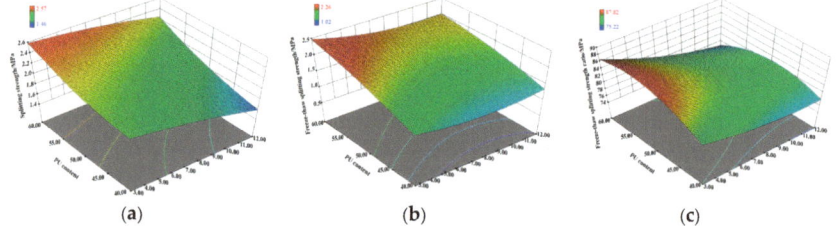

Figure 13. Response surface: (**a**) splitting strength; (**b**) freeze-thaw splitting strength; and (**c**) TSR.

Table 10. Variance and reliability analysis of the three response values.

Response Value		Quadratic Sum	Degree of Freedom	Mean Square	F Ratio	p-Value (>F) [1]
7d Splitting Strength	Model	0.926	6	0.003	14.88	0.0002
	PU content	0.411	2	0.205	22.83	0.0003
	Preparation method	0.501	2	0.251	15.42	0.0017
	Dosage of CA	0.007	2	0.004	6.38	0.0253
Freeze-thaw splitting strength	Model	1.87	6	0.21	19.67	0.0001
	PU content	0.60	2	0.60	56.86	0.0002
	Preparation method	0.78	2	0.78	73.79	0.0001
	Dosage of CA	0.23	2	0.23	21.62	0.0109
TSR	Model	199.36	6	22.15	35.52	0.0001
	PU content	8.41	2	8.41	13.49	0.0079
	Preparation method	11.98	2	11.98	19.21	0.0032
	Dosage of CA	111.30	2	111.30	178.48	0.0001

[1] When the p-value is greater than 0.05, the model significance is low (indicating no statistical significance), and the regression model is not available; if the converse is true, the regression model significance is high.

From Table 10, it can be seen that the p-values of the three types of models were less than 0.05, which indicates that the models fit well and can accurately reflect the effects of each factor on the 7d splitting strength, freeze-thaw splitting strength, and TSR.

Furthermore, the experimental results were found to be consistent with the results of the orthogonal analysis, indicating that the response surface optimization analysis has a good reliability.

To determine the optimal ratio of PU asphalt, the maximum splitting strength and freeze-thaw splitting strength were taken as the optimization objectives. In Design-Expert, the optimal factor level obtained using the numerical tool in the optimization module is shown in Table 11.

Table 11. Optimal composition.

PU Content	Preparation Method	Dosage of CA
56.64%	Prefabrication method	3.58%

Five groups of specimens were prepared according to the optimal scheme, and the base asphalt was used as the control group to verify its mechanical properties. First, the polyurethane was heated to 60 °C in the oven and the base asphalt was heated to 120 °C. After heating was completed, it was sheared at 120 °C and 4000 rpm for 10 min. Then, a 3.5% curing agent was added, and shearing was continued for 5 min. Finally, the PU asphalt was mixed with the aggregate to prepare the mixture, and the Marshall test specimen was formed for testing. The specific procedures conducted are exhibited in Figure 14. The property indexes of the comparison epoxy asphalt mixture are shown in Figure 15.

Figure 14. Optimal scheme flow chart.

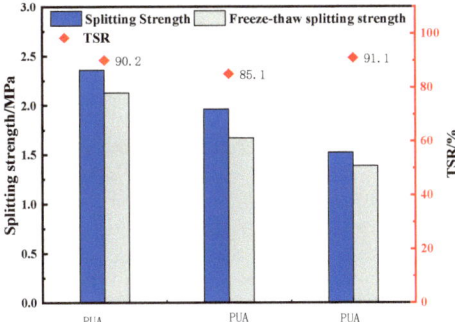

Figure 15. Performance comparison of the mixture.

It can be concluded from Figure 15 that the splitting strength, freeze-thaw splitting strength, and TSR of a PU asphalt mixture are 12.9%, 21.6%, and 5.7% higher than those of an epoxy asphalt mixture, and 55.2% and 53.7% higher than those of an asphalt mixture, respectively, while the TSR does not increase. This indicates that the mechanical properties and low-temperature performance of PU asphalt are markedly better than those of epoxy asphalt and SBS-modified asphalt, while the water stability has also been improved; it has excellent mechanical properties, low-temperature performance, and water stability.

3.5. Tensile Properties

PU asphalt is a thermosetting material, and its mechanical properties are usually evaluated using direct tensile tests. The effects of various PU and CA contents on the mechanical performance of PU asphalt were determined through conducting tensile tests. The samples are shown in Figure 16, and the test results are presented in Figure 17. When the amount of PU was constant, an increase in the CA content made the tensile strength increase initially and then decrease, and the elongation at the break also decreased, which indicates that a certain amount of CA can improve the mechanical strength of PU asphalt but will adversely affect its flexibility. Therefore, the mass fraction of CA was set to 3%.

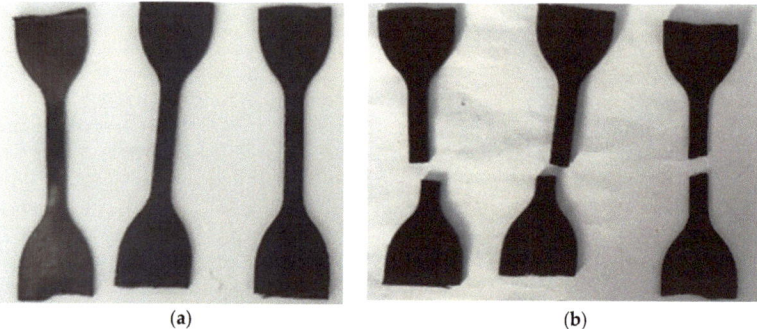

Figure 16. Tensile test specimen: (**a**) shape before tensile test; and (**b**) shape after tensile test.

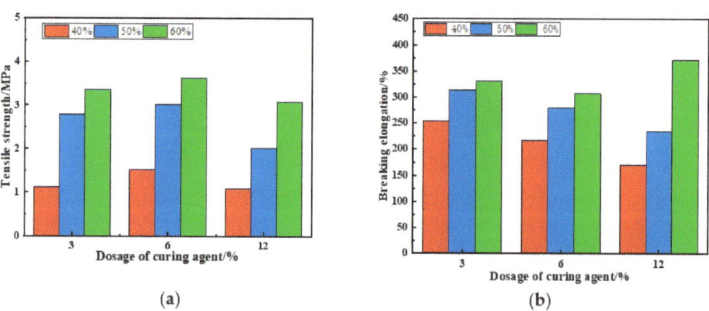

Figure 17. Tensile properties of PU asphalt: (**a**) tensile strength; and (**b**) breaking elongation.

It can be seen from Figure 17 that when the mass fraction of CA was set to 3%, and with the increase in PU content from 40% to 50%, the tensile strength of the specimen increased from 1.13 MPa to 2.78 MPa. The increment was 1.65 MPa, while the elongation at break increased from 254% to 314%, representing an increase of 60%. With an increase in PU content from 50% to 60%, the tensile strength of the specimen further improved from 2.78 MPa to 3.37 MPa, with an increment of 0.59 MPa, and the elongation at break increased from 314% to 331%, with an increment of 17%.

It can be concluded that, with increases in the PU content, the cross-linking network inside PU asphalt becomes denser. After the complete curing reaction, a stable three-dimensional network structure was formed inside PU asphalt, which greatly improves its mechanical strength. When the PU content exceeds 50%, the tensile performance increment decreases, indicating that a PU content of about 50% is the critical point for the formation of the three-dimensional network structure inside PU asphalt.

3.6. Failure Surface Analysis

The PU mixture specimens were prepared as according to the optimal scheme, and the failure surfaces of the specimens were analyzed. The splitting failure surface of the specimen is shown in Figure 18a. The splitting failure surface after freeze-thaw is shown in Figure 18b.

Figure 18. (**a**) Splitting surface diagram of the PU asphalt mixture; and (**b**) freeze-thaw splitting surface of the PU asphalt mixture.

It can be seen in Figure 18 that after the splitting test of the specimen without freeze-thaw, there were many broken aggregates on the failure surface of the specimen, indicating that failure mainly occurs in the aggregate, while the fracture form is aggregate failure, indicating that PU-modified asphalt has an excellent bond strength. The formation of a thicker binder film enhances the strength of the binder–aggregate interface and improves the mechanical strength of the mixture.

In comparison to the failure surface of the specimen before freeze-thaw, there were fewer aggregates broken on the failure surface after freeze-thaw, while the crack extended along the interface between the binder and the aggregate, indicating that in the low-temperature environment, the effect of water accelerates the failure of the binder–aggregate interface, the damage occurs mainly along the interface between the binder and the aggregate, and the fracture form is interface damage. The splitting failure simulation diagram of the specimen is shown in Figure 19.

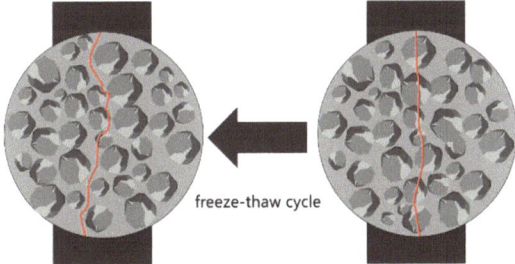

Figure 19. Damage simulation diagram of the PU asphalt mixture specimen.

Under the action of water erosion, water molecules entered the interface between the binder and the aggregate, which reduced the interface's strength. When the bonding ability of the binder was low, water erosion greatly reduced the bonding strength of the interface, and the binder and the aggregate were therefore separated. Therefore, it is of particularly importance to improve the bonding ability of the binder.

As shown in Figure 19, the PU asphalt–aggregate interface strength was decreased in water and low-temperature environments, which adversely affects the interface's strength; however, some aggregate damage still remains in the polyurethane asphalt specimens, and the splitting strength after freeze-thaw was still significantly better than that of the asphalt mixture indicating that PU asphalt has an excellent bonding performance and water stability.

4. Conclusions

The main conclusions of this paper are summarized as follows:

(1) Under the same preparation conditions and 10% PU content, the selected PU-1, PU-2, PU-3, and PU-4 materials can improve the mechanical properties of asphalt to varying degrees. Among them, PU-1 has the most significant improvement, at 95% higher than that of base asphalt.

(2) Based on the orthogonal tests and variance analysis, the influence of the related factors on the splitting tensile strength and freeze-thaw splitting strength of PU-modified asphalt at different ages was obtained, and it was found that the PU content was the most significant.

(3) The optimal ratio parameters of PU-modified asphalt are determined as follows: the PU content is 56.64%, the preparation process is the prefabricated method, and the CA content is 3.58%. The establishment of the response surface was found to be highly consistent, which can therefore guide the optimization of the PU-modified asphalt ratio.

(4) Through conducting tensile tests, it was found that the PU asphalt prepared by the best scheme has good tensile properties and flexibility.

(5) Through the analysis of failure surface morphology, it was found that PU-modified asphalt significantly improves the strength of the binder–aggregate interface, and the material performance was found to be excellent. PU-modified asphalt exhibits obvious strengthening and toughening effects on the mixture.

Author Contributions: Conceptualization, M.S. and Y.Y.; methodology, M.S.; software, S.J.; validation, M.S. and H.W.; formal analysis, S.J. and Q.X.; investigation, S.J.; resources, M.S. and H.W.; data curation, J.Z.; writing—original draft preparation, S.J.; writing—review and editing, M.S.; visualization, Y.Y.; supervision, M.S. and Y.Z.; project administration, M.S.; funding acquisition, M.S. and Y.Z. All authors have read and agreed to the published version of the manuscript.

Funding: This work was supported by the Shandong Natural Science Foundation (ZR2022QE054), the Key Scientific Research Projects in the Transportation Industry of the Ministry of Transport (2019MS2028), the Shandong Expressway Group Project (HSB 2021-72), and the Open Project of Shandong Key Laboratory of Highway Technology and Safety Assessment (SH202107).

Institutional Review Board Statement: Not applicable.

Data Availability Statement: This original copy does not include distributed figures and tables, thus all figures and tables in this original copy are unique.

Acknowledgments: We would like to recognize numerous co-workers, students, and research facility associates for giving specialized assistance on instrument use.

Conflicts of Interest: Author Min Sun was employed by the School of Transportation Engineering, Shandong Jianzhu University. The remaining authors declare that the research was conducted in the absence of any commercial or financial relationships that could be construed as a potential conflict of interest.

References

1. Cong, L.; Yang, F.; Guo, G.H.; Ren, M.D.; Shi, J.C.; Tan, L. The use of polyurethane for asphalt pavement engineering applications: A state-of-the-art review. *Constr. Build. Mater.* **2019**, *225*, 1012–1025. [CrossRef]
2. Zhao, Q.M.; Jing, S.; Lu, X.J.; Liu, Y.; Wang, P.; Sun, M.; Wang, L. The Properties of Micro Carbon Fiber Composite Modified High-Viscosity Asphalts and Mixtures. *Polymers* **2022**, *14*, 2718. [CrossRef] [PubMed]
3. Hou, X.D.; Xiao, F.P.; Wang, J.G.; Amirkhanian, S. Identification of asphalt aging characterization by spectrophotometry technique. *Fuel* **2018**, *226*, 230–239. [CrossRef]
4. Ryms, M.; Denda, H.; Jaskula, P. Thermal stabilization and permanent deformation resistance of LWA/PCM-modified asphalt road surfaces. *Constr. Build. Mater.* **2017**, *142*, 328–341. [CrossRef]
5. Chen, J.S.; Wang, T.J.; Lee, C.T. Evaluation of a highly-modified asphalt binder for field performance. *Constr. Build. Mater.* **2018**, *171*, 539–545. [CrossRef]

6. Polaczyk, P.; Weaver, S.C.; Ma, Y.T.; Zhang, M.M.; Jiang, X.; Huang, B.S. Laboratory investigation of graphene modified asphalt efficacy to pavement performance. *Road Mater. Pavement Des.* **2023**, *24*, 587–607. [CrossRef]
7. Xiao, R.; Huang, B.S. Moisture Damage Mechanism and Thermodynamic Properties of Hot-Mix Asphalt under Aging Conditions. *ACS Sustain. Chem. Eng.* **2022**, *10*, 14865–14887. [CrossRef]
8. Wang, Y.H.; Polaczyk, P.; He, J.X.; Lu, H.; Xiao, R.; Huang, B.S. Dispersion, compatibility, and rheological properties of graphene-modified asphalt binders. *Constr. Build. Mater.* **2022**, *350*, 128886. [CrossRef]
9. Chen, B.; Dong, F.; Yu, X.; Zheng, C. Evaluation of Properties and Micro-Characteristics of Waste Polyurethane/Styrene-Butadiene-Styrene Composite Modified Asphalt. *Polymers* **2021**, *13*, 2249. [CrossRef]
10. Li, S.; Xu, W.Y.; Zhang, F.F.; Wu, H.; Ge, Q.X. Effect of Graphene Oxide on Aging Properties of Polyurethane-SBS Modified Asphalt and Asphalt Mixture. *Polymers* **2022**, *14*, 3496. [CrossRef]
11. Wu, J.P.; Herrington, P.R.; Alabaster, D. Long-term durability of epoxy-modified open-graded porous asphalt wearing course. *Int. J. Pavement Eng.* **2019**, *20*, 920–927. [CrossRef]
12. Lu, Q.; Bors, J. Alternate uses of epoxy asphalt on bridge decks and roadways. *Constr. Build. Mater.* **2015**, *78*, 18–25. [CrossRef]
13. Qian, Z.D.; Wang, Y.Q.; Song, X. Mechanical Properties on Thermosetting Epoxy Asphalt Waterproof Bond Material for Cement Concrete Pavement Based on Composite Structures. In Proceedings of the International Conference on Civil Engineering and Transportation (ICCET 2011), Jinan, China, 14–16 October 2011; pp. 948–954.
14. Akindoyo, J.O.; Beg, M.D.H.; Ghazali, S.; Islam, M.R.; Jeyaratnam, N.; Yuvaraj, A.R. Polyurethane types, synthesis and applications—A review. *RSC Adv.* **2016**, *6*, 114453–114482. [CrossRef]
15. Engels, H.W.; Pirkl, H.G.; Albers, R.; Albach, R.W.; Krause, J.; Hoffmann, A.; Casselmann, H.; Dormish, J. Polyurethanes: Versatile Materials and Sustainable Problem Solvers for Today's Challenges. *Angew. Chem. Int. Ed.* **2013**, *52*, 9422–9441. [CrossRef] [PubMed]
16. Huang, G.; Yang, T.H.; He, Z.Y.; Yu, L.; Xiao, H.X. Polyurethane as a modifier for road asphalt: A literature review. *Constr. Build. Mater.* **2022**, *356*, 129058. [CrossRef]
17. Xu, Y.; Li, Y.Z.; Duan, M.X.; Ji, J.; Xu, S.F. Compaction Characteristics of Single-Component Polyurethane Mixtures. *J. Mater. Civ. Eng.* **2021**, *33*, 04021221. [CrossRef]
18. Xu, Y.; Wu, L.; Lv, X.P.; Chou, Z.J.; Li, W.J.; Ji, J.; Xu, S.F. Improvement of Water Stability of Macroporous Polyurethane Mixture. *J. Mater. Civ. Eng.* **2022**, *34*, 04022285. [CrossRef]
19. Gao, G.H.; Sun, M.; Xu, C.C.; Qu, G.Z.; Yang, Y.H. Interlaminar Shear Characteristics of Typical Polyurethane Mixture Pavement. *Polymers* **2022**, *14*, 3827. [CrossRef]
20. Sun, M.; Zheng, M.L.; Qu, G.Z.; Yuan, K.; Bi, Y.F.; Wang, J. Performance of polyurethane modified asphalt and its mixtures. *Constr. Build. Mater.* **2018**, *191*, 386–397. [CrossRef]
21. Jin, X.; Guo, N.S.; You, Z.P.; Wang, L.; Wen, Y.K.; Tan, Y.Q. Rheological properties and micro-characteristics of polyurethane composite modified asphalt. *Constr. Build. Mater.* **2020**, *234*, 117395. [CrossRef]
22. Yang, F.; Cong, L.; Li, Z.L.; Yuan, J.J.; Guo, G.H.; Tan, L. Study on preparation and performance of a thermosetting polyurethane modified asphalt binder for bridge deck pavements. *Constr. Build. Mater.* **2022**, *326*, 126784. [CrossRef]
23. Lu, P.Z.; Huang, S.M.; Shen, Y.; Zhou, C.H.; Shao, L.M. Mechanical performance analysis of polyurethane-modified asphalt using molecular dynamics method. *Polym. Eng. Sci.* **2021**, *61*, 2323–2338. [CrossRef]
24. Yu, R.; Zhu, X.; Zhang, M.; Fang, C. Investigation on the Short-Term Aging-Resistance of Thermoplastic Polyurethane-Modified Asphalt Binders. *Polymers* **2018**, *10*, 1189. [CrossRef]
25. Jin, X.; Guo, N.S.; You, Z.P.; Tan, Y.Q. Research and Development Trends of Polyurethane Modified Asphalt. *Mater. Rep.* **2019**, *33*, 3686–3694.
26. Zhang, Z.P.; Sun, J.; Jia, M.; Ban, X.Y.; Wang, L.; Chen, L.Q.; Huang, T.; Liu, H. Effects of Polyurethane Thermoplastic Elastomer on Properties of Asphalt Binder and Asphalt Mixture. *J. Mater. Civ. Eng.* **2021**, *33*, 04020477. [CrossRef]
27. Salas, M.Á.; Pérez-Acebo, H.; Calderón, V.; Gonzalo-Orden, H. Bitumen modified with recycled polyurethane foam for employment in hot mix asphalt. *Rev. Ing. E Investig.* **2018**, *38*, 60–66. [CrossRef]
28. Jin, X.; Sun, S.W.; Guo, N.S.; Huang, S.L.; You, Z.P.; Tan, Y.Q. Influence on Polyurethane Synthesis Parameters Upon the Performance of Base Asphalt. *Front. Mater.* **2021**, *8*, 656261. [CrossRef]
29. Zhang, Z.P.; Sun, J.; Huang, Z.G.; Wang, F.; Jia, M.; Lv, W.J.; Ye, J.J. A laboratory study of epoxy/polyurethane modified asphalt binders and mixtures suitable for flexible bridge deck pavement. *Constr. Build. Mater.* **2021**, *274*, 122084. [CrossRef]
30. Yan, W.; Ou, Y.J.; Xie, J.; Huang, T.; Peng, X.H. Study on Properties of Bone Glue/Polyurethane Composite Modified Asphalt and Its Mixture. *Materials* **2021**, *14*, 3769. [CrossRef]
31. Li, C.X. Preparation of polyurethane modified asphalt and road performance evaluation of mixture. *J. Wuhan Univ. Technol. Transp. Sci. Eng.* **2017**, *41*, 958–963.
32. Zhang, Z.P.; Sun, J.; Jia, M.; Qi, B.; Zhang, H.L.; Lv, W.J.; Mao, Z.Y.; Chang, P.T.; Peng, J.; Liu, Y.C. Study on a thermosetting polyurethane modified asphalt suitable for bridge deck pavements: Formula and properties. *Constr. Build. Mater.* **2020**, *241*, 118122. [CrossRef]
33. Jia, M.; Zhang, Z.P.; Yang, N.; Qi, B.; Wang, W.T.; Huang, Z.G.; Sun, J.; Luo, F.K.; Huang, T. Performance Evaluation of Thermosetting and Thermoplastic Polyurethane Asphalt Mixtures. *J. Mater. Civ. Eng.* **2022**, *34*, 04022097. [CrossRef]

34. *JTG E20-2011*; Standard Test Methods of Bitumen and Bituminous Mixtures for Highway Engineering. Ministry of Transport of the People's Republic of China: Beijing, China, 2011.
35. *ISO 37:2017*; Rubber, Vulcanized or Thermoplastic—Determination of Tensile Stress-Strain Properties. International Organization for Standardization: Geneva, Switzerland, 2009.

Disclaimer/Publisher's Note: The statements, opinions and data contained in all publications are solely those of the individual author(s) and contributor(s) and not of MDPI and/or the editor(s). MDPI and/or the editor(s) disclaim responsibility for any injury to people or property resulting from any ideas, methods, instructions or products referred to in the content.

Article

Dynamic Processes and Mechanical Properties of Lipid–Nanoparticle Mixtures

Fan Pan [1,*], Lingling Sun [2] and Shiben Li [2,*]

1 School of Data Science and Artificial Intelligence, Wenzhou University of Technology, Wenzhou 325035, China
2 Department of Physics, Wenzhou University, Wenzhou 325035, China
* Correspondence: panfan@wzu.edu.cn (F.P.); shibenli@wzu.edu.cn (S.L.)

Abstract: In this study, we investigate the dynamic processes and mechanical properties of lipid nanoparticle mixtures in a melt via dissipation particle dynamic simulation. By investigating the distribution of nanoparticles in lamellar and hexagonal lipid matrices in equilibrium state and dynamic processes, we observe that the morphology of such composites depends not only on the geometric features of the lipid matrix but also on the concentration of nanoparticles. The dynamic processes are also demonstrated by calculating the average radius of gyration, which indicates the isotropic conformation of lipid molecules in the x–y plane and that the lipid chains are stretched in the z direction with the addition of nanoparticles. Meanwhile, we predict the mechanical properties of lipid–nanoparticle mixtures in lamellar structures by analyzing the interfacial tensions. Results show that the interfacial tension decreased with the increase in nanoparticle concentration. These results provide molecular-level information for the rational and a priori design of new lipid nanocomposites with ad hoc tailored properties.

Keywords: dynamic process; mechanical property; lipid–nanoparticle mixture; dissipative particle dynamic simulation

1. Introduction

With the development of nanotechnology, nanoparticles of various shapes and sizes have been extensively used in biomedical, liquid sensing, fuel cell, packaging and other fields because they can impart different properties such as mechanical, optical, thermal and rheological properties of materials [1–6]. However, the effects of nanoparticles on material properties are varied, and these potential applications require the efficient control of the distribution of nanoparticles in the matrix. Amphiphilic molecules such as lipids and block copolymers, which possess both hydrophilic and hydrophobic moieties, can self-assemble into a variety of periodic microstructures (spheres, cylinders and lamellae), making them ideal scaffolds for organizing nanoparticles into well-defined nanostructures, such as nanoplates, nanowires or nanospheres [2,5,7–12]. Due to the easy preparation of block copolymers, the phase behavior of nanoparticles in block copolymers has been studied extensively, both experimentally [13,14] and theoretically [7–9,15].

In contrast to block copolymers, lipid molecules, which are one of the major constituents of biological systems, as well as various commercial products, such as foods and cosmetics, are basic components in nature. More than 3,000 unique lipid structures have been reported according to the Lipid Maps Structure Database (LMSD). Lipid molecules commonly contain one or more polar hydrophilic functional groups and one or more flexible hydrophobic fatty acid chains [11,16–18]. Given the amphipathicity of head and tail chains, lipids can self-assemble into many different structures, either in aqueous solutions or in melts [11,12,16,18–21]. Phospholipids, usually with one head hydrophilic group and two tail hydrophobic chains, are the most familiar lipid structures, since they are the main components of biofilms and often exist in the liquid environment. Other types of common lipids that are composed of one hydrophilic group and one hydrophobic chain, such as monoelaidin (ME), mono-olein (MO), monovaccenin (MV) and monolinolein (ML), can self-assemble into a series of bicontinuous cubic structures in the water environment or form crystalline phases, usually of lamellar structures and with zero or low hydration, which are quite different from phospholipids [16,20]. A very accurate phase diagram of glycerol a mono-oleate/water system was reported by Qiu et al. based on X-ray diffraction measurements, which was described in the water concentration range from the dry state to full hydration and in the temperature range from -15 to $55\ ^\circ$C. It shows that with no water or in the presence of a small amount of water, glycerol mono-oleate forms a lamellar crystal phase or a liquid crystal phase at temperatures under 37 $^\circ$C, while with water content of 20–40%, it forms a gyroid cubic crystalline phase and that with the addition of more water to a concentration of more than 40%, a diamond cubic liquid crystalline phase is formed [20]. The liquid crystalline phases of these polar lipids have high solubilization capacity for lipophilic, hydrophilic and amphiphilic guest molecules and can protect molecules against oxidation or hydrolysis. Thus, they are expected to have more applications than polymers and have received sufficient research interest with respect to their applications in various pharmaceutical, food and biotechnical areas, such as for use as a drug delivery matrix for peptides, proteins, vitamins and amino acids [12,16,21–23].

The full use of these liquid crystal phases in such applications calls for understanding of their structures and mechanical properties at the molecular level. Computer simulations offer a unique approach to explore the microstructures and have a certain predictability. For example, self-consistent field theory (SCFT), which is the most widely used method to estimate phases, is usually adopted to predict the phase of polymers and the distribution of nanoparticles in the polymer matrix [24–26]. However, SCFT can only observe the probability distribution of configurations in a potential field because it is a method based on phase-field theories; it cannot capture polymer dynamics and the packing structure of polymer chains in detail. Compared to SCFT, molecular dynamics (MD), which simulate the motion of molecular systems following Newton's equations with an initial configuration and velocity, can resolve the limitations of SCFT [27–30]. However, the system size and the time scale of classical MD are not sufficient for investigating large-scale systems. To overcome the shortcomings of both SCFT and MD, dissipative particle dynamics (DPD) based on a coarse-grained (CG) model was developed to investigate the equilibrium or nonequilibrium properties of such systems. DPD is the inheritance and development of MD and LGA(lattice-gas automata) simulation; it can obtain the spatial position and velocity distribution information of each particle in the time evolution process just like MD. It also eliminates the concept of lattice, which greatly reduces the systematic error and allows a large time scale to cause the system to be in equilibrium with a limited computational load. Due to these advantages, DPD has been applied to a number of applications and been proven to reproduce the expected behavior successfully, such as in the study of polymer solutions [31], block-copolymer–nanoparticle composites [32] and lipid bilayer membranes [33]. For example, Cai et al. studied the self-assembly behavior of poly(γ-benzyl-L-glutamate)-block-poly (ethylene glycol) (PBLG-b-PEG) block copolymer blended with gold nanoparticles both by experiment and by DPD simulation and showed that the PBLG-b-PEG block copolymer self-assembled into cylindrical micelles in pure form. However, when introducing gold nanoparticles, the formed aggregate morphology transformed from long, cylindrical micelles to spherical micelles. Moreover, the nanoparticles were mostly found near the core/shell interface and in the core center of the micelles. Their DPD simulation results were found to be in good agreement with their experimental observations [34]. Kranenburg and Smit performed DPD simulations on a mesoscopic model of a phospholipid molecular structure containing a head group of three hydrophilic beads and two tail chains varying in length from four to seven beads. They investigated the phase behavior of double-tail lipids with varied tail length, headgroup interactions and temperature, which can reproduce the experimentally observed phases [35]. However, to the best of our knowledge, DPD simulation of lipid–nanoparticle mixtures is limited.

Hence, we can address the issue by systematically modeling equilibrium structures of a binary mixture of lipid molecules and nanoparticles in a melt by using a DPD simulation based on a coarse-grained (CG) model. In this work, we select lipid molecules with one head chain and one tail chain to investigate the phase behavior of lipid–nanoparticle mixtures under different concentrations of nanoparticles. We are also interested in the dynamic processes and mechanical properties of lipid–nanoparticle mixtures.

2. Method and Model

2.1. Method

The simulation was based on the DPD method, which was originally developed to simulate hydrodynamic interactions during molecular dynamic simulation [36–40]. In the DPD method, three categories of forces were introduced to describe the movement of beads, in accordance with the Newton's law, where conservation force ($\mathbf{F}_{ij}^C$) represents the force required to exclude the volume effect, dissipative fore ($\mathbf{F}_{ij}^D$) represents the viscous resistance among moving beads and a random force ($\mathbf{F}_{ij}^R$) typifies a stochastic force. These forces occur between a pair of beads: the i-th and j-th beads. Thus, the total force on the i-th bead can be expressed as follows:

$$\mathbf{F}_i = \sum_{i \neq j} \left(\mathbf{F}_{ij}^C + \mathbf{F}_{ij}^D + \mathbf{F}_{ij}^R \right). \tag{1}$$

The conservative force is the soft repulsion acting along the intermolecular vector, which is presented as follows:

$$\mathbf{F}_{ij}^C = a_{ij} w(r_{ij}) \hat{\mathbf{r}}_{ij}, \tag{2}$$

where a_{ij} indicates the maximum repulsive force between the i-th and j-th beads, and $r_{ij} = |\mathbf{r}_i - \mathbf{r}_j|, \hat{\mathbf{r}}_{ij} = (\mathbf{r}_i - \mathbf{r}_j)/r_{ij}$. The weight function ($w(r_{ij})$) can be expressed as follows:

$$w(r_{ij}) = \begin{cases} 1 - \frac{r_{ij}}{r_c} & r_{ij} < r_c \\ 0 & r_{ij} > r_c \end{cases}, \tag{3}$$

where r_c is the cutoff radius. The dissipative force is a hydrodynamic drag force, which is presented as follows:

$$\mathbf{F}_{ij}^D = -\gamma w^2(r_{ij})(\hat{\mathbf{r}}_{ij} \cdot \mathbf{v}_{ij})\hat{\mathbf{r}}_{ij}. \tag{4}$$

Here, $\mathbf{v}_{ij} = \mathbf{v}_i - \mathbf{v}_j$. The random force corresponds to the thermal noise, which is represented as follows:

$$\mathbf{F}_{ij}^R = \sigma w(r_{ij}) \zeta_{ij} \Delta t^{-1/2} \hat{\mathbf{r}}_{ij}, \tag{5}$$

γ is the friction coefficient, and σ is the noise amplitude. γ and σ are related as $\sigma^2 = 2\gamma k_B T$, where T is the absolute temperature, and k_B is the Boltzmann constant. ζ_{ij} denotes a random number from a uniform random distribution with unit variance and Gaussian distribution. $\sigma = 3.0$ and $\gamma = 4.5$ are usually used as standard values in the simulation.

2.2. Model

Our simulation is based on a coarse-grained (CG) model that treats a small group of atoms as a single bead located at the center of mass of the group. The atomistic lipid corresponding to our simulation model is glycerol mono-oleate(1-(cis-9-Octadecenoyl)-rac-glycerol), which contains a hydrocarbon chain, an ester bond and a glycerol backbone. The two remaining carbons of the glycerol moiety are free and confer polar characteristics to this part of the molecule. Thus, this part is hydrophilic and we commonly called it the head chain. In contrast, the C18 hydrocarbon chain, which formed a cis double bond at the 9,10 positions, is strongly hydrophobic, and we commonly referred this part as the tail chain. In our work, we adopt a single type of coarse grain to model head-chain or tail-chain beads according to their amphipathicity of atom groups, which is similar to the work of other researchers [35,41]. The mapping of atomistic lipids and the corresponding model used in this paper is shown in Figure 1, where the hydrophilic beads (H) and hydrophobic tails (T) are shown as red and yellow beads, respectively. Two neighboring beads in one lipid chain are connected by a harmonic spring force with spring constant k_s and equilibrium bond length r_s:

$$\mathbf{F}_{ij} = k_s \left(1 - \frac{r_{ij}}{r_s}\right) \hat{\mathbf{r}}_{ij}. \tag{6}$$

Here, we set the parameters $k_s = 120.0$ and $r_s = 0.7 r_c$ to make the bonded chain structure, which are similar to previous worcks [42–44]. In this paper, the rigidity of the head particles in the lipid structure is greater than that of the tail particles; thus, an additional angular harmonic potential energy based on angular potential constant (k_θ) and equilibrated angle (θ_0) is set to make a rod-like chain structure. The angular harmonic bending force is presented as follows:

$$\mathbf{F}^\theta = -\nabla \left[k_\theta (\theta - \theta_0)^2\right], \tag{7}$$

where θ is the angle between two adjacent bonds from the center particle (i). We set $k_\theta = 6.0$ and $\theta_0 = \pi$ for three consecutive particles to ensure a stiff rod chain with enough mobility for self-assembly. In addition, the nanoparticles were modeled as a single bead and shown as blue beads in Figure 1.

There are also several works providing more accurate coarse-grained models. For example, Shelley et al. described a method for developing a CG model for phospholipids by fitting some potential parameters on the basis of comparisons, which can semiquantitatively reproduce the density profile of an aqueous dimyristoylphosphatidylcholine (DMPC) bilayer [45]. Marrink et al. optimized the current CG models in terms of four aspects: speed, accuracy, applicability and versatility. Their CG model was proven to be versatile in studying nonlamellar systems [46]. Compared to their models, it seems that our approach is less refined. Nevertheless, the coarse-grained models we cite here are qualitative rather than quantitative in their predictions, so they still have certain validity.

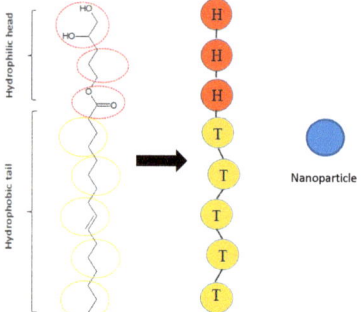

Figure 1. DPD model of the lipid molecule and the nanoparticle. The atomistic representation of a lipid molecular structure consisting of one head group and one tail group is demonstrated in the leftmost column, and the corresponding coarse-grained model is shown in the middle column. Hydrophilic head beads are indicated in red, and hydrophobic tail beads are represented in yellow. The rightmost side shows a schematic representation of the nanoparticle, which is indicated by blue color.

2.3. Parameters

In the DPD method, reduced units were usually used for convenience. The cut-off radius (r_c) represents units of simulated length; bead mass (m) defines units of simulated mass, and $k_B T$ defines the unit of simulated energy. In addition, the time was scaled as normalized units (τ). The cutoff radius can be estimated as $r_c = (\rho V_b)^{1/3}$, where V_b and ρ

represent the volume and particle density of a DPD bead, respectively. In our simulation, the modified version of the velocity–Verlet algorithm proposed by Groot and Warren was used to integrate the motion process [47]:

$$\begin{aligned}
\mathbf{r}_i(t + \Delta t) &= \mathbf{r}_i(t) + \Delta t \mathbf{v}_i(t) + \tfrac{1}{2}(\Delta t)^2 \mathbf{f}_i(t) \\
\tilde{\mathbf{v}}_i(t + \Delta t) &= \mathbf{v}_i(t) + \lambda \Delta t \mathbf{f}_i(t) \\
\mathbf{f}_i(t + \Delta t) &= \mathbf{f}_i(\mathbf{r}(t + \Delta t), \tilde{\mathbf{v}}(t + \Delta t)) \\
\mathbf{v}_i(t + \Delta t) &= \mathbf{v}_i(t) + \tfrac{1}{2}\Delta t (\mathbf{f}_i(t) + \mathbf{f}_i(t + \Delta t)).
\end{aligned} \quad (8)$$

We selected the time step $\Delta t = 0.01\tau$, where the time unit (τ) is defined as follows [48]:

$$\tau = \sqrt{mr_c^2/k_B T}. \quad (9)$$

We set repulsive interaction parameters as $a_{ii} = 25$ for the same types of particles and as $a_{ij} = 100$ for different types of particles in the simulation, which were applied extensively in previous works [47,49,50]. The Flory–Huggins parameter (χ) can be calculated from the relationship between a_{ii} and a_{ij}, that is, $\chi = 0.286(a_{ij} - a_{ii})$. This formula has been adopted in previous simulations [51]. We listed the interaction parameters between different DPD beads in Table 1. All simulations were performed in a cubic box with a volume of $V = L \times L \times L$ under periodic boundary conditions. We performed the calculations for box sizes ranging from $L = 25r_c$ to $L = 35r_c$ to avoid the finite size effect [52]. Then, we optimized the size of the simulation box as $L = 30r_c$. The whole system was implemented with the NVT ensemble using a large-scale atomic/molecular massively parallel simulator (LAMMPS).

Table 1. The interaction parameter introduced in this article. H and T represent head and tail beads of lipid chain, N denotes nanoparticle beads.

a_{ij}	H	T	N
H	25		
T	100	25	
N	100	100	25

In general, the equilibrium state can be achieved after about 200,000 DPD time steps in the dynamic process. In this state, the energy decreased to the lowest value. Taking the system with pure melt lipid molecules as an example (Figure 2), we observed that the energy of the microstructure eventually decreased to a gentle state. Since the initial structures have a strong influence on the final results of simulations, we usually input several different initial structures to compare the energy of their stable structures, such as the preassembled lamellar and hexagonal structures, as well as random inputting, and we selected the equilibrium state with the lowest energy in all studied systems. After determining the initial structure of the lipid matrix, we randomly placed nanoparticles in the initial structures. In our present work, we performed about 300,000 DPD time steps and 6 rounds of DPD simulations with different random seeds in all simulations to ensure the acquisition of equilibrium structures and the repeatability of the findings. The physical quantities are presented as the six averaged parallel simulations dates Section 3 to minimize error and obtain a better representation.

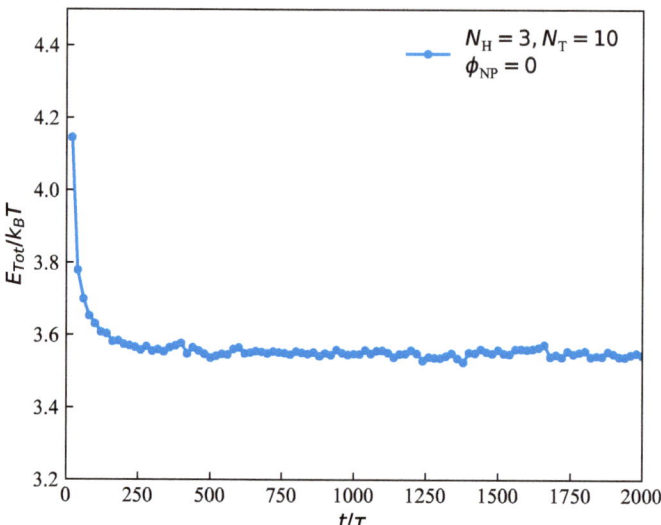

Figure 2. An example of obtaining the equilibrium state in the dynamic process with parameters $N_H = 3$, $N_T = 10$, the total energy E_{Tot}/k_BT as a function of time steps.

3. Results and Discussion

In this section, we analyze and discuss the results from the DPD simulation for the self-assembly of pure lipid molecules and lipid–nanoparticle composites in a melt. In order to systematically study the influences of nanoparticles on the lipid matrix, different nanoparticle concentrations were set, including ϕ_{NP} =0.03, 0.05 and 0.15, to represent low concentration, medium concentration and high concentration, respectively. The nanoparticles selected in this study are neutral, which indicates the absence of preferable interaction between nanoparticles and hydrophobic or hydrophilic parts of lipid molecules. The equilibrium structures are displayed in Figures 3 and 4, the dynamic processes are demonstrated in Figures 5 and 6 and the mechanical properties are illustrated in Figures 7 and 8.

3.1. Equilibrium Structures

In general, the equilibrium structure of lipid–nanoparticle composites depends on several factors, such as the different block ratios of lipid molecules, that is, the number of head particles and tail particles of lipids (N_H, N_T), the interaction between nanoparticles and lipid molecules (a_{ij}), the concentration of nanoparticles (ϕ_{NP}), etc. This is similar to diblock copolymer–nanoparticle composites, which have been proven experimentally and theoretically [2,5].

First, we perform the DPD simulation on the phase separation of lipid molecules in a pure melt with $N_H = 3$ and $N_T = 10$, which shows the lamellar microphase-separated morphology (Figure 3a1,a2). These lamellae have multilayer structures because of the amphiphilicity of lipid molecules in bulk. Figure 3a3 demonstrates the density profiles of the head particles and tail particles of lipids along the z-direction. Several peaks are found in the density distribution: the red curve corresponds to the head particles, and the yellow curve corresponds to the tail particles. The z-coordinate positions corresponding to the peaks in these curves coincide with the centers of the microdomain of head and tail particles (Figure 3a1). This density distribution diagram exhibits the lamellar structure of pure molten lipid molecules.

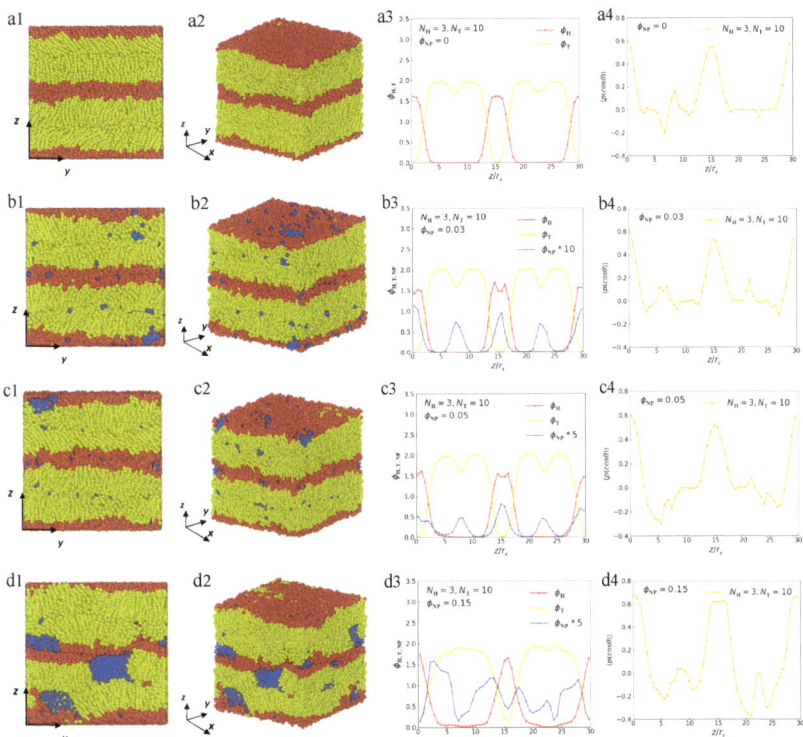

Figure 3. Representative lamellar microstructures of lipid–nanoparticle mixtures with $N_H = 3$ and $N_T = 10$. The microstructures for different views, density profiles and order parameter profiles are listed on the left side, middle and rightmost side, respectively. (**a1**) The front view of pure lipid molecule with $\phi_{NP} = 0$. (**a2**) The side view of pure lipid molecule. (**a3**) The density distribution of pure lipid molecule along the z direction. (**a4**) Order parameter profile of pure lipid molecule. (**b1**) The front view of lipid–nanoparticle mixtures with $\phi_{NP} = 0.03$. (**b2**) The side view of lipid–nanoparticle mixtures with $\phi_{NP} = 0.03$. (**b3**) The density distribution of lipid–nanoparticle mixtures with $\phi_{NP} = 0.03$ along the z direction. (**b4**) Order parameter profile of lipid–nanoparticle mixtures with $\phi_{NP} = 0.03$. (**c1**) The front view of lipid–nanoparticle mixtures with $\phi_{NP} = 0.05$. (**c2**) The side view of lipid–nanoparticle mixtures with $\phi_{NP} = 0.05$. (**c3**) The density distribution of lipid–nanoparticle mixtures with $\phi_{NP} = 0.05$ along the z direction. (**c4**) Order parameter profile of lipid–nanoparticle mixtures with $\phi_{NP} = 0.05$. (**d1**) The front view of lipid–nanoparticle mixtures with $\phi_{NP} = 0.15$. (**d2**) The side view of lipid–nanoparticle mixtures with $\phi_{NP} = 0.15$. (**d3**) The density distribution of lipid–nanoparticle mixtures with $\phi_{NP} = 0.15$ along the z direction. (**d4**) Order parameter profile of lipid–nanoparticle mixtures with $\phi_{NP} = 0.15$.

The microstructure can only display the spatial distribution but cannot provide orientation properties. In quantitatively monitoring the orientational ordering of the lipid chains, we introduce the following order parameter [53,54]:

$$\langle P(\cos\theta)\rangle = \left\langle \frac{3}{2}\cos^2\theta - \frac{1}{2} \right\rangle, \tag{10}$$

where θ is the angle between the chain direction and the z-axis, and the bracket represents an ensemble average. When the chain direction is parallel to the z direction, the order parameter takes a value of 1. When the chain direction is perpendicular to the z direction, the order parameter takes a value of -0.5, whereas a value of 0 represents a complete disorder in the distribution [17]. Figure 3a4 shows that the order parameter of lipid chains is about 0.6 at the positions where the head particles are concentrated, which indicates that the lipid chains are nearly parallel to the z direction at these places because the rigidity of the head chains in the lipids in our study is greater than that of the tail chains. This phenomenon indicates that the lamellar structure has a liquid–crystalline characteristic with well-orientational orders.

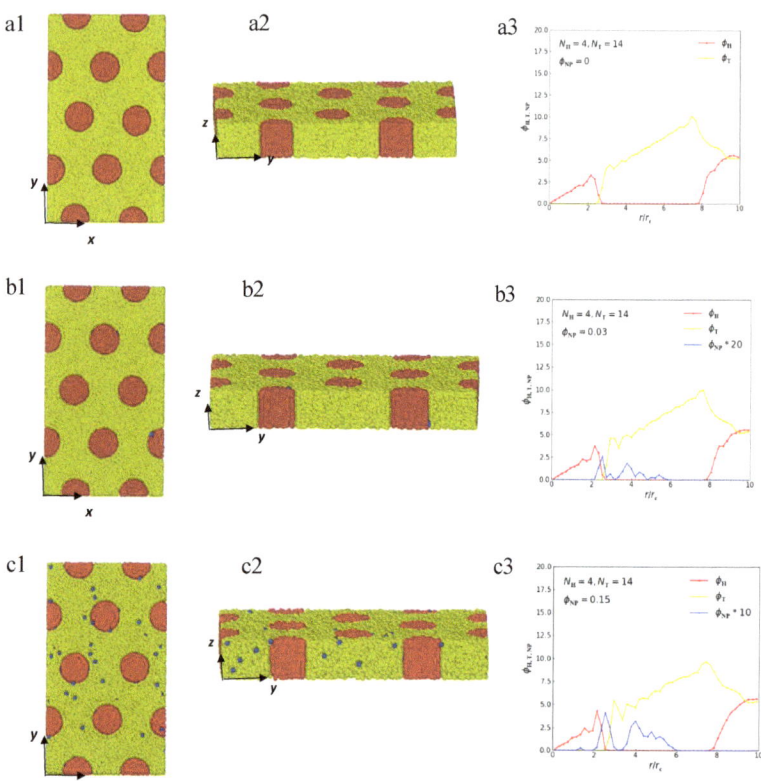

Figure 4. Representative hexagonal microstructures of lipid–nanoparticle mixtures with $N_H = 4$ and $N_T = 14$. The microstructures for different views are listed on the left and middle, and the radial density profiles are listed on the rightmost side. (**a1**) The front view of pure lipid molecule with $\phi_{NP} = 0$. (**a2**) The side view of pure lipid molecule. (**a3**) The radial density distribution of pure lipid molecule. (**b1**) The front view of lipid–nanoparticle mixtures with $\phi_{NP} = 0.03$. (**b2**) The side view of lipid–nanoparticle mixtures with $\phi_{NP} = 0.03$. (**b3**) The radial density distribution of lipid–nanoparticle mixtures with $\phi_{NP} = 0.03$. (**c1**) The front view of lipid–nanoparticle mixtures with $\phi_{NP} = 0.15$. (**c2**) The side view of lipid–nanoparticle mixtures with $\phi_{NP} = 0.15$. (**c3**) The radial density distribution of lipid–nanoparticle mixtures with $\phi_{NP} = 0.15$.

Then, we randomly add nanoparticles into the pure melt to study the effects of nanoparticles on the lamellar phase separation of lipid molecules. The equilibrium self-assembled structures; density profiles of the microphase separation; and distributions of nanoparticles with $\phi_{NP} = 0.03$, $\phi_{NP} = 0.05$ and $\phi_{NP} = 0.15$ are shown in Figure 3b–d, respectively. Here, the concentration of nanoparticles (ϕ_{NP}) is defined as follows:

$$\phi_{NP} = \frac{N_{NP}}{D^3}, \tag{11}$$

where N_{NP} represents the number of nanoparticles, and D^3 is the volume of the simulation box. As shown in Figure 3b1,b2, these snapshots of equilibrium self-assembled structures indicate that the nanoparticles tend to concentrate at the center of the microdomains, forming nanosheets within the lipid matrix, and the lamellar structure of the matrix can be preserved well when $\phi_{NP} = 0.03$. The density profile of the nanoparticles, which is magnified by a factor of 10 to make it more obvious, displays several peaks at the center of head-particle domains and tail-particle domains (Figure 3b3). Meanwhile, we observed a significant 'crater' in the center of the density profile of the hydrophilic head particles (Figure 3b3), indicating an exclusion of head particles from this region, which shows that the nanoparticles are localized within this cavity. The entire system is organized into a well-ordered 'core–shell' structure. These results are consistent with previous studies by Thompson et al. on diblock–nanoparticle mixtures to predict ordered phases based on mean field theory [24]. The phenomenon of $\phi_{NP} = 0.05$ is similar to that of $\phi_{NP} = 0.03$, as seen from Figure 3c. However, when the concentration of nanoparticles increases to $\phi_{NP} = 0.15$, the nanoparticles tend to form clusters (Figure 3d1,d2), and we can seen from Figure 3d3 that the layered distribution of nanoparticles significantly weakens where the peaks in the density distribution of nanoparticles are weaker than that shown in Figure 3b3,c3. However, the order parameters (Figure 3b4–d4) show that the orientational order of lipid molecules is almost not affected by the nanoparticles, even at ϕ_{NP} of 0.15.

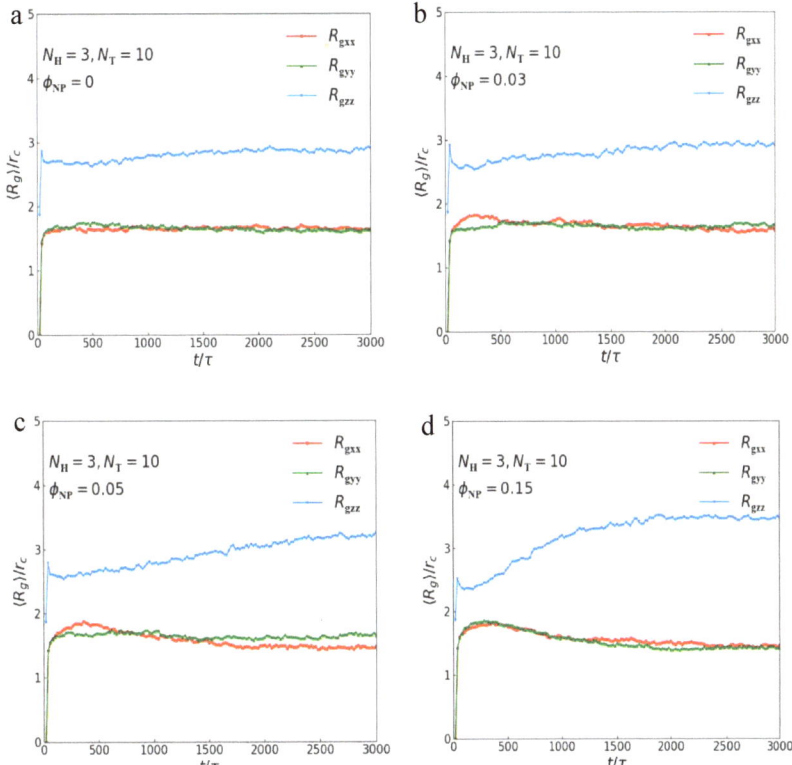

Figure 5. The average radius of gyration of lipid–nanoparticle mixtures with lamellar structure. There are three components $\langle R_{gxx} \rangle$, $\langle R_{gyy} \rangle$ and $\langle R_{gzz} \rangle$ of $\langle R_g \rangle$ at different concentrations of nanoparticles as functions of time steps with (**a**) $\phi_{NP} = 0$, (**b**) $\phi_{NP} = 0.03$, (**c**) $\phi_{NP} = 0.05$ and (**d**) $\phi_{NP} = 0.15$.

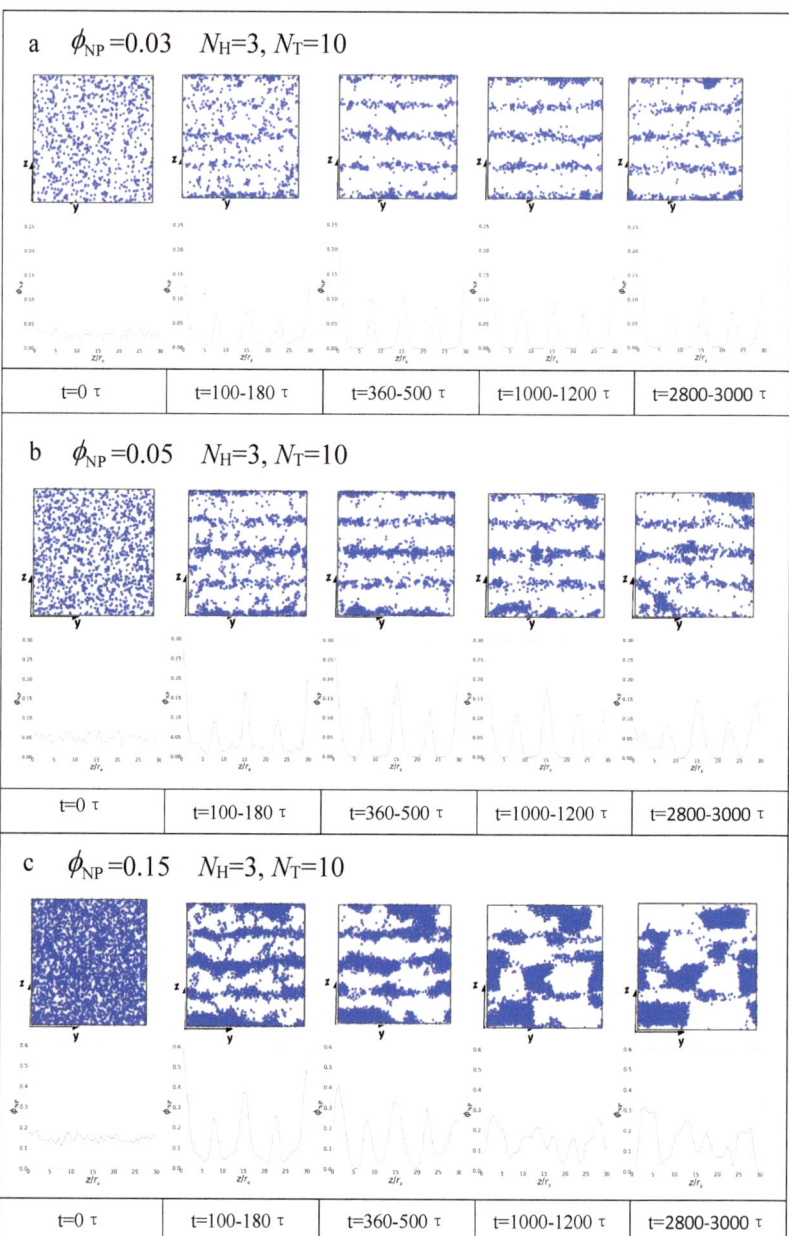

Figure 6. The dynamic processes for position distributions of nanoparticles in lamellar structures with (**a**) ϕ_{NP} = 0.03, (**b**) ϕ_{NP} = 0.05 and (**c**) ϕ_{NP} = 0.15. The upper line shows the schematic diagram of nanoparticles at each time stage, and the lower line displays the corresponding density profiles of nanoparticles.

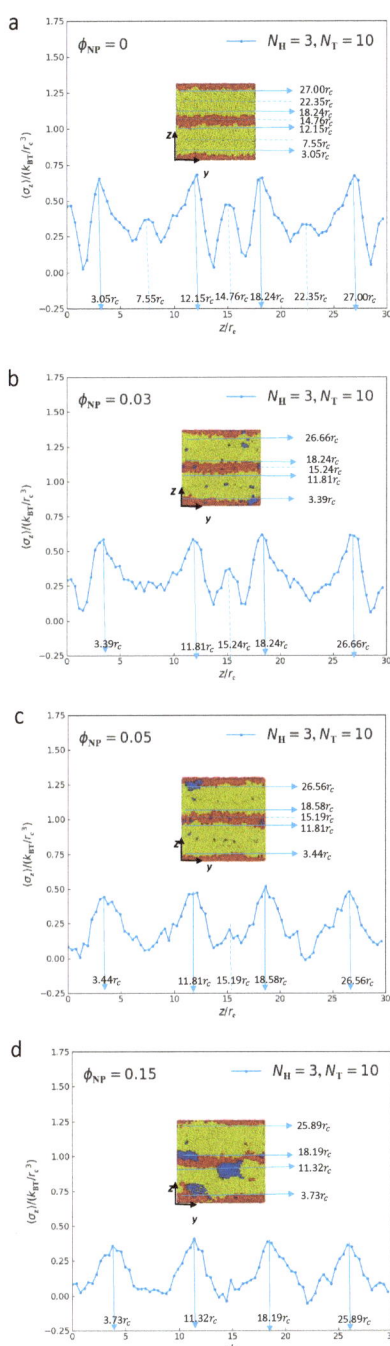

Figure 7. The average tensions ($\langle \sigma_z \rangle$) as a function of distance along the z-axis for lamellar structures with (**a**) $\phi_{NP} = 0$, (**b**) $\phi_{NP} = 0.03$, (**c**) $\phi_{NP} = 0.05$ and (**d**) $\phi_{NP} = 0.15$.

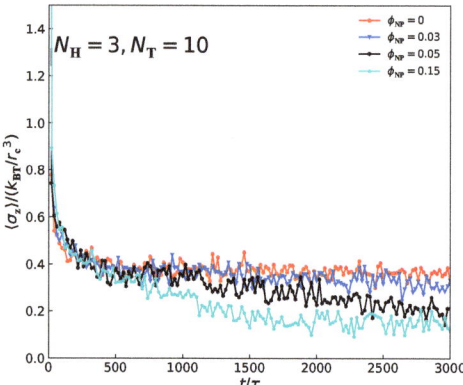

Figure 8. The average tensions ($\langle \sigma_z \rangle$) along the z-direction as a function of time for lamellar structures with $\phi_{NP} = 0$, $\phi_{NP} = 0.03$, $\phi_{NP} = 0.05$ and $\phi_{NP} = 0.15$.

In the case of hexagonal lipid–nanoparticle composites with parameters $N_H = 4$ and $N_T = 14$ (Figure 4), we can observe that the head particles form cylindrical cores, and the tail particles are gathered around the column boundaries. The lipid molecules can self-assemble into a hexagonal structure in the bulk, which is consistent with the phase diagram of glycerol mono-oleate [16,20]. Figure 4a1,a2 display the front and side views of the hexagonal structure of a pure lipid molecule. The red and yellow curves in Figure 4a3 show the radial density profiles of head and tail particles, respectively. In Figure 4a3, the red curve reaches the peak value at the interface between two microdomains and then decreases to zero. Meanwhile, the yellow curve begins to increase gradually at the interface until it reaches the boundaries of other cylinders. This density distribution diagram is consistent with the hexagonal structure. When randomly distributed nanoparticles are loaded, the nanoparticles prefer to drive to the surface of the head-particle cylinders, as shown in Figure 4b1,b2, where blue beads represent nanoparticles. In addition, the peak of the density profiles of nanoparticles, which is magnified by a factor of 20 (Figure 4b3), appears at the position where the head and tail chains join. When the concentration of nanoparticles increases to $\phi_{NP} = 0.15$, the distribution of nanoparticles is not very regular according to the snapshot of microstructure (Figure 4c1,c2); however, the density profile of nanoparticles, which is magnified by a factor of 10 (Figure 4c3), shows an obvious peak value at the position where the head and tail chains join. That is to say that the nanoparticles tend to be segregated at the interface between two microdomains in the equilibrium structure in a hexagonal structure. Previous works have reported that nanoparticles segregate to the interface when the interfacial tension between two different domains is sufficiently large and the particles are neutral in block copolymer–nanoparticle composites [55–57]. Here, we find that nanoparticles segregate to the interface between two microdomains in the hexagonal structure but localize at the center of microdomains in the lamellar structure. This finding indicates that the distribution of nanoparticles depends on the structure of the lipid matrix, which has important guiding significance for the production of new lipid nanomaterials.

3.2. Dynamic Processes

In this subsection, we concentrate on the dynamic processes of lipid–nanoparticle mixtures to understand their formation mechanism. We investigate the distinct dynamic processes of lipid–nanoparticle mixtures using different concentrations of nanoparticles by calculating the average radius of gyration and the distribution of nanoparticles with time steps. The physical quantity of the average radius of gyration $\langle R_g \rangle$ [58] is an important parameter to describe the polymer size. Thus, the three components of average radius of gyration ($\langle R_{gxx}\rangle$, $\langle R_{gyy}\rangle$ and $\langle R_{gzz}\rangle$), as functions of time steps, can reflect dynamic information about chain size in the three axes. The radius of gyration tensor can be expressed as follows:

$$\mathbf{R}_g^2 = \begin{pmatrix} R_{gxx}^2 & R_{gxy}^2 & R_{gxz}^2 \\ R_{gyx}^2 & R_{gyy}^2 & R_{gyz}^2 \\ R_{gzx}^2 & R_{gzy}^2 & R_{gzz}^2 \end{pmatrix}. \tag{12}$$

Therefore, the element $R_{g\alpha\beta}^2$ is presented as follows:

$$\langle R_{g\alpha\beta}^2 \rangle = \frac{1}{N}\sum_i \langle (r_{i,\alpha} - r_{c,\alpha})(r_{i,\beta} - r_{c,\beta})\rangle, \tag{13}$$

where $\alpha, \beta \in \{x, y, z\}$, $r_{i,\alpha}$ and $r_{c,\alpha}$ represent the α coordinate of the i-th bead and the center of mass, respectively, and N refers to the number of chains. Since we can obtain spatial positions of each bead at every time step from simulation, we can determine the center of mass of each chain and calculate the element of the average radius of gyration tensor ($\langle R_{gxx}^2\rangle$, $\langle R_{gyy}^2\rangle$ and $\langle R_{gzz}^2\rangle$) according to Equation (13); then, $\langle R_{gxx}\rangle$, $\langle R_{gyy}\rangle$ and $\langle R_{gzz}\rangle$ are the square root of $\langle R_{gxx}^2\rangle$, $\langle R_{gyy}^2\rangle$ and $\langle R_{gzz}^2\rangle$, respectively.

Previous studies have shown that the effect of nanoparticles on the size of polymer matrix is complex and that it is related to the type, size and dispersion state of nanoparticles. Mackay et al. [59] found that the polystyrene chain expanded and increased with the increase in nanoparticle concentration by neutron scattering of polystyrene/polystyrene nanoparticles, whereas a polystyrene/silica nanoparticle experiment showed that the size of the polystyrene chain was not affected [60,61]. In the current simulation, we focus on the average radius of gyration of lipid–nanoparticle mixtures with lamellar structures, where the parameters are $N_H = 3$ and $N_T = 10$. The average radius of gyration for pure lipid molecules and lipid–nanoparticle mixtures with different nanoparticle concentrations are shown in Figure 5a–d. The three components ($\langle R_{gxx}\rangle$, $\langle R_{gyy}\rangle$ and $\langle R_{gzz}\rangle$) of $\langle R_g\rangle$ are listed during the dynamic processes to clearly show the effect of nanoparticles on the microstructures. In pure lipid, we find that the three components converge to a stable value in sufficient time, as shown in Figure 5a. During the self-assembly period, the average values of $\langle R_{gxx}\rangle$ and $\langle R_{gyy}\rangle$ increase rapidly initially and then reach a stable value; we observe that the average values of $\langle R_{gxx}\rangle$ and $\langle R_{gyy}\rangle$ are basically the same during the whole simulation process, and the average values of $\langle R_{gzz}\rangle$ are much larger than $\langle R_{gxx}\rangle$ and $\langle R_{gyy}\rangle$. These results indicate the isotropic conformation of lipid molecules in the x–y plane, and the chains are arranged along the z axis, which is consistent with the microstructural morphologies and the order parameter profiles displayed in Figure 3. When nanoparticles are loaded (Figure 5b–d), we find that the values of $\langle R_{gzz}\rangle$ increase with the increase in nanoparticle concentration, while the values of $\langle R_{gxx}\rangle$ and $\langle R_{gyy}\rangle$ do not change obviously. Moreover, we calculated the mean and standard deviation values of these parameters in equilibrium states to compare them quantitatively, namely $\langle R_{gxx}\rangle = 1.65 \pm 0.17 r_c$, $\langle R_{gyy}\rangle = 1.62 \pm 0.18 r_c$ and $\langle R_{gzz}\rangle = 2.87 \pm 0.07 r_c$ for $\phi_{NP} = 0$; $\langle R_{gxx}\rangle = 1.60 \pm 0.12 r_c$, $\langle R_{gyy}\rangle = 1.64 \pm 0.12 r_c$ and $\langle R_{gzz}\rangle = 2.92 \pm 0.13 r_c$ for $\phi_{NP} = 0.03$; $\langle R_{gxx}\rangle = 1.46 \pm 0.06 r_c$, $\langle R_{gyy}\rangle = 1.59 \pm 0.12 r_c$ and $\langle R_{gzz}\rangle = 3.15 \pm 0.16 r_c$ for $\phi_{NP} = 0.05$; and $\langle R_{gxx}\rangle = 1.47 \pm 0.12 r_c$, $\langle R_{gyy}\rangle = 1.42 \pm 0.11 r_c$ and $\langle R_{gzz}\rangle = 3.47 \pm 0.28 r_c$ for $\phi_{NP} = 0.15$. These results indicate that the lipid chains are stretched in the z direction with the addition of nanoparticles.

We also investigate the dynamic processes of lamellar structures with different concentrations of nanoparticles by illustrating the distribution of nanoparticles with time steps. Figure 6 shows the variation in the distribution of nanoparticles with different concentrations of nanoparticles over time with $N_H = 3$ and $N_T = 10$. The upper part shows the distribution of nanoparticles at different time stages, and the lower part shows the corresponding density profiles of nanoparticles. Figure 6a illustrates the concentration of nanoparticles at $\phi_{NP} = 0.03$. We can observe that the nanoparticles are randomly distributed during the initial stage and then gradually present a layered distribution along the z axis with time. The appearance of several peaks corresponding to the center of microdomains on the corresponding density profiles of nanoparticles from $t = 100\tau$ confirm this. When ϕ_{NP} increases to 0.05, we find that the nanoparticles begin to form clusters from $t = 1000\tau$ (Figure 6b); however, the corresponding density profiles of nanoparticles still show layered distribution during the whole simulation process at this concentration. When ϕ_{NP} reaches 0.15, as shown in Figure 6c, the nanoparticles are distributed from random to layered; then, from $t = 1000\tau$, a large number of nanoparticles form clusters, and the layered distribution of nanoparticles disappears, which can be seen from the corresponding density profiles of nanoparticles.

Because the nanoparticles are amphiphobic in our simulation, at low concentrations, the entropy of nanoparticles dominates; therefore, they tend to disperse in the system. However, owing to the close packing of rod-like lipid chains, the free volume is much greater in the central zone of each layer, providing more space for nanoparticles [62,63]. Localizing nanoparticles in these spaces sacrifices some translational entropy of the nanoparticles but avoids an even larger chain stretching penalty incurred by distributing the nanoparticles throughout the domain (Figures 3b and 6a). When the nanoparticle concentration is increased, the distribution of nanoparticles around the center of domains narrows, and nanoparticles are densely packed around the center of the lipid domain. The nanoparticle dispersion within the domain becomes progressively more unfavorable as the lipid chains stretch farther to accommodate more particles. This increase in stretching penalty cannot be offset by nanoparticle translation entropy, thereby preventing nanoparticles from spreading throughout the domains. More nanoparticles are localized near the center of the lipid domain to avoid exceedingly large stretching penalty. Finally, above the threshold concentration of nanoparticles, the excess nanoparticles cannot assemble in the lipid domain, and the system presents an ordered lipid/nanoparticle phase coexisting with a macrophase separation of nanoparticles. As shown in Figures 3d and 6c, for $\phi_{NP} = 0.15$, large clusters of nanoparticles form, and the lipid chains are stretched in the z direction. This behavior is similar to that reported in the experimental study conducted by Kim et al. in symmetric PS-b-P2VP block copolymers blended with gold nanoparticles [64].

3.3. Mechanical Properties

In this subsection, we present the mechanical properties of lipid–nanoparticle composites with different concentrations of nanoparticles by calculating the interfacial tensions. The interfacial tension of the lipid membrane has elicited considerable interest in recent years [65–67].

Based on the Irving–Kirkwood definition, the formulation of tension (σ_z) along the z direction is presented as follows [68–72]:

$$\langle \sigma_z \rangle = \langle p_{zz} \rangle - \frac{1}{2}(\langle p_{xx} \rangle + \langle p_{yy} \rangle), \tag{14}$$

where the component of the pressure tensor (P_{zz}) can be achieved as follows:

$$\langle p_{xx} \rangle = \frac{1}{V} \left\langle \sum_{i=1}^{N} m_i v_{ix} v_{ix} + \sum_{i=1}^{N} \sum_{j>i}^{N} F_{ijx} x_{ij} \right\rangle. \tag{15}$$

where V and N are the volume and the number of DPD beads in the simulated box, respectively; x_{ij} and F_{ijx} denote the relative position and force between i-th and j-th particles along the x-axis, respectively; and the components P_{yy} and P_{zz} have the same formula as P_{xx}, only with changes in the corresponding subscripts.

Figure 7 displays the interfacial tension ($\langle \sigma_z \rangle$) as a function of distance along the z direction for the lamellar structure of lipid–nanoparticle mixtures with different nanoparticle concentrations. In pure lipid molecules (Figure 7a), there several large peaks appear at positions of $z = 3.05r_c$, $12.15r_c$, $18.24r_c$ and $27.0r_c$. We find that these locations are almost located at the interfaces between microdomains, as shown in the illustration in Figure 7a. In addition, we also observe that there are several small peaks that appear at positions of $z = 7.55r_c$, $14.76r_c$ and $22.35r_c$. These locations are the joins of two neighboring lipid chains arranged along the z direction. At positions away from these interfaces, the values of $\langle \sigma_z \rangle$ tend to be zero. The distributions of the interfacial tension ($\langle \sigma_z \rangle$) for $\phi_{NP} = 0.03$, $\phi_{NP} = 0.05$ and $\phi_{NP} = 0.15$ are similar to those of pure lipid molecules, except the values of large peaks and small peaks decrease with the increase in nanoparticle concentration. We can calculate that the average values of large peaks for pure molten lipid molecules, $\phi_{NP} = 0.03$, $\phi_{NP} = 0.05$ and $\phi_{NP} = 0.15$ are 0.654, 0.593, 0.501 and 0.390, respectively. The average values of small peaks for pure molten lipid molecules are about 0.390, and the small peaks disappear gradually when the nanoparticles are added, as shown in Figure 7b–d. This means the average interfacial tension ($\langle \sigma_z \rangle$) decreases with the increase in nanoparticles concentration. Moreover, we can determine the size of domain space based on the difference of the z-axis positions corresponding to these large peaks; the results are as follows: the size of domain spaces for tail beads are $9.1r_c$, $8.42r_c$, $8.37r_c$ and $7.59r_c$ for pure lipid molecules, $\phi_{NP} = 0.03$, $\phi_{NP} = 0.05$ and $\phi_{NP} = 0.15$, respectively, and the size of domain spaces for head beads are $6.09r_c$, $6.43r_c$, $6.70r_c$ and $6.87r_c$, respectively. That is to say that the size of domain space for head beads increases with the addition of nanoparticles, while the size of domain space for tail beads decreases with the addition of nanoparticles. This phenomenon is reasonable because the rigid head chain of lipid move apart to provide enough space for nanoparticles, which makes their thickness increase, while the flexible tail chains free up space for nanoparticles by curling them, which causes a reduction in their thickness.

In addition, we investigate the interfacial tension ($\langle \sigma_z \rangle$) changes with time steps for the lamellar structures of pure molten lipid molecules and lipid–nanoparticle mixtures (Figure 8). The red curve corresponds to pure molten lipid molecules; the blue curves correspond to $\phi_{NP} = 0.03$; and the black and green curves correspond to $\phi_{NP} = 0.05$ and $\phi_{NP} = 0.15$, respectively. In these cases, tension evolution can be divided into three parts: first, between the time of 0 and 500τ, the tension decreases rapidly, which corresponds to the random generation stage. From 500 to 2000τ, the tension shows a weak declining trend, which corresponds to the adjustment perforation stage. At this stage, the structure is still undergoing minor adjustments and evolution. Finally, the value of interfacial tensions reaches a stable value, which means the structures have already reached an equilibrium state. In addition, after reaching equilibrium states, the mean values of internal tensions are 0.36, 0.31, 0.23 and 0.15 for pure lipid molecules, $\phi_{NP} = 0.03$, $\phi_{NP} = 0.05$ and $\phi_{NP} = 0.15$, respectively. This means that the addition of nanoparticles seems to remove the interfacial tension of lipid molecules. This phenomenon is consistent with the experimental results reported by Chung et al., who showed that the introduction of silica nanoparticles into a binary polymer system can prevent the phase separation of the two polymers by gathering the nanoparticles at the interfaces [73].

4. Summary

In this work, the self-assembly behavior of lipid–nanoparticle mixtures are simulated by the DPD method. Two common structures, that is, lamellar and hexagonal structures, are observed in equilibrium states. By analyzing bead density distributions and order parameters of the equilibrium structure, we observe that nanoparticles tend to concentrate at the center of microdomains, forming nanosheets within the lipid matrix at a low nanoparticle concentration and forming clusters with the increase in nanoparticle concentration in lamellar structures. In contrast, in hexagonal structures, we find that nanoparticles segregate to the interface between two microdomains. This result indicates that the distribution of nanoparticles depends on the structure of the lipid matrix, which has important guiding significance for the production of new nanomaterials.

Then, we investigate the dynamic process of lamellar structures by calculating the average radius of gyration and the distribution of nanoparticles with time steps. Three components ($\langle R_{gxx} \rangle$, $\langle R_{gyy} \rangle$ and $\langle R_{gzz} \rangle$) of $\langle R_g \rangle$ are listed during the dynamic process to clearly show the effect of nanoparticles on the microstructure of the lipid matrix. The average values of $\langle R_{gzz} \rangle$ are much larger than $\langle R_{gxx} \rangle$ and $\langle R_{gyy} \rangle$, and the average values of $\langle R_{gxx} \rangle$ and $\langle R_{gyy} \rangle$ are almost the same, indicating the isotropic conformation of lipid molecules in the x–y plane and that the chains are arranged along the z axis. The increase in $\langle R_{gzz} \rangle$ with the increase in nanoparticle concentration indicates that the lipid chains are stretched in the z direction with the addition of nanoparticles.

We also investigate the mechanical properties of lipid–nanoparticle composites with different concentrations of nanoparticles by calculating the interfacial tensions. We observe that there several large peaks appear at positions of the interfaces between microdomains, and several small peaks appear in the middle of microdomains where the two neighboring lipid chains arranged along the z direction join. By determining the z axis of these large peaks, we can calculate the size of the microdomain space, and we find that the size of the domain space for head beads increases with the addition of nanoparticles, while the size of the domain space for tail beads decrease with the addition of nanoparticles. In addition, the average interfacial tension decreases with the increase in nanoparticle concentration after reaching equilibrium states. That is to say that the addition of nanoparticles can remove the interfacial tension between lipids molecules.

Author Contributions: Data curation, F.P.; Methodology, L.S.; Project administration, S.L.; Supervision, S.L.; Writing—original draft, F.P. All authors have read and agreed to the published version of the manuscript.

Funding: This research was funded by a Program of the National Natural Science Foundation of China (Grant No. 21973070).

Institutional Review Board Statement: Not applicable.

Informed Consent Statement: Not applicable.

Data Availability Statement: The data presented in this study are available in the article.

Conflicts of Interest: The authors declare no conflict of interest.

References

1. Kongkanand, A.; Kuwabata, S.; Girishkumar, G.; Kamat, P. Single-wall carbon nanotubes supported platinum nanoparticles with improved electrocatalytic activity for oxygen reduction reaction. *Langmuir* **2006**, *22*, 2392–2396. [CrossRef] [PubMed]
2. Yan, L.T.; Xie, X.M. Computational modeling and simulation of nanoparticle self-assembly in polymeric systems: Structures, properties and external field effects. *Prog. Polym. Sci.* **2013**, *38*, 369–405. [CrossRef]
3. Soenen, S.J.; Parak, W.J.; Rejman, J.; Manshian, B. (Intra)cellular stability of inorganic nanoparticles: Effects on cytotoxicity, particle functionality, and biomedical applications. *Chem. Rev.* **2015**, *115*, 2109–2135. [CrossRef] [PubMed]
4. Blanco, E.; Shen, H.; Ferrari, M. Principles of nanoparticle design for overcoming biological barriers to drug delivery. *Nat. Biotechnol.* **2015**, *33*, 941–951. [CrossRef]
5. Sarkar, B.; Alexandridis, P. Block copolymer–nanoparticle composites: Structure, functional properties, and processing. *Prog. Polym. Sci.* **2015**, *40*, 33–62. [CrossRef]
6. Chen, G.Y.; Roy, I.; Yang, C.H.; Prasad, P.N. Nanochemistry and nanomedicine for nanoparticle-based diagnostics and therapy. *Chem. Rev.* **2016**, *116*, 2826–2885. [CrossRef]
7. Kim, J.U.; O'Shaughnessy, B. Morphology selection of nanoparticle dispersions by polymer media. *Phys. Rev. Lett.* **2002**, *89*, 238301. [CrossRef]
8. Buxton, G.A.; Lee, J.Y.; Balazs, A.C. Computer simulation of morphologies and optical properties of filled diblock copolymers. *Macromolecules* **2003**, *36*, 9631–9637. [CrossRef]
9. Schultz, A.J.; Hall, C.K.; Genzer, J. Computer simulation of block copolymer/nanoparticle composites. *Macromolecules* **2005**, *38*, 3007–3016. [CrossRef]
10. Balazs, A.C.; Emrick, T.; Russell, T.P. Nanoparticle polymer composites: Where two small worlds meet. *Science* **2006**, *314*, 1107–1110. [CrossRef]
11. Kulkarni, C.V. Lipid crystallization: From self-assembly to hierarchical and biological ordering. *Nanoscale* **2012**, *4*, 5779–5791. [CrossRef] [PubMed]
12. Tan, J.Y.B.; Yoon, B.K.; Cho, N.J.; Lovric, J.; Jug, M.; Jackman, J.A. Lipid nanoparticle technology for delivering biologically active fatty acids and monoglycerides. *Int. J. Mol. Sci.* **2021**, *22*, 9664. [CrossRef] [PubMed]

13. Zhang, Q.L.; Gupta, S.; Emrick, T.; Russell, T.P. Surface-functionalized CdSe nanorods for assembly in diblock copolymer templates. *J. Am. Chem. Soc.* **2006**, *128*, 3898–3899. [CrossRef] [PubMed]
14. Warren, S.C.; Disalvo, F.J.; Wiesner, U. Nanoparticle-tuned assembly and disassembly of mesostructured silica hybrids. *Nat. Mater.* **2007**, *6*, 156–161. [CrossRef]
15. Zhang, J.; Chen, X.F.; Wei, H.B.; Wan, X.H. Tunable assembly of amphiphilic rod-coil block copolymers in solution. *Chem. Soc. Rev.* **2013**, *42*, 9127–9154. [CrossRef]
16. Milak, S.; Zimmer, A. Glycerol monooleate liquid crystalline phases used in drug delivery systems. *Int. J. Pharm.* **2015**, *478*, 569–587. [CrossRef]
17. Qiang, X.W.; Wang, X.H.; Ji, Y.Y.; Li, S.B.; He, L.L. Liquid-crystal self-assembly of lipid membranes on solutions: A dissipative particle dynamic simulation study. *Polymer* **2017**, *115*, 1–11. [CrossRef]
18. Shi, A.; Claridge, S.A. Lipids: An atomic toolkit for the endless frontier. *ACS Nano* **2021**, *15*, 15429–15445. [CrossRef]
19. Larsson, K. Aqueous dispersions of cubic lipid-water phases. *Curr. Opin. Colloid. Interface Sci.* **2000**, *5*, 64–69. [CrossRef]
20. Qiu, H.; Caffrey, M. The phase diagram of the monoolein/water system: Metastability and equilibrium aspects. *Biomaterials* **2000**, *21*, 223–234. [CrossRef]
21. Kulkarni, C.V.; Wachter, W.; Iglesias-Salto, G.; Engelskirchen, S.; Ahualli, S. Monoolein: A magic lipid. *Phys. Chem. Chem. Phys.* **2011**, *13*, 3004–3021. [CrossRef] [PubMed]
22. Boge, L.; Bysell, H.; Ringstad, L.; Wennman, D.; Umerska, A.; Cassisa, V.; Eriksson, J.; Joly-Guillou, M.L.; Edwards, K.; Andersson, M. Lipid-based liquid crystals as carriers for antimicrobial peptides: Phase behavior and antimicrobial effect. *Langmuir* **2016**, *32*, 4217–4228. [CrossRef] [PubMed]
23. Matougui, N.; Boge, L.; Groo, A.C.; Umerska, A.; Ringstad, L.; Bysell, H.; Saulnier, P. Lipid-based nanoformulations for peptide delivery. *Int. J. Pharm.* **2016**, *502*, 80–97. [CrossRef] [PubMed]
24. Thompson, R.B.; Ginzburg, V.V.; Matsen, M.W.; Balazs, A.C. Predicting the mesophases of copolymer-nanoparticle composites. *Science* **2001**, *292*, 2469–2472. [CrossRef] [PubMed]
25. Lauw, Y. Equilibrium morphologies of nonionic lipid-nanoparticle mixtures in water: A self-consistent mean-field prediction. *J. Colloid Interface Sci.* **2009**, *332*, 491–496. [CrossRef]
26. Tang, J.Z; Jiang, Y.; Zhang, X.H.; Yan, D.D.; Chen, J.Z.Y. Phase diagram of rod-coil diblock copolymer melts. *Macromolecules* **2015**, *48*, 9060–9070. [CrossRef]
27. Esteban-Martin, S.; Salgado, J. Self-assembling of peptide/membrane complexes by atomistic molecular dynamics simulations. *Biophys. J.* **2007**, *92*, 903–912. [CrossRef]
28. Poger, D.; Van Gunsteren, W.F. ; Mark, A.E. A new force field for simulating phosphatidylcholine bilayers. *J. Comput. Chem.* **2010**, *31*, 1117–1125. [CrossRef]
29. Shinoda, W.; DeVane, R.; Klein, M.L. Zwitterionic lipid assemblies: Molecular dynamics studies of monolayers, bilayers, and vesicles using a new coarse grain force field. *J. Phys. Chem. B* **2010**, *114*, 6836–6849. [CrossRef]
30. Skjevik, A.A.; Madej, B.D.; Dickson, C.J.; Teigen, K.; Walker, R.C.; Gould, I.R. All-atom lipid bilayer self-assembly with the AMBER and CHARMM lipid force fields. *Chem. Commun.* **2015**, *51*, 4402–4405. [CrossRef]
31. Wu, H.H.; He, L.L.; Wang, X.H.; Wang, Y.W.; Jiang, Z.T. Liquid crystalline assembly of rod-coil diblock copolymer and homopolymer blends by dissipative particle dynamics simulation. *Soft Matter.* **2014**, *10*, 6278–6285. [CrossRef] [PubMed]
32. Nam, C.; Lee, W.B.; Kim, Y. Self-assembly of rod-coil diblock copolymer-nanoparticle composites in thin films: Dissipative particle dynamics. *Soft Matter* **2021**, *17*, 2384–2391. [CrossRef] [PubMed]
33. Hu, S.W.; Huang, C.Y.; Tsao, H.K.; Sheng, Y.J. Hybrid membranes of lipids and diblock copolymers: From homogeneity to rafts to phase separation. *Phys. Rev. E* **2019**, *99*, 012403. [CrossRef] [PubMed]
34. Cai, C.H.; Wang, L.Q.; Lin, J.P.; Zhang, X. Morphology transformation of hybrid micelles self-assembled from rod-coil block copolymer and nanoparticles. *Langmuir* **2012**, *28*, 4515–4524. [CrossRef]
35. Kranenburg, M.; Smit, B. Phase behavior of model lipid bilayers. *J. Phys. Chem. B* **2005**, *109*, 6553–6563. [CrossRef]
36. Hoogerbru, P.J.; Koelman, J.M.V.A. Simulating microscopic hydrodynamic phenomena with dissipative particle dynamics. *Europhys. Lett.* **1992**, *19*, 155–160. [CrossRef]
37. Koelman, J.M.V.A.; Hoogerbru, P.J. Dynamic Simulations of Hard-Sphere Suspensions Under Steady Shear. *Europhys. Lett.* **1993**, *21*, 363–368. [CrossRef]
38. Espanol, P.; Warren, P.B. Perspective: Dissipative particle dynamics. *J. Chem. Phys.* **2017**, *146*, 150901. [CrossRef]
39. Basan, M.; Prost, J.; Joanny, J.F.; Elgeti, J. Dissipative particle dynamics simulations for biological tissues: Rheology and competition. *Phys. Biol.* **2011**, *8*, 026014. [CrossRef]
40. Sevink, G.J.A.; Fraaije, J.G.E.M. Efficient solvent-free dissipative particle dynamics for lipid bilayers. *Soft Matter.* **2014**, *10*, 5129–5146. [CrossRef]
41. Lin, C.M.; Li, C.S.; Sheng, Y.J.; Wu, D.T.; Tsao, H.K. Size-dependent properties of small unilamellar vesicles formed by model lipids. *Langmuir* **2012**, *28*, 689–700. [CrossRef] [PubMed]
42. Groot, R.D.; Rabone, K.L. Mesoscopic simulation of cell membrane damage, morphology change and rupture by nonionic surfactants. *Biophys. J.* **2001**, *81*, 725–736. [CrossRef] [PubMed]
43. Venturoli, M.; Smit, B.; Sperotto, M.M. Simulation studies of protein-induced bilayer deformations, and lipid-induced protein tilting, on a mesoscopic model for lipid bilayers with embedded proteins. *Biophys. J.* **2005**, *88*, 1778–1798. [CrossRef] [PubMed]

44. Zhang, L.Y.; Becton, M.; Wang, X.Q. Designing nanoparticle translocation through cell membranes by varying amphiphilic polymer coatings. *J. Phys. Chem. B* **2015**, *119*, 3786–3794. [CrossRef]
45. Shelley, J.C.; Shelley, M.Y.; Reeder, R.C.; Bandyopadhyay, S.; Klein, M.L. A coarse grain model for phospholipid simulations. *J. Phys. Chem. B* **2001**, *105*, 4464–4470. [CrossRef]
46. Marrink, S.J.; de Vries, A.H.; Mark, A.E. Coarse grained model for semiquantitative lipid simulations. *J. Phys. Chem. B* **2004**, *108*, 750–760. [CrossRef]
47. Groot, R.D.; Warren, P.B. Dissipative particle dynamics: Bridging the gap between atomistic and mesoscopic simulation. *J. Chem. Phys.* **1997**, *107*, 4423–4435. [CrossRef]
48. Groot, R.D.; Madden, T.J. Dynamic simulation of diblock copolymer microphase separation. *J. Chem. Phys.* **1998**, *108*, 8713–8724. [CrossRef]
49. Ding, H.M.; Ma, Y.Q. Design maps for cellular uptake of gene nanovectors by computer simulation. *Biomaterials* **2013**, *34*, 8401–8407. [CrossRef]
50. Ding, H.M.; Ma, Y.Q. Theoretical and computational investigations of nanoparticle-biomembrane interactions in cellular delivery. *Small* **2015**, *11*, 1055–1071. [CrossRef]
51. Maiti, A.; McGrother, S. Bead-bead interaction parameters in dissipative particle dynamics: Relation to bead-size, solubility parameter, and surface tension. *J. Chem. Phys.* **2004**, *120*, 1594–1601. [CrossRef] [PubMed]
52. Velazquez, M.E.; Gama-Goicochea, A.; Gonzalez-Melchor, M.; Neria, M.; Alejandre, J. Finite-size effects in dissipative particle dynamics simulations. *J. Chem. Phys.* **2006**, *124*, 084104. [CrossRef]
53. Hadley, K.R.; McCabe, C. A simulation study of the self-assembly of coarse-grained skin lipids. *Soft Matter.* **2012**, *8*, 4802–4814. [CrossRef] [PubMed]
54. Skjevik, A.A.; Madej, B.D.; Walker, R.C.; Teigen, K. LIPID11: A modular framework for lipid simulations using amber. *J. Phys. Chem. B* **2012**, *116*, 11124–11136. [CrossRef] [PubMed]
55. Matsen, M.W.; Thompson, R.B. Particle Distributions in a Block Copolymer Nanocomposite. *Macromolecules* **2008**, *41*, 1853–1860. [CrossRef]
56. Walther, A.; Matussek, K.; Muller, A.H.E. Engineering nanostructured polymer blends with controlled nanoparticle location using Janus particles. *ACS Nano* **2008**, *2*, 1167–1178. [CrossRef]
57. Kim, J.U.; Matsen, M.W. Positioning Janus nanoparticles in block copolymer scaffolds. *Phys. Rev. Lett.* **2009**, *102*, 078303. [CrossRef]
58. Huang, W.W.; Zaburdaev, V. The shape of pinned forced polymer loops. *Soft Matter.* **2019**, *15*, 1785–1792. [CrossRef]
59. Tuteja, A.; Duxbury, P.M.; Mackay, M.E. Polymer chain swelling induced by dispersed nanoparticles. *Phys. Rev. Lett.* **2008**, *100*, 077801. [CrossRef]
60. Sen, S.; Xie, Y.P.; Kumar, S.K.; Yang, H.; Bansal, A.; Ho, D.L.; Hall, L.; Hooper, J.B.; Schweizer, K.S. Chain conformations and bound-layer correlations in polymer nanocomposites. *Phys. Rev. Lett.* **2007**, *98*, 128302. [CrossRef]
61. Crawford, M.K.; Smalley, R.J.; Cohen, G.; Hogan, B.; Wood, B.; Kumar, S.K.; Melnichenko, Y.B.; He, L.; Guise, W.; Hammouda, B. Chain conformation in polymer nanocomposites with uniformly dispersed nanoparticles. *Phys. Rev. Lett.* **2013**, *110*, 196001. [CrossRef] [PubMed]
62. Teng, C.Y.; Sheng, Y.J.; Tsao, H.K. Boundary-induced segregation in nanoscale thin films of athermal polymer blends. *Soft Matter.* **2016**, *12*, 4603–4610. [CrossRef] [PubMed]
63. Cheng, Y.T.; Tsao, H.K.; Sheng, Y.J. Interfacial assembly of nanorods: Smectic alignment and multilayer stacking. *Nanoscale* **2021**, *13*, 14236–14244. [CrossRef] [PubMed]
64. Kim, B.J.; Fredrickson, G.H.; Bang, J.; Hawker, C.J.; Kramer, E.J. Tailoring Core-Shell Polymer-Coated Nanoparticles as Block Copolymer Surfactants. *Macromolecules* **2009**, *42*, 6193–6201. [CrossRef]
65. Petelska, A.D. Interfacial tension of bilayer lipid membranes. *Cent. Eur. J. Chem.* **2012**, *10*, 16–26. [CrossRef]
66. Takei, T.; Yaguchi, T.; Fujii, T.; Nomoto, T.; Toyota, T.; Fujinami, M. Measurement of membrane tension of free standing lipid bilayers via laser-induced surface deformation spectroscopy. *Soft Matter.* **2015**, *11*, 8641–8647. [CrossRef]
67. Lamour, G.; Allard, A.; Pelta, J.; Labdi, S.; Lenz, M.; Campillo, C. Mapping and Modeling the Nanomechanics of Bare and Protein-Coated Lipid Nanotubes. *Phys. Rev. X* **2020**, *10*, 011031. [CrossRef]
68. Irving, J.H.; Kirkwood, J.G. The statistical mechanical theory of transport processes. IV. the equations of hydrodynamics. *J. Chem. Phys.* **1950**, *18*, 817–829. [CrossRef]
69. Heinz, H.; Paul, W.; Binder, K. Calculation of local pressure tensors in systems with many-body interactions. *Phys. Rev. E* **2005**, *72*, 066704. [CrossRef]
70. Chang, C.C.; Sheng, Y.J.; Tsao, H.K. Wetting hysteresis of nanodrops on nanorough surfaces. *Phys. Rev. E* **2016**, *94*, 042807. [CrossRef]
71. Ting, C.L.; Muller, M. Membrane stress profiles from self-consistent field theory. *J. Chem. Phys.* **2017**, *146*, 104901. [CrossRef] [PubMed]

72. Chu, K.C.; Tsao, H.K.; Sheng, Y.J. Spontaneous spreading of nanodroplets on partially wetting surfaces with continuous grooves: Synergy of imbibition and capillary. *J. Mol. Liq.* **2021**, *339*, 117270. [CrossRef]
73. Chung, H.J.; Ohno, K.; Fukuda, T.; Composto, R.J. Self-regulated structures in nanocomposites by directed nanoparticle assembly. *Nano Lett.* **2005**, *5*, 1878–1882. [CrossRef] [PubMed]

Disclaimer/Publisher's Note: The statements, opinions and data contained in all publications are solely those of the individual author(s) and contributor(s) and not of MDPI and/or the editor(s). MDPI and/or the editor(s) disclaim responsibility for any injury to people or property resulting from any ideas, methods, instructions or products referred to in the content.

Article

Fiber Orientation Quantification for Large Area Additively Manufactured Parts Using SEM Imaging

Rifat Ara Nargis and David Abram Jack *

Department of Mechanical Engineering, Baylor University, Waco, TX 76798, USA; rifat_nargis1@baylor.edu
* Correspondence: david_jack@baylor.edu

Abstract: Polymer-based additively manufactured parts are increasing in popularity for industrial applications due to their ease of manufacturing and design form freedom, but their structural and thermal performances are often limited to those of the base polymer system. These limitations can be mitigated by the addition of carbon fiber reinforcements to the polymer matrix, which enhances both the structural performance and the dimensional stability during cooling. The local fiber orientation within the processed beads directly impacts the mechanical and thermal performances, and correlating the orientation to processing parameter variations would lead to better part quality. This study presents a novel approach for analyzing the spatially varying fiber orientation through the use of scanning electron microscopy (SEM). This paper presents the sample preparation procedure including SEM image acquisition and analysis methods to quantify the internal fiber orientation of additively manufactured carbon fiber-reinforced composites. Large area additively manufactured beads with 13% by weight large aspect ratio carbon fiber-reinforced acrylonitrile butadiene styrene (ABS) pellets are the feedstock used in this study. Fiber orientation is quantified using the method of ellipses (MoE), and the spatial change in fiber orientation across the deposited bead cross-section is studied as a function of process parameters including extrusion speed, raster height, and extrusion temperature zones. The results in the present paper show the results from the novel use of SEM to obtain the local fiber orientation, and results show the variation in alignment within the individual processed bead as well as an overall aligned orientation state along the direction of deposition.

Keywords: large area additive manufacturing; carbon fiber-reinforced polymer matrix composite; fiber orientation characterization; method of ellipses; scanning electron microscopy

Citation: Nargis, R.A.; Jack, D.A. Fiber Orientation Quantification for Large Area Additively Manufactured Parts Using SEM Imaging. *Polymers* **2023**, *15*, 2871. https://doi.org/10.3390/polym15132871

Academic Editors: Francesco Galiano, Mustafa Özcanli, Phuong Nguyen-Tri and Ana Pilipović

Received: 27 May 2023
Revised: 16 June 2023
Accepted: 22 June 2023
Published: 29 June 2023

Copyright: © 2023 by the authors. Licensee MDPI, Basel, Switzerland. This article is an open access article distributed under the terms and conditions of the Creative Commons Attribution (CC BY) license (https://creativecommons.org/licenses/by/4.0/).

1. Introduction

Polymer composites are seeing increased use in the manufacturing industry because of their enhanced mechanical and thermal properties, low densities, formability, and ability to mass-produce components. Of interest in the present scope is the use of short fiber-reinforced thermoplastic polymer composites to enhance structures fabricated through an additive manufacturing (AM) process. Additive manufacturing (AM), which is often termed freeform fabrication or 3D printing, is a process in which parts are built by depositing materials such as plastic, composite, metal, ceramic, and concrete, layer by layer, according to digital 3D design data. There are many variations of different additive manufacturing processes, among which fused deposition modeling (FDM) is one of the more commonly employed AM methods. Fused deposition modeling, also known as fused filament fabrication (FFF), is an additive manufacturing process that often uses thermoplastic materials such as acrylonitrile butadiene styrene (ABS), nylon, and polylactic acid (PLA), often in the form of continuous filament, fed from a heated nozzle or printer extruder head and methodically deposited to form layers. The large area additive manufacturing (LAAM) system used for this study would be considered within the family of fused deposition modeling. This system was first studied at Oak Ridge National Laboratory, where Duty et al. [1] demonstrated the ability to take raw fiber-filled pellets and fabricate large-volume AM structures. A review by Vicente et al. [2] studied recent applications for LAAM systems, showed their use in the transportation industry, and concluded that the LAAM process may be the most suitable approach for the production of large-scale components. Viciente et al. highlight that LAAM is used in the aerospace industry for reducing complex assemblies, large-scale mock-ups, and speeding the design process for new systems. In the automotive industry, Viciente et al. show use cases for LAAM for electric vehicles, prototyping, and tooling for large components. The system utilized in the present scope of work utilizes a pellet-fed system with a high-flow-rate extruder capable of ~25 lbs/h deposition. These pellet-fed systems allow for higher deposition rates while simultaneously reducing the cost of the raw materials. Recently, Pricci et al. [3] presented a sophisticated model to correlate the screw speed for a pellet-fed system similar to that of Duty et al. to that of the desired mass flow rate. Similarly, Barera et al. [4] studied the impact of the temperature and extrusion pressure on the final part performance for carbon fiber-filled polyamide 6. Yeole [5] used the LAAM process to fabricate large-scale molds and showed the anisotropy of the individual extrudate as well as the sensitivity of the final part performance to the choice of printing path. In the present study, we select ranges of screw speeds and adjust the translation speed of the extruder to obtain the desired cross-sectional area. A large area additive manufacturing (LAAM) printer is used to melt the thermoplastic pellet, and the molten material is deposited through an extruder. The material is then extruded to a heated build plate or platform where the resulting anisotropic material cools and hardens.

The polymer used in the present study is acrylonitrile butadiene styrene as it is broadly used in additively manufactured part production. ABS is a terpolymer that consists of butadiene rubber dispersed in a poly styrene-co-acrylonitrile matrix [6]. ABS has an improved toughness relative to many mass-produced polymer blends that require processing temperatures below 250 °C. The structural and thermal performance of ABS is limited, but the addition of fiber reinforcements, specifically carbon fibers, allows for enhanced structural performance while simultaneously reducing dimensional variations during thermal changes. Specifically, Love et al. [7] showed that the printed neat ABS had a CTE of 87 µm/°C, whereas their 13% carbon fiber-filled system had a CTE of 10 µm/°C along the deposition direction but 106 µm/°C transverse to the print direction. Lobov et al. [8] demonstrated that the incorporation of carbon fiber in their ABS filament improved the tensile strength by 30% and the fracture toughness by 20% in their 3D printed system. Wang et al. [9] studied 3D printed ABS systems with both carbon and Kevlar fiber reinforcements. They observed an improvement in both the strength and stiffness upon the incorporation of carbon fibers, whereas the Kevlar fibers had little to no effect on the stiffness and strength. Actually, the Kevlar fiber reduced the performance relative to the neat ABS system. Carbon fiber-filled thermoplastics allow for design tailoring ability at a relatively low cost, as demonstrated in [10]. Velez-Garcia et al. [11] demonstrated for an injection molded composite that the quantification of the fiber orientation and any induced porosity due to the addition of the fibers is essential to understanding the mechanical, structural, and thermal properties of the final manufactured part.

One of the most prevalent methods to measure the fiber orientation in a polymer matrix is the method of ellipses (MoE) (see, e.g., [12]). Using MoE, Velez-Garcia [13] showed the characteristic values of complete and incomplete elliptical footprints, and the resulting fiber orientations were quantified from micrographs. Hofmann et al. [14] extended the method to look at fiber flexure within the section region for fibers with an aspect ratio significantly larger than those in the present study. The method of ellipses for rigid fibers has its own limitations, specifically, the ambiguity problem where a pair of orientation states can be represented by a given elliptical footprint (see, e.g., [15]). Typically, the standard method of ellipses for circular fibers produces an ambiguity of the fiber orientation angle relative to the surface normal, causing an undetermined state for the off-diagonal components of the orientation tensor. However, two ambiguities can exist for elliptical cross-section fibers because of the additional degree of freedom of the roll about the axis of the fiber [16], which is not considered with the assumed axisymmetric fibers used in the present research.

As discussed by Goldstein [17], scanning electron microscopy (SEM) generates images of a sample by scanning the surface with electron beams. There are two types of electrons that can be captured with the system used in the present research: high-energy backscattered electrons and low-energy secondary electrons. From the reflected signal, SEM provides data on the surface topology and composition of the sample (see, e.g., [17]) and can generate three-dimensional-like images of the surface with the help of a large depth of field and shadow relief effects of the secondary and backscattered electron contrast. The JEOL SEM 6610-LV (USA headquarters in Peabody, Massachusetts) used in this study can achieve up to a 300,000× magnification in a high vacuum mode, but in the present study, it is the use of the low vacuum mode that allows for the investigation of features just below the surface of the polished sample without the need to sputter coat the surface with an electrically conductive material. The present work utilizes the electrical mismatch between the semi-conducting carbon fibers and the electrically insulating polymer matrix to yield high-resolution images of the boundary between individual carbon fibers and an ABS matrix. Additional non-optical methods were considered by the authors, such as optical-FTIR and AFM, and may be worth future studies. For example, the mismatch in stiffness between the polymer matrix and the carbon fiber may be detectable by AFM. Similarly, the difference in the IR spectrum of the polymer matrix and the carbon fiber may be identifiable using high-resolution optical-FTIR. Due to the availability of SEM to the researchers, SEM was used in this study to produce electrical surface maps to quantify the direction of each fiber in the etched surface, and the results in the present paper provide a potential alternative method to remove the orientation ambiguity.

In fiber-reinforced polymers, the volume fraction of fiber, type of fiber, fiber size, and fiber orientation directly influence the performance of the material (see, e.g., [18]). Fiber orientation in the matrix may vary due to the motion of the fiber compared to the surrounding matrix in the material flow from the nozzle during processing. Love et al. [7] demonstrated that the addition of carbon fibers into the acrylonitrile butadiene styrene (ABS) polymer matrix showed a significant increase in the stiffness and strength of their FDM 3D printed parts. The final 3D printed part bears most of the loads in the print direction where the fibers are mostly aligned, as demonstrated by Tekinalp et al. [19]. The tensile strength and tensile modulus were shown in [19,20] with the addition of carbon fibers. Specifically, in [19], the authors found that the addition of carbon fibers enhanced the tensile strength and tensile stiffness by 115% and 700%, respectively. Russell and Jack [21] obtained a stiffness of 2.25 GPa for their neat ABS system, which was increased to 3.2 GPa for their lightly loaded, moderately aligned carbon fiber-reinforced system. Similarly, Duty et al. [1] found that through the addition of carbon fibers and enhancing the elongational effects to enhance the fiber alignment, they were able to achieve a stiffness of 8.2 GPa for their carbon fiber-filled ABS system. In addition to impacting the structural performance, the fiber orientation will influence the thermal properties such as thermal conductivity and coefficient of thermal expansion. The thermal conductivity was shown by Chen and Wang [22] to be substantially higher where the fibers have a higher degree of fiber orientation for their injection-molded composite. Hassan et al. [23] showed that through the addition of carbon fibers to their ABS system, the thermal conductivity increased by a factor of 3 to 7 times that of the neat polymer as a function of orientation. In some cases, they observed that the conductivity along the aligned fiber direction was 3 to 4 times that transverse to the alignment. Tezvergil et al. [24] showed that in the direction of fiber alignment, the coefficient of thermal expansion is lower than that of the base polymer, whereas transverse to the fiber alignment direction, the coefficient of thermal expansion is closer to that of the base polymer.

In the present study, a method is presented to quantify the local fiber orientation of an additively manufactured carbon fiber-reinforced composite through the use of scanning electron microscopy. In addition, the industrially relevant LAAM-deposited beads, with a cross-sectional area greater than 20 mm^2, are investigated, whereas most literature focuses on beads with a cross-sectional area less than 1 mm^2. The approach in the present study also differs from that of the literature which typically relies upon optical microscopy methods. Samples are formed using the large area additive manufacturing process and are then sectioned, polished, and imaged using SEM. The SEM approach leverages the mismatch in electrical conductivity between that of the semi-conducting carbon fiber and the electrically insulating polymer matrix. In addition to presenting a new method for quantifying the local changes in orientation within the LAAM-deposited beads, several process parameters unique to the LAAM approach, specifically screw speed, processing temperature zones, and bead cross-sectional area, are studied for their impact on the final orientation state of fibers.

2. Materials and Methods

The following sections discuss the selection of various process parameters studied for the LAAM system, the sample preparation procedure, and the image acquisition. In total, 27 different sets of processing parameters were investigated over a range of temperature profiles in the extruder, extrusion speed, and cross-sectional area of the deposited bead. The LAAM system was custom designed by the researchers to process fiber filled pellets with the full specifications and part component listing given in [25]. In addition, six additional specimens were fabricated to correlate the sensitivity of the developed fiber orientation quantification method to the sample size.

2.1. Sample Fabrication

In this research, to study the effects of different process parameters on the fiber orientation, three process parameters were selected: nozzle temperature, screw speed within the extruder by varying the RPM, and the area of the cross-section of the deposited extrudate. The extruder had three different heated zones. The three set point temperatures for zones 1, 2, and 3, were set to three different scenarios, specifically 190/195/200 °C, 200/205/210 °C, and 210/215/220 °C, with the last temperature zone listed being closest to the extrusion nozzle. Three different RPM values were selected for the single screw extruder: 1500 RPM, 1875 RPM, and 2250 RPM. Lastly, the three cross-sectional areas studied were 22 mm^2, 27 mm^2, and 32 mm^2, where the cross-sectional area was of the deposited bead. The extruder height relative to the print bed was kept fixed for all specimens in the present study at 3 mm from the height of the previously deposited layer. The cross-sectional area of a printed bead for a certain screw speed and a certain temperature was controlled by adjusting the gantry speed. This resulted in a total of $3^3 = 27$ samples in this study to analyze for fiber orientation. All samples in the present study were composed of 3 layers of deposited beads, with the interior bead being used for all the orientation analyses. This interior bead would be indicative of the majority of the deposited beads in a layered structure, whereas the top and bottom beads would potentially have a different orientation state due to edge effects.

To control the cross-sectional area of the final print part, the G-code controlling the gantry speed was modified. The mass flow rate $\dot{m}_{ABS}$ was measured using a standard laboratory scale and stopwatch for each of the 9 variations of extruder zone temperatures and extruder screw speed. The volumetric flow rate, $\dot{V}_{ABS}$, is found from the mass flow rate $\dot{m}_{ABS}$ divided by the density ρ_{ABS} as follows:

$$\dot{V}_{ABS} = \frac{\dot{m}_{ABS}}{\rho_{ABS}} = \frac{m_{ABS}}{t\,\rho_{ABS}} \qquad (1)$$

where $\dot{m}_{ABS}$ is the measured mass of extruded polymer over the duration of time, t. This can be converted to the speed of the gantry by dividing it by the desired cross-sectional area A_{ABS} as follows:

$$Gantry\ Speed = \frac{\dot{V}_{ABS'}}{A_{ABS}} \qquad (2)$$

The actual cross-sectional areas of the layers of the printed sample were validated against the estimate from Equation (2) with the help of optical microscopy of sectioned samples. The 27 samples are categorized according to the process parameters and shown in Table 1.

Table 1. Characterization of sample number according to the process parameters.

		Screw Speed 1500 RPM	Screw Speed 1875 RPM	Screw Speed 2250 RPM
Temperature Zones 190/195/200 (°C)	Area 22 mm^2	10	13	16
	Area 27 mm^2	11	14	17
	Area 32 mm^2	12	15	18
Temperature Zones 200/205/210 (°C)	Area 22 mm^2	1	4	7
	Area 27 mm^2	2	5	8
	Area 32 mm^2	3	6	9
Temperature Zones 210/215/220 (°C)	Area 22 mm^2	19	22	25
	Area 22 mm^2	10	13	16
	Area 27 mm^2	11	14	17

2.2. Surface Preparation for Imaging

Proper surface preparation of the sectioned samples is critical to obtain the desired characteristic material and topological data for both optical and scanning electron microscopy imaging, specifically obtaining a high contrast between the fiber edge and the surrounding polymer matrix. Sample preparation steps include 3D printing the chopped carbon fiber-filled ABS on the available LAAM system, sectioning samples using a low-speed saw, placing samples into a thermoset mold, polishing the molds to expose the sample surface, cleaning the sample surface during polishing, and etching the polished samples to expose the fiber edges.

For polishing the samples, a proper recipe is very important for obtaining the best micrographs while using microscopes. Velez-Garcia et al. [11] presented a methodology for polishing composites with glass fibers that was not effective for the carbon fiber-reinforced ABS samples used in the present study. A parametric study was performed in the present research, and a suitable polishing technique for the carbon fiber-reinforced composites was identified. The carbon fiber samples were polished with a Buehler EcoMat 3000 (headquarters, Lake Bluff, Illinois, USA) variable-speed grinder automatic polisher, with 120-, 320-, 400-, 600-, and 1200-grade silicon carbide electro-coated abrasive sandpaper. Samples were polished in a wet environment with 120-grade sandpaper for 2 min, 320-grade for 4 min, 400-grade for 6 min, 600-grade for 6 min, and 1200-grade for 12 min. Wet conditions while polishing are essential to avoid excessive fiber breakage as carbon fibers are very lightweight and the grinding from the polisher can easily break the fibers and embed them into the surrounding polymer matrix, as was observed in the present study along with the earlier work of Velez-Garcia et al. [11]. The continuous flow of the fluid, in this case, water, helps to eliminate the debris.

Next, the sample was polished with a micro-cloth with an alumina polishing compound in three stages. In each stage, a successively smaller grit of the polishing compound was used: 5-micron, 1-micron, and then 0.5-micron. Each step was performed for 30 min for each of the three polishing steps, and polishing compounds were added every 10 min. The sequence of polishing was repeated once to enhance the surface's smoothness. In all polishing steps, a wet environment was maintained continuously to avoid excessive breakage of fibers on the surface and to enhance the exposed fiber footprint border.

High-frequency and high-intensity sound waves are used for removing foreign contaminants from the surface of a sample, as performed in [26]. During polishing, the sample was cleaned for 30 s with a Branson 1510 ultrasound sonicator (St. Louis, MO, USA), every 10 min during sanding, to remove any debris that could accumulate and scratch the surface.

During sonication, an ultrasound polishing compound was mixed with water in a ratio of 6:1 to serve as a cleaning agent and rinsed in water after sonication.

Typical SEM produces images in grayscale and often produces images of a degraded low contrast [27]. There are many image processing techniques for enhancing degraded images (see, e.g., [16,25]). In the present work, plasma etching, similar to that described in [11], was used to remove a small layer of material from the polished surface to enhance the contrast without the use of image processing. After etching, the fibers slightly stick out of the matrix, which provides a clearer distinction from the matrix. The sample was etched with a Plasma-Etch PS-50 (Carson City, NV, USA) for 40–50 min in an environment of oxygen mixed with carbon tetrafluoride (CF_4). The ratio of oxygen and CF_4 gas was 3:1. The before and after images taken from different regions of the same sample are provided in Figure 1, and the enhanced contrast from the small removal of the polymer matrix from the fiber perimeter can be observed. Notice that in Figure 1a there is little contrast between the carbon fibers and the surrounding polymer matrix. However, after etching, the fibers in Figure 1b are clearly evident, and their surface area is easier to quantify. For all SEM images, the system settings, including the working distance, spot size, electron voltage, and magnification, are shown alongside the scale bar. All samples in the present study were etched after the completion of the final polishing step.

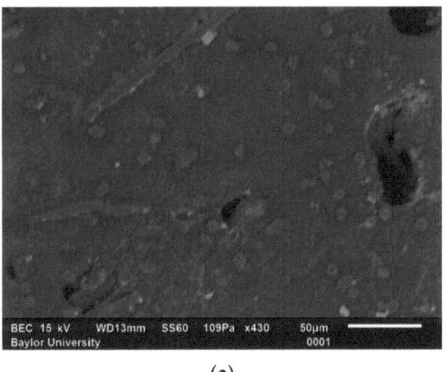

(a)

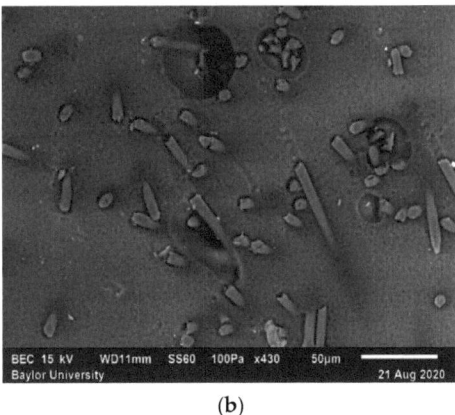

(b)

Figure 1. SEM images (**a**) before plasma etching and (**b**) after plasma etching a sample.

2.3. Image Acquisition Process from Scanning Electron Microscope

A JEOL JSM-6610LV Scan Electron Microscope was used to capture high-resolution images with an increased depth of field of the polished sample using the available backscatter detection mode.

In Figure 2, an optical image of the polished sample is shown at 20× magnification. Observe in the figure that there are three deposited layers, termed beads. The sample is from a custom large area additively manufactured system as described in Russell [25]. The typical fiber length produced in this system is on the order of 200~300 μm with a nominal aspect ratio ranging from 20 to 40, much larger than that of standard fused filament systems with fiber aspect ratios ranging from 5 to 10. The larger-aspect-ratio fibers provide a significant improvement in the final processed part performance. In the present study, we will focus on the middle bead as that would be representative of the core region of the fabricated component. This middle bead is further subdivided into 9 sub-regions as depicted in Figure 2, and the results for the orientation state within each region will be quantified in the following results. Later results will show that the center section of the bead, Region 5, has the lowest alignment state, whereas the corner regions of the deposited bead, Regions 1, 3, 7, and 9, tend to have the highest alignment state.

In Figure 3, the image acquired from the scanning electron microscope of Region 1 from Figure 2 is shown. In Region 1, the fibers are mostly oriented towards the out-of-plane direction, defined as x_3 (into and out of the page). The fibers are the regions of the circular or near-circular ellipses. Notice in the figure several small areas where the signal appears to be saturated, indicated by white. These regions are caused by poor surface conductivity and were identified to be the loose alumina polishing compound entrapped within voids that were exposed during polishing. The dark region in the center of the image is a large void within the extrudate. The smaller voids that are shown by the entrapped alumina and the macro voids will both contribute to a reduction in the overall structural performance, whereas the larger voids tend to randomize the final alignment state.

Shown in Figure 4 is the SEM image of the center-middle region of the sample extruded bead depicted in Figure 2 and identified as Region 5. In contrast to Region 1, in Region 5, the fibers are not uniformly dispersed but have a banded nature to the dispersion. In addition, the ellipsoids observed have a larger major-to-minor axis ratio indicating less alignment in the flow direction. This latter observation is made based on viewing the variations in the cross-sectioned ellipses. It is also clear that there are several significant voids measurably larger than the fibers themselves as well as multiple voids on the order of the fiber diameters present throughout the observed region. The SEM images of all 9 regions for each of 27 samples defined in Table 1 were collected as part of this study and were used in the present work for the fiber orientation quantification presented in a later section.

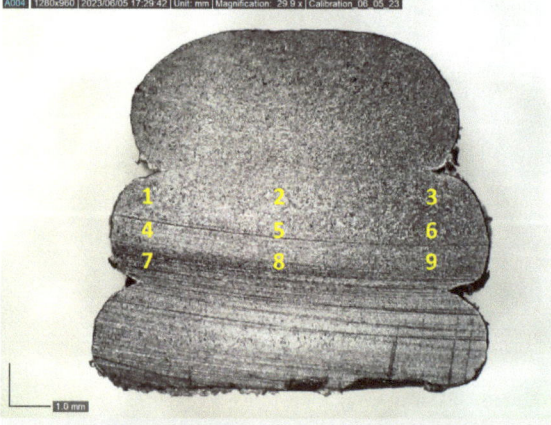

Figure 2. Optical image of the surface of the polished sample taken at ~30× magnification.

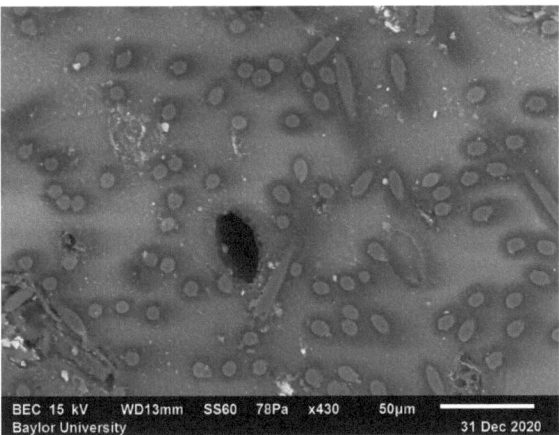

Figure 3. SEM Image of Region 1.

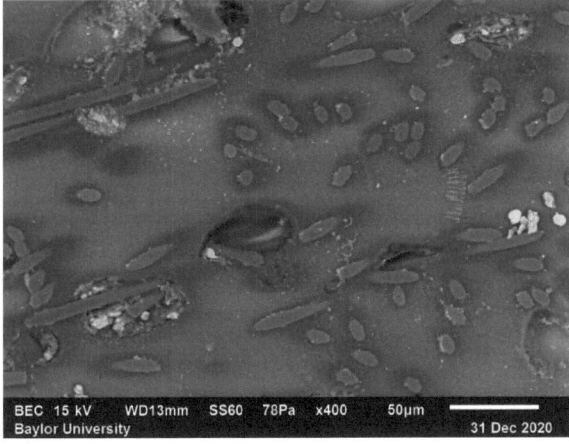

Figure 4. SEM Image of Region 5.

3. Fiber Orientation Identification

3.1. Fiber Orientation Measurement

Fiber orientation measurements in the composite material are very important for a reliable assessment of the physical properties. Reflective optical microscopy and scanning electron microscopy are two common methods for fiber orientation measurement (see, e.g., [28]). For fibers with a circular cross-section, the most common method for the quantification of fiber orientation is the method of ellipses (MoE) (see, e.g., [4,19,20]). For the circular fibers, at the cross-section which appears on the intercepting plane, an elliptical image can be observed, as shown in Figure 5. The directional angles $\left(\theta_f, \phi_f\right)$ of the ellipse are shown in Figure 5 along with the center of the ellipse (x_c, y_c) and the major M and minor m axes of the ellipse.

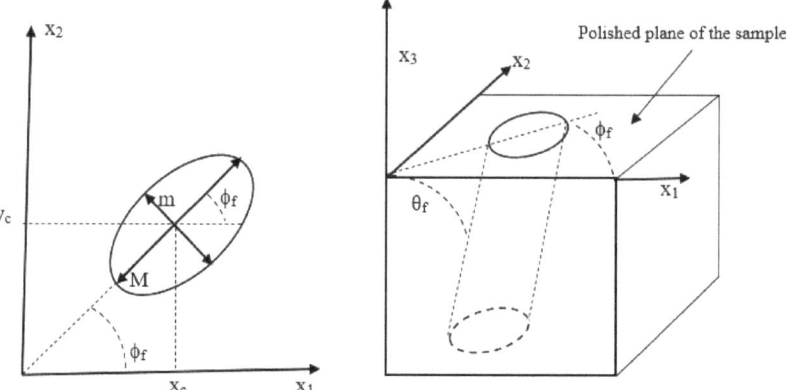

Figure 5. Definition of geometrical parameters measured in the method of ellipses.

The out-of-plane angle, θ_f, can be obtained from the geometry as

$$\theta_f = \cos^{-1}\left(\frac{m}{M}\right) \tag{3}$$

where M is the major axis, m is the minor axis, and θ_f and is the out-of-plane angle (see, e.g., [15]). Short or chopped fibers in a composite matrix flow behave as rigid cylindrical rods. Short fibers can translate and rotate in the polymer melt in any of the three coordinate directions. For the LAAM system considered, with aspect ratios ranging between 20 and 40, the fibers were not observed to have any flexure. For a single fiber with a high aspect ratio, the spatial orientation can be described using spherical coordinates with the in-plane angle ϕ_f, where $\phi_f \in \{0, 2\pi\}$, and out-of-plane angle θ_f, where $\theta_f \in \{0, \pi\}$, as shown in Figure 6. The unit vector **p**, obtained from Figure 6, has the following components:

$$\mathbf{p} = \left\{ \sin\theta_f \cos\phi_f, \sin\theta_f \sin\phi_f, \cos\theta_f \right\} \tag{4}$$

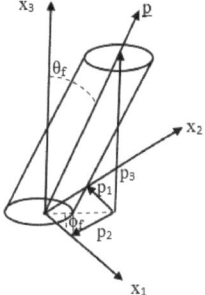

Figure 6. Rigid fiber particle orientation in spherical coordinates along with the polished plane from sectioning indicated by the top surface.

Orientation tensors were popularized by Advani and Tucker [29] to represent the orientation of a population of fibers composed of the individual orientations of each individual fiber. The local orientations of the fibers in a part are commonly represented by the orientation tensors. For example, the second-order orientation tensor A_{ij} is the probability density of the inner product of direction vector **p** with itself. For a continuous distribution of fibers defined by the orientation probability density function $\psi\left(\phi_f,\theta_f\right)$, this is defined as follows:

$$A_{ij} = \int_S p_i\left(\theta_f,\phi_f\right) p_j\left(\theta_f,\phi_f\right) \psi\left(\theta_f,\phi_f\right) dS \qquad (5)$$

where S is the unit sphere. By construction, orientation tensors are symmetric, $A_{ij} = A_{ji}$, and their trace is unity, $A_{11} + A_{22} + A_{33} = 1$. For a discrete selection of fiber angles randomly selected from the orientation probability density function $\psi\left(\phi_f,\theta_f\right)$, such as the case when measuring the orientation from the SEM images in the present study, the orientation tensor A_{ij} may be expressed as follows (see, e.g., [15]):

$$A_{ij} \approx \frac{\sum_{n=1}^{N_f}(p_ip_j)_n L_n F_n}{\sum_{n=1}^{N_f} L_n F_n} \qquad (6)$$

where L_n is the length of the nth fiber and N_f is the total number of fibers in the sample. The expression F_n is a weighing function, defined in terms of L_n and the diameter of the nth fiber d_n, which relates the orientation per unit area to the orientation per unit volume and is defined as

$$F_n = \frac{1}{L_n \cos\left(\theta_f\right)_n + d_n \sin\left(\theta_f\right)_n} \qquad (7)$$

In the present study, the value for the length is 250 μm and the diameter is 7 μm; both values are taken from the companion study of Russell [25].

The method of ellipses has some experimental and geometrical limitations that can lead to inaccurate measurements (see, e.g., [28,30]) which may lead to inaccurate interpretations of the fiber orientation state. When using standard optical techniques for the data collection, there will be a problem of ambiguity for the out-of-plane angle θ_f due to the fiber symmetry, but by construction, $\psi\left(\phi_f,\theta_f\right)$ retains the same symmetries, so this is not an issue. Of experimental concern is the sensitivity in the measurement of θ_f which increases as the cross-section of fiber in the prepared specimen approaches that of a perfect circle, resulting in measurement errors (see, e.g., [28]). This error can be mitigated by increasing the resolution of the images used for analysis. A third source of potential error is termed the ambiguity problem and is depicted in Figure 7 and shown by two fibers, one with ϕ_f and a second with $\phi_f + \pi$. Notice in this case that both fibers have identical cross-sections in the (x_1,x_2) plane, and this ambiguity for ϕ_f prevents the tabulation of off-diagonal components of A_{ij}, A_{13} and A_{23}, from being computed. As the diagonal components of A_{ij} are products of the unit vector p_i with itself, the choice of ϕ_f or $\phi_f + \pi$ does not change the result.

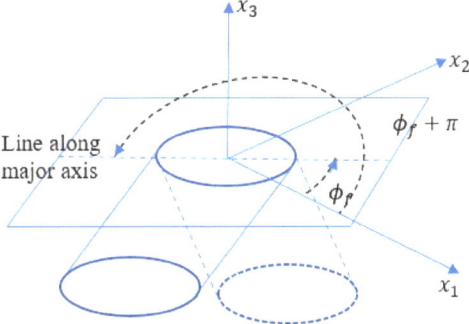

Figure 7. Depiction of sectioned surface demonstrating the fiber ambiguity problem.

Velez-Garcia et al. [11] developed a method for solving the ambiguity issue by exposing the polished fiber tips from underneath the polished surface by plasma etching of glass fiber samples. This exposes a characteristic shadow in the optical reflection micrograph in the under-surface region and was used to determine the direction of angle alignment [11]. With SEM, a similar shadow or tail will be shown below both with and without plasma etching for semi-conducting fibers, such as carbon fibers. Effective visualization of the shadow is very important to determine the orientation of the fiber in the matrix.

Another important aspect for measuring the fiber orientation is the flexibility of carbon fibers in the ABS matrix. The flexibility is described as the tendency of the fibers to bend in the material flow (see, e.g., [31,32]). After analyzing all the SEM images in this study and in the companion study of Russell [25], no bending of the fiber in the investigated system was found.

3.2. Fiber Orientation Quantification

As shown in Equation (3) the orientation of the fiber can be determined in terms of the semi-major and semi-minor axis lengths. Each fiber is traced from the image created using SEM by capturing multiple points (x_1, x_2) on the perimeter of each fiber. These points are then used to fit, in a least-squares sense, the equation for an ellipse as

$$ax_1^2 + 2bx_1x_2 + cx_2^2 + 2dx_1 + 2fx_2 + g = 0 \qquad (8)$$

The ellipses are drawn around the cross-section from the data as shown in Figure 8 by clicking at least 6 points to define the perimeter, and for the objects with higher ellipticity, up to 15 points were used to define the perimeter. An optimization algorithm using least-squares regression was created in MATLAB (Headquarters at Natlick, MA, USA) to estimate the parameters $a, b, \ldots, g$. Once they are tabulated, the center point $(x_{1,0}, x_{2,0})$ along with the semi-major axis M and semi-minor axis m, can be calculated as

$$x_{1,0} = \frac{cd - bf}{b^2 - ac} \qquad (9)$$

$$x_{2,0} = \frac{af - bd}{b^2 - ac} \qquad (10)$$

$$M = \sqrt{\frac{2(af^2 + cd^2 + gb^2 - 2bdf - acg)}{(b^2 - ac)\left[\sqrt{(a-c)^2 + 4b^2} - (a+c)\right]}} \qquad (11)$$

$$m = \sqrt{\frac{2(af^2 + cd^2 + gb^2 - 2bdf - acg)}{(b^2 - ac)\left[-\sqrt{(a-c)^2 + 4b^2} - (a+c)\right]}} \qquad (12)$$

Lastly, the rotation from the x_1 axis to the major axis, specifically angle ϕ_f from Figure 7, is cast in terms of the elliptical parameters as

$$\phi_f = \begin{cases} 0 & \text{for } b = 0 \text{ and } a < c \\ \frac{1}{2}\pi & \text{for } b = 0 \text{ and } a > c \\ \frac{1}{2}\cot^{-1}\left(\frac{a-c}{2b}\right) & \text{for } b \neq 0 \text{ and } a < c \\ \frac{\pi}{2} + \frac{1}{2}\cot^{-1}\left(\frac{a-c}{2b}\right) & \text{for } b \neq 0 \text{ and } a > c \end{cases} \qquad (13)$$

Figure 8 shows the elliptical curve fits over each individual cross-sectioned fiber from Region 1, the same data previously shown in Figure 3. The semi-major and semi-minor axis lengths along with the direction of the major axis are extracted from the best-fit ellipse to Equation (8), and this information is then used to evaluate the out-of-plane angle θ_f from Equation (3) for each individual fiber. With the in-plane and out-of-plane angle values, the unit vector **p** is evaluated from Equation (4), and then the orientation tensor A_{ij} is evaluated from Equation (6).

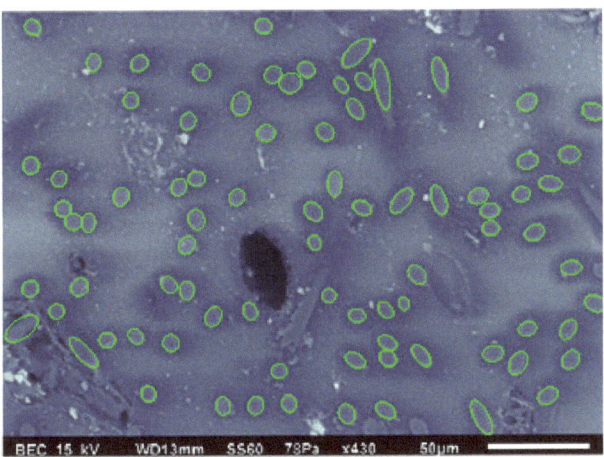

Figure 8. Best-fit ellipses from SEM image of Region 1.

4. Orientation Results and Discussion

SEM imaging and the elliptical analysis for orientation were performed on all 27 samples from Table 2 for each of the nine regions shown in Figure 2 for a total of 243 unique datasets, with each dataset containing between 50 and 95 identifiable elliptical cross-sections. One sample, sample 15, was randomly selected for an expanded analysis. The average for the orientation parameters A_{11}, A_{22}, and A_{33} along with the respective fiber count is provided in Table 2 for just sample 15. The orientation data from Table 2 are also plotted in Figure 9.

Table 2. Orientation tensor for different regions of sample 15.

Region	1	2	3	4	5	6	7	8	9
Fiber Count	83	64	77	86	53	75	80	54	91
A_{11}	0.21	0.18	0.3	0.1	0.41	0.26	0.2	0.43	0.14
A_{22}	0.2	0.26	0.12	0.26	0.24	0.2	0.13	0.07	0.2
A_{33}	0.59	0.56	0.58	0.64	0.36	0.54	0.67	0.5	0.66

From the fiber orientation measurement given in Table 2 and shown in Figure 9, it is evident that fibers are most aligned towards the x_3 direction as A_{33} is generally the highest parameter, but there are significant spatial inhomogeneities of the orientation state within the individual deposited bead. Notice that for regions along the vertical centerline, Regions 2, 5, and 8, the alignment tends to be lower (lowest A_{33}), as is the case for the regions along the horizontal centerline, Regions 4, 5, and 6, with the center region of the extrudate, Region 5, experiencing the lowest alignment along the print direction. Although Region 2 has a relatively high alignment in this particular sample, Region 2 has a relatively lower fiber alignment in general for the majority of the samples investigated, which will be shown later in this text.

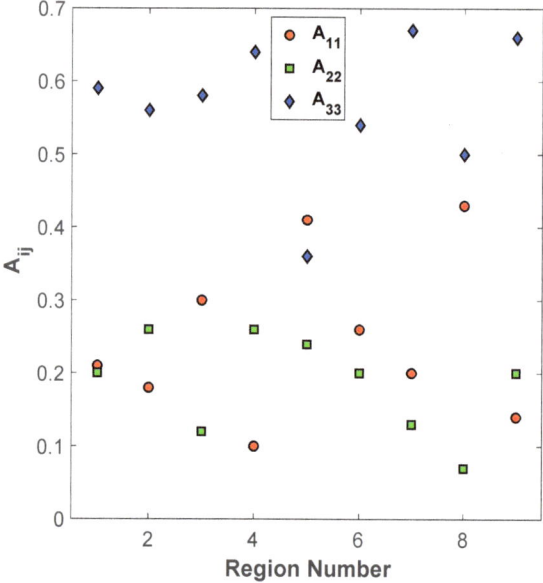

Figure 9. Change in A_{11}, A_{22}, and A_{33} in 9 different regions of a layer of sample 15 in the deposited bead.

The above process of orientation extraction from each of the nine regions for the remaining 27 samples is shown in Table 1 for the three different temperatures, screw speeds, and cross-sectional areas. The orientation results from all 27 samples for each of the nine regions are shown in Figure 10. In general, the trend that the alignment is highest along the print direction is shown by the increased value of A_{33}. The observation that the alignment is highest on the corners, Regions 1, 3, 7, and 9, is generally true, as is the observation that the x_3 alignment is lowest in the center of the deposited bead, Region 5. It is worth noting that there is an observed correlation between the fiber count and the fiber alignment along x_3 of a particular region.

4.1. Effect of Nozzle Temperature on Fiber Orientation

The sensitivity of the alignment to the processing parameter of temperature is investigated in the present section. This was accomplished by fixing the screw speed to 1500 RPM and the cross-sectional area to 22 mm^2 (samples 1, 10, and 19 from Table 1). The variation in orientation tensors A_{11}, A_{22}, and A_{33} in each of the nine regions of Figure 2 for the changing nozzle temperature was computed, and A_{33} is shown in Figure 11. Notice that the three temperature zones of 190/195/200 °C, 200/205/210 °C, and 210/215/220 °C have been noted as 195 °C, 205 °C, and 215 °C, respectively, in Figure 11. Overall, the temperature state of $T = 205$ °C has the highest overall alignment, but this is by a nominal amount.

4.2. Effect of Screw Speed on Fiber Orientation

The sensitivity of the alignment to the processing parameter of screw speed is investigated in the present section. This was accomplished by fixing the temperature to 190/195/200 °C and the cross-sectional area to 22 mm^2 (samples 10, 13, and 16 from Table 1). The variation in the orientation tensors A_{11}, A_{22}, and A_{33} in each of the nine regions depicted in Figure 2 by changing the screw speed to 1500 RPM, 1875 RPM, and 2250 RPM was computed, and the value for A_{33} is shown in Figure 12. From the figure, it is not clear which of the three extrusion speeds results in the highest orientation state.

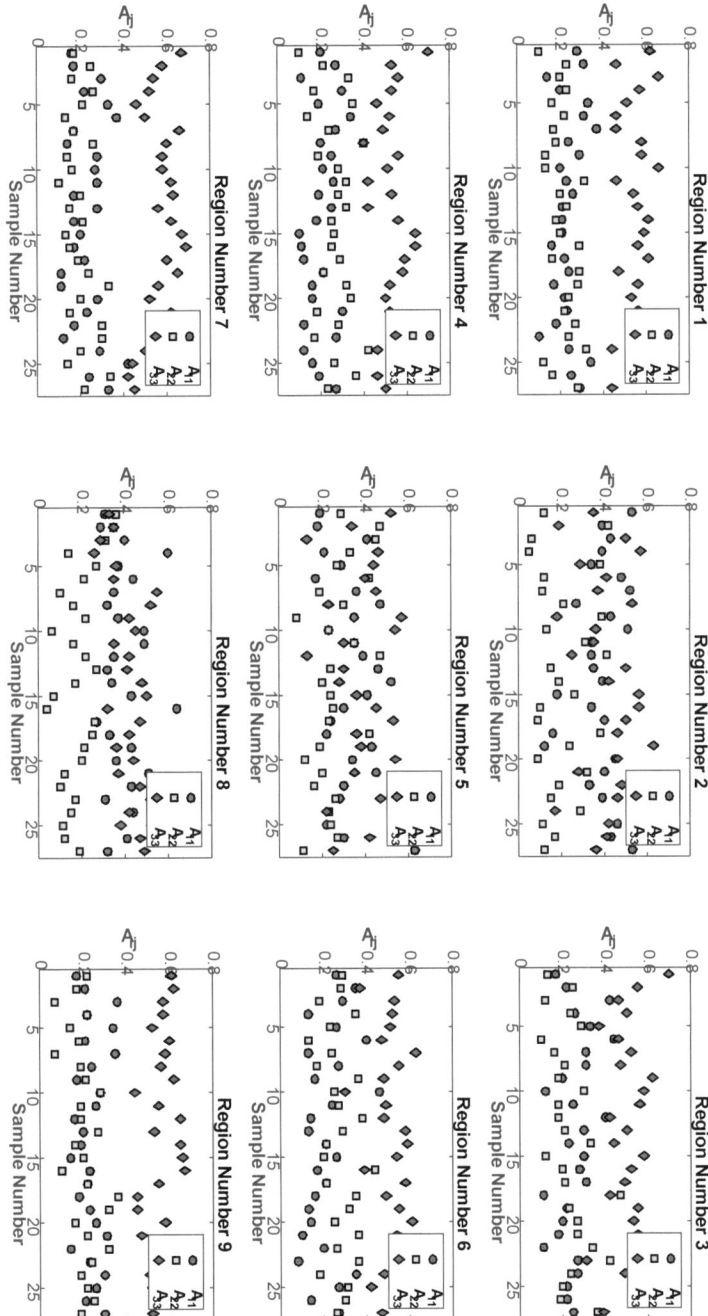

Figure 10. Plots of A_{11}, A_{22}, and A_{33} for each sample by region.

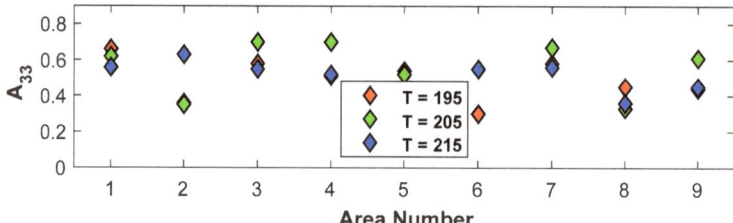

Figure 11. Change in orientation tensors for 1500 RPM and $A = 22$ mm^2 for varying temperature.

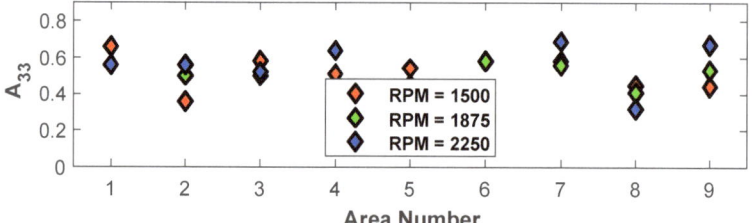

Figure 12. Change in orientation tensors for temperature = 190/195/200 °C and $A = 22$ mm^2 when varying the extrusion speed.

4.3. Effect of Cross-Sectional Area on Fiber Orientation

The sensitivity of the alignment to the processing parameter of the deposition area is investigated in the present section. This was accomplished by fixing the temperature to 190/195/200 °C and the extruder RPM to 1500 RPM (samples 10, 11, and 12 from Table 1). The variation in the orientation tensors A_{11}, A_{22}, and A_{33} in each of the nine regions depicted in Figure 2 by changing the deposition area from 22 mm^2, 27 mm^2, and 32 mm^2 was computed, and the value for A_{33} is shown in Figure 13. From the figure, it is not clear which of the three cross-sectional areas yields the highest orientation state.

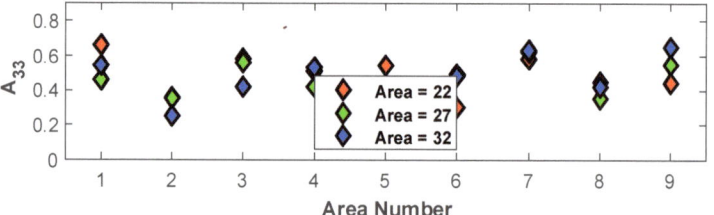

Figure 13. Change in orientation tensors at 9 regions with varying cross-sectional area with RPM = 1500 at nozzle temperature = 190/195/200 °C.

4.4. Correlation among All Three Process Parameters with the Orientation Tensors

The orientation state from all 27 samples from Table 1 is presented in Figure 14. The value presented is the orientation taken from a single sample by averaging the orientation from the nine regions within the sample. The error bars are not plotted in the figure as they provide confusion, but for each sample, the typical standard deviation is ~0.074.

From Figure 14, it can be seen that the two highest alignments of fibers in the x_3 direction occur for the print condition with the temperature of 200/205/210 °C, screw speed of 1500 RPM, and cross-sectional area of 22 mm^2 and the print condition with the temperature of 190/195/200 °C, screw speed of 1875 RPM, and cross-sectional area of 32 mm^2.

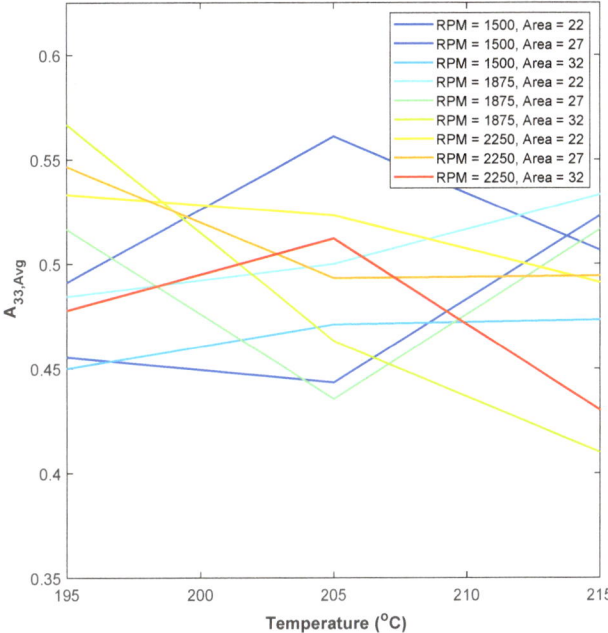

Figure 14. Change in A_{33} value for different print conditions.

Figure 14 shows a trend of the fiber orientation being the greatest towards the print direction at lower nozzle temperatures generally at higher screw speeds. Again, for the lowest RPM and lowest cross-sectional area, and the highest RPM and highest cross-sectional area, the fiber orientation towards the print direction can be obtained at the nozzle temperature of 200/205/210 °C.

Studying and analyzing all 27 samples by changing three process parameters, it can be concluded that there is not a statistically significant correlation between the investigated process parameters and the orientation tensors. This is unfortunately not a strong conclusion as to the sensitivity of the fiber orientation to processing parameter variation.

4.5. Fiber Orientation Measurement for Different Samples with Identical Processing Conditions

This final study takes multiple cross-sections of the same deposited bead to identify any inhomogeneities within the bead itself. A single deposited bead is sectioned at six locations along the bead, and then the orientation tensors A_{11}, A_{22}, and A_{33} are measured using the aforementioned approach for SEM image construction. In Figure 15, the average of the orientation tensors for the nine regions along the six samples is shown. These six samples are printed at a nozzle temperature of 190/195/200 °C, a screw speed of 1875 RPM, and a cross-sectional area of 32 mm^2, the same parameters as sample 15 from Table 1. The mean orientation from each region is plotted in Figure 15, with the error bars defined as one standard deviation about the mean. There is a significant variation in the value of the tensors along the length of the printed part, in contradiction to the assumed state of uniformity along the deposited bead length. The A_{33} component within each region varies across each of the six samples.

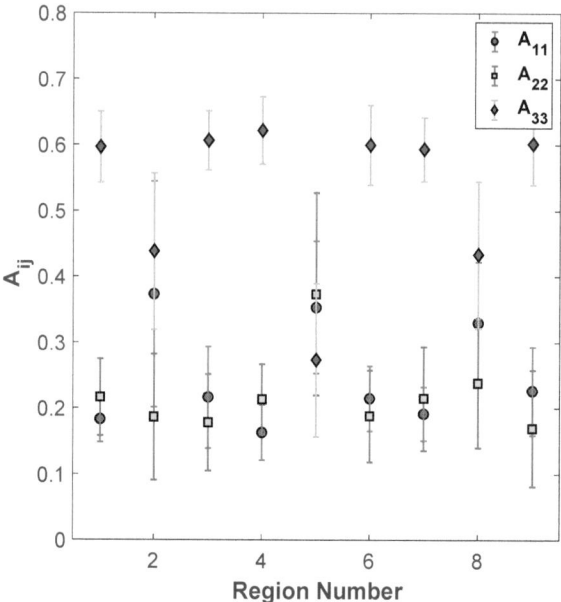

Figure 15. Plots of the average of A_{11}, A_{22}, and A_{33} for 6 samples taken from deposited beads fabricated with the properties of Part #15.

It can be observed from Figure 15 that the orientation tensor values follow the same trend as discussed before of having the highest values in the print direction on the corners (Regions 1, 3, 7, and 9) and on the edges (Regions 4 and 6), with the near-the-surface regions along the deposition center line (Regions 2 and 8) having a reduced alignment and the center of the deposited bead (Region 5) having the lowest alignment state. As shown by the error bars, taken from the respective standard deviations, the orientation tensor values in the same region for the samples taken along the same bead have considerable variation. Thus, there is an expected considerable amount of part-to-part variation. This is in alignment with the results from the preceding sections where there are few conclusions that can be made in regard to the processing parameter sensitivity. Thus, to properly correlate the sensitivity of processing parameters to the orientation state, additional sectioned samples would be required along the length of individual beads, and this was beyond the scope of the present study. Based on the results in Figure 15, it is clear there is significant variability along the extruded bead for the same process parameters, and before a quantification can be made to identify the optimal set of process parameters, the manufacturing process itself must be made more consistent.

5. Conclusions and Future Work

The present paper presented a method to quantify the spatial variation in the orientation state of carbon fibers within an additively manufactured deposited bead. For this study, 13% weight fraction carbon fiber-reinforced ABS pellets were printed using a custom large area additive manufacturing (LAAM) system over a range of processing parameters. The resulting extrudate was sectioned, polished, and subsequently analyzed with SEM for studying spatial variations of the local fiber orientation. The presented work introduced a novel approach using scanning electron microscopy to image sectioned samples to quantify the orientation state. Methods for proper cutting, polishing, cleaning, and plasma etching of the sample are presented in this work. The proper preparation of the samples, including etching, was shown to be helpful in providing a high contrast of the fiber–matrix interface.

The variation in fiber alignment was analyzed by changing three different process parameters: nozzle temperature, screw speed, and cross-sectional area of the printed beads. A total of 27 samples were printed, and each sample was subdivided into nine regions to study the spatial variation in fiber orientation. The results presented in this study indicate that fibers are most aligned at the corners of the extrudate, with the fibers in the center region of the extrudate being near a random orientation state. There are subtle variations in the sensitivity of the orientation to variations in the process parameters from scan to scan, but drawing conclusions is made difficult due to the finite sample sizes and the variability within the same sample caused by the limited available data. To quantity this latter issue, six samples from the same processing parameter set were fabricated, sectioned, imaged, and characterized. From this large dataset, it was seen that a better estimation of the mean values could be obtained with a standard deviation between samples of greater than 0.05 for the longitudinal component of the orientation tensor A_{33}. This value is significant as the range of the orientation tensor component of the average A_{33} over the 27 permutations of process parameters is between 0.42 and 0.57. Thus, for example, the process parameter set of $RPM = 1500$, $A = 22 \text{ mm}^2$, and $T = 205\ °C$ had the highest value of alignment with $A_{33} = 0.57$, but 9 of the remaining 26 sets of investigated process parameters exhibited an alignment within one standard deviation of the highest average alignment measured.

As part of the developed method, an interesting future study was revealed by the nature of the electron charge distribution build-up on the surface. This may provide a solution to the ambiguity problem inherent to the method of ellipses. For example, Region 5 of sample 15 is selected to present this potential solution and is shown in Figure 16. Observe that SEM generates a three-dimensional-like image of the surface. In the SEM image, an under-surface shadow of each fiber can be traced as noted by the non-uniform darkened regions around the fiber; thus, the actual orientation of each fiber in the matrix from Figure 7 can be determined. In the figure, several fibers are highlighted, and the resulting dark-gray tail can be observed. It is hypothesized that this dark-gray region is caused by the semi-conducting nature of the carbon fibers, thus allowing the surface change to dissipate in a mechanism different from that of the polymer surface, an insulating material. This is especially obvious in fibers with a higher elliptical nature for which the fiber under the surface remains closer to the surface for a larger distance.

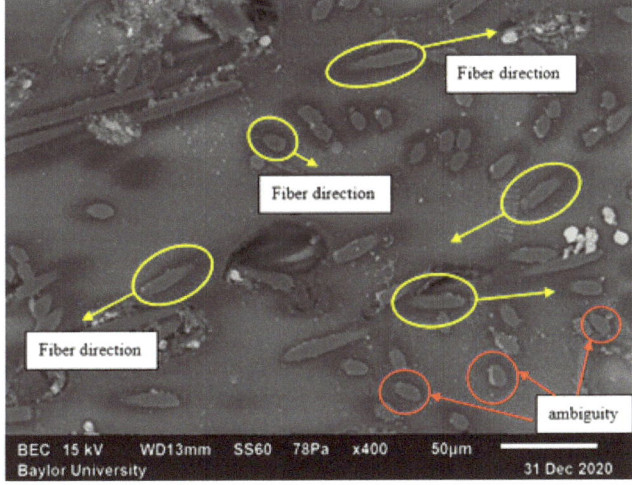

Figure 16. Tracing under the surface shadow of fiber to determine the orientation (Region 5 of sample 15). Fibers highlighted in yellow indicate representative fiber alignment that might be properly characterized using optical microscopy, whereas the fibers highlighted in red have a clear SEM shadow but would be ambiguous using optical microscopy and standard polishing techniques.

In the future, more samples need to be studied to establish the correlations of fiber orientation to the investigated processing parameters. Regardless, the orientation state for the investigated printed composite was quantified, and a moderate degree of alignment was observed. Additional work will be required in manufacturing to optimize the alignment state as well as to promote consistency within the deposited beads themselves. It is noted that there were a considerable number of voids in all samples, and it is well established that void formation in fiber-filled systems tends to reduce the structural performance while randomizing the orientation state. Thus, it is suggested that an investigation of the sensitivity of the final void content to processing parameters may yield more promise in achieving the goal of increasing the alignment state.

Author Contributions: Conceptualization, D.A.J.; methodology, D.A.J. and R.A.N.; software, D.A.J. and R.A.N.; validation, R.A.N.; formal analysis, R.A.N.; investigation, D.A.J. and R.A.N.; resources, D.A.J.; data curation, R.A.N.; writing—original draft preparation, R.A.N.; writing—review and editing, D.A.J. and R.A.N.; visualization, D.A.J. and R.A.N.; supervision, D.A.J.; project administration, D.A.J.; funding acquisition, D.A.J. All authors have read and agreed to the published version of the manuscript.

Funding: This research was supported by Baylor University through a research assistantship.

Institutional Review Board Statement: Ethical review and approval was not applicable as the studies performed did not involve human or animal subjects.

Data Availability Statement: SEM images can be provided upon request to the corresponding author.

Acknowledgments: The authors would like to acknowledge Timothy Russell for aiding in the fabrication of samples from LAAM and Ben Blandford and Velez Garcia for their guidance in polishing techniques.

Conflicts of Interest: The authors declare no conflict of interest.

References

1. Duty, C.E.; Kunc, V.; Compton, B.; Post, B.; Erdman, D.; Smith, R.; Lind, R.; Lloyd, P.; Love, L. Structure and mechanical behavior of Big Area Additive Manufacturing (BAAM) materials. *Rapid Prototyp. J.* **2017**, *23*, 181–189. [CrossRef]
2. Vicente, C.M.S.; Sardinha, M.; Reis, L.; Ribeiro, A.; Leite, M. Large-format additive manufacturing of polymer extrusion-based deposition systems: Review and applications. *Prog. Addit. Manuf.* **2023**. [CrossRef]
3. Pricci, A.; de Tullio, M.D.; Percoco, G. Modeling of extrusion-based additive manufacturing for pelletized thermoplastics: Analytical relationships between process parameters and extrusion outcomes. *CIRP J. Manuf. Sci. Technol.* **2023**, *41*, 239–258. [CrossRef]
4. Barera, G.; Dul, S.; Pegoretti, A. Screw Extrusion Additive Manufacturing of Carbon Fiber Reinforced PA6 Tools. *J. Mater. Eng. Perform.* **2023**. [CrossRef]
5. Yeole, P.S. Thermoplastic Additive Manufacturing for Composites and Molds. Ph.D. Thesis, University of Tennessee Knoxville, Tennessee, TX, USA, 2020. Available online: https://trace.tennessee.edu/utk_graddiss/6099/ (accessed on 27 June 2023).
6. Meincke, O.; Kaempfer, D.; Weickmann, H.; Friedrich, C.; Vathauer, M.; Warth, H. Mechanical properties and electrical conductivity of carbon-nanotube filled polyamide-6 and its blends with acrylonitrile/butadiene/styrene. *Polymer* **2004**, *45*, 739–748. [CrossRef]
7. Love, L.J.; Kunc, V.; Rios, O.; Duty, C.E.; Elliott, A.M.; Post, B.K.; Smith, R.J.; Blue, C.A. The importance of carbon fiber to polymer additive manufacturing. *J. Mater. Res.* **2014**, *29*, 1893–1898. [CrossRef]
8. Lobov, E.; Dobrydneva, A.; Vindokurov, I.; Tashkinov, M. Effect of Short Carbon Fiber Reinforcement on Mechanical Properties of 3D-Printed Acrylonitrile Butadiene Styrene. *Polymers* **2023**, *15*, 2011. [CrossRef]
9. Wang, K.; Li, S.; Rao, Y.; Wu, Y.; Peng, Y.; Yao, S.; Zhang, H.; Ahzi, S. Flexure Behaviors of ABS-Based Composites Containing Carbon and Kevlar Fibers by Material Extrusion 3D Printing. *Polymers* **2019**, *11*, 1878. [CrossRef]
10. Blok, L.G.; Longana, M.L.; Yu, H.; Woods, B.K.S. An investigation into 3D printing of fibre reinforced thermoplastic composites. *Addit. Manuf.* **2018**, *22*, 176–186. [CrossRef]
11. Vélez-García, G.; Wapperom, P.; Kunc, V.; Baird, D.; Zink-Sharp, A. Sample preparation and image acquisition using optical-reflective microscopy in the measurement of fiber orientation in thermoplastic composites. *J. Microsc.* **2012**, *248*, 23–33. [CrossRef]
12. Fischer, G.; Eyerer, P. Measuring spatial orientation of short fiber reinforced thermoplastics by image analysis. *Polym. Compos.* **1988**, *9*, 297–304. [CrossRef]
13. Velez-Garcia, G.M. Experimental Evaluation and Simulations of Fiber Orientation in Injection Molding of Polymers Containing Short Glass Fibers. 2012. Available online: https://vtechworks.lib.vt.edu/handle/10919/27335 (accessed on 20 February 2023).

14. Hofmann, J.T.; Vélez-Garcia, G.M.; Baird, D.G.; Whittington, A.R. Application and evaluation of the method of ellipses for measuring the orientation of long, semi-flexible fibers. *Polym. Compos.* **2013**, *34*, 390–398. [CrossRef]
15. Vélez-García, G.M.; Wapperom, P.; Baird, D.G.; Aning, A.O.; Kunc, V. Unambiguous orientation in short fiber composites over small sampling area in a center-gated disk. *Compos. Part Appl. Sci. Manuf.* **2012**, *43*, 104–113. [CrossRef]
16. Sharp, N.D.; Goodsell, J.E.; Favaloro, A.J. Measuring Fiber Orientation of Elliptical Fibers from Optical Microscopy. *J. Compos. Sci.* **2019**, *3*, 23. [CrossRef]
17. Goldstein, J.I.; Newbury, D.E.; Echlin, P.; Joy, D.C.; Lyman, C.E.; Lifshin, E.; Sawyer, L.; Michael, J.R. *Scanning Electron Microscopy and X-ray Microanalysis*, 3rd ed.; Springer: Boston, MA, USA, 2003. [CrossRef]
18. Jahan, N. Fiber Orientation Prediction and Strength Evaluation of Composite Spur Gears Reinforced by Discontinuous Fiber. Master's Thesis, South Dakota University, Vermillion, SD, USA, 2016. Available online: https://openprairie.sdstate.edu/etd/1044 (accessed on 27 June 2023).
19. Tekinalp, H.L.; Kunc, V.; Velez-Garcia, G.M.; Duty, C.E.; Love, L.J.; Naskar, A.K.; Blue, C.A.; Ozcan, S. Highly oriented carbon fiber–polymer composites via additive manufacturing. *Compos. Sci. Technol.* **2014**, *105*, 144–150. [CrossRef]
20. Jiang, D.; Smith, D.E. Anisotropic mechanical properties of oriented carbon fiber filled polymer composites produced with fused filament fabrication. *Addit. Manuf.* **2017**, *18*, 84–94. [CrossRef]
21. Russell, T.; Jack, D. Stiffness Prediction and Validation of Large Volume 3D Printed, Short-Fiber-Filled Polymer Composites. In Proceedings of the SAMPE 2019, Charlotte, NC, USA, 20–23 May 2019. [CrossRef]
22. Chen, C.H.; Wang, Y.-C. Effective thermal conductivity of misoriented short-fiber reinforced thermoplastics. *Mech. Mater.* **1996**, *23*, 217–228. [CrossRef]
23. Hassen, A.A.; Dinwiddie, R.B.; Kim, S.; Tekinalp, H.L.; Kumar, V.; Lindahl, J.; Yeole, P.; Duty, C.; Vaidya, U.; Wang, H.; et al. Anisotropic thermal behavior of extrusion-based large scale additively manufactured carbon-fiber reinforced thermoplastic structures. *Polym. Compos.* **2022**, *43*, 3678–3690. [CrossRef]
24. Tezvergil, A.; Lassila, L.V.; Vallittu, P.K. The effect of fiber orientation on the thermal expansion coefficients of fiber-reinforced composites. *Dent. Mater. Off. Publ. Acad. Dent. Mater.* **2003**, *19*, 471–477. [CrossRef]
25. Russell, T.D. Mechanical and Thermal Property Prediction in Single Beads of Large Area Additive Manufactured Short-Fiber Polymer Composites. Master's Thesis, Baylor University, Waco, TX, USA, 2021. Available online: https://baylor-ir.tdl.org/handle/2104/11726 (accessed on 23 May 2023).
26. Rudresh, B.M.; Kumar, B.N.R. Influence of Experimental Parameters on Friction and Wear Mechanisms of PA66/PTFE Blend Reinforced with Glass Fiber. *Trans. Indian Inst. Met.* **2018**, *71*, 339–349. [CrossRef]
27. Al-Ameen, Z. An improved contrast equalization technique for contrast enhancement in scanning electron microscopy images. *Microsc. Res. Tech.* **2018**, *81*, 1132–1142. [CrossRef]
28. Lee, K.S.; Lee, S.W.; Chung, K.; Kang, T.J.; Youn, J.R. Measurement and numerical simulation of three-dimensional fiber orientation states in injection-molded short-fiber-reinforced plastics. *J. Appl. Polym. Sci.* **2003**, *88*, 500–509. [CrossRef]
29. Advani, S.G.; Tucker, C.L. The Use of Tensors to Describe and Predict Fiber Orientation in Short Fiber Composites. *J. Rheol.* **1987**, *31*, 751–784. [CrossRef]
30. Toll, S.; Andersson, P.-O. Microstructural characterization of injection moulded composites using image analysis. *Composites* **1991**, *22*, 298–306. [CrossRef]
31. Switzer, L.H. Simulating Systems of Flexible Fibers. Ph.D. Thesis, University of Wisconsin, Madison, MD, USA, 2002. Available online: https://www.proquest.com/docview/305533244 (accessed on 27 May 2023).
32. Meyer, K.J.; Hofmann, J.T.; Baird, D.G. Initial conditions for simulating glass fiber orientation in the filling of center-gated disks. *Compos. Part Appl. Sci. Manuf.* **2013**, *49*, 192–202. [CrossRef]

Disclaimer/Publisher's Note: The statements, opinions and data contained in all publications are solely those of the individual author(s) and contributor(s) and not of MDPI and/or the editor(s). MDPI and/or the editor(s) disclaim responsibility for any injury to people or property resulting from any ideas, methods, instructions or products referred to in the content.

Article

Technologies for Mechanical Recycling of Carbon Fiber-Reinforced Polymers (CFRP) Composites: End Mill, High-Energy Ball Milling, and Ultrasonication

Enrique Martínez-Franco [1], Victor Alfonzo Gomez Culebro [2] and E. A. Franco-Urquiza [2,*]

[1] Manufacturing Department, Center for Engineering and Industrial Development, Av. Pie de la Cuesta 702-No. 702, Desarrollo San Pablo, Santiago de Querétaro 76125, Mexico; enrique.martinez@cidesi.edu.mx

[2] Aerospace Section, Center for Engineering and Industrial Development, Carretera Estatal 200, km 23, Queretaro 76270, Mexico; victor.gomez@cidesi.edu.mx

* Correspondence: edgar.franco@cidesi.edu.mx

Citation: Martínez-Franco, E.; Gomez Culebro, V.A.; Franco-Urquiza, E.A. Technologies for Mechanical Recycling of Carbon Fiber-Reinforced Polymers (CFRP) Composites: End Mill, High-Energy Ball Milling, and Ultrasonication. *Polymers* **2024**, *16*, 2350. https://doi.org/10.3390/polym16162350

Academic Editors: Phuong Nguyen-Tri, Ana Pilipović and Mustafa Özcanli

Received: 23 May 2024
Revised: 9 August 2024
Accepted: 14 August 2024
Published: 20 August 2024

Copyright: © 2024 by the authors. Licensee MDPI, Basel, Switzerland. This article is an open access article distributed under the terms and conditions of the Creative Commons Attribution (CC BY) license (https://creativecommons.org/licenses/by/4.0/).

Abstract: Carbon fiber reinforced polymer (CFRP) composites have very high specific properties, which is why they are used in the aerospace, wind power, and sports sectors. However, the high consumption of CFRP compounds leads to a high volume of waste, and it is necessary to formulate mechanical recycling strategies for these materials at the end of their useful life. The recycling differences between cutting-end mills and high-energy ball milling (HEBM) were evaluated. HEBM recycling allowed us to obtain small recycled particles, but separating their components, carbon fiber, epoxy resin, and CFRP particles, was impossible. In the case of mill recycling, these were obtained directly from cutting a CFRP composite laminate. The recycled materials resulted in a combination of long fibers and micrometric particles—a sieving step allowed for more homogeneous residues. Although long, individual carbon fibers can pass through the sieve. Ultrasonication did not significantly affect HEBM recyclates because of the high energy they are subjected to during the grinding process, but it was influential on end mill recyclates. The ultrasonication amplitude notably impacted the separation of the epoxy resin from the carbon fiber. The end mill and HEBM waste production process promote the presence of trapped air and electrostatics, which allows recyclates to float in water and be hydrophobic.

Keywords: mechanical recycling; carbon fiber-reinforced polymers; end milling; high-energy ball milling; ultrasonication

1. Introduction

Carbon fiber composite materials are manufactured by impregnating carbon fiber fabrics with epoxy resins to obtain carbon fiber-reinforced polymer (CFRP) composites. CFRPs have very high specific properties, so they play an essential role in the growth of various high-tech sectors such as aerospace, wind, and sports. However, the high consumption of CFRP leads to a high volume of waste, so it is necessary to formulate viable recycling strategies for these technological materials at the end of their useful life [1]. Academia and industry combine efforts to improve the efficiency of resource consumption for global sustainable development and environmental conservation, aligned with the circular economy [2–5]. An example of the above is the European Community, which aims at the recovery, reuse, and recycling of almost all components in vehicles at the end of their life cycle [6–8], and similar decisions have been made in the aerospace industry [9].

CFRP currently constitutes more than 50% of the structural mass of commercial aircraft and will reach the end of life (EoL) in 20–30 years [10]. Therefore, the question arises about how these technological materials will be recycled.

The main drawback of CFRP is that thermosetting resins form cross-linked structures during their curing process, so they cannot be melted and reprocessed [11]. Additionally, carbon fibers are abrasive. The management of CFRP waste is mainly carried out through disposal and recycling. Disposal is the most common method because waste is taken to landfills [12,13]. Thermal recycling, which involves energy recovery, commonly known as pyrolysis, consists of incinerating waste to take advantage of the gases produced by combustion [14,15]. Chemical recycling, known as solvolysis, uses chemical solutions to remove the epoxy resin from CFRP. This process is relatively cheaper than heartburn, although it is more polluting and risky since it uses typically dangerous chemicals. In both recycling processes, the thermosetting resin (which has a high thermomechanical potential) is eliminated, and only the carbon fiber is recovered, which is damaged by the harsh processes that considerably reduce its mechanical properties.

Mechanical recycling recovers both the thermosetting resin and the carbon fiber without damaging them, as it involves a primary low-speed cutting and grinding process to reduce the size of the CFRP into pieces in the range of 50 to 100 mm [16,17]. Subsequently, these crushed pieces are ground to obtain particles with a size distribution of 10 to 1 mm [18,19]. M. Durante and co-workers [1] reported that mechanical recycling represents the most promising way to recover carbon fibers from the polymeric matrix of the CFRP components since this methodology does not require burning at high temperatures or chemical substances to decompose the polymeric matrix.

Mechanical recycling allows the recovery of three types of waste: short carbon fiber, epoxy resin powder, and CFRP microparticles [18]. Mechanical recycling of CFRP does not induce structural damage to the carbon fiber and provides high expectations of reuse and sustainable savings to develop new materials or products, taking advantage of the intrinsic properties of carbon fiber and epoxy resin.

The most used and well-known mechanical recycling technology is the one that uses end mill tools coated with carbides (carbide end mills) because of the high abrasion of CFRP. This technology is relatively the most accessible, requiring only a conventional small helix end mill. The tool follows a programmed trajectory using computer-aided manufacturing systems. Nonetheless, the main restriction is tool wear and difficulty milling curved parts. Norbert Geier and collaborators [19] conducted a complete review of edge-cutting technologies in CFRP composites. The review article is oriented toward cutting efficiency in composites and is helpful in guiding efforts toward mechanical recycling of CFRP. Other research explored various configurations of milling tools to increase end mill tool life [17,20]. Process conditions such as orientation, speed, and cutting feed have also been evaluated, with feed speed having the most significant effect, followed by cutting speed, on tool wear [18,21]. Hosokawa et al. [22] examined the effect of tool coatings. The force analysis showed that the tangential and normal force components decrease with increasing helix angle.

Mechanical milling is a technique used to convert long carbon fibers into short fibers [23]. Ball and hammer mills are the most common methods for converting dry carbon fiber into powder. In the ball milling process, carbon fiber is placed in a drum that houses high-hardness (usually stainless steel) and high-density spheres. Through the rotation of the drum, the balls begin to move in a cascade fashion and collide with the surrounding material, causing localized high-pressure impacts that grind the fibers into powder.

High energy ball milling (HEBM) is a complex process that involves a high speed of ball movement, also causing collisions and friction between them that favor homogeneous mixing, promote morphology changes, and influence the evolution of defects of the crystal lattice and formation of new phases. Thus, HEBM mainly synthesizes powder products to develop materials and coatings [24]. Although HEBM has been employed to fabricate ultrafine graphite waste [25], the technique has not been explored for recycling CFRP composites.

Another technique that can be used to provide an additional step in the mechanical recycling of CFRP composites is ultrasound. Ultrasonic treatment can alter the surface roughness characteristics of particles and promote their efficient separation by flotation [26].

On the other hand, the sonication process uses ultrasonic sound waves that produce thousands of microscopic vacuum bubbles in the solution because of the applied pressure. The bubbles collapse in solution during the cavitation process, leading to the generation of enormous energy. This way, ultrasonic cavitation leads to solid particles' dispersion, homogenization, and disintegration. Ultrasonication has been used to disperse nanoparticles in a polymer matrix [27]. The platelets that make up the mineral clays [28,29] can be exfoliated using the ultrasonication technique to increase the thermal and mechanical properties of the polymers [30,31]. Ultrasonication is also used to disperse the nanoclay platelets that reinforce CFRP composites by adding small percentages by weight (<5 wt %) to the epoxy resin [27]. Other nanoparticles have reinforced fiber-reinforced polymer composites [32]. Hiroaki Miyagawa et al. [33] investigated the mechanical and thermophysical properties of biobased epoxy nanocomposites reinforced with organo-montmorillonite clay and carbon fibers. They used sonication to process the organically modified clay into bio-based glassy epoxy networks, which resulted in the clay nanoplatelets being homogeneously dispersed and thoroughly exfoliated in the matrix. There is no evidence of ultrasonication reducing the particle size of recycled CFRP (rCFRP) composites at the moment.

This work uses different technologies to obtain recycled materials from CFRP composites. The first is the most widely used to cut this composite type using end milling cutters. End milling is used more to finish composites than to destroy them. This is because the cutter edge wears down quickly. However, using end mills to evaluate the mechanical recycling capacity is convenient. The recycled materials obtained comprise short carbon fiber, epoxy resin powder, and rCFRP particles. The second technique is unconventional since the HEBM technique is used in this work to pulverize the recyclates. This technique is complementary to assessing the efficiency of obtaining finer and smaller recycled particles. The recycled materials obtained by end milling and grinding are subjected to ultrasonication to separate epoxy resin residues from the carbon fibers. The morphology of microscopic CFRP particles was inspected after cutting processes with end mills and high-energy ball milling. Recycled materials could be used as efficient reinforcements since mechanical recycling does not alter the structural capacity of CFRP composites [18], opening up a range of possibilities for research and development of new products, including large-scale recycling technology.

2. Materials and Methods

The CFRP composites consisted of a 1×1 3K taffeta carbon fiber fabric weighing 198 g/m^2, supplied by Quintum. The carbon fabric has a tensile strength of 2137 MPa and an elastic modulus of 227 GPa. The epoxy resin Epolam 2015 from Sika was used as the polymer matrix. According to the manufacturer's technical sheet, Epolam 2015 has a viscosity (25 °C) of 1680 mPa·s and a density of 1.15 g/cc. The hardener Epolam 2015, with amine groups, was used as a catalyst in a mixing ratio of 32 by weight with a useful life of 140 min at room temperature (25 °C).

Carbon fiber and epoxy resin laminates were manufactured in a controlled environment using the vacuum-assisted resin infusion (VARI) method. The negative pressure was 2670 Pa, and the resin curing reaction took approximately 24 h (~25 °C). The orientation of the carbon fiber fabrics was quasi-isotropic, with a stacking sequence of $[\{0/90/ \pm 45/0/90/ \pm 45\}_2/\{0/90\}]_s$. The CFRP laminates had 18 layers of carbon fiber plain weave fabric and nominal dimensions of 300 mm × 300 mm × 4 mm. More information on manufacturing CFRP laminates can be found in previous work [18].

For mechanical recycling technology using end mill cutters, CFRP laminates of 90 mm × 250 mm were placed horizontally on a Challenger CNC model MM-430. Carbide end mill cutters of 10 mm diameter and four flutes were used to cut the lateral part of the laminate. Each section had one cutter, and the laminates were firmly clamped to prevent vibrations or slipping during cutting. In a previous work [18], the amount of mechanically recycled CFRP (rCFRP) collected was measured as a function of the cutting speed, ranging from 1100 rpm, 1800 rpm, and 2500 rpm. The rCFRP was separated on a Ro-Tap laboratory sieve equipped with W.S. sieves to obtain the recyclates composed of short carbon fiber, epoxy resin powder, and rCFRP microparticles.

The recyclates were ground using HEBM technology. For this, a Simoloyer CM01-2l horizontal milling was used. Samples were taken at different grinding times as in previous experimental tests. The final processing time was determined to be 40 min.

In order to separate the recyclates, two experimental tests were carried out by taking samples of 1 and 2 g of recyclates in a beaker containing 500 mL of tap water. Samples were placed on a hot plate magnetic stirrer (Thermo Fisher Scientific, Waltham, MA, USA) and subjected to 350 rpm and 510 rpm, respectively. The tests were carried out at room temperature for 40 min. Subsequently, the samples were placed inside an Autoscience AS2060B (Pontiac, MI, USA) ultrasonic bath at 47 kHz for 40 min.

The ultrasonication technique was used using a QSONICA Q700 (Newton, CT, USA) ultrasonicator to evaluate the separation capacity of the rCFRP particles. The test involved placing 1 g and 2 g of rCFRP in 200 mL of tap water inside a beaker. A 24.5 mm diameter probe was used, and tests were performed varying the amplitude of 50 and 75% for 30 min to avoid excessive water heating.

The morphology of the rCFRP processed by the various mechanical recycling technologies such as end mill, HEBM, and ultrasonication was observed in a JEOL 6610LV scanning electron microscope (SEM) at 10 kV. For the SEM characterization, powders were (Dearborn Road Peabody, MA, USA) metalized using a JEOL Smart coater DII-29010SCTR that is operated under vacuum from a mechanical pump and provides a gold coating. This equipment generates a cloud ion from a current source of 5 A and a power of 500 VA. The parameter in the equipment is the sputtering time, which was adequate for the best observation of powders under SEM, and it was 5 min.

The chemical composition of the rCFRPs was determined by X-ray fluorescence (XRF) on a Spectro Xepos (Kleve, Germany). Five different samples were measured, and the average value obtained was presented.

3. Results and Discussions

Figure 1 shows the rCFRP obtained after the end milling process applying different velocities of the tool. The cut was made in the lateral zone of the CFRP laminates (Figure 1a) [18]. It is relevant to highlight that only the recyclates deposited on the white sheet of paper were collected and analyzed. Particles ejected into the air or deposited outside the leaf were not analyzed. Furthermore, it is highly recommended that cutting and grinding of the compounds be carried out in a closed room to avoid the volatility of micro and nanometric-sized recyclates that could cause damage to the health of the workers. Particle detectors in the air are also recommended to check the air quality of the cutting area and detect the concentration of particles to which the workers would be subjected.

The end mill recyclates obtained are small particles of different sizes. However, cutting speeds seem to influence end mill recyclates. For example, cutting speeds of 1100 and 1800 rpm promote chip detachment or fiber pullout (Figure 1b,c), while at 2500 rpm, small pieces of sheets are obtained, as seen in Figure 1d.

The presence of waste in the form of fiber pullout or delamination during the milling of CFRP composites has been reported by other authors and is attributable to defects induced by the cutting process [34]. It is relevant to highlight that most of the research evaluates the effect of cutting on the machining quality of CFRP composites and not on mechanical recycling [21,35]. In a previous work [18], the presence of fiber pullout and delaminations indicate the beginning of tool wear. However, delamination also seems to occur due to the effect of cutting speed. In their research, Pascual et al. [36] concluded that the axial cutting force is mainly responsible for the formation of delamination by debonding and expulsion induced by machining. Therefore, the evaluation of recycling technologies in this work will be carried out using only the recycling obtained at 1100 rpm.

The end mill recyclates at 1100 rpm were sieved using two meshes of 90–425 microns. Figure 2 presents the SEM micrographs corresponding to the end mill recyclates.

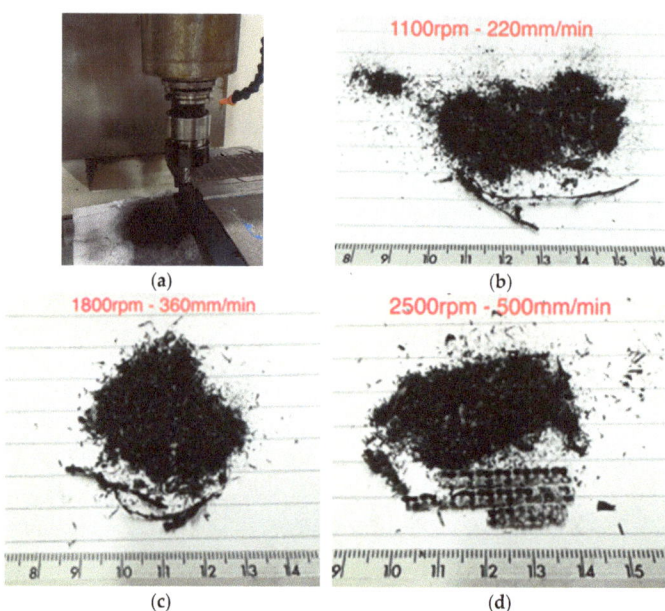

Figure 1. Photographs of the end milling recyclates obtained at different operation conditions: (**a**) edge milling, (**b**) 1100 rpm, (**c**) 1800 rpm, and (**d**) 2500 rpm.

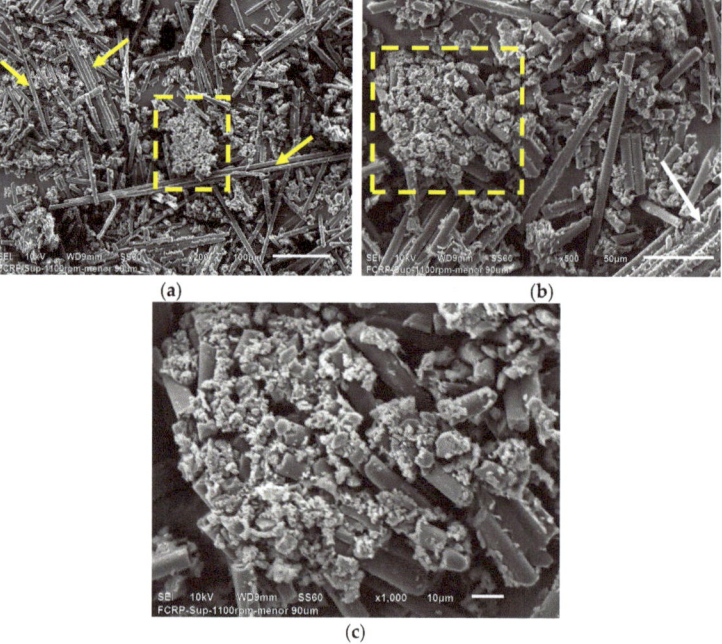

Figure 2. SEM micrographs corresponding to the sieved end mill recyclates. Features of recyclates can be observed at different magnifications: (**a**) 200×, (**b**) 500×, (**c**) 1000×. The yellow arrows indicate long fibers, the dashed box highlights agglomerated particles, and the white arrow indicates rCFRP particles.

It is possible to see long carbon fibers indicated by yellow arrows, which managed to pass through the 90 and 425-micron mesh during the sieving process. Agglomerated particles framed in the yellow dashed box are also seen (Figure 2a). At higher magnifications (Figure 2b), rCFRP particles indicated by a withe arrow can be seen. The rCFRP particles are made up of aligned carbon fibers joined together with epoxy resin. Meanwhile, the agglomerates are composed of shorter carbon fiber with a large amount of epoxy resin, as observed in Figure 2c. Figure 2c shows the characteristic features of the agglomerated particles, where epoxy resin particle residues agglomerate a considerable amount of short carbon fibers with a heterogeneous distribution.

Figure 3 presents the HEBM procedure performed. The end mill recyclates (1100 rpm, without sieving, and containing large fibers pullout) were milled using HEBM technology at a rotor speed of 860 rpm and a processing time of 40 min.

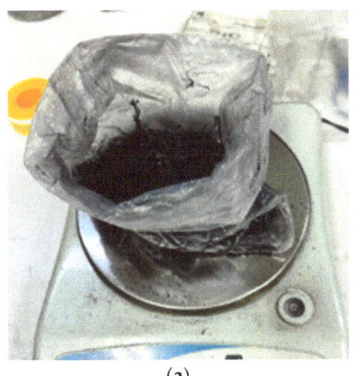
(a)

(b)

(c)

Figure 3. Photographs corresponding to the HEBM processing: (**a**) material weighing, (**b**) before the start of HEBM, (**c**) after the HEBM procedure.

The as-is recyclate obtained from side milling of the CFRP laminates was weighed on a laboratory scale (Figure 3a), then the recyclate was placed inside the mill, and the HEBM process was started (Figure 3b). After milling, the recyclate powder was extracted, as shown in Figure 3c. Figure 4 presents the SEM micrographs that reveal the morphology of the HEBM recyclates.

From HEBM recycling, it is possible to observe three types of waste: single short carbon fibers or graphite particles, and epoxy particles. The latter can be seen at low magnifications, indicated with yellow circles in Figure 4c. The HEBM process produces more agglomerated epoxy resin particles than the end milling process, where the agglomerates are micrometric resin particles containing randomly distributed short carbon fibers.

At high magnifications (Figure 4b), individual long carbon fibers, graphite particles, and agglomerated epoxy particles are visible. After the HEBM process, the carbon fibers' length is dramatically reduced compared with recycled end mills, obtaining graphite powder of fewer than 10 microns. Figure 4c shows that the short carbon fibers and the graphite particles contain traces of epoxy resin on their surface, which confirms the high adhesion between both constituents produced by the molecular compatibility between the carbon fiber and the epoxy matrix in the CFRP compounds. Particle size evaluation of powder obtained by HEBM was performed using the ImageJ free software (https://ij.imjoy.io/ access on 8 July 2024) using five SEM images, and two types of particles were considered: C-fiber and epoxy. Results are shown in Figure 5.

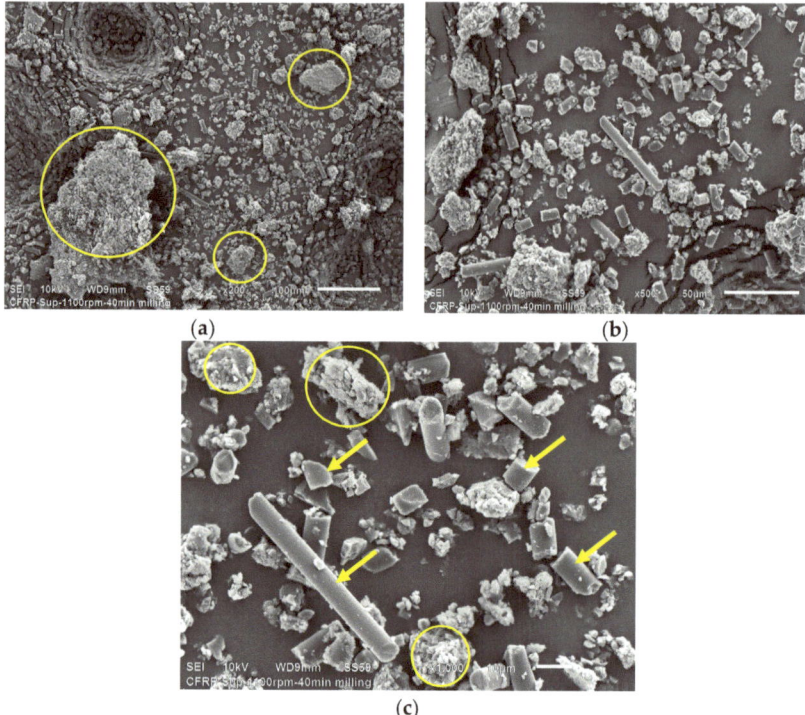

Figure 4. SEM micrographs corresponding to the HEBM recyclates. Features of recyclates can be observed at different magnifications: (**a**) 200×, (**b**) 500×, (**c**) 1000×. The yellow circles highlight agglomerated particles, and the yellow arrows indicate single carbon fibers and graphite particles.

Figure 5 shows a significant difference in particle size compared with C-fibers, some up to 80 microns, vs. epoxy, which has particles near 900 microns. However, the average size of particles is 7.06 and 32.86 microns of C-fibers and epoxy, respectively. The average of C-fibers is considered to be in the range of submicrometric size.

On the other hand, mechanical recycling of CFRP compounds entails contamination promoted by the wear of cutting and grinding tools. The elements mostly present in tools are chromium (Cr), iron (Fe), and cobalt (Co). The XRF technique was used to detect these elements. Table 1 presents the chemical composition of the recyclates obtained through end milling and the HEBM process.

Table 1. Chemical composition of recyclates.

Process	End Milling	HEBM
Element	Concentration (Wt.%)	Concentration (wt.%)
Cr	0.0023	0.213
Fe	0.0123	1.566
Co	0.0017	0.132

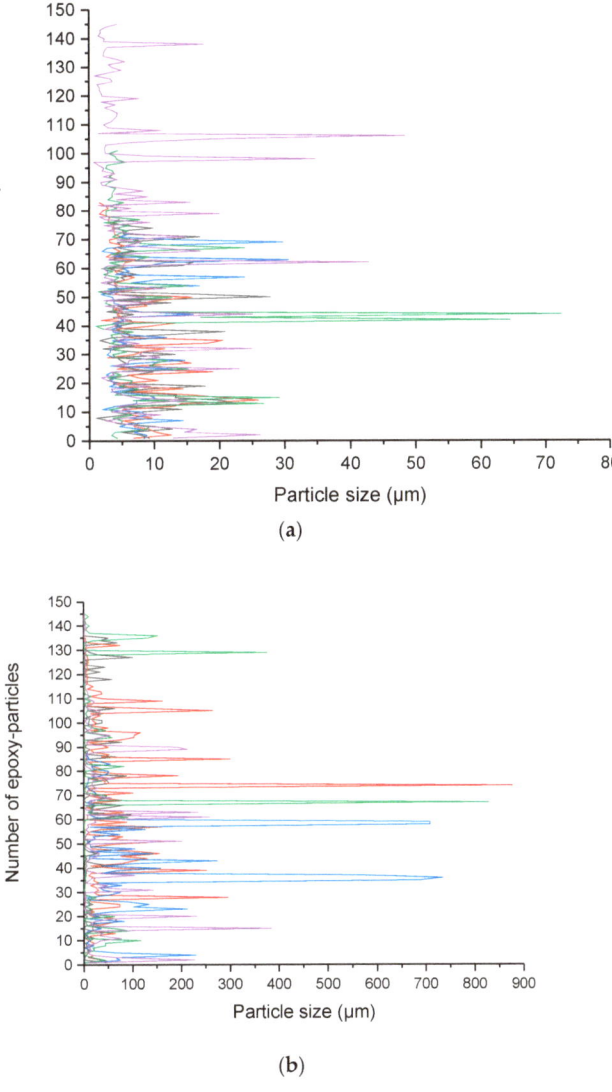

Figure 5. Particle size distribution of powder obtained after HEBM by using ImageJ free software, (**a**) C-particles and (**b**) epoxy particles. Colors are related to different SEM images.

As expected, the most significant elemental contaminants come from iron, which is the base manufacturing material for cutting tools and grinding balls. Curiously, the end milling-cutting process does not entail excessive contamination. In contrast, the HEBM process produces considerable elemental contaminants, even more so if mechanical recycling is considered to be evaluated at the laboratory level (that is, grams of material). Therefore, it is essential to emphasize that the presence of these elements, mainly Fe, should not be ignored when using recyclates to manufacture new materials through melt welding processes since it could alter the properties of the materials.

Specific punctual chemical analysis using an Energy Dispersive Spectrometer (EDS) of SEM equipment was performed directly on the powder obtained after HEBM. The evaluation was performed to confirm epoxy resin particles as part of the product. Additional/complimentary results are located in the Supporting Information section. Then, Figure 6 shows the punctual EDS evaluation in an epoxy particle.

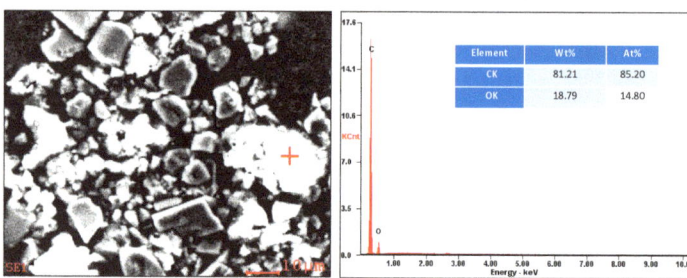

Figure 6. SEM-EDS punctual chemical analysis showing the presence of epoxy resin in milled powder.

SEM-EDS chemical analysis results confirm the presence of epoxy particles separated from carbon fibers. It clearly identifies the peculiar morphology of carbon fibers and the small particles of this material after HEBM.

Continuing with the effort to separate the components of the recyclates (carbon fiber, epoxy resin, and rCFRP particles), a magnetic stirring test followed by ultrasound was carried out using the recyclates HEBM. Magnetic stirring was performed at 350 rpm for 10 min and 510 rpm for another 10 min (Figure 7a,b, respectively). During the test, the formation of a vortex, a product of magnetic stirring, was more pronounced as the stirring speed increased.

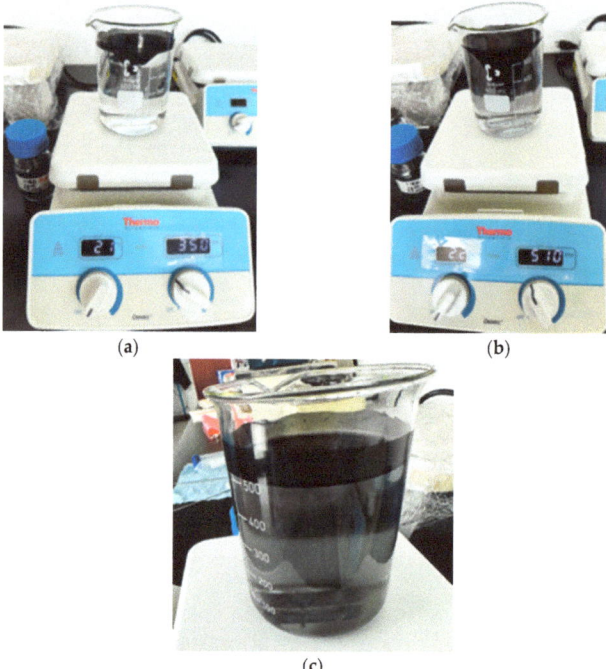

Figure 7. Photographs corresponding to the magnetic stirring process: (**a**) 350 rpm, (**b**) 510 rpm, (**c**) after stirring.

At the end of the test, and after 10 min, most of the HEBM recyclates remained on the surface, although a small amount of the sample precipitated, as shown in Figure 7c.

According to the literature, carbon fiber and epoxy resin density is 1.75–2.00 g/cm^3 and 1.2–1.3 g/cm^3, respectively. In our case, the Epolam 2015 has a 1.15 g/cm3 density. Therefore, the recyclates were expected to sink to the bottom of the beaker.

During the magnetic stirring test, it was observed that the recyclates showed highly hydrophobic behavior, which could explain their ability to float in water.

The same batch of HEBM recyclates was subjected to an ultrasonic bath for 40 min (Figure 8a) to separate the components of the recyclates through cavitation bubbles induced by high-frequency pressure waves.

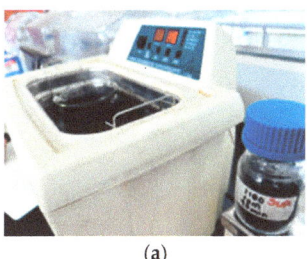
(a)

(b)
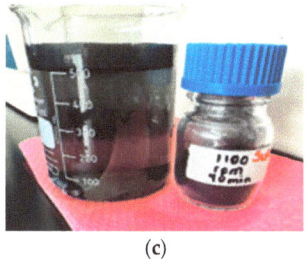
(c)

Figure 8. Photographs corresponding to the (**a**) ultrasonic bath process, (**b**) floating recyclates, (**c**) precipitated HEBM recyclates after 30 min.

At the end of the test, the recyclates remained on the water surface (Figure 8b). However, precipitates and some particles suspended in the water inside the beaker were observed, as seen in Figure 8c.

To elucidate the flotability of HEBM recyclates, flotation tests of the recyclate constituents were performed in distilled water. First, the epoxy resin was evaluated. For this purpose, a 5 cm × 5 cm × 1 cm epoxy resin plate (Epolam 2015) was prepared. The resin plate was wrapped in a cloth and manually crushed into small pieces using a hammer. Approximately 10 g of ground epoxy resin was placed in the beaker, as shown in Figure 9a.

Figure 9. Photographs corresponding to (**a**) crushed epoxy resin, (**b**) short carbon fiber, and (**c**) CFRP samples submerged in distilled water.

Most resin initially sinks, although some resin remaining floats. When stirring with a glass rod, most of the epoxy resin sinks to the bottom, and only a tiny proportion floats. A closer look (image box in Figure 9a) shows that some resin chunks floating in water are tiny and appear to have internal bubbles induced during the epoxy board manufacturing process. These bubbles and the aspect ratio of the particles could promote floating in water. Subsequently, the same test was performed using short carbon fiber (5–10 mm), as presented in Figure 9b. The carbon fiber was cut from the plain weave fabric used in this work. Approximately 10 g of short carbon fiber was placed in the beaker. The behavior was similar to that observed with the epoxy resin. A significant proportion of the short carbon fiber sinks, but some short carbon fibers float because of surface tension, so when pushing the fiber with laboratory tweezers or stirring it with a glass rod, the surface tension breaks, and the fiber sinks to the bottom of the beaker (image box in Figure 9b).

The epoxy particles and the short carbon fibers float because of the high aspect ratio of the particles. Edward Bormashenko [37] published that capillary forces better support more elongated bodies of a fixed volume because of an increase in the perimeter of the triple line, which is what occurs in the flotation of slender bodies, similar to what has been observed with particles (flakes) of epoxy resin and carbon fibers.

It is important to note that when the resin particles and carbon fiber were removed from the beaker, both were completely soaked in water.

On the other hand, some pieces of CFRP composites were cut and immersed in distilled water, as shown in Figure 9c. As expected, the carbon fiber composite pieces sank immediately. However, several bubbles were observed to form. The bubbles could indicate that the composites have imperfections (small holes) that contain air.

Figure 10 shows the HEBM recyclate (powder) floating in distilled water. Contrary to what might be expected, the recyclate floats noticeably. The recyclate was stirred with a glass rod and remained afloat. Manual shaking was also carried out, observing that the recyclate did not become wet and remained floating. Only small particles sink after manual shaking. Finally, a PVC plastic rod was introduced to the beaker, showing how the recyclate adheres to the PVC rod, as shown in Figure 10b. The recyclate remains adhered even after stirring the PVC rod a little while immersed in water. Continuing with the stirring, some agglomerated pieces of recyclate powder detach from the rod and return to the surface, as seen in Figure 10c. A more vigorous stirring allows all the adhered and agglomerated recyclate to detach from the PVC rod and return to the surface (Figure 10d). However, a considerable proportion of recyclate adheres to its surface when the PVC rod is removed, as presented in Figure 10e. It is relevant to consider the appearance of shiny reflective surfaces (Figure 10b), which could be attributable to the occurrence of water–air interfaces.

The recyclate seems to have an electrostatic charge induced during the HEBM process, which promotes the agglomeration of the recyclate. If we also consider what is observed in Figure 2, where the particles of the composites seem to melt because of the friction of the end-mill cutting process, then we can suggest the particles contain trapped air, which promotes the floating of the recyclate.

Based on the above results, the floating of the recyclate would be due to the combination of three situations: surface tension, trapped air, and the electrostatics induced by the processing [38].

It seems that the induced electrostatics would be influencing the hydrophobicity of the recyclate. Mateusz Kruszelnicki et al. [39] indicated that flotation can be derived from the wettability of particles. The results showed that for the flotation process to occur, the minimum (critical) value of the contact angle must be exceeded, which was determined to be approximately 25° when the electrostatic interactions were attractive or weakly repulsive and up to 62° for strongly repulsive electrostatic interactions.

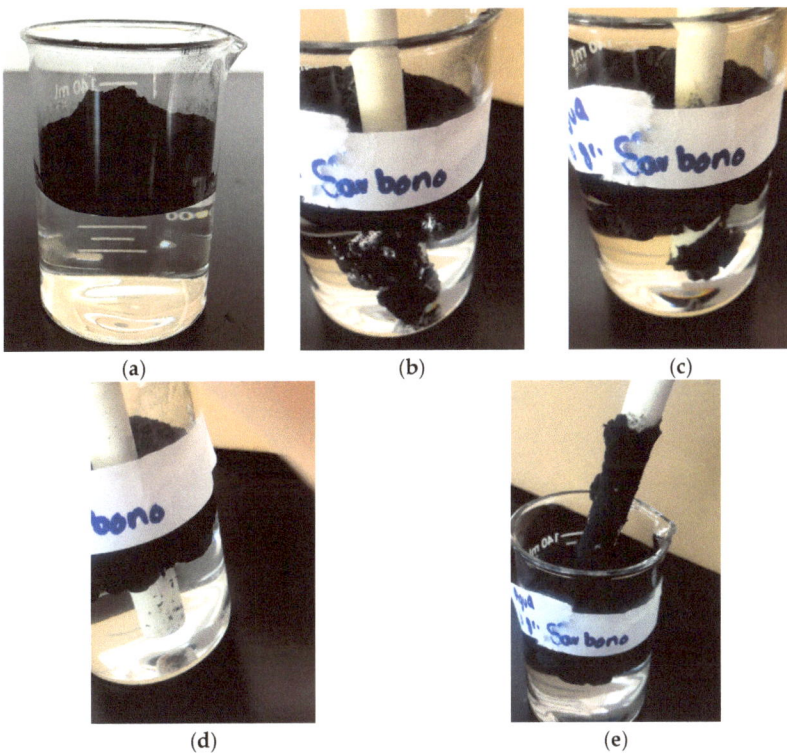

Figure 10. Photographs corresponding to (**a**) recyclates floating in distilled water, (**b**) recyclates adhering to PVC road submerged in distilled water, (**c**) recyclates detached from PVC road after stirring, (**d**) completely detached recyclates that return to the surface, and (**e**) recyclates adhered to PVC road outside the beaker.

Other authors [40] experimented with floating metals, taking advantage of the energy of the water surface. The authors attribute this effect to the enhanced flotation ability of nanostructures on the surfaces of copper and stainless steel sheets. Sufficiently thin hydrophobic metals can slowly float underwater by trapping air on the surface.

It is well known that the ability of materials to sink or float depends on their density, shape, and the amount of water they displace [41,42]. Guo et al. [43] mentioned that this ability is further enhanced by superhydrophobic surfaces that exhibit water-repellent behavior as it provides a layer of air to replace volumes of water for increased buoyancy. Furthermore, attached air bubbles or layers can enhance flotation ability [44]. Edward Bormashenko [37] studied the flotation of bodies with different geometries, showing that more elongated bodies float thanks to capillary forces, taking into account the effect of the aspect ratio of the bodies. Thus, when the lateral dimension of the floating bodies is much smaller than the length of the capillary, the buoyancy is negligible, and the flotation is determined by the surface tension, according to what was studied by Dominic Vella [45].

Figure 11 presents the SEM micrographs corresponding to the morphological analysis of the precipitated HEBM recyclates.

When comparing the micrographs corresponding to the HEBM recyclates with the HEBM ultrasonic recyclates at high magnifications (Figures 4c and 11c), the HEBM ultrasonic recyclates contain a more significant amount of epoxy resin particles than the HEBM recyclates. The above would indicate that the precipitates are mostly epoxy resin, which allows progress toward a process of separation of the components of the recyclates [35]. In both cases, carbon fibers with some epoxy resin can be seen, which would indicate that the ultrasonic does not produce significant changes in the morphology of the recyclates, referring to the dispersion or separation of the epoxy resin adhered to the surface of the carbon fiber.

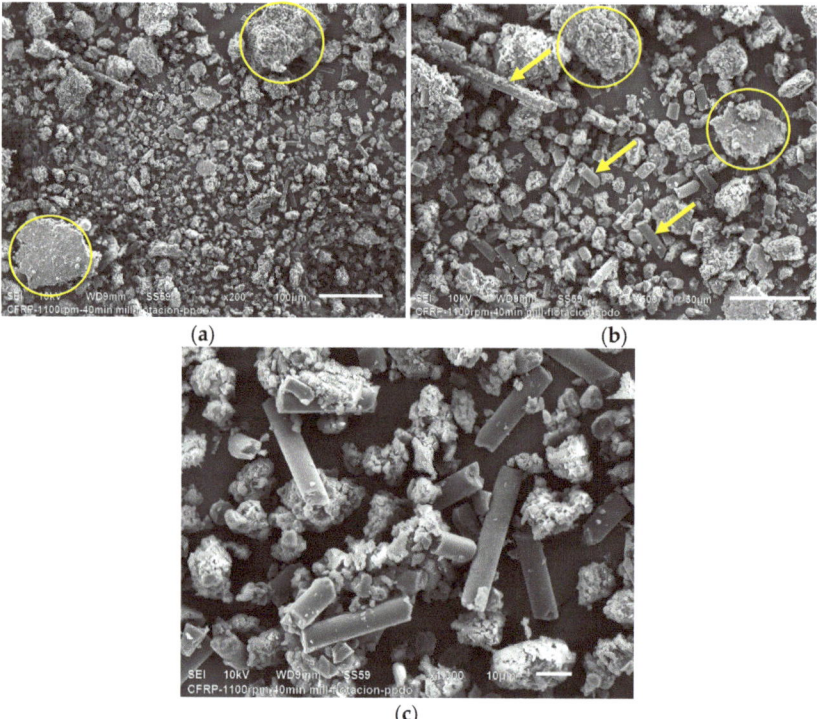

Figure 11. SEM micrographs corresponding to the precipitated HEBM recyclates after ultrasonic bath. Features of recyclates can be observed at different magnifications: (a) 200×, (b) 500×, (c) 1000×. The yellow circles highlight agglomerated particles, and the yellow arrows indicate single carbon fibers and graphite particles.

Resin particles are agglomerates that may contain graphite particles and short carbon fibers because of the strong molecular interaction between both constituents. The interface between the carbon fiber and the resin matrix plays a fundamental role in controlling the overall properties of the composites. Polyacrylonitrile (PAN) is the most common precursor for manufacturing carbon fibers, so the mechanical and electrochemical properties of the fibers largely depend on the nature of the preparation methods and subsequent processing treatments. The modification of carbon fibers is expected to improve their structural performance. Thus, their reinforcing performance often depends on the interfacial bonding strength between the fibers and the resin matrix. Unmodified carbon fibers compromise the mechanical properties of the composite [46].

Ultrasonication is the technique adopted to reduce the size of clay particles and disperse carbon nanotubes [47]. Ultrasonication produces cavitation bubbles that collapse rapidly, leading to impact shock waves and interparticle collisions. In this way, sonication induces the disaggregation of particles and favors their dispersion depending on the ultrasonication energy [48].

Ultrasonication was carried out using samples of HEBM recyclates and sieved end mill recyclates (<90 microns). Figure 12 presents the sequence of the experimental ultrasonication procedure performed on the HEBM recyclates and the sieved end mill recyclates.

Four experimental ultrasonication tests were run that consisted of varying the amplitude (50–75%) and the content of recyclates (1–2 g). The temperature was monitored during the experimental tests. For an amplitude of 50%, the test exceeded 60 °C after 15 min of ultrasonication, while for an amplitude of 75%, the test exceeded 60 °C after 10 min of operation. Observations in the recipient of water/powder solution were that it evaporated and spread the material that floats on the surface. This solution behavior could not be controlled during ultrasonication, and cooling was performed by placing the recipient in a vessel with cold water, which required approximately 10 or 15 min, depending on the energy sonication used to reach 25–27 °C in the solution.

(a)

(b)

(c)

(d)

Figure 12. Photographs corresponding to the ultrasonication process: (**a**) sample 1 g, (**b**) sample 2 g, (**c**) sample 2 g after sonication, (**d**) sample 2 g after 30 min that reveals the separation of floated and precipitated recyclates.

Figure 12a corresponds to the 1 g sample of HEBM recyclates, while Figure 12b corresponds to the 2 g sample of HEBM recyclates. The physical difference in using 1 or 2 g of recyclates is seen. Using 2 g of recyclates subjected to an amplitude of 75% produced a considerable amount of bubbles, which could promote the effective separation of the components of the recyclates. After ultrasonication, the suspension remains completely black (Figure 12c). After 30 min, the suspension cools, and the floating recyclates and precipitates are separated, as shown in Figure 12d. As the content of recyclates increases, the precipitates increase. A similar behavior was observed in the sieved end mill recyclates. Following the methodological sequence of the research, the morphology of the precipitated recyclates was evaluated by SEM.

Figure 13 presents the SEM micrographs corresponding to the ultrasonicated precipitates of HEBM recyclates, while Figure 14 presents the SEM micrographs of the ultrasonicated precipitates of sieved end mill recyclates.

When comparing the SEM micrographs of the ultrasonicated HEBM recyclates, it does not seem—at first glance—that there are evident changes in the morphology of the precipitates. However, significant morphological variations can be observed if we compare the morphologies of the recyclates subjected to a lower amplitude and low residue content (Figure 13a) with the tests carried out with recyclates of 2 g and higher amplitude (Figure 13d). Ultrasonication breaks down the hard epoxy resin particles and reduces their size. Cracks are observed in some larger epoxy particles, as highlighted in a dashed square indicated with arrows in Figure 13. With 2 g of recyclates and 75% amplitude, a greater content of graphite particles and a considerable amount of epoxy resin microparticles can be observed, much smaller than those observed in Figure 13a.

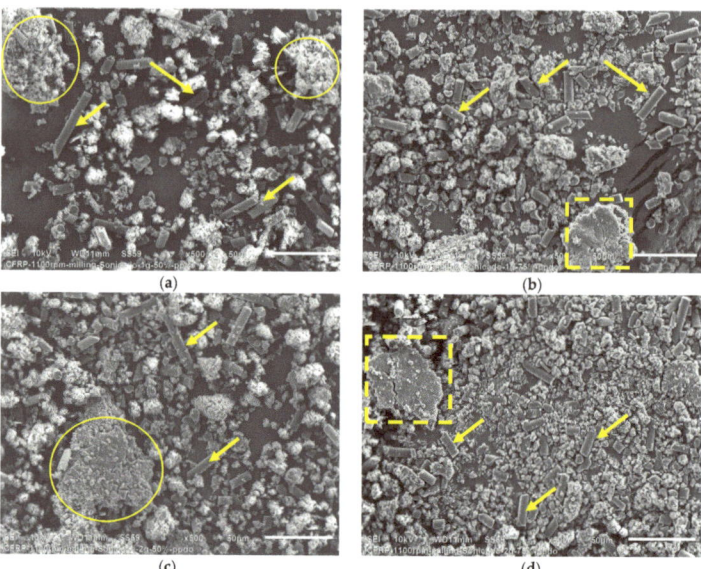

Figure 13. SEM micrographs corresponding to the ultrasonicated precipitates of HEBM recyclates: (**a**) 1 g and 50% amplitude, (**b**) 1 g and 75% amplitude, (**c**) 2 g and 50% amplitude, (**d**) 2 g and 75% amplitude. The yellow circles highlight agglomerated particles, and the yellow arrows indicate single carbon fibers and graphite particles. The yellow dashed square indicates fractured particles.

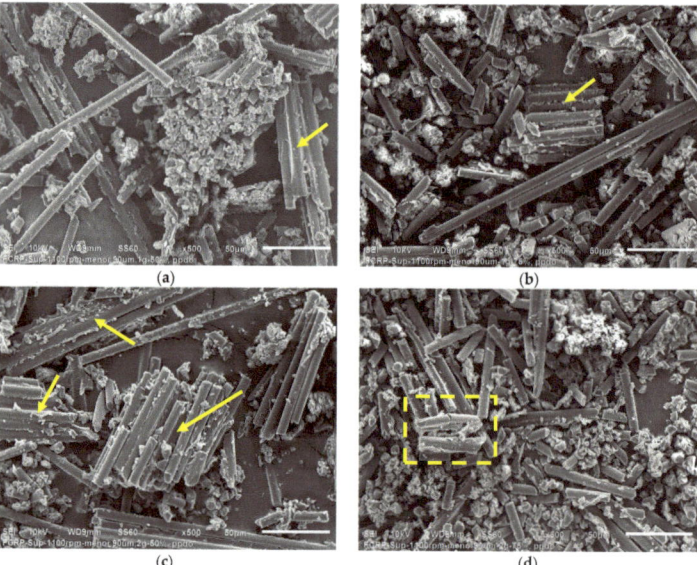

Figure 14. SEM micrographs corresponding to the ultrasonicated precipitates of sieved end mill recyclates: (**a**) 1 g and 50% amplitude, (**b**) 1 g and 75% amplitude, (**c**) 2 g and 50% amplitude, (**d**) 2 g and 75% amplitude. The yellow arrows indicate rCFRP particles and the yellow dashed square indicates fractured carbon fiber.

It is possible to appreciate that the morphology of the precipitates corresponding to the sieved and ultrasonicated end mill recyclates differs from the HEBM recyclates. The first difference is that individual long carbon fibers, agglomerated CFRP particles, and some epoxy resin particles are observed. Most of the epoxy resin is attached to the carbon fibers. Contrary to what is observed in Figure 2c, the particles present carbon fibers aligned and homogeneously distributed (Figure 14b,c). Figure 14d, which represents the morphology of 2 g recyclates subjected to high amplitude ultrasonication, shows individual carbon fibers and dispersed epoxy resin particles. Ultrasonication separates the carbon fibers and releases the epoxy resin from their surface, so a higher content of micrometric epoxy resin particles distributed between the individual carbon fibers is observed. In addition, the carbon fibers are less long because ultrasonication breaks the agglomerates and fragments the fibers, as seen in Figure 14d. The results confirm that ultrasonication is effective in sifted end mill recyclates since the carbon fibers are released from the epoxy resin and break into tiny agglomerated particles.

The strategies evaluated in the mechanical recycling of CFRP composites allow us to assess the differences between end mills and HEBM recycling. HEBM recycling requires small samples to enter the ball mill; a prior cutting or grinding step is necessary. HEBM recycling allows us to obtain small particles of recyclates, but it is impossible to separate their components: carbon fiber, epoxy resin, and CFRP particles.

Mill recyclates are obtained directly by cutting large CFRP parts or components. The recyclates obtained are a combination of long fibers and micrometric particles. A sieving step allows for particle homogeneity. However, large carbon fibers can pass the sieve.

Ultrasonication does not significantly affect HEBM recyclates because of the high energy they are subjected to during the grinding process, but it is effective on end-mill recyclates. The ultrasonication amplitude has a greater impact than the content of the samples in terms of the ability to separate the epoxy resin from the carbon fiber.

4. Conclusions

The morphology of the recyclates obtained through different mechanical recycling techniques and subsequent treatments was analyzed to separate the constituents of the recyclates, which are carbon fiber, epoxy powder, and CRFP particles.

The sieving of the recycle end mill separates the waste, although individual carbon fibers manage to pass through the wire cloth.

The ultrasonic bath did not present a significant difference in separating the epoxy resin from the carbon fiber.

The end mill produces CFRP particles, epoxy resin agglomerated waste, and some individual fibers with epoxy resin on their surface.

The HEBM process produces tiny particles of graphite and epoxy resin.

Ultrasonication separates CFRP into individual long carbon fibers, breaks down the epoxy resin particles to make them smaller, fragments the carbon fibers, and separates the epoxy resin from the surface of the fibers.

Ultrasonication allows effective separation of the constituents from the recyclates. However, the procedure is time-consuming and at a laboratory level. At the moment, it is impossible to collect the recyclates' constituents separately.

CFRP constituents do not float in distilled water or tap water. However, the recyclates float. The end mill and HEBM waste production process promote the presence of trapped air and electrostatics, which allows recyclates to float in water and be hydrophobic. The appearance of shiny reflective surfaces could be attributable to the occurrence of water–air interfaces.

The results of this research present progress towards developing technology for the mechanical recycling of CFRP and open a range of opportunities in reusing recyclates to create new products.

Supplementary Materials: The following supporting information can be downloaded at: https://www.mdpi.com/article/10.3390/polym16162350/s1, Figure S1: Particle size evaluation: (a) Carbon fiber, (b) epoxy resin particles. Figure S2: Several SEM microgprahs taken from different zones. Figure S3: Punctual analyses where we observed peaks of carbon and oxygen. Figure S4: Process of ultrasonication. Figure S5: Photographs corresponding to (a) crushed epoxy resin,

(b) 10 mg of chopped carbon fiber, (c) 5 mg of chopped carbon fiber submerged in tap water. Figure S6: Photograph corresponding to the resin chunks that float in tap water. Video S1: Adherence rCFRP by electrostatic.

Author Contributions: Conceptualization, E.A.F.-U. and E.M.-F.; methodology, E.A.F.-U., E.M.-F. and V.A.G.C.; validation, E.A.F.-U. and V.A.G.C.; formal analysis, E.A.F.-U.; investigation, E.A.F.-U., E.M.-F. and V.A.G.C.; resources, E.A.F.-U.; writing—original draft preparation, E.M.-F.; writing—review and editing, E.A.F.-U.; supervision, E.A.F.-U.; project administration E.A.F.-U.; funding acquisition, E.A.F.-U. All authors have read and agreed to the published version of the manuscript.

Funding: This work was funded by Centro de Ingenieria y Desarrollo Industrial (CIDESI). Grant No. QID028.

Institutional Review Board Statement: Not applicable.

Data Availability Statement: The original contributions presented in the study are included in the article/Supplementary Materials, further inquiries can be directed to the corresponding author.

Acknowledgments: The authors would like to acknowledge the technical support provided by Cecilia Zarate and Sergio Lopez from the CIDESI campus Airport.

Conflicts of Interest: The authors declare no conflict of interest.

References

1. Durante, M.; Boccarusso, L.; De Fazio, D.; Formisano, A.; Langella, A. Investigation on the Mechanical Recycling of Carbon Fiber-Reinforced Polymers by Peripheral Down-Milling. *Polymers* **2023**, *15*, 854. [CrossRef] [PubMed]
2. Zhang, C.; Wang, C.; Gao, M.; Liu, C. Emergy-based sustainability measurement and evaluation of industrial production systems. *Environ. Sci. Pollut. Res.* **2023**, *30*, 22375–22387. [CrossRef] [PubMed]
3. Pauliuk, S.; Koslowski, M.; Madhu, K.; Schulte, S.; Kilchert, S. Co-design of digital transformation and sustainable development strategies-What socio-metabolic and industrial ecology research can contribute. *J. Clean. Prod.* **2022**, *343*, 130997. [CrossRef]
4. Bengtsson, M.; Alfredsson, E.; Cohen, M.; Lorek, S.; Schroeder, P. Transforming systems of consumption and production for achieving the sustainable development goals: Moving beyond efficiency. *Sustain. Sci.* **2018**, *13*, 1533–1547. [CrossRef]
5. Omer, A.M. Energy, environment and sustainable development. *Renew. Sustain. Energy Rev.* **2008**, *12*, 2265–2300. [CrossRef]
6. Soo, V.K.; Doolan, M.; Compston, P.; Duflou, J.R.; Peeters, J.; Umeda, Y. The influence of end-of-life regulation on vehicle material circularity: A comparison of Europe, Japan, Australia and the US. *Resour. Conserv. Recycl.* **2021**, *168*, 105294. [CrossRef]
7. Karagoz, S.; Aydin, N.; Simic, V. End-of-life vehicle management: A comprehensive review. *J. Mater. Cycles Waste Manag.* **2020**, *22*, 416–442. [CrossRef]
8. Giannouli, M.; de Haan, P.; Keller, M.; Samaras, Z. Waste from road transport: Development of a model to predict waste from end-of-life and operation phases of road vehicles in Europe. *J. Clean. Prod.* **2007**, *15*, 1169–1182. [CrossRef]
9. Vieira, D.R.; Vieira, R.K.; Chang Chain, M. Strategy and management for the recycling of carbon fiber-reinforced polymers (CFRPs) in the aircraft industry: A critical review. *Int. J. Sustain. Dev. World Ecol.* **2017**, *24*, 214–223. [CrossRef]
10. Vo Dong, A.; Azzaro-Pantel, C.; Boix, M. A multi-period optimisation approach for deployment and optimal design of an aerospace CFRP waste management supply chain. *Waste Manag.* **2019**, *95*, 201–216. [CrossRef]
11. Rodríguez-González, J.A.; Renteria-Rodríguez, V.; Rubio-González, C.; Franco-Urquiza, E.A. Spray-coating of graphene nanoplatelets on sisal fibers and its influence on electromechanical behavior of biocomposite laminates. *J. Reinf. Plast. Compos.* **2024**, *43*, 3–15. [CrossRef]
12. Petrakli, F.; Gkika, A.; Bonou, A.; Karayannis, P.; Koumoulos, E.P.; Semitekolos, D.; Trompeta, A.F.; Rocha, N.; Santos, R.M.; Simmonds, G.; et al. End-of-life recycling options of (nano)enhanced CFRP composite prototypes waste-a life cycle perspective. *Polymers* **2020**, *12*, 2129. [CrossRef] [PubMed]
13. Gopalraj, S.K.; Deviatkin, I.; Horttanainen, M.; Kärki, T. Life cycle assessment of a thermal recycling process as an alternative to existing cfrp and gfrp composite wastes management options. *Polymers* **2021**, *13*, 4430. [CrossRef] [PubMed]
14. Sukanto, H.; Raharjo, W.W.; Ariawan, D.; Triyono, J. Carbon fibers recovery from CFRP recycling process and their usage: A review. *IOP Conf. Ser. Mater. Sci. Eng.* **2021**, *1034*, 12087. [CrossRef]
15. Naqvi, S.R.; Prabhakara, H.M.; Bramer, E.A.; Dierkes, W.; Akkerman, R.; Brem, G. Resources, Conservation & Recycling A critical review on recycling of end-of-life carbon fi bre/glass fi bre reinforced composites waste using pyrolysis towards a circular economy. *Resour. Conserv. Recycl.* **2018**, *136*, 118–129. [CrossRef]
16. Pickering, S.J. Recycling technologies for thermoset composite materials—Current status. *Compos. Part A Appl. Sci. Manuf.* **2006**, *37*, 1206–1215. [CrossRef]
17. Oliveux, G.; Dandy, L.O.; Leeke, G.A. Current status of recycling of fibre reinforced polymers: Review of technologies, reuse and resulting properties. *Prog. Mater. Sci.* **2015**, *72*, 61–99. [CrossRef]

18. Vega-Leal, C.; Zárate-Pérez, C.; Gomez-Culebro, V.A.; Burelo, M.; Franco-Urquiza, E.A.; Treviño-Quintanilla, C.D. Mechanical recycling of carbon fibre reinforced polymers. Part 1: Influence of cutting speed on recycled particles and composites properties. *Int. J. Sustain. Eng.* **2024**, *17*, 1–10. [CrossRef]
19. Geier, N.; Xu, J.; Poór, D.I.; Dege, J.H.; Davim, J.P. A review on advanced cutting tools and technologies for edge trimming of carbon fibre reinforced polymer (CFRP) composites. *Compos. Part B Eng.* **2023**, *266*, 111037. [CrossRef]
20. Prakash, R.; Krishnaraj, V.; Zitoune, R.; Sheikh-Ahmad, J. High-Speed Edge Trimming of CFRP and Online Monitoring of Performance of Router Tools Using Acoustic Emission. *Materials* **2016**, *9*, 798. [CrossRef]
21. Slamani, M.; Chatelain, J.F.; Hamedanianpour, H. Influence of machining parameters on surface quality during high speed edge trimming of carbon fiber reinforced polymers. *Int. J. Mater. Form.* **2019**, *12*, 331–353. [CrossRef]
22. Hosokawa, A.; Hirose, N.; Ueda, T.; Furumoto, T. High-quality machining of CFRP with high helix end mill. *CIRP Ann.* **2014**, *63*, 89–92. [CrossRef]
23. Newman, B.; Creighton, C.; Henderson, L.C.; Stojcevski, F. A review of milled carbon fibres in composite materials. *Compos. Part A Appl. Sci. Manuf.* **2022**, *163*, 107249. [CrossRef]
24. Dudina, D.V.; Bokhonov, B.B. Materials Development Using High-Energy Ball Milling: A Review Dedicated to the Memory of M.A. Korchagin. *J. Compos. Sci.* **2022**, *6*, 188. [CrossRef]
25. Kanlayakan, W.; Lhosupasirirat, S.; Amnatsin, C.; Verojpipath, N.; Pragobjinda, B.; Srikhirin, T. Improvement of Ball Mill Performance in Recycled Ultrafine Graphite Waste Production for Carbon Block Applications. *ACS Omega* **2023**, *8*, 27312–27322. [CrossRef] [PubMed]
26. Gong, X.; Yao, J.; Yin, W.; Yin, X.; Ban, X.; Wang, Y. Effect of acid corrosion on the surface roughness and floatability of magnesite and dolomite. *Green Smart Min. Eng.* **2024**, *404*, 125002. [CrossRef]
27. Agnihotri, S.N.; Thakur, R.K.; Singh, K.K. Influence of nanoclay filler on mechanical properties of CFRP composites. *Mater. Today Proc.* **2022**, *66*, 1734–1738. [CrossRef]
28. Franco-Urquiza, E.A. Clay-Based Polymer Nanocomposites: Essential Work of Fracture. *Polymers* **2021**, *13*, 2399. [CrossRef] [PubMed]
29. Franco-Urquiza, E.A.; Cailloux, J.; Santana, O.; Maspoch, M.L.; Velazquez Infante, J.C. The influence of the clay particles on the mechanical properties and fracture behavior of PLA/o-MMT composite films. *Adv. Polym. Technol.* **2015**, *34*, 21470. [CrossRef]
30. Franco-Urquiza, E.; Perez, J.G.; Sánchez-Soto, M.; Santana, O.O.; Maspoch, M.L. The effect of organo-modifier on the structure and properties of poly[ethylene-(vinyl alcohol)]/organo-modified montmorillonite composites. *Polym. Int.* **2010**, *59*, 778–786. [CrossRef]
31. Maspoch, M.L.; Franco-Urquiza, E.; Gamez-Perez, J.; Santana, O.O.; Sánchez-Soto, M. Fracture behaviour of poly[ethylene-(vinyl alcohol)]/organo-clay composites. *Polym. Int.* **2009**, *58*, 648–655. [CrossRef]
32. Franco-Urquiza, E.; Rentería-Rodríguez, A. Effect of nanoparticles on the mechanical properties of kenaf fiber-reinforced bio-based epoxy resin. *Text. Res. J.* **2021**, *91*, 1313–1325. [CrossRef]
33. Miyagawa, H.; Jurek, R.J.; Mohanty, A.K.; Misra, M.; Drzal, L.T. Biobased epoxy/clay nanocomposites as a new matrix for CFRP. *Compos. Part A Appl. Sci. Manuf.* **2006**, *37*, 54–62. [CrossRef]
34. Sadek, A.; Meshreki, M.; Attia, M.H. Characterization and optimization of orbital drilling of woven carbon fiber reinforced epoxy laminates. *CIRP Ann.* **2012**, *61*, 123–126. [CrossRef]
35. Ferraris, M.; Ventrella, A.; Salvo, M.; Avalle, M.; Pavia, F.; Martin, E. Comparison of shear strength tests on AV119 epoxy-joined carbon/carbon composites. *Compos. Part B Eng.* **2010**, *41*, 182–191. [CrossRef]
36. Pascual, V.; San-Juan, M.; Santos, F.J.; Martín, Ó.; de Tiedra, M.P. Study of axial cutting forces and delamination phenomenon in drilling of carbon fiber composites. *Procedia Manuf.* **2017**, *13*, 67–72. [CrossRef]
37. Bormashenko, E. Surface tension supported floating of heavy objects: Why elongated bodies float better? *J. Colloid Interface Sci.* **2016**, *463*, 8–12. [CrossRef] [PubMed]
38. Novin, N.; Shameli, A.; Balali, E.; Zomorodbakhsh, S. How electrostatic and non-electrostatic interactions play a role in water wettability of possible nanostructure surfaces. *J. Nanostruct. Chem.* **2020**, *10*, 69–74. [CrossRef]
39. Kruszelnicki, M.; Polowczyk, I.; Kowalczuk, P.B. Insight into the influence of surface wettability on flotation properties of solid particles—Critical contact angle in flotation. *Powder Technol.* **2024**, *431*, 119056. [CrossRef]
40. Tsai, J.Y.; Huang, G.F.; Shieh, J.; Hsu, C.C.; Ostrikov, K. (Ken) Harvesting water surface energy: Self-jumping nanostructured hydrophobic metals. *iScience* **2021**, *24*, 102746. [CrossRef]
41. Keller, J.B. Surface tension force on a partly submerged body. *Phys. Fluids* **1998**, *10*, 3009–3010. [CrossRef]
42. Naylor, D.; Tsai, S.S.H. Archimedes' principle with surface tension effects in undergraduate fluid mechanics. *Int. J. Mech. Eng. Educ.* **2021**, *50*, 749–763. [CrossRef]
43. Guo, J.; Yu, S.; Li, J.; Guo, Z. Fabrication of functional superhydrophobic engineering materials via an extremely rapid and simple route. *Chem. Commun.* **2015**, *51*, 6493–6495. [CrossRef] [PubMed]
44. Pan, Q.; Wang, M. Miniature Boats with Striking Loading Capacity Fabricated from Superhydrophobic Copper Meshes. *ACS Appl. Mater. Interfaces* **2009**, *1*, 420–423. [CrossRef]
45. Vella, D. Floating Versus Sinking. *Annu. Rev. Fluid Mech.* **2015**, *47*, 115–135. [CrossRef]
46. Casimero, C.; Hegarty, C.; Mcglynn, R.J.; Davis, J. Ultrasonic exfoliation of carbon fiber: Electroanalytical perspectives. *J. Appl. Electrochem.* **2020**, *50*, 383–394. [CrossRef]

47. Poli, A.L.; Batista, T.; Schmitt, C.C.; Gessner, F.; Neumann, M.G. Effect of sonication on the particle size of montmorillonite clays. *J. Colloid Interface Sci.* **2008**, *325*, 386–390. [CrossRef]
48. Chen, S.J.; Zou, B.; Collins, F.; Zhao, X.L.; Majumber, M.; Duan, W.H. Predicting the influence of ultrasonication energy on the reinforcing efficiency of carbon nanotubes. *Carbon* **2014**, *77*, 1–10. [CrossRef]

Disclaimer/Publisher's Note: The statements, opinions and data contained in all publications are solely those of the individual author(s) and contributor(s) and not of MDPI and/or the editor(s). MDPI and/or the editor(s) disclaim responsibility for any injury to people or property resulting from any ideas, methods, instructions or products referred to in the content.

Article

Use of Recycled Additive Materials to Promote Efficient Use of Resources While Acting as an Effective Toughness Modifier of Wood–Polymer Composites

Luísa Rosenstock Völtz [1,2], Linn Berglund [1] and Kristiina Oksman [1,2,3,*]

1 Division of Materials Science, Department of Engineering Sciences and Mathematics, Luleå University of Technology, SE-97187 Luleå, Sweden; luisa.voltz@ltu.se (L.R.V.); linn.berglund@ltu.se (L.B.)
2 Wallenberg Wood Science Center (WWSC), Luleå University of Technology, SE-97187 Luleå, Sweden
3 Department of Mechanical & Industrial Engineering (MIE), University of Toronto, Toronto, ON M5S 3G8, Canada
* Correspondence: kristiina.oksman@ltu.se

Abstract: Wood–polymer composites (WPCs) with polypropylene (PP) matrix suffer from low toughness, and fossil-based impact modifiers are used to improve their performance. Material substitution of virgin fossil-based materials and material recycling are key aspects of sustainable development and therefore recycled denim fabric, and elastomer were evaluated to replace the virgin elastomer modifier commonly used in commercial WPCs. Microtomography images showed that the extrusion process fibrillated the denim fabric into long, thin fibers that were well dispersed within the WPC, while the recycled elastomer was found close to the wood fibers, acting as a soft interphase between the wood fibers and PP. The fracture toughness (K_{IC}) of the WPC with recycled denim fabric matched the commercial WPC which was 1.4 MPa m$^{1/2}$ and improved the composite tensile strength by 18% and E-modulus by 54%. Recycled elastomer resulted in slightly lower K_{IC}, 1.1 MPa m$^{1/2}$, as well as strength and modulus while increasing elongation and contributing to toughness. The results of this study showed that recycled materials can potentially be used to replace virgin fossil-based elastomeric modifiers in commercial WPCs, thereby reducing the CO_2 footprint by 23% and contributing to more efficient use of resources.

Keywords: wood–polymer composites; recycled modifiers; impact properties; fracture toughness; microtomography

Citation: Völtz, L.R.; Berglund, L.; Oksman, K. Use of Recycled Additive Materials to Promote Efficient Use of Resources While Acting as an Effective Toughness Modifier of Wood–Polymer Composites. *Polymers* **2024**, *16*, 2549. https://doi.org/10.3390/polym16182549

Academic Editors: Phuong Nguyen-Tri, Ana Pilipović and Mustafa Özcanli

Received: 20 July 2024
Revised: 26 August 2024
Accepted: 5 September 2024
Published: 10 September 2024

Copyright: © 2024 by the authors. Licensee MDPI, Basel, Switzerland. This article is an open access article distributed under the terms and conditions of the Creative Commons Attribution (CC BY) license (https://creativecommons.org/licenses/by/4.0/).

1. Introduction

The addition of wood flour/fibers in thermoplastics has many benefits, such as reduced material cost, increased mechanical properties, and decreased weight compared with other fillers, and wood is derived from renewable resources [1,2]. Residual wood from industries, as well as recycled thermoplastics, has been employed to produce WPCs, making these composites partially or completely produced of recycled materials. However, the addition of wood flour or fibers into polypropylene results in a decrease in their impact strength and fracture toughness [1–4]. Few studies have investigated the impact and fracture properties of WPCs [3,5], the use of modifiers [1,4,6–14], or examples where recycled materials were used as impact modifiers in WPCs are challenging to find.

The most common strategy for improving the toughness of WPCs includes the addition of soft elastomeric modifiers to make the matrix softer or create a soft interface [1,9,12,15]. The fracture mechanism of soft inclusions is that these are more ductile than the polymer matrix, and act as stress concentrators, resulting in improved impact strength via crazing, shear yielding, microcracks, and/or cavitation [15–18]. It has been found that the presence of elastomers may form two structures: (1) elastomers are dispersed in the polymeric matrix as a third phase, and/or (2) elastomers encapsulating the rigid reinforcement, resulting in improved fiber–matrix adhesion [1,9,11]. Elastomers, such as ethylene–propylene–diene–monomer (EPDM) [1,9,10], styrene–ethylene–butylene–styrene (SEBS) [1,6], and polyolefin elastomers (POE) [8,11,12], have been used as impact and toughness modifiers for WPCs. For example, Oksman and Clemons [1] added 10 wt% of elastomers, EPDM, EPDM–g–maleic anhydride (MA), and SEBS–g–MA to polypropylene (PP)–wood flour composites, where the addition of EPDM decreased the tensile strength and modulus, but increased the notched Izod impact strength from 26 to 38 J/m, and EPDM–g–MA and SEBS–g–MA further increased the impact strength to 48 and 51 J/m, respectively. They suggested that the elastomers–g–MA formed a flexible interphase around the wood particles and matrix, resulting in a higher impact strength compared with EPDM alone.

Recently, tough, and long fibers have been studied to improve composite toughness and impact properties, these are often referred to as hybrid composites [13,14,19–21]. The fracture mechanism in these composites is the higher absorption energy capacity from debonding and pull-outs of long fibers resulting in better impact strength and toughness. However, its effectiveness depends on several factors, including the fiber properties, fiber aspect ratio, and the interface between the matrix and fibers [19,20]. It has been found that longer fibers in composites bridge a higher number of cracks, resisting crack opening and crack propagation, whereas short fibers act as flaws, reducing the aforementioned properties [19,21].

Várdai et al. have studied the use of long fibers in WPCs, such as poly(ethylene terephthalate) (PET) and poly(vinyl alcohol) (PVA) fibers in the WPCs [13,14]. In one of the studies, they added short PET fibers with a length of 4 mm up to 40 wt% for hybridization of WPCs with 20 wt% wood flour [13]. They showed that the addition of approx. 40 wt% PET fibers in the WPC resulted in the highest impact resistance when maleic anhydride grafted polypropylene (MAPP) was used as a coupling agent, the impact strength increased from 2 to 15 kJ/m^2 and they suggested that the PET fibers hindered crack initiation and propagation, preventing catastrophic failure by increasing the plastic deformation of the matrix. In another study, the same authors used PVA fibers, with a length of 3 mm for hybridization of PP-wood flour composites (20 wt%) and showed that the addition of 20 wt% PVA fibers had a positive effect resulting in an impact resistance of approx. 25 kJ/m^2 [14]. The results of these studies showed that polymer fibers are excellent impact modifiers for WPCs; however, the feeding of short polymer fibers can be challenging in the continuous extrusion process.

Although the use of elastomers as matrix modifiers in WPCs is an effective and well-known practice in industry the soft elastomer has a negative impact on strength and stiffness, and it will also increase the cost. In addition, virgin fossil-based materials are increasing competition for limited resources [22]. To provide a resource-efficient solution for fossil-based modifiers and reduce waste generation, targeting the United Nations Sustainable Development Goals (SDGs 8 and 12) [23], there is a need to study the effects of recycled materials as modifiers in composites with low toughness. With the use of recycled polymers or/and waste wood to make WPCs [24–28], and the studies presented in this work, companies can design impact-modified WPCs based on recycled materials, contributing to a more sustainable circular production approach. When considering materials that are currently disposed of in landfills or incinerated but could potentially be used as modifiers for enhancing toughness and impact properties, two types could be selected: recycled textiles (long fibers) and elastomers (soft inclusions). For example, every year, 12.6 million tons of textile waste is generated in Europe, approximately 77% of which goes to landfills or incinerators [29]. Textiles are woven from spun yarns, such as cotton or polycotton fibers, using an appropriate method, these recycled textiles can be converted back into fibers, turning them into excellent additives or reinforcements in composites. Another noteworthy factor is the estimated 185,000 tons of EPDM dumped into landfills every year [30]. As mentioned above, the use of EPDM elastomers has proven to be an excellent toughness modifier in WPC; therefore, recycled EPDM elastomers have the potential to provide similar behavior. These substantial quantities of waste textiles and elastomers can offer an opportunity to function as impact modifiers and thus contribute to the resource-efficient use of materials, promote recycling, and increase the circular economy.

Hence, this study focuses on WPCs with polypropylene as a matrix and wood fibers as reinforcement with the addition of recycled materials as impact modifiers. The recycled materials were denim fabric and recycled elastomer that were added targeting improved impact behavior and toughness reaching those of commercial WPCs modified with virgin fossil-based elastomer. The WPC microstructures were studied using scanning electron microscopy and X-ray microtomography. Mechanical characterizations, including impact tests, fracture toughness with digital image correlation, and tensile tests, were conducted.

2. Materials and Methods
2.1. Materials

WPC40: Commercially available WPC-grade DuraSense Pure 40 Food, with a melt flow index of 0.32 g/10 min (190 °C/2.16 kg), consisting of 60 wt% polypropylene (PP), 40 wt% wood fibers (WF), and a minor amount of maleic anhydride-grafted polypropylene (MAPP) was kindly supplied by Stora Enso (Hylte Mill, Hylte, Sweden), and it was used as base material for the compounding with the recycled materials.

Recycled-textile masterbatch (PP-rT): A denim fabric from used jeans (composition of 71% cotton, 27% polyester, and 2% elastane) was cut into 10–15-mm-wide strips, as shown in Figure 1a. Two different yarns were separated from the fabric shown in Figure 1b, the blue yarn is cotton fibers, and the white one is polyester fibers with thick elastane fibers as a minor component. The three separated fiber types are shown in a polarized optical microscopy image in Figure 1c. Struktol TPW 113 lubricant was purchased from the Struktol Company (St. Louis, MO, USA). Both the PP matrix and MAPP coupling agent were supplied by Stora Enso (Hylte Mill, Hylte, Sweden) and used for the preparation of the recycled textile masterbatch.

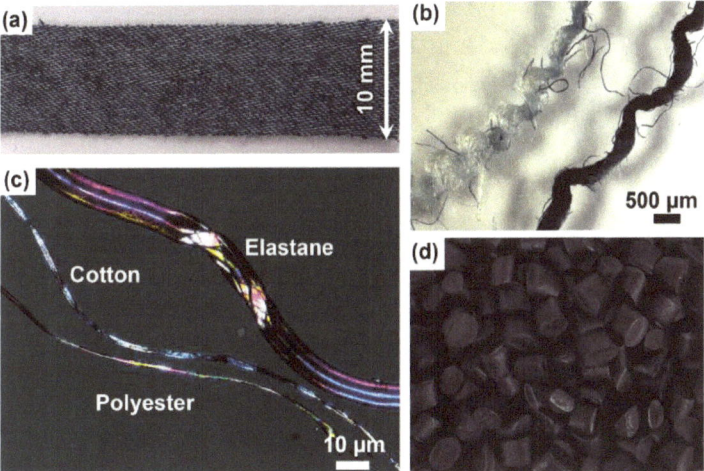

Figure 1. Used recycled materials in this study: (**a**) a strip of denim fabric; (**b**) two different types of yarns separated from the denim fabric, where the dark blue is dyed cotton and the white is polyester/elastane yarn; (**c**) polarized optical microscopy image showing the size difference of elastane, cotton, and polyester fibers separated from the yarns; and (**d**) image of the received recycled elastomer in granulated form.

Recycled elastomer (rE): Consisting of 60 wt% PP and high-density polyethylene (PP-HDPE) and 40 wt% recycled EPDM elastomer with additives. It was kindly provided by Ecorub AB (Lövånger, Sweden), the rE pellets are shown in Figure 1d. According to the datasheet, the tensile modulus of rE is 0.34 GPa, strength 11 MPa, elongation at break 23%, and Charpy impact strength 21 kJ/m^2.

WPC-E: A commercial impact-modified WPC, grade DuraSense Pure 30 Flex G, was used as a reference, consisting of 60 wt% PP, 30 wt% WF, 10 wt% elastomeric modifier, and a small amount of MAPP coupling agent. It was kindly supplied by Stora Enso (Hylte Mill, Hylte, Sweden).

Chemicals: Xylene purchased from Sigma-Aldrich (98%, Darmstadt, Germany) and ethanol 95% (Clean Chemical Sweden AB, Borlänge, Sweden) were used for Soxhlet extraction of the fibers from the matrix polymer for size measurements.

2.2. Processing

2.2.1. PP-rT Masterbatch

A co-rotating twin screw extruder (TSE) (ZSK 18 MEGALab, Coperion GmbH, Stuttgart, Germany) was used to manufacture the PP-rT masterbatch, which consisted of 57.5 wt% PP, 40 wt% used denim fabric, and 2.5 wt% MAPP and lubricant combined. The coupling agent MAPP content was adjusted to be the same in the WPCs as in the WPC-E, and a small amount of lubricant was used to improve processability due to the high fiber content. PP, MAPP, and the lubricant were premixed and fed into the main inlet of the extruder using a K-SFS-24 gravimetric feeder (Coperion K-Tron GmbH, Stuttgart, Germany). The denim fabric was fed using the recycled textile long-fiber thermoplastic process (RT-LFT) [31]. The denim strip was fed into the side extruder, and the content was controlled by the side extruder screw speed, knowing the targeted fiber content, weight per meter of the fabric strip, and length per revolution of the screw used, according to Equation (1)

$$\text{Screw speed} = \frac{\text{targeted fiber content}}{(\text{screw length/rev}) \times \text{time} \times (\text{weight/ meter denim strip})} \quad (1)$$

The total throughput was set to 4 kg/h and the targeted textile fiber content was 40 wt%, resulting in a feeding rate of 1.6 kg/h denim fabric. The length per revolution is 0.05 m, time 1 h (60 min), the weight of the fabric strip was 0.0044 kg/m, and the resulting speed was 122 rpm, which was set as the screw speed of the side extruder. A schematic of the PP-rT process is shown in Figure 2, where the temperature profile set between 180 and 195 °C is shown. The feeding rate of the polymer was 2.4 kg/h and the main screw speed for the extruder was 200 rpm. The extruded strands were cooled in a water bath, pelletized into granulates and dried overnight in an oven (60 °C) before further processing.

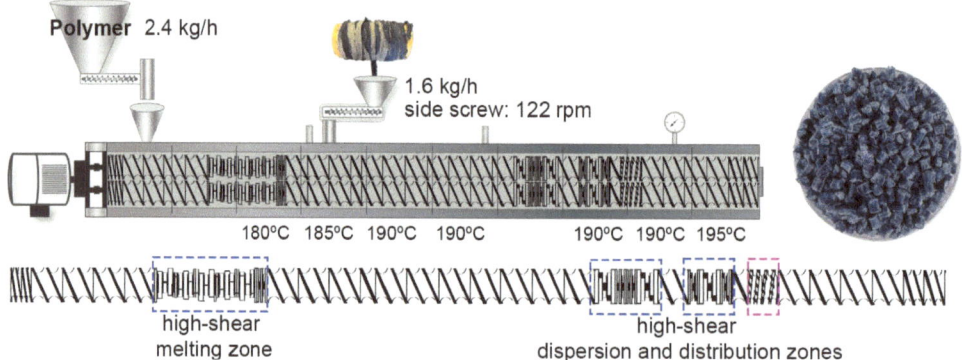

Figure 2. A schematic of the processing of textile masterbatch where the denim fabric was fed using a side extruder, the PP-rT masterbatch granulates, and the screw-configuration.

2.2.2. Manufacturing of the Modified WPCs

WPCs with recycled modifiers were compounded using the same co-rotating TSE as used for the masterbatch production, described in the previous section. The textile masterbatch (PP-rT) was mixed with commercial WPC40 to obtain the final WPC-rT, and the recycled elastomer (rE) was mixed with WPC40 to obtain the final WPC-rE. The prepared WPCs had the same MAPP content as the commercially modified WPC (WPC-E). The materials are listed in Table 1.

Table 1. Coding and material compositions, commercial grade of WPC40, manufactured masterbatch PP-rT, recycled textile-modified WPC (WPC-rT), recycled elastomer-modified WPC (WPC-rE), and commercial impact-modified elastomer WPC-E.

Material Codes	Polymer (wt%)	WF (wt%)	Impact Modifier (wt%)	Processing Aids (wt%)
WPC40	60	40	-	-
PP-rT	57.5	-	40	2.5
WPC-rT	60	30	10	-
WPC-rE	60	30	10	-
WPC-E	60	30	10	-

For both WPCs, the processing parameters and screw design are shown in Figure 3, with a temperature profile between 180 and 200 °C, total throughput of 3 kg/h, and screw speed of 300 rpm. The materials were cooled in a water bath, pelletized to granulates using the strand pelletizer, and dried overnight in an oven (60 °C).

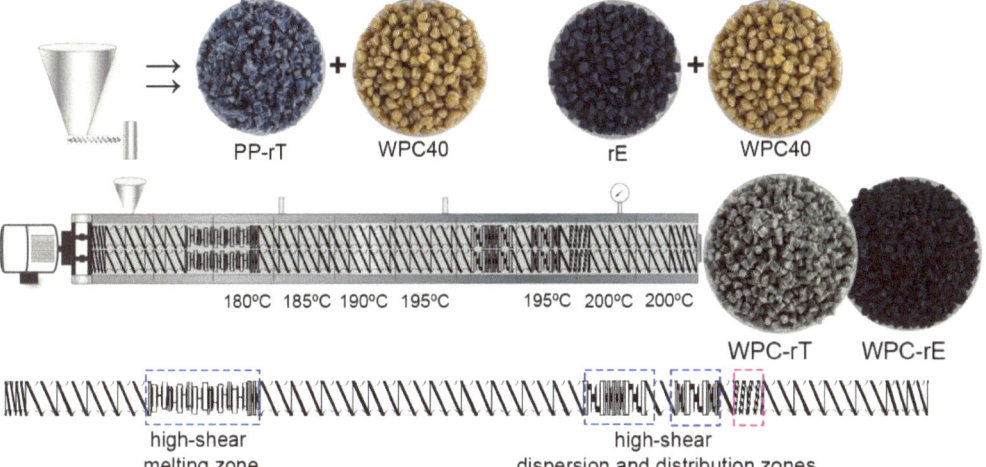

Figure 3. Extrusion profile temperatures at different zones. Granulated PP-rT was compounded with WPC40 to produce WPC-rT, and rE was also compounded with WPC40 to form WPC-rE. The screw profile with different elements is shown on the bottom of the picture, with the high-shear zones highlighted.

Samples for X-ray microtomography (μ-CT) and impact tests were produced using a compression-molding LabEcon 300 press (Fontijine Presses BV, Rotterdam, The Netherlands). The WPC granulates (15 g) were placed in a metal frame (130 × 130 × 1.5 mm³) between Mylar® films (Lohmann Technologies, Milton Keynes, UK) and steel plates. The preheating time was 5 min at 200 °C, compression molding pressure was 5.9 MPa for 60 s, and cooling was conducted under the same pressure at 25 °C for 4 min. The samples for the µ-CT study were cut from the molded sheets using a CMA0604-B-A laser (Han's Yueming Laser Group Co. Ltd., Dongguan City, China) to a size of 4 × 4 mm^2.

For the fracture toughness test, the samples were injection molded into a rectangular shape (44 × 10 × 5 mm^3), using a Haake II MiniJet Pro Piston Injection-Moulding System (JK Lab Instrument AB, Åkersberg, Sweden). The cylinder temperature was 200 °C, the mold temperature was 75 °C, the injection pressure was 400 bar and the holding time was 40 s, and the back pressure was 150 bar for 30 s. The single-edge-notch bending (SENB) specimen geometry is shown in Figure 4a, a notch with a depth of 3.5 (±0.5) mm was made with a saw with a blade thickness of 0.5 mm, and the pre-crack with a new microtome blade in the notch using compression mode in a test rig (Instron 4411 Series, Instron, Norwood, MA, USA) at a rate of 0.1 mm/min until a maximum displacement of −1.5 mm was reached (0.45 < a/W > 0.55), where a/W is the crack length across the width of the specimen. Both the notch and the pre-crack are shown in Figure 4b. All samples were painted white and speckled with a black dye to provide contrast for the digital image correlations (Figure 4c) and the test setup is shown in Figure 4d.

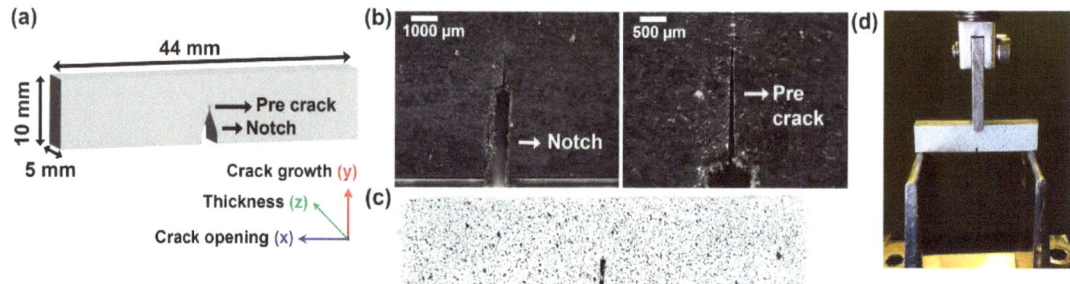

Figure 4. Fracture toughness test: (**a**) schematic of the used specimen's shape; (**b**) stereomicroscope image of the notch and pre-crack; (**c**) specimen with speckle pattern for contrasting; and (**d**) fracture toughness test set-up in bending mode.

For tensile testing, WPC-rT, WPC-rE, and WPC-E granulates were injection-molded into a type-V (ASTM D638) [32] specimen, using the same Haake II MiniJet, with a cross-section area of 3.18 × 3.20 mm^2. The cylinder temperature was 200 °C and the mold temperature was 75 °C, the injection pressure was 400 bar, the hold time was 40 s, followed by back pressure of 150 bar for 30 s.

2.3. Characterizations

2.3.1. Microstructure of Raw Materials

A stereomicroscope (SM) (Nikon SMZ1270, BergmanLabora, Danderyd, Sweden) was used to study the yarns separated from the denim fabric, and single fibers separated from the textile were studied using a polarized optical microscope (POM) (Nikon Eclipse LV100POL, BergmanLabora, Danderyd, Sweden).

2.3.2. Microstructure of Composites

The injection-molded specimens were fractured in liquid nitrogen, and the fracture surfaces were analyzed using a scanning electron microscope (SEM) (JSM-IT300LV, JEOL Ltd., Tokyo, Japan) at an acceleration voltage of 10 kV. The specimen surface was sputter-coated (Leica EM ACE200, Leica Microsystems, Wetzlar, Germany) with a 10-nm-thick layer of platinum to avoid charging.

X-ray microtomography (μ-CT) (Zeiss Xradia 510 Versa, Carl Zeiss, Pleasanton, CA, USA) was used to study the microstructure of the composite materials and dispersion of the textile fibers. Specimens were scanned using a 20× objective, which performed interior tomography with a field of view of 0.56 mm and voxel size of 0.56 μm to visualize the 3D structure of the materials. The scanned region of interest was positioned at the center of each sample and scanning was performed with an X-ray tube voltage of 50 kV, output power of 4 W, and without any X-ray filters. Reconstruction was performed using filtered back-projection with Zeiss Scout-and-Scan Reconstructor software (version 11.1), and 3D visualization and analysis of the structure were performed using Dragonfly Pro Software version 2022.2.0.1409 (Object Research Systems ORS Inc., Montreal, QC, Canada).

Measurement of fiber dimensions, the textile- and wood fibers were extracted from the composite (WPC-rT) pellets by boiling them in xylene at 170 °C for 2 h, followed by Soxhlet extraction in xylene (185 °C) for 24 h. Extracted fibers were filtered and washed with ethanol and distilled water and dispersed in water (0.1 wt%) using magnetic stirring. Textile fiber dimensions were measured using optical microscopy (OM) (Eclipse LV100PO, BergmanLabora, Danderyd, Sweden) and an image analysis program (ImageJ, 1.54d National Institutes of Health, University of Wisconsin, Madison, WI, USA). At least 100 fibers were measured.

2.3.3. Falling Weight Impact Testing and Fracture Toughness

An Instron Dynatup Minitower Drop Weight Impact Testing device (Instron, High Wycombe, UK) was used for the impact load and energy measurements, and the sample was clamped onto a round sample holder (diameter 75 mm). The test was performed at 1.6 m/s and with 3.6 kg drop mass, resulting in 4.7 J impact energy. The composite materials were conditioned in RH 50% at 25 °C for 48 h prior to the tests. Images of the impact specimens were captured, and the damaged areas were measured using ImageJ software (National Institutes of Health, University of Wisconsin, WI, USA).

The fracture toughness tests were made following ASTM D5045 [33] standard in the bending mode (with a span length of 40 mm) for samples with sharp pre-cracks using an Instron 4411 Series (Instron, Norwood, MA, USA), (Figure 4d), with a crosshead speed of 10 mm/min and load cell of 500 N. Three to four samples were measured, and the average was calculated. Crack propagation was monitored using a JAI GO-5000M camera (Stemmer Imaging AB, Stockholm, Sweden) connected to a digital image correlation (DIC) system ZEISS ARAMIS GOM Correlate Software, version 2.0.1 (Carl Zeiss GOM Metrology GmbH, Braunschweig, Germany). After the test, crack propagation and length were analyzed using a stereomicroscope (Nikon SMZ1270, BergmanLabora, Danderyd, Sweden). Fracture surfaces along the x–y and y–z planes were also analyzed using SEM (JSM-IT300LV, JEOL Ltd., Tokyo, Japan) at an acceleration voltage of 10 kV, and the specimens were sputter-coated (Leica EM ACE200, Leica Microsystems, Wetzlar, Germany) with platinum (10-nm-thick layer). The critical-stress-intensity factor (K_{IC}) for the SENB specimens was calculated following, Equations (2)–(4)

$$K_Q = \left(\frac{P_Q}{B\sqrt{W}}\right) f(x) \qquad (2)$$

$$x = a/W \qquad (3)$$

P_Q is the determined load in the load-displacement diagram, B is the thickness, W is the width of the specimen, a is the crack length, and $f(x)$ is defined as

$$f(x) = 6x^{1/2} \frac{\left[1.99 - x(1-x)\left(2.15 - 3.93x + 2.7x^2\right)\right]}{(1+2x)(1-x)^{3/2}} \qquad (4)$$

This standard assumes linear elastic behavior of the specimen, and plane-strain at the crack tip, which can be validated by two conditions: Condition 1: The test is valid only if Equation (5) is satisfied, as follows

$$P_{MAX}/P_Q < 1.1 \tag{5}$$

where P_{MAX} is the maximum load that the specimen was able to sustain during the test.

Condition 2: The K_Q is only valid if the size criteria (Equation (6)), which ensure plane strain fracture toughness, is satisfied if

$$B, a, (W - a) > 2.5 \left(K_Q/\sigma_y\right)^2 \tag{6}$$

where σ_y is the yield stress of the material, which was taken from the maximum stress obtained by the tensile test. If both conditions are satisfied that K_Q is equal to K_{IC}.

The strain energy release rate, G_{IC}, at fracture initiation was calculated using Equation (7)

$$G_{IC} = \frac{K_{IC}^2}{E} \times (1 - v^2) \tag{7}$$

where E is the modulus of the WPCs from tensile test conducted at the same strain rate as the fracture toughness and v is the Poisson's ratio of the WPCs, calculated using Equation (8)

$$v = v_f \times V_f + v_m \times (1 - V_f) \tag{8}$$

where v_f is the Poisson's ratio of the fiber, estimated to be 0.38; v_m is the Poisson's ratio of the matrix, estimated to be 0.42; and V_f is the fiber volume fracture [20].

One-way analysis of variance (ANOVA) and Tukey's honesty significant difference tests, with a significance level of 5%, were used for the results from the impact test, and fracture toughness tests using Past4 software 4.12b (Natural History Museum, Oslo, Norway).

2.3.4. Mechanical Testing

The mechanical testing was performed using a universal testing machine (Autograph AG-X, JK Lab Instrument AB, Åkersberg, Sweden). Testing was performed according to the ASTM D638 standard, with a load cell of 5 kN, 25 mm between the grips, and a strain rate of 2.5 mm/min. A noncontact video extensometer model DVE-201 (JK Lab Instrument AB, Åkersberg, Sweden) was used to measure strain for at least five specimens. Samples were conditioned in RH 50% at 25 °C for 48 h prior to testing. In addition, the fracture surfaces from the tensile test were analyzed using SEM (JSM-IT300LV) at an acceleration voltage of 10 kV. The specimen surface was sputter-coated (Leica EM ACE200) with a 10 nm thick layer of platinum. The E modulus for G_{IC}, was calculated from the tensile test with a crosshead speed of 10 mm/min (ASTM D5045) using an AGX-V series testing machine (BergmanLabora, Danderyd, Sweden) equipped with a TRViewX digital video extensometer (BergmanLabora, Danderyd, Sweden). ANOVA and Tukey's honesty significant difference in both tests, with a significance level of 5%, was used for the results from the tensile test using Past4 software 4.12b.

2.3.5. Eco-Audit Analysis Tool

The eco-audit tool in Granta EduPack (Level 3 Sustainability, Version 22.2.2, 2022, ANSYS, Canonsburg, PA, USA) was used for a simplified analysis of the environmental impact. In this analysis, the individual materials are analyzed in terms of their mass, and both energy and CO_2 footprint only take into consideration virgin materials. In contrast, the values for recycled materials are zero.

3. Results and Discussion

The results of the different characterizations performed on the WPCs are discussed in several subsections. First, a section of the microstructure and morphology was used to understand how the different constituents (polymer, wood fibers, and modifiers) were dispersed and distributed within the composites. Secondly, results were reported from the different mechanical characterizations, mainly focusing on impact, fracture toughness, and toughness properties. This study focused on the effect of recycled materials as modifiers in the replacement of fossil-based modifiers presented in the commercially modified WPC, where comparable properties should be achieved for all WPCs.

3.1. Microstructural Analysis of the Composites

Figure 5 shows cross-sections of the WPCs, fractured in liquid nitrogen. Figure 5a shows that the denim fabric is separated into long and thin fibers, it is also possible to see some holes in the matrix from fiber pullouts, in addition, some thicker fibers that may be wood fibers are visible. Figure 5b shows the thin fiber at higher magnification, the surface is smooth and clean indicating that the adhesion with the polymer has been poor. One reason for the poor interaction may be the dye present in the recycled textile. Serra et al. [34] showed in a study that the presence of the dye in recycled cotton reduced the interaction between polypropylene matrix and cotton fibers. The cross-section of WPC-rE is shown in Figure 5c, it is difficult to distinguish the wood fibers from the fracture surface, this could indicate good adhesion between the wood fibers and the polymer or that the wood fibers are covered by the elastomer and therefore not clearly visible. Figure 5d a higher magnification shows round elastomeric particles in the matrix, also some holes that may be places where wood fibers have been before the break, and some wood fibers coated by rE. The fracture surface of WPC-E is shown in Figure 5e, some wood fibers can be distinguished, but these are broken, the fracture has occurred in wood fibers and no fiber pull-outs are visible, this indicates very good adhesion between the matrix and the wood fibers. Figure 5f shows a magnified view of broken wood fibers. Furthermore, the elastomeric phase is not visible in WPC-E, indicating good compatibility between the matrix and the elastomer.

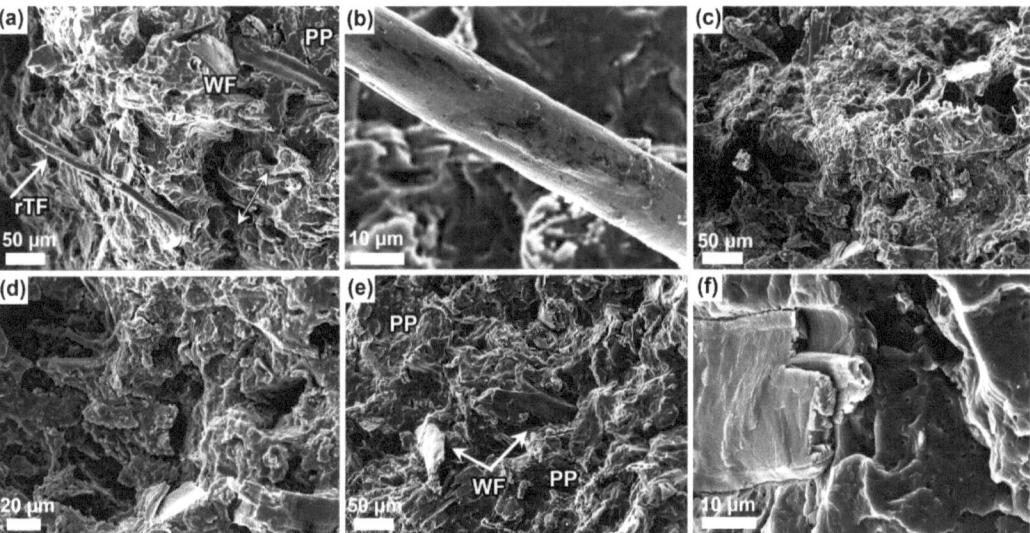

Figure 5. Fracture cross-sections of the WPCs: (**a**) WPC-rT; (**b**) higher magnification of the rTF with a clean surface; (**c**) WPC-rE; (**d**) magnified view of recycled elastomer (rE) covering the wood fibers; (**e**) WPC-E with some broken wood fibers; and (**f**) magnified broken wood fiber.

Three-dimensional reconstructions from μ-CT are shown in Figure 6 (videos and higher magnifications are available in Supporting Information), where different components of the composites are seen. The fibers are shown with a purple color as well as with a natural color. As expected, WPC-rT contains more fibers than the other WPCs, and it is possible to distinguish fine and long fibers that are separated from the denim fabric and no remnants of the fabric are visible. If comparing the WPC-rE and WPC-E, fewer fibers are visible in WPC-rE; however, there is a more green color which is the elastomeric modifier, this covers the fibers as discussed in the previous section and therefore a lower number of purple fibers are visible. Larger spots of purple particles/fibers are also visible, these are wood particles/fibers. The modifier in the WPC-E is more homogeneously distributed in the composite and has a smaller size compared with rE, and it appeared to have a better interaction with the matrix than with the fibers. Higher magnification of the 2D reconstruction is shown in Supporting Information Figure S1, where the recycled elastomer covers the fibers, whereas the commercial elastomer is evenly dispersed in the matrix.

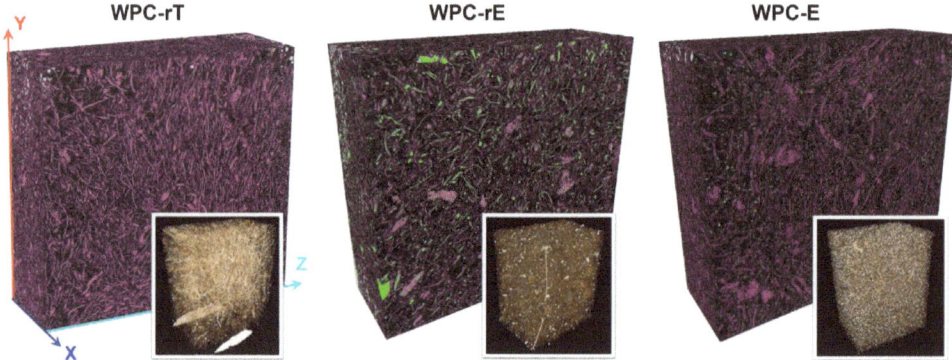

Figure 6. Three-dimensional reconstructions μ-CT of WPCs. Different colors represent different components in the composites. PP-matrix is black, fibers are purple, and elastomers are green; and bottom right figures show the same WPCs which are taken from the videos (available in Supporting Information).

Soxhlet extraction of the WPC-rT compounded material showed that the recycled textile fiber length was considerably reduced during the process, as shown in Supporting Information Figure S2. Recycled denim textile fibers are longer and have a smaller diameter than wood fibers; their length ranges from 65 to 2500 μm, with most in the range of 100–1100 μm. The aspect ratio for recycled denim textile fibers was found to be in the range of 5–130, with the majority ranging between 10 and 50. A previous study [35] reported that the average aspect ratio of the same wood fibers was lower than that of the rTF. Considering that previous studies used much longer fibers to provide toughness [19,20], rTFs are longer than wood fibers and may act as modifiers for WPCs via fiber pull-out. Further investigations will be conducted based on the fracture-toughness results. Regarding rE, it was assumed that the rE particle size would be close to that reported by Oksman and Clemons [1], ranging between 0.1 and 1 μm, with an average size of 0.3 μm.

3.2. Impact Strength and Fracture Toughness

Typical load–time and energy–time graphs for the drop-weight impact tests are presented in Figure 7. The WPC-rT presented the highest load-time as well as energy-time graphs, the higher load needed to initiate cracks while the WPC-rE showed the lowest load and energy-time. The WPC-E is performing between these two composites. All samples were completely perforated with the impactor. The weak interface between textile fibers and WPC and the length of the textile fibers resulted in a fiber pull-out mechanism while the addition of elastomers acted as stress concentrations, improving impact properties.

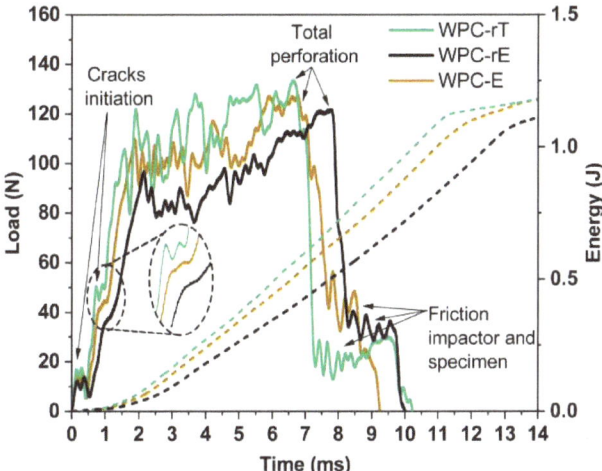

Figure 7. Load–time (continuous lines) and energy–time (dotted lines) curves from the drop-weight impact test, where crack initiation phases, total perforation of the impactor, and friction between impact and specimen are indicated with arrows.

The WPCs exhibited similar impact behavior, including disturbance under the load, which is associated with crack initiation in the first milliseconds and the subsequent increase in load. The abrupt drop in the load was due to the total perforation of the impactor and complete breakage of the specimen (indicated by the arrows in the graph). The maximum impact load for WPC-rT was approximately 129 ± 3 N, the WPC-rE load was slightly lower (120 ± 6 N), and the WPC-E load was approximately 125 ± 5 N. The energy-time shows the energy at the peak load, which was found similar for all materials 1.18–1.20 J. The absorbed energy was calculated by integrating the load-deflection curves, as shown in Supporting Information Figure S3a, where a higher average impact energy absorbed by the materials was found for WPC-rE, with a value of 1.5 (±0.3) J, followed by 1.3 (±0.2) J for WPC-E and 1.2 (±0.3) J for WPC-rT. However, the differences were small and not statistically significant according to ANOVA (5%) and Tukey's tests, indicating that the energy absorption was similar for the WPCs. As mentioned previously, all specimens were broken during the impact test in the presence of fragments owing to the total perforation of the impactor. In addition, for WPC-rT and WPC-E, the area around the crack became white, which can be attributed to the crazing and cracking of the matrix, as also reported by Puech et al. [36]. However, it was not possible to observe a similar color change in WPC-rE, possibly because of the black color of the specimen (Figure S3b). The calculated average area of damage was higher for WPC-rT (1715 mm^2) and lower for WPC-E (1351 mm^2); however, the values among all composites were not statistically different.

Table 2 shows the plane-strain fracture toughness (K_{IC}) and energy release rate (G_{IC}) of the WPCs, and the most representative load-displacement curves are shown in Figure 8, together with the DIC images. All materials exhibited linear behavior with a sudden load drop after the maximum load, indicating unstable crack growth [37] (represented by events 1 and 2 in Figure 8), and Figure S4 in Supporting Information. For both WPC-rT and WPC-E, the cracks exhibited a small deviation, as seen in the DIC images after the test, until complete failure was reached. For WPC-rE, the crack propagated longitudinal to the applied stress.

Table 2. Fracture toughness (K_{IC}), energy (G_{IC}), and toughness for the WPCs. (Marked with the same letter within the same column are not significantly different at a 5% significant level based on ANOVA and Tukey's test).

WPCs	K_{IC} (MPa m$^{1/2}$)	G_{IC} (kJ m^{-2})	Toughness (MJ m^{-3})
WPC-rT	1.44 ± 0.08 [A]	0.48 ± 0.1 [A]	1.1 ± 0.1 [A]
WPC-rE	1.13 ± 0.10 [B]	0.45 ± 0.1 [A]	1.1 ± 0.2 [A]
WPC-E	1.44 ± 0.09 [A]	0.73 ± 0.1 [B]	1.0 ± 0.1 [A]

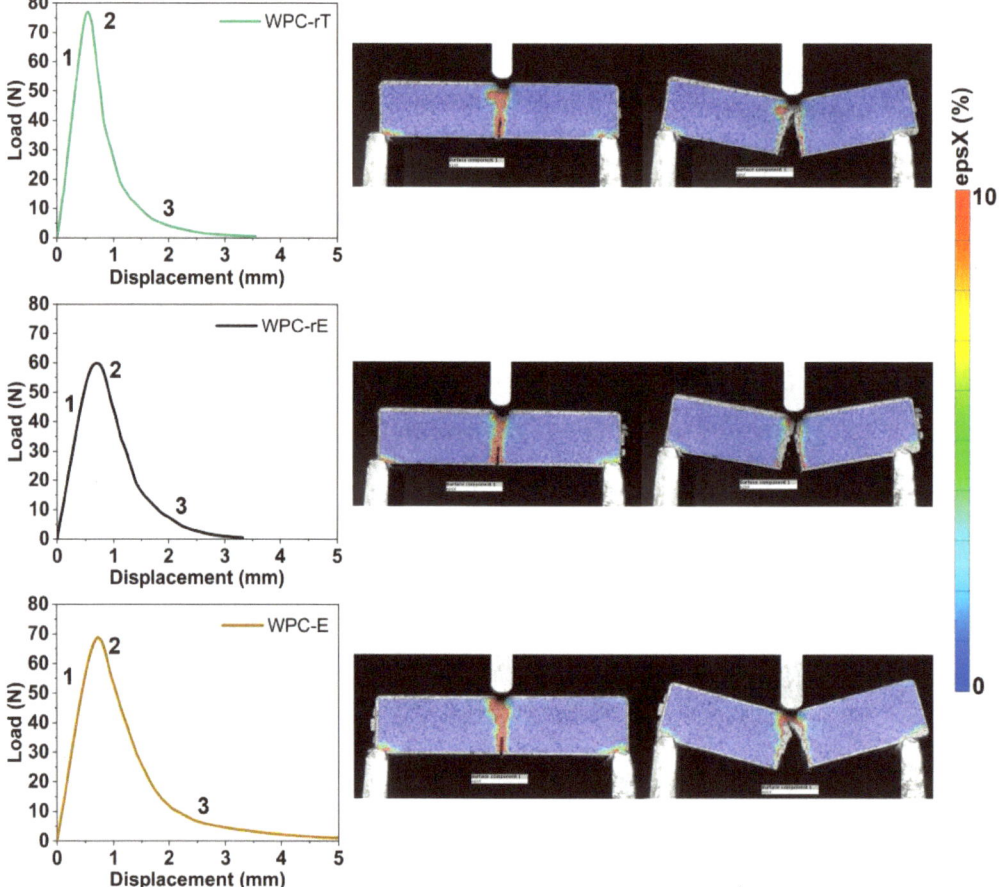

Figure 8. Load–displacement curves from fracture toughness with their respective DIC images (at the beginning of the test, and the end of the test); the color bar in DIC images represents the strain on the x-axis.

The average total time for the crack to propagation until complete for WPC-E was 31 s, followed by WPC-rE at 25 s, and WPC-rT at 17 s. The K_{IC} values were the same for WPC-rT and WPC-E, and a slight decrease was observed for WPC-rE. Even though the crack propagation had no deviation and a lower K_{IC} was found for WPC-rE, the presence of the modifier did not lead to a catastrophic and fast failure. The energy (G_{IC}) was found similar for all WPCs made from recycled materials, while WPC-E presented higher energy.

By observing the fracture specimens along the crack propagation in SEM and stereomicroscope, the suggested fracture toughness mechanisms of the modifiers used in this study are shown in Figures 9–11. For WPC-rT, long textile fibers play an important role, the energy is mainly dissipated by debonding, fiber pull-out, and bridging mechanisms (Figure 9), as suggested elsewhere [21]. The fiber pull-out will occur if the fibers are long, facilitating dissipation of energy along their length [21], and if the interaction between these fibers and the matrix is poor, as shown in Figure S5a. In addition, the crack path can change when the crack hits the fibers [38]. A few microcracks were also observed in the specimens perpendicular to the crack path (Figure S5b), in addition to the white lines observed in the stereomicroscope, which could be attributed to stress-whitening zones, as also observed in the impact specimens.

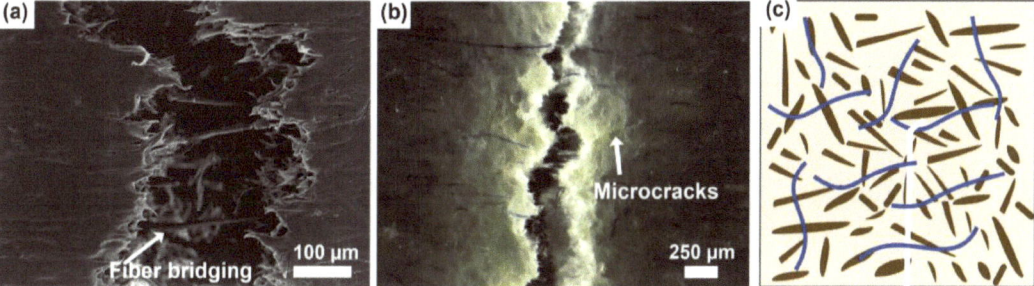

Figure 9. WPC-rT: (**a**) micrograph of a detailed view of crack from fracture toughness specimen, where fiber pull-outs and fiber bridging are visible; (**b**) stereomicroscope image of the crack, where fiber pull-outs and microcracks are also visible, as well as stress-whitening zone around the crack; (**c**) fiber pullout and fiber bridging are suggested fracture toughness mechanisms. Textile fibers are visualized as blue and wood fibers as brown.

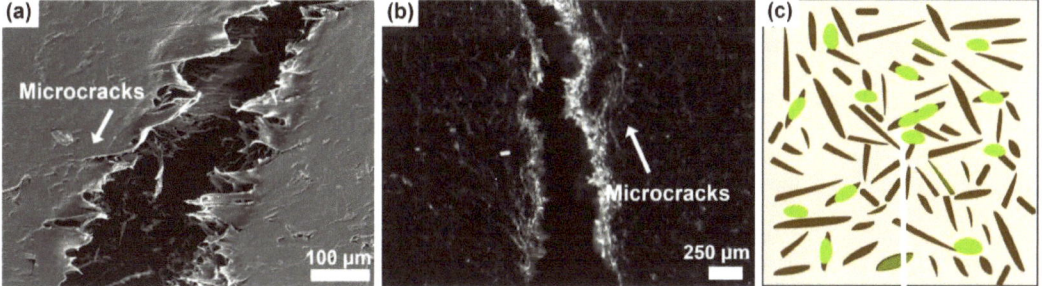

Figure 10. WPC-rE: (**a**) micrograph of the crack from the fracture toughness test showing crack propagation in the matrix with crazing mechanism; (**b**) a stereo-microscopy image of the crack; and crazing; and (**c**) the suggested fracture toughness mechanism. Wood fibers are visualized as brown and recycled elastomer as green.

Based on the microstructure of the WPC-rE in Figure 10a,b, it is seen that the interaction is very strong between the recycled elastomer and wood fibers, in addition, the crack is propagating along the applied stress in the matrix. Few microcracks are visible, and the applied stress forms micro voids perpendicular to the crack at points with high-stress concentrations creating crazing in the polymer, and due to the continuous crack propagation. Those elongated fibrils were clearly seen in WPC-rE, but not in WPC-rT. The rE is present in large size around the harder wood fibers, aiding in the schematic depiction in Figure 10c. The main mechanism for WPC-rE is the soft elastomer between the fibers and matrix (Figure S5d), this is improving the impact and toughness properties [1] while reducing the mechanical properties compared to WPC-E. As seen in SEM images and µ-CT the recycled elastomer was located near the wood fibers or their interface, with the formation of micro voids and crazing being the main toughening mechanism.

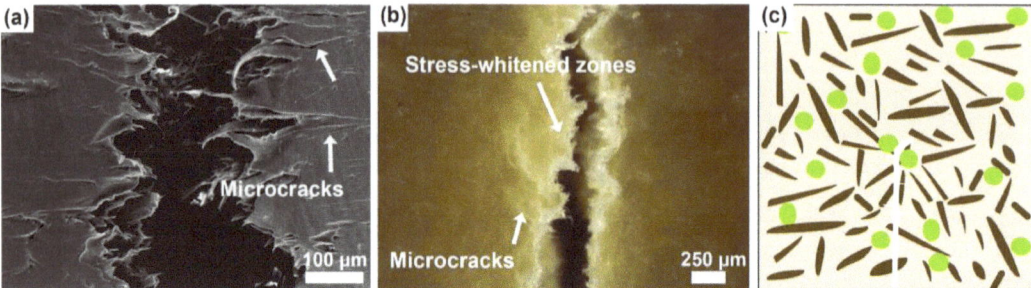

Figure 11. WPC-E: (**a**) micrograph of a detailed view of a crack from fracture toughness test where few microcracks are visible; (**b**) stereomicroscope image of the crack where microcracks are visible, along with a stress-whitened area; and (**c**) suggested fracture toughness mechanism. Wood fibers are visualized as brown and elastomer as green.

As discussed before, the commercial elastomer was homogeneously distributed in the PP matrix in a single phase, because of the good interaction between the polymer and the elastomer, and the fracture mechanism could be explained by the presence of microcracks perpendicular to the crack propagation (see Figure 11a,b). Broken fibrils and microcracks (Figure S5e,f) are also observed in WPC-E, which could be the result of crazing. Additionally, crack deflection was observed for WPC-E, this mechanism aided in reducing the stress intensity at the crack tip [15]. Moreover, by performing a detailed examination using a stereomicroscope, it became apparent that stress-whitened zones developed along the crack, which may have been associated with craze-like features [18]. In summary, the fracture mechanisms could be observed as microcrack formation and failed fibrils due to crazing during crack propagation (Figure 11c).

3.3. Tensile Properties

Figure 12 shows the stress–strain curves of the WPCs and a summary of the mechanical properties is shown in Table S1 (Supporting Information). It is seen that the WPC-rT has the highest strength and modulus followed by WPC-E and WPC-rE. The tensile strength of WPC-rT improved from 39 (WPC-E) to 46 MPa, and the tensile modulus, from 2.4 to 3.7 GPa, the improvement is because overall fiber content increased, from 30 to 40 wt% and the absence of soft elastomer. The average strain of 3.4% is slightly lower than the WPC-E 3.7% but the difference is not significant (seen in Table S1). The tensile strength (30 MPa) and modulus (2.2 GPa) of WPC-rE are lower compared to WPC-E; the reason is that the soft recycled elastomer is, at least partially, on the wood fiber surfaces, reducing the strength and modulus. A soft interface can reduce the effectiveness of stress transfer from the matrix to the fibers [1]. The elastomeric encapsulation of the wood fibers slightly improved the elongation at break (to 4.1%), but it is not a significant improvement (Table S1). This was previously discussed by Oksman and Clemons [1], where the elastomer-covered wood particle surfaces were more effective as toughness modifiers for WPCs but this reduced both strength and modulus more than an elastomer present as a soft phase in the matrix polymer. Fracture surface images of samples from tensile testing, are seen in Supporting Information Figure S6a shows several places where the denim fibers have been pulled out from the matrix, it is difficult to see the wood fibers, which indicates better interaction between the wood fibers and polymer matrix. Figure S6b shows WPC-rE, it is difficult to see the wood fibers and that indicates good adhesion between the fibers and the matrix, therefore it is suggested that the wood fibers are covered by the rE. The toughness, indicated by the area under the stress–strain curves, showed similar values for all WPCs of approximately 1.0–1.1 MJ m^{-3} (shown in Table 2), with no statistically significant difference.

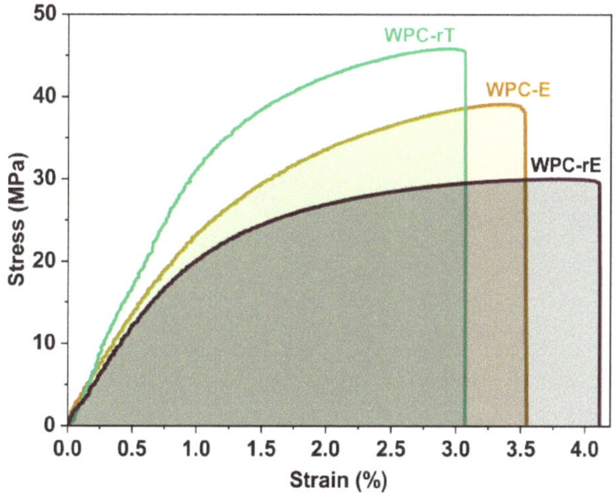

Figure 12. Representative stress–strain curves from tensile testing, where the area under the graph represents the toughness.

3.4. Eco-Audit Analysis

The eco-audit showed that WPC-E needs 58 MJ/kg and the WPCs with recycled modifiers reduced it to 46 MJ/kg. This shows that using only 10% recycled materials will have a positive effect on energy by 26%. The WPC-E had a CO_2 footprint of 2.4 kg/kg material this was reduced to 1.8 kg/kg with recycled modifiers. It is important to point out that only the contribution of the virgin elastomer impacts 23% of the total CO_2 footprint. The details of the analysis are shown in Table S2 in the supporting information.

4. Conclusions

This study aimed to assess the potential of recycled textiles and recycled elastomeric polymers in improving the properties of PP-based WPCs, especially impact and (fracture) toughness properties, to be on a similar level with the commercial impact-modified WPCs (WPC-E). The developed WPCs consisted of 60 wt% polypropylene, 30 wt% wood fibers, 10 wt% recycled impact modifiers, and a small amount of compatibilizer, manufactured using a co-rotating twin-screw extruder.

The microtomography results showed the recycled denim textile strip was separated into fibers that were well dispersed and distributed within the WPC. The manufactured WPCs did not show significant differences in terms of their impact properties and toughness. The composites with recycled denim fabric as a modifier (WPC-rT) exhibited fracture toughness like that of WPC-E, whereas the composite with recycled elastomer (WPC-rE) exhibited slightly lower fracture toughness. The toughening mechanism investigated in this study showed that the recycled denim fabric, which was successfully separated into fibers during the compounding, lacking interaction with the matrix, led to a fiber pull-out mechanism maintaining the fracture toughness. For WPC-rE, the recycled elastomer covered the wood fibers, suggesting a good interface between the fiber and the matrix, but with decreased strength and stiffness, and increased failure at break. In WPC-E, the elastomer was well-distributed in the polymer matrix, and the main toughening mechanisms were crazing and microcracks formation, with cracks slightly deviating from the original path.

Understanding the behavior of recycled materials is important for designing sustainable materials. In this study, the recycled denim fabric and elastomer were used as impact modifiers in WPCs. These modifiers especially the denim fabric resulted in impact properties and toughness like the commercial WPC but with better mechanical properties. This promotes the circular economy—where used materials are reintroduced into the economy, instead of being landfilled—reducing waste generation and better use of resources, and these findings could potentially lead to new WPCs based on 100% recycled materials.

Supplementary Materials: The following supporting information can be downloaded at: https://www.mdpi.com/article/10.3390/polym16182549/s1, Figure S1: μ-CT 2D reconstructions of WPCs. Different colors represent different components inside the composites. Black background is the PP-matrix, presented in all WPCs, purple indicates the fibers, and green indicates the elastomeric modifier presented only in WPC-rE and WPC-E; Figure S2: (a) Polarized optical microscope images of the extracted fibers from WPC-rT where WF is wood fibers and rTF is recycled textile fibers, and (b) rTF length and aspect ratio distribution after extraction; Figure S3: (a) Drop weight impact curves load-deflection for all tested specimens, (b) fractographs of the specimens for all the WPCs, where the average damage areas are: 1715 (±280) mm^2, 1448 (±325) mm^2, and 1351 (±475) mm^2 for WPC-rT, WPC-rE and WPC-E, respectively; Figure S4: Load-displacement curves from the fracture toughness test with their respective DIC of the fractured specimens; Figure S5: SEM micrographs of SENB samples: WPC-rT (a) Fracture surface micrograph (yz plane) showing fiber imprint and weak interaction between recycled textile and matrix, (b) Surface micrograph (yx plane) showing microcracks formations; WPC-rE: (c) Surface micrograph (xy plane) where failed fibrils are seen which could be suggested to a craze-like formation, and (d) Fracture surface micrograph (yz plane) where the wood fibers is covered by recycled elastomer and polymer matrix, indicating good interface; WPC-E: (e) Surface micrograph (xy plane) where failed fibrils are seen which could be suggested to a craze-like formation, and (f) Surface micrograph (xy plane) showing microcracks and void formations; Figure S6: Micrographs of fracture surfaces from tensile test: (a) WPC-rT with fiber pull-outs, and (b) WPC-rE where wood fibers are difficult to be seen because rE covers the fibers; Table S1: Mechanical properties summarized for the WPCs; Table S2: Detailed analysis from the eco-audit tool. Video S1: WPC-rT microCT animation; Video S2: WPC-rE microCT animation; Video S3: WPC-E microCT animation.

Author Contributions: Conceptualization, L.R.V. and K.O.; methodology, L.R.V. and K.O.; validation, L.R.V. and K.O.; formal analysis, L.R.V.; investigation, L.R.V., L.B. and K.O.; resources, K.O.; writing—original draft preparation, L.R.V., L.B. and K.O.; writing—review and editing, L.R.V., L.B. and K.O.; visualization, L.R.V. and K.O.; supervision, L.B. and K.O.; project administration, K.O.; funding acquisition, K.O. All authors have read and agreed to the published version of the manuscript.

Funding: This research was funded by Knut and Alice Wallenberg Stiftelse (KAW), Stora Enso (Project KAW 2018.0451), and Bio4Energy, a national strategic research program.

Data Availability Statement: The original contributions presented in the study are included in the article/Supplementary Materials, further inquiries can be directed to the corresponding author.

Acknowledgments: The authors would like to thank Knut and Alice Wallenberg Stiftelse (KAW), Stora Enso, and Bio4Energy, a national strategic research program, for their financial support. The authors would also like to thank Stora Enso AB and Ecorub AB for providing materials. We also thank the Wallenberg Wood Science Center (WWSC). The authors would like to thank Henrik Lycksam at Fluid and Experimental Mechanics, Luleå University of Technology, Sweden, for the help with X-ray microtomography.

Conflicts of Interest: The authors declare no conflicts of interest.

References

1. Oksman, K.; Clemons, C. Mechanical Properties and Morphology of Impact Modified Polypropylene-Wood Flour Composites. *J. Appl. Polym. Sci.* **1998**, *67*, 1503–1513. [CrossRef]
2. Klyosov, A.A. *Wood-Plastic Composites*; John Wiley & Sons: Hoboken, NJ, USA, 2007.
3. Pérez, E.; Famá, L.; Pardo, S.G.; Abad, M.J.; Bernal, C. Tensile and Fracture Behaviour of PP/Wood Flour Composites. *Compos. Part B Eng.* **2012**, *43*, 2795–2800. [CrossRef]

4. Ferdinánd, M.; Jerabek, M.; Várdai, R.; Lummerstorfer, T.; Pretschuh, C.; Gahleitner, M.; Faludi, G.; Móczó, J.; Pukánszky, B. Impact Modification of Wood Flour Reinforced PP Composites: Problems, Analysis, Solution. *Compos. Part A Appl. Sci. Manuf.* **2023**, *167*, 107445. [CrossRef]
5. Guo, Y.; Zhu, S.; Chen, Y.; Liu, D.; Li, D. Acoustic Emission-Based Study to Characterize the Crack Initiation Point of Wood Fiber/HDPE Composites. *Polymers* **2019**, *11*, 701. [CrossRef] [PubMed]
6. Oksman, K.; Lindberg, H. Influence of Thermoplastic Elastomers on Adhesion in Polyethylene–Wood Flour Composites. *J. Appl. Polym. Sci.* **1998**, *68*, 1845–1855. [CrossRef]
7. Hristov, V.N.; Lach, R.; Grellmann, W. Impact Fracture Behavior of Modified Polypropylene/Wood Fiber Composites. *Polym. Test.* **2004**, *23*, 581–589. [CrossRef]
8. Jiang, F.; Qin, T. Toughening Wood/Polypropylene Composites with Polyethylene Octene Elastomer (POE). *J. For. Res.* **2006**, *17*, 312–314. [CrossRef]
9. Clemons, C. Elastomer Modified Polypropylene–Polyethylene Blends as Matrices for Wood Flour–Plastic Composites. *Compos. Part A Appl. Sci. Manuf.* **2010**, *41*, 1559–1569. [CrossRef]
10. Sudár, A.; Renner, K.; Móczó, J.; Lummerstorfer, T.; Burgstaller, C.; Jerabek, M.; Gahleitner, M.; Doshev, P.; Pukánszky, B. Fracture Resistance of Hybrid PP/Elastomer/Wood Composites. *Compos. Struct.* **2016**, *141*, 146–154. [CrossRef]
11. Yi, S.; Xu, S.; Li, Y.; Gan, W.; Yi, X.; Liu, W.; Wang, Q.; Wang, H.; Ou, R. Synergistic Toughening Effects of Grafting Modification and Elastomer-Olefin Block Copolymer Addition on the Fracture Resistance of Wood Particle/Polypropylene/Elastomer Composites. *Mater. Des.* **2019**, *181*, 107918. [CrossRef]
12. Mazzanti, V.; Malagutti, L.; Santoni, A.; Sbardella, F.; Calzolari, A.; Sarasini, F.; Mollica, F. Correlation between Mechanical Properties and Processing Conditions in Rubber-Toughened Wood Polymer Composites. *Polymers* **2020**, *12*, 1170. [CrossRef] [PubMed]
13. Várdai, R.; Lummerstorfer, T.; Pretschuh, C.; Jerabek, M.; Gahleitner, M.; Bartos, A.; Móczó, J.; Anggono, J.; Pukánszky, B. Improvement of the Impact Resistance of Natural Fiber–Reinforced Polypropylene Composites through Hybridization. *Polym. Adv. Technol.* **2021**, *32*, 2499–2507. [CrossRef]
14. Várdai, R.; Ferdinánd, M.; Lummerstorfer, T.; Pretschuh, C.; Jerabek, M.; Gahleitner, M.; Faludi, G.; Móczó, J.; Pukánszky, B. Impact Modification of Hybrid Polypropylene Composites with Poly(Vinyl Alcohol) Fibers. *J. Reinf. Plast. Compos.* **2022**, *41*, 399–410. [CrossRef]
15. Kinloch, A.J.; Young, J.R. *Fracture Behaviour of Polymers*; Springer: Dordrecht, The Netherlands, 2013.
16. Katz, H.S.; Mileski, J.V. *Handbook of Fillers and Reinforcements for Plastics*; Van Nostrand Reinhold Company: New York, NY, USA, 1978.
17. Liang, J.Z.; Li, R.K.Y. Rubber Toughening in Polypropylene: A Review. *J. Appl. Polym. Sci.* **2000**, *77*, 409–417. [CrossRef]
18. Zebarjad, S.M.; Bagheri, R.; Lazzeri, A.; Serajzadeh, S. Fracture Behaviour of Isotactic Polypropylene under Static Loading Condition. *Mater. Des.* **2003**, *24*, 105–109. [CrossRef]
19. Ranganathan, N.; Oksman, K.; Nayak, S.K.; Sain, M. Structure Property Relation of Hybrid Biocomposites Based on Jute, Viscose, and Polypropylene: The Effect of the Fibre Content and the Length on the Fracture Toughness and the Fatigue Properties. *Compos. Part A Appl. Sci. Manuf.* **2016**, *83*, 169–175. [CrossRef]
20. Ranganathan, N.; Oksman, K.; Nayak, S.K.; Sain, M. Regenerated Cellulose Fibers as Impact Modifier in Long Jute Fiber Reinforced Polypropylene Composites: Effect on Mechanical Properties, Morphology, and Fiber Breakage. *J. Appl. Polym. Sci.* **2015**, *132*, 41301. [CrossRef]
21. Hajiha, H.; Sain, M. High Toughness Hybrid Biocomposite Process Optimization. *Compos. Sci. Technol.* **2015**, *111*, 44–49. [CrossRef]
22. Tang, S.; Li, J.; Wang, R.; Zhang, J.; Lu, Y.; Hu, G.-H.; Wang, Z.; Zhang, L.; Key, S. Current Trends in Bio-Based Elastomer Materials. *SusMat* **2022**, *2*, 2–33. [CrossRef]
23. United Nations Sustainable Development Goals Report. 2022. Available online: https://www.un.org/sustainabledevelopment/progress-report/ (accessed on 18 September 2023).
24. Dairi, B.; Djidjelli, H.; Boukerrou, A.; Migneault, S.; Koubaa, A. Morphological, Mechanical, and Physical Properties of Composites Made with Wood Flour-Reinforced Polypropylene/Recycled Poly(Ethylene Terephthalate) Blends. *Polym. Compos.* **2017**, *38*, 1749–1755. [CrossRef]
25. Tang, W.; Xu, J.; Fan, Q.; Li, W.; Zhou, H.; Liu, T.; Guo, C.; Ou, R.; Hao, X.; Wang, Q. Rheological Behavior and Mechanical Properties of Ultra-High-Filled Wood Fiber/Polypropylene Composites Using Waste Wood Sawdust and Recycled Polypropylene as Raw Materials. *Constr. Build. Mater.* **2022**, *351*, 128977. [CrossRef]
26. Martinez Lopez, Y.; Paes, J.B.; Gustave, D.; Gonçalves, F.G.; Méndez, F.C.; Theodoro Nantet, A.C. Production of Wood-Plastic Composites Using *Cedrela Odorata* Sawdust Waste and Recycled Thermoplastics Mixture from Post-Consumer Products—A Sustainable Approach for Cleaner Production in Cuba. *J. Clean. Prod.* **2020**, *244*, 118723. [CrossRef]
27. Keskisaari, A.; Kärki, T. The Use of Waste Materials in Wood-Plastic Composites and Their Impact on the Profitability of the Product. *Resour. Conserv. Recycl.* **2018**, *134*, 257–261. [CrossRef]
28. Khamseh, M.; Maroufkhani, M.; Moghanlou, S.; Lotfi, A.; Pourabbas, B.; Razavi Aghjeh, M.K. Development of Sustainable Cellulose-Based Composite of Polypropylene Reinforced by Recycled Microfibrillar Poly (Ethylene Terephthalate). *Polym. Compos.* **2023**, *44*, 7058–7069. [CrossRef]

29. European Commission Circular Economy for Textiles Report. 2023. Available online: https://ec.europa.eu/commission/presscorner/detail/en/ip_23_3635 (accessed on 18 September 2023).
30. West, R. The Benefits of Using Recycled EPDM. 2009. Available online: https://www.sprayfoammagazine.com/foam-news/the-benefits-of-using-recycled-epdm/1021 (accessed on 18 September 2023).
31. Völtz, L.R.; Berglund, L.; Oksman, K. Resource-Efficient Manufacturing Process of Composite Materials: Fibrillation of Recycled Textiles and Compounding with Thermoplastic Polymer. *Compos. Part A Appl. Sci. Manuf.* **2023**, *175*, 107773. [CrossRef]
32. *ASTM D638-22*; Standard Test Method for Tensile Properties of Plastics. American Society for Testing and Materials International: West Conshohocken, PA, USA, 2022.
33. *ASTM D5045-14*; Standard Test Methods for Plane-Strain Fracture Toughness and Strain Energy Release Rate of Plastic Materials. American Society for Testing and Materials International: West Conshohocken, PA, USA, 2022.
34. Serra, A.; Tarrés, Q.; Claramunt, J.; Mutjé, P.; Ardanuy, M.; Espinach, F.X. Behavior of the Interphase of Dyed Cotton Residue Flocks Reinforced Polypropylene Composites. *Compos. Part B Eng.* **2017**, *128*, 200–207. [CrossRef]
35. Völtz, L.R.; Di Guiseppe, I.; Geng, S.; Oksman, K. The Effect of Recycling on Wood-Fiber Thermoplastic Composites. *Polymers* **2020**, *12*, 1750. [CrossRef]
36. Puech, L.; Ramakrishnan, K.R.; Le Moigne, N.; Corn, S.; Slangen, P.R.; Duc, A.L.; Boudhani, H.; Bergeret, A. Investigating the Impact Behaviour of Short Hemp Fibres Reinforced Polypropylene Biocomposites through High Speed Imaging and Finite Element Modelling. *Compos. Part A Appl. Sci. Manuf.* **2018**, *109*, 428–439. [CrossRef]
37. Kinloch, A.J.; Shaw, S.J.; Tod, D.A.; Hunston, D.L. Deformation and Fracture Behaviour of a Rubber-Toughened Epoxy: 1. Microstructure and Fracture Studies. *Polymer* **1983**, *24*, 1341–1354. [CrossRef]
38. Naebe, M.; Abolhasani, M.M.; Khayyam, H.; Amini, A.; Fox, B. Crack Damage in Polymers and Composites: A Review. *Polym. Rev.* **2016**, *56*, 31–69. [CrossRef]

Disclaimer/Publisher's Note: The statements, opinions and data contained in all publications are solely those of the individual author(s) and contributor(s) and not of MDPI and/or the editor(s). MDPI and/or the editor(s) disclaim responsibility for any injury to people or property resulting from any ideas, methods, instructions or products referred to in the content.

MDPI AG
Grosspeteranlage 5
4052 Basel
Switzerland
Tel.: +41 61 683 77 34

Polymers Editorial Office
E-mail: polymers@mdpi.com
www.mdpi.com/journal/polymers

Disclaimer/Publisher's Note: The statements, opinions and data contained in all publications are solely those of the individual author(s) and contributor(s) and not of MDPI and/or the editor(s). MDPI and/or the editor(s) disclaim responsibility for any injury to people or property resulting from any ideas, methods, instructions or products referred to in the content.

www.ingramcontent.com/pod-product-compliance
Lightning Source LLC
LaVergne TN
LVHW072322090526
838202LV00019B/2337